D0904183

Effective Logic Computation

Effective Logic Computation

KLAUS TRUEMPER

A Wiley-Interscience Publication

JOHN WILEY & SONS, INC.

New York • Chichester • Weinheim • Brisbane • Singapore • Toronto

Sci
T
57.6
.T75
1998

Copyright © 1998 by John Wiley & Sons, Inc.

All rights reserved. Published simultaneously in Canada.

Library of Congress Cataloging in Publication Data:

Truemper, K., 1942–
    Effective logic computation / Klaus Truemper.
       p. cm.
    "A Wiley-Interscience publication."
    Includes bibliographical references and index.
    ISBN 0-471-23886-4 (cloth : alk. paper)
    1. System analysis. 2. Operations research. 3. Logic, Symbolic
and mathematical. I. Title.
    T57.6.T75 1998
    003—dc21

                                   97-13438
                                       CIP

Printed in the United States of America

10 9 8 7 6 5 4 3 2 1

Contents

National Science Foundation, the Office of Naval Research, and the author's home institution, the University of Texas at Dallas.

The implementation of the theory in the Leibniz System was carried out initially with private funds, and it was later funded with revenues produced by the system.

Several people supported the development of the theory and the implementation of the Leibniz System. First and foremost, we should mention M. Grötschel, who, believing in our vision of a new computational tool for logic computation, helped in numerous ways. A major portion of the implementation effort was accomplished by R. Karpelowitz with exemplary dedication and precision. Substantial help in various forms was provided by J. N. Barrer, R. E. Bixby, M. Jünger, C. Ratliff, G. Rinaldi, and D. K. Wagner.

Many persons assisted with the search for references—in particular, E. Boros, R. Chandrasekaran, G. Cornuéjols, W. H. Cunningham, U. Derigs, J. V. Franco, G. Gallo, M. X. Goemans, P. L. Hammer, P. Hansen, F. Harche, J. N. Hooker, B. Jaumard, H. Kleine Büning, B. Korte, O. Kullmann, D. W. Loveland, J. A. Makowsky, I. Mitterreiter, D. Pretolani, G. Qian, G. Rago, A. Sassano, I. Schiermeyer, E. Speckenmeyer, I. H. Sudborough, R. P. Swaminathan, A. Tamir, and H. Zhang. The University of Bonn and New York University assisted in the search for and verification of reference material.

Typesetting was done in TeX. Spell and syntax checking was accomplished with the Laempel System for Intelligent Text Processing in an efficient and untiring manner. We should mention that the latter system was constructed in joint work with Y. Zhao, using the Leibniz System for the processing of the numerous logic modules.

Alas, the checking of content still required human intellect, with much assistance provided by G. Felici, G. Qian, E. Speckenmeyer, M. Stoer, J. Straach, I. Truemper, and Y. Zhao.

The drawings were expertly done by Y. Zhao. Further help with the processing of the drawings was provided by R. L. Brooks, G. Qian, G. Rinaldi, and F.-S. Sun.

To all who gave so much support in so many ways, I express my sincere thanks. This book and the Leibniz System could not have been created without their help.

Preface

The engineering advances of the second half of the 20th century have created an avalanche of new technology. Control and use of that technology require, among many things, effective computational methods for logic.

This book proposes one such method. It is based on a new theory of logic computation. Main features of the theory are an extension of propositional logic, an analysis of logic formulas via combinatorial structures, and a construction of logic solution algorithms based on that analysis.

The research for this project began in 1985. Most results were established during the period 1985–1989. To prove practical utility of the new theory, we implemented the ideas in a commercially available software system for logic programming called the Leibniz System, in honor of the mathematician, physicist, engineer, and philosopher G. W. Leibniz (1646–1716).

The first version of the Leibniz System was completed in 1990. Since then, students in a course on expert systems taught by the author have used the Leibniz System to construct numerous expert systems for a wide variety of real-world problems.

In addition, in work with G. Felici, G. Rinaldi, G. Qian, J. Straach, and Y. Zhao, we have developed several large, intelligent systems for complex settings.

These applications of the Leibniz System to real-world problems have verified our prediction that the new theory would yield very fast solution algorithms for real-world logic problems. We began the writing of this book once that fact had been established.

While we created the Leibniz System, other researchers obtained and published a few results that we had found independently but had not published. In this book, we shall not attempt to establish who had done what

ix

first. Instead, we simply cite all references that in any way are related to the material of this book.

Uses of the Book

The material of this book may be used as a blueprint for the design of logic programming software—for example, for expert system shells, for Prolog-type systems, or for systems similar to the Leibniz System.

The book may also be used as the main text for a graduate course in computational logic. We have done so repeatedly with preliminary versions.

Overview

This book is completely self-contained. In particular, it does not require any background in logic or combinatorics.

Chapter 1 reviews the history of logic and explains the basic ideas of the new theory of logic computation.

Chapter 2 provides basic definitions and results of logic and combinatorics. The chapter needs only to be scanned on first reading.

Chapter 3 reviews matroid theory. Here, too, scanning of the material suffices initially.

Chapter 4 defines an extension of propositional logic called system IB and introduces a number of related concepts.

Chapter 5 deals with special classes of logic problems.

Chapter 6 presents characterizations of a certain special class of logic problems. The chapter may be skipped without loss of continuity.

Chapter 7 introduces a property called Boolean closedness that is employed in Chapters 8 and 10.

Chapters 8–12 describe several decompositions for logic problems. The chapters may be read in any order.

Chapter 13 synthesizes the results of the preceding chapters and establishes the so-called analysis algorithm, which constructs solution algorithms for logic problems.

Chapter 14 shows that the analysis algorithm produces efficient solution algorithms for large classes of logic problems. The classes include a great many real-world logic problems.

Acknowledgments

The development of the theory was funded in Germany by the German Science Foundation (Deutsche Forschungsgemeinschaft) and the Alexander von Humboldt-Foundation, and it was funded in the United States by the

Effective Logic Computation

Chapter 1

Introduction

1.1 Overview

Theoretical results in computational complexity predict that computing in logic is very difficult in general, even on very fast machines. However, that negative conclusion leaves the door open for the existence of attractive or at least reasonably effective solution algorithms for particular classes of logic problems. A number of research programs have aimed at such solution algorithms. Each one of those programs belongs to one of two categories.

In programs of the first category, algorithms and theories are desired that handle broad problem classes. The size and complexity of the problem instances is such that researchers are willing to employ heuristic and empirically based methods as part of the computing arsenal. Accordingly, it typically is difficult to predict the performance of the algorithms. So one accepts that a given algorithm may solve one instance very quickly, but may not perform so well on another. There is nothing wrong with that compromise. After all, if such an algorithm produces an important result after extensive or even extraordinary computing effort, then this is good news indeed.

Research programs of the second category focus on particular problem classes and aim at results and theories that produce fast solution algorithms, with guaranteed performance. Such algorithms are needed whenever the time available for the solution of the problem instances is severely limited. Real-world logic situations often impose that constraint. In fact, for real-time systems, it is mandatory that a solution be derived within a guaranteed time interval.

The work of this book falls into the second category. The book presents a new theory for logic computation that supports the construction of fast solution algorithms, with guaranteed performance, for large classes of real-world problems.

Prior work in this area concentrated almost exclusively on a few special problem classes, plus some simple extensions. Real-world applications, however, often lead to logic problems that do not fall into these classes or extensions.

Much of the theory of this book has been used to create a commercially available software system for logic programming. That software system is called the *Leibniz System*, to honor the mathematician, physicist, engineer, and philosopher G. W. Leibniz (1646–1716), who first proposed that logic computations should be employed to solve real-world problems. To date, the Leibniz System has been used to construct a large number of intelligent computer systems. The excellent performance of these systems constitutes empirical proof that the theory presented in this book is practical and useful.

Logic Problems SAT and MINSAT

The logic problems considered in this book involve logic variables and clauses. The variables can take on the values *True* and *False*. The clauses follow a format called *conjunctive normal form* (CNF), where variables, possibly negated, are joined by the logical "or." Suppose certain *True/False* values have been assigned to the variables. Declare a clause to have the value *True* if some variable in the clause has the value *True* or some negated variable in the clause has the value *False*. A clause with the value *True* is also said to be *satisfied*. We consider two problem classes.

For logic instances of the first class, one must decide whether the clauses can be satisfied. If this is so, one must provide an assignment of *True/False* values to the variables proving that conclusion. The class of these instances is called the *satisfiability problem*, abbreviated SAT.

The instances of the second class have variables and clauses like the SAT instances, but also involve, for each variable, a rational nonnegative number called the *cost* for that variable. The cost is incurred when one assigns the value *True* to the variable. The assignment of *False* does not entail any cost. Given the variables, clauses, and costs, one must decide whether the clauses are satisfiable. If this is so, one must obtain an assignment of *True/False* values for the variables such that all clauses are satisfied and the total cost of the assignment is minimum. The class of these logic instances is called the *minimum cost satisfiability problem*, abbreviated MINSAT.

As an aside, the MINSAT condition that the costs for *True* values be nonnegative and the costs for *False* be zero poses no substantive restriction, since any instance with arbitrary costs for *True* and *False* values can be readily transformed to an equivalent one satisfying the stated condition.

Encoding of SAT and MINSAT

The relationships between the variables and clauses of a SAT or MINSAT instance can be conveniently encoded in a $\{0, \pm 1\}$ matrix, say, A, where the columns of A represent the variables and the rows represent the clauses. The entry of A in row i and column j is equal to 1 if the jth variable occurs in the ith clause, is equal to -1 if the jth variable occurs negated in the ith clause, and is 0 otherwise.

In the SAT case, the matrix A completely specifies the instance. In the MINSAT case, we also need a vector, say, c, that contains the cost values. The jth element of c is the cost of *True* for the jth variable. We combine A and c to a matrix/vector pair (A, c) that contains all information about the MINSAT instance.

SAT and MINSAT Problem Classes

A real-world logic application typically gives rise to a few, or at most several dozen, SAT or MINSAT problem classes, each of which may be generated by a two-step process. First, one defines one SAT instance A or MINSAT instance (A, c). Second, one declares the class to consist of A or (A, c) as well as all instances that may be obtained from A or (A, c) by the deletion of rows and columns. We denote by $\overline{A}$ or $(\overline{A}, \overline{c})$ a typical instance of the class.

Given the particular structure of the classes, it would seem possible, indeed prudent, that we handle a given logic application by creating a special solution algorithm for each SAT or MINSAT problem class of that application. The theory for logic computation described in this book is based on that idea.

Analysis Algorithm

A main result of this book is a scheme that investigates the structure of the SAT instance A or MINSAT instance (A, c) defining a given class. Since the scheme analyzes the combinatorial/logic structure of the given A or (A, c), we call it the *analysis algorithm*. Based on the insight gained into the structure of A or (A, c), the analysis algorithm assembles a *solution algorithm* that correctly processes all instances $\overline{A}$ or $(\overline{A}, \overline{c})$ of the class. The analysis algorithm also computes an upper bound on the run time of the solution algorithm. That bound is valid regardless of the case of $\overline{A}$

or $(\overline{A}, \overline{c})$ being solved, and it constitutes a performance guarantee for the solution algorithm.

In the language of computer science, one may call the analysis algorithm a *compiler* that accepts A or (A, c) as input and that outputs a solution algorithm, together with a performance guarantee, for the SAT or MINSAT problem class defined by A or (A, c).

Computational Complexity Versus Practical Utility

Computational complexity theory classifies SAT and MINSAT as difficult. The precise term is $\mathcal{NP}$-*complete*. For the moment, we leave that term undefined and also do not go into the definition of *polynomial time* algorithms. The classifications "$\mathcal{NP}$-complete" and "polynomial time" are based on an asymptotic viewpoint where the performance of algorithms is evaluated as the size of the instances grows arbitrarily large. From that position, polynomial time algorithms are considered to be attractive. On the other hand, according to the present state of knowledge, $\mathcal{NP}$-complete problems are difficult.

In this book, we classify all algorithms according to these and related concepts. We point out, however, that these classifications are not very helpful when one evaluates the practical utility of solution algorithms. This is due to the fact that the real-world classes of SAT and MINSAT instances that one wants to solve often are finite. Accordingly, the asymptotic viewpoint of computational complexity theory does not apply. For example, consider the class of SAT or MINSAT instances having at most 10,000 variables and 10,000 clauses. A great many SAT and MINSAT instances of real-world applications fall into that class. In fact, if one could produce a fast solution algorithm for all such instances, it would be an extraordinary achievement.

So how should one judge the practical utility of the theory described in this book—in particular, the utility of the analysis algorithm? We know of just one way. One should implement the analysis algorithm, apply it to many real-world SAT and MINSAT problem classes, and evaluate the performance bounds of the solution algorithms so produced. A number of years ago we began that effort, and in the process we created the earlier-mentioned Leibniz System. Using that system, we have produced solution algorithms for hundreds of SAT and MINSAT problem classes arising from practical applications. In almost all cases, the solution algorithm generated by the Leibniz System turned out to have a small upper time bound and thus was guaranteed to be fast. The implementation effort for the Leibniz System had the negative effect of delaying publication of this book for a number of years. But given the test results, we publish our work in confidence, knowing that the ideas, results, and algorithms proposed here are practical and useful.

1.2 History

This section points out key facts and results of the past that paved the way for computing in logic or *computational logic*, as it is now called. Excellent references are Newman (1956) and Kneale and Kneale (1984).

In the discussion below, we repeatedly refer to an "or" operator being *exclusive* or *inclusive*. The exclusive "or" means "either ... or ...," and the inclusive "or" means "... or ... or both."

We shall not attempt to even outline the roots of logic. Suffice it to say that the Greeks—in particular, Aristotle (384–322 B.C.)—developed logic to clarify discourse and to verify the correctness of mathematical arguments. The results formulated at that time stood virtually unchanged right up to the time of G. W. Leibniz (1646–1716), who developed several versions of a calculus for logic.

Leibniz

In the first version of the logic calculus, Leibniz employed $+$ and $-$ as operators. In today's terminology, the $+$ is the exclusive "or." The $-$ operator is defined indirectly, by declaring that $C = A - B$ if B and C have nothing in common and $A = B + C$. These definitions cause difficulties. For example, $A + A$ makes sense only if A is nil, and $A - B$ requires that A subsumes B.

Leibniz eventually dropped the $-$ operator and moved from $+$ to a more general $\oplus$ that supports two interpretations. In today's language, the axioms formulated by Leibniz for $\oplus$ permit interpretation as the logical "and" or as the inclusive "or." It is unfortunate that Leibniz permitted one symbol to have these two interpretations. Had he employed two symbols, he probably would have developed axioms linking the two concepts and thus would have created the foundation for modern logic. As it was, that development had to wait one hundred and fifty years.

Nevertheless, the results obtained by Leibniz for the $\oplus$ operator, as well as his other far-ranging investigations and efforts that, for example, produced a calculating machine for addition, subtraction, multiplication, and division, propelled him to envision an encoding of human knowledge and a computational process that would reduce rational discussions to mere calculations. Thus, it is appropriate to consider Leibniz to be the father of computational logic.

Boole

We move forward by one hundred and fifty years, to the time of G. Boole (1815–1864). He created a calculus for logic that, in today's terms, uses the logical "and," the exclusive "or," and a certain inverse of that "or."

We sketch Boole's approach. He starts with classes having certain features. Let x, y, z denote such classes. The universe of discourse is denoted by 1, and the null class is denoted by 0. The class having the features of both x and y is xy. For x and y with disjoint features, $x + y$ is the class where each element has either the features of x or those of y. Finally, if $z = x + y$, then $x = z - y$ is declared to hold.

Since 1 is the universe and 0 is the null class, $1x = x$ and $0x = 0$. The complement of x is $1 - x$. Just as in the first calculus of Leibniz, these definitions cause problems. But then Boole defines a specialized version by adding the condition that each class must be equal to 0 or 1. This solves the problems arising from subtraction. Boole handles the undefined $x + x$ by allowing integers larger than 1 during computations. The latter change has the drawback that intermediate steps of calculations may not have an interpretation in terms of classes. Nevertheless, Boole's system makes logic calculations possible.

Subsequently, Boole's system was simplified. The symbol $+$ was declared to stand for the inclusive "or," and the $-$ operator was dropped. That system is known today as *Boolean algebra* or *propositional logic*.

Frege, Russell, Whitehead, Hilbert, and Gödel

In terms of time, it is a short step from Boole to G. Frege (1848–1925), who created a complete notation for mathematical logic. He used that notation in far-ranging investigations. Today, Frege's formulas would be called *parse trees* of logic expressions. Such trees may be a bit cumbersome to typeset, and they might have seemed strange to contemporaries. But they exhibit the structure of logic expressions clearly and at least as well as the logic notation in use today.

Frege envisioned that mathematics could be based on logic. He set out to prove that conjecture in an extraordinary effort spanning many years. His work contains many novel ideas. But later, B. Russell (1872–1970) showed that the assumptions made by Frege contained a flaw.

A. N. Whitehead (1861–1947) and Russell successfully carried out much of Frege's plan and argued in a three-volume treatise called *Principia Mathematica* that all of mathematics can be derived from logic. At that time, it seemed that in principle one could derive all theorems of mathematics by computation.

Subsequently, D. Hilbert (1862–1943) refined the *axiomatic method*, introduced *meta-mathematics*, and hoped that with these tools one could carry out a program that eventually would establish the consistency of most if not all mathematical systems.

In 1931, K. Gödel (1906–1978) proved that the construction of Whitehead and Russell cannot build all of mathematics and that Hilbert's program generally cannot be carried out. That result, called the *incomplete-*

ness theorem, implies that one cannot systematically compute all theorems of mathematics starting from some given axioms.

The availability of computers in the second half of the 20th century raised the hope that computations in logic could be used to solve real-world problems. Loveland (1984) gives a concise account of the developments since the 1950s. The vision of effective logic computation is gradually being translated into reality. We give an overview of the various approaches later, once we have defined the logic problems to be considered.

1.3 Logic Problems

We discuss propositional logic, give precise definitions of the problems SAT and MINSAT, and introduce first-order logic and related material.

Propositional Logic

Propositional logic consists of *Boolean variables*, which can take on the values *True* and *False*, and the operators ¬ (denoting "not"), ∧ (denoting "and"), and ∨ (denoting inclusive "or"). We skip the axioms of propositional logic, since they are discussed in Chapter 2.

CNF System

A *conjunctive normal form system*, abbreviated CNF *system*, is, for some $n \geq 1$, a logic expression of the form $S_1 \wedge S_2 \wedge \ldots \wedge S_n$, where each S_i consists of possibly negated Boolean variables that are joined by ∨. An example for S_i is $s_1 \vee \neg s_2$. Each S_i is called a CNF *clause*, or simply a *clause* for present purposes.

SAT and MINSAT

Suppose we assign *True/False* values to the variables of a CNF system S. Declare a clause of S to be *satisfied* if the clause contains at least one variable with the value *True* or at least one negated variable with the value *False*. The CNF system S is *satisfied* if all clauses are satisfied.

Collectively, the CNF systems constitute the instances of the *satisfiability problem* SAT. Given an instance S, one must find *True/False* values for the variables of S such that S is satisfied or declare that no such values exist. In the former case, the *True/False* values constitute a *satisfying solution* for S. In the latter case, S is *unsatisfiable*.

Suppose we assign to each variable of a CNF system a rational non-negative cost, which is interpreted to be the cost of assigning *True* to the variable. The assignment of *False* does not result in any cost. The CNF systems with such costs are the instances of the *minimum cost satisfiability problem* MINSAT. Given an instance of MINSAT, one must find a satisfying solution such that the total cost of the assignment is minimum or declare the instance to be unsatisfiable.

Computational complexity theory classifies both SAT and MINSAT, as well as several subclasses, as difficult; see Cook (1971), Garey and Johnson (1979), Blass and Gurevich (1982), Lichtenstein (1982), Tovey (1984), Hunt and Stearns (1990), Kratochvíl, Savický, and Tuza (1993), Boros, Crama, Hammer, and Saks (1994), and Kratochvíl (1994).

First-Order Logic

We move from propositional logic to *first-order logic*, defined as follows. Let U be some nonempty set, called the *universe*. For any $k \geq 1$, denote the set $\{(u_1, u_2, \ldots, u_k) \mid u_i \in U, \text{ for all } i\}$, which is the *product* of k copies of U, by $\prod_{i=1}^{k} U$. When k is small, we may denote $\prod_{i=1}^{k} U$ by $U \times U \times \ldots \times U$, with k U-terms. A *predicate* or *truth function* is a function p that, for some $k \geq 1$, takes the elements of $\prod_{i=1}^{k} U$ to the set $\{True, False\}$.

Predicates are employed in connection with the *universal quantifier* $\forall$, denoting "for all" or "for each," and the *existential quantifier* $\exists$, denoting "there exists." Example statements involving these quantifiers and a predicate p that takes the pairs of $U \times U$ to $\{True, False\}$ are $\forall x (\forall y [p(x, y)])$ and $\forall x (\exists y [p(x, y)])$, meaning "for all $x \in U$ and for all $y \in U$, $p(x, y)$ holds" and "for each $x \in U$, there exists $y \in U$ such that $p(x, y)$ holds." Note the implicit use of U in each quantification.

We skip details of the construction of logic formulas of first-order logic. It is covered in Chapter 2. Suffice it to say that the quantifiers $\forall$ and $\exists$ are combined with the operators $\neg$, $\wedge$, and $\vee$ of propositional logic.

Analogously to SAT and MINSAT, one may define a *satisfiability problem* and a *minimum cost satisfiability problem* of first-order logic.

In general, the universe U is considered to be infinite. When U is restricted to be finite, the logic formulas can be restated as expressions of propositional logic. Chapter 2 contains details. In this book, we only consider the case of finite U. The satisfiability problem and the minimum cost satisfiability problem of first-order logic then become the problems SAT and MINSAT of propositional logic.

Other Types and Problems of Logic

There are many other types of logic—for example, nonmonotonic logic, probabilistic logic, and logic under uncertainty. There are also a number of logic problems besides SAT and MINSAT. Space constraints prohibit even a cursory discussion of these logic types and problems. Details may be found in Chang and Lee (1973), Loveland (1978, 1984), Hailperin (1986), Genesereth and Nilsson (1987), Pearl (1988), Fitting (1990), Guan and Bell (1991), Wos, Overbeek, Lusk, and Boyle (1992), Bibel (1993), and Nerode and Shore (1993).

In this book, we shall not describe applications of computational logic, since even a terse summary would require many pages and even at that would fail to depict the richness and range of possibilities just for the uses of SAT and MINSAT. Indeed, the word "applications" itself is too restrictive. A better view is that, in almost any rational endeavor, some reasoning is required and can be carried out by tools of logic.

1.4 Prior Results

Section 1.1 classifies research programs according to two categories. Programs of the first category deal with relatively large classes of logic problems. They aim at solution algorithms that need not have a good performance guarantee. In contrast, programs of the second category only consider solution algorithms with such a guarantee. In this section, we outline prior approaches and results of the programs of both categories. Since this book is concerned with the SAT and MINSAT problems, we restrict ourselves to solution methods for these problems. We begin with the SAT case.

SAT Algorithms

The methods for SAT employed in research programs of the first category can be classified as follows: resolution, enumeration, polyhedral techniques, and encoding of solution space. For a comprehensive coverage of these methods and variations, see Kleine Büning and Lettmann (1994).

We do not include here randomized or heuristic methods—for example, the schemes proposed in Kamath, Karmarkar, Ramakrishnan, and Resende (1990, 1992), Dubois and Carlier (1991), Koutsoupias and Papadimitriou (1992), or Franco and Swaminathan (1997b)—since they may fail to solve a given instance.

We first discuss resolution.

Resolution

It is convenient that we view each clause of a CNF system as a set which contains the variables of the clause. For example, the CNF system (s_1 ∨

$\neg s_2$) $\wedge \neg s_3 \wedge (s_1 \vee s_4)$ becomes in set notation $\{s_1, \neg s_2\}$, $\{\neg s_3\}$, $\{s_1, s_4\}$. Note that the empty set corresponds to a clause without variables, called the *empty clause*. Clearly, that clause cannot be satisfied.

Suppose y is a variable of a CNF system defined by sets C_1, C_2, $\ldots$, C_n. Since any clause containing both y and $\neg y$ always evaluates to *True* and thus does not impose any restriction, we may assume that no such clause is present. Hence, some of the clauses C_1, C_2, $\ldots$, C_n contain y and not $\neg y$, others contain $\neg y$ and not y, and the remaining clauses do not contain y or $\neg y$.

Suppose a clause C_i contains y, and a clause C_j contains $\neg y$. Using $D_i = C_i - \{y\}$ and $D_j = C_j - \{\neg y\}$, we define a new clause $D_i \cup D_j$. The derivation of $D_i \cup D_j$ from C_i and C_j is called *resolution*.

We claim that any satisfying solution for C_i and C_j also satisfies $D_i \cup D_j$. The proof is as follows.

If a satisfying solution for C_i and C_j has $y = True$, then the term $\neg y$ of C_j is *False*, and the fact that C_j is satisfied implies that $D_j = C_j - \{\neg y\}$ is satisfied. Similarly, if y has the value *False*, then D_i is satisfied. Thus, no matter which value is assigned to y, we must have $D_i \cup D_j$ satisfied.

One may use resolution to reduce a given CNF system to one with fewer variables that is satisfiable if and only if this is so for the original CNF system. To eliminate one variable, say, y, we replace all clauses C_i containing y and all clauses C_j containing $\neg y$ by all possible clauses $D_i \cup D_j$ derivable by resolution. The above arguments prove that the new CNF system is satisfiable if this is so for the original one. To show the converse, let us assume that we have a satisfying solution for the new system. Thus, all $D_i \cup D_j$ are satisfied. But this implies that all D_i or all D_j are satisfied. Assume the former case. Then all clauses $C_i = D_i \cup \{y\}$ are satisfied. Using $y = False$, all clauses $C_j = D_j \cup \{\neg y\}$ are satisfied as well. The case where all D_j are satisfied is handled by $y = True$. Thus, in both cases, the original CNF system is satisfied.

We combine the above resolution step for the elimination of variables with elementary reduction steps to get a solution algorithm for SAT that is usually called the *resolution algorithm*. The reductions are as follows.

If there is no clause at all, arbitrarily assign *True/False* values to the variables, and declare the CNF system to be satisfiable. If there is an empty clause, then declare the system to be unsatisfiable. If there is a clause with just one variable y (resp. $\neg y$), then any satisfying solution must have $y = True$ (resp. $y = False$); hence, we assign the appropriate value to y and reduce the system by deleting all satisfied clauses and all occurrences of y or $\neg y$ from the remaining clauses.

Davis and Putnam (1960) first proposed the above resolution algorithm. For this reason, it is sometimes called the *Davis–Putnam algorithm*. We should mention, though, that some authors use the name "Davis–Putnam algorithm" for the enumerative method of the next subsection and cite the Davis and Putnam (1960) reference. That use of the name seems inappropriate, since the reference does not contain any enumerative algorithms.

It is well known that the resolution method may be viewed as a specialization of an algorithm for solving linear inequalities called the *Fourier–Motzkin elimination method*. For details about the latter method, see Schrijver (1986).

The performance of the resolution algorithm generally is not good. The main reason is that the elimination of variables may increase the number of clauses rather substantially. For details, see Davis, Logemann, and Loveland (1962), Tseitin (1968), Galil (1977a, 1977b), Ben-Ari (1980), Haken (1985), Urquhart (1987), Buss and Turán (1988), Chvátal and Szemerédi (1988), Hooker (1988a), Cook and Pitassi (1990), and Fouks (1992).

For additional material about resolution, see Chang (1970), Cook, Coullard, and Turán (1987), Hooker (1988c, 1992, 1996), Kleine Büning and Löwen (1989), Avron (1993), Goerdt (1992a, 1992b, 1993), Heusch (1994), and Kullmann (1997a).

We should emphasize that Davis and Putnam (1960) do not employ the term "resolution" in the description of their algorithm. That term and the generalization of the above-described resolution step to the *resolution principle* of first-order logic are due to Robinson (1965a, 1965b). One cannot possibly overstate the importance of that seminal contribution. It started an avalanche of research in computation for first-order logic. Since in this book we confine ourselves to the SAT and MINSAT problems of propositional logic, we omit discussion of the resolution principle and of the many results based on it. The interested reader should consult the books cited at the end of Section 1.3.

We digress for a moment. The resolution step is implicit in the system of *syllogisms* of the Greeks. Various laws of propositional logic—to wit, the laws of *conjunction argument, constructive dilemma, destructive dilemma, detachment* (= *modus ponens*), and *disjunctive inference*, as well as the *chain rule* (= law of the *syllogism*) and *modus ponens*—are nothing but manifestations of resolution. Given that fact, we marvel at the persistence with which some texts on logic or discrete mathematics ask the reader to understand and differentiate among the cited laws, when it would suffice to just introduce the unifying resolution step.

We turn to enumeration.

Enumeration

Algorithms based on enumeration use the same elementary reduction steps as the resolution algorithm. However, the elimination of variables in the latter method is replaced by a search that involves a rooted binary tree where the two descendants of any node represent the fixing of some variable to *True* and *False*. Thus, for any node other than the root, the unique path from that node to the root corresponds to the fixing of certain variables to *True/False* values.

With each node of the tree, we associate a CNF system. For the root node, it is the original CNF system. For any other node, it is the CNF system that is obtained from the original one by fixing variables as specified by the path from the root to the given node.

In general, enumerative algorithms explicitly or implicitly construct the tree, starting with the root node, until a satisfying solution is found or unsatisfiability can be proved. There are numerous ways to carry out that construction task, with attendant differences of computational efficiency. The first such method is due to Davis, Logemann, and Loveland (1962) and is sometimes called the *Davis–Putnam–Loveland algorithm*. Since that time, numerous versions have been created. Recent results and related material are in Purdom (1984), Monien and Speckenmeyer (1985), Blair, Jeroslow, and Lowe (1986), van Gelder (1988), Bugrara, Pan, and Purdom (1989), Gallo and Urbani (1989), Iwama (1989), Jeroslow and Wang (1990), Dubois (1991), Tanaka (1991), Billionnet and Sutter (1992), Larrabee (1992), Letz, Schumann, Bayerl, and Bibel (1992), Hooker (1993), Pretolani (1993a, 1996), Vlach (1993), Zhang (1993), Schiermeyer (1993, 1996), Gallo and Pretolani (1995), Hooker and Vinay (1995), Böhm (1996), Böhm and Speckenmeyer (1996), Freeman (1996), Rodošek (1996), Zhang (1996), Kullmann (1997b, 1997c), Kullmann and Luckhardt (1997), and Rodošek and Schiermeyer (1997).

For probabilistic analyses of enumerative algorithms, see Brown and Purdom (1981), Goldberg, Purdom, and Brown (1982), Franco (1983, 1986, 1991, 1993), Franco and Paull (1983), Purdom (1984, 1990), Purdom and Brown (1985a, 1985b, 1987), Speckenmeyer, Monien, and Vornberger (1988), Bugrara and Purdom (1988), Franco and Ho (1988), Bugrara, Pan, and Purdom (1989), Iwama (1989), Dubois and Carlier (1991), and Speckenmeyer, Böhm, and Heusch (1997). For a comprehensive coverage of probabilistic results, see Franco and Swaminathan (1997a).

We describe polyhedral techniques next.

Polyhedral Techniques

Section 1.1 refers to a matrix encoding of SAT and a matrix/vector encoding of MINSAT. We need details of that encoding to discuss polyhedral methods.

Consider the single clause

(1.4.1) $$\neg x_1 \lor x_2 \lor x_3 \lor \neg x_4$$

It evaluates to *True* if and only if $x_1 = $ *False*, or $x_2 = $ *True*, or $x_3 = $ *True*, or $x_4 = $ *False*. Compare the clause of (1.4.1) with the inequality

$$(1.4.2) \qquad (-1) \cdot (r_1 - 1) + (+1) \cdot r_2 + (+1) \cdot r_3 + (-1) \cdot (r_4 - 1) \geq 1$$

where r_1, r_2, r_3, r_4 are $\{0, 1\}$ variables. That inequality is satisfied if and only if $r_1 = 0$, or $r_2 = 1$, or $r_3 = 1$, or $r_4 = 0$.

Suppose that, for $j = 1, 2, 3, 4$, we affiliate $x_j = $ *True* (resp. $x_j = $ *False*) with $r_j = 1$ (resp. $r_j = 0$). Then any solution satisfying the clause (1.4.1) corresponds to a solution for the inequality (1.4.2), and vice versa.

We move the constant terms of the left-hand side of (1.4.2) to the right-hand side and get

$$(1.4.3) \qquad (-1) \cdot r_1 + (+1) \cdot r_2 + (+1) \cdot r_3 + (-1) \cdot r_4 \geq 1 - 2$$

Let r be the column vector containing r_1, r_2, r_3, r_4. Define a matrix A by

$$(1.4.4) \qquad A = [-1, +1, +1, -1]$$

With that notation, the inequality (1.4.3) can be written as

$$(1.4.5) \qquad A \cdot r \geq 1 - 2$$

Note that the 2 on the right-hand side corresponds to the number of -1s in the single row of A.

We move from a single clause to a CNF system S with variables x_1, $x_2, \ldots, x_n$ and with m clauses. Define A to be the $m \times n$ matrix where the entry in row i and column j is 1 if clause i contains x_j, is -1 if clause i contains $\neg x_j$, and is 0 otherwise. Let r be the column vector containing $\{0, 1\}$ variables $r_1, r_1, \ldots, r_n$. Declare $q(A)$ to be the integer $m \times 1$ vector whose entry in position i is equal to the number of -1s in row i of A. Let $\underline{1}$ denote any column vector containing just 1s.

Analogously to the earlier example, we associate the value *True* (resp. *False*) for any x_j with the value 1 (resp. 0) for r_j. It is then easy to see that *True*/*False* values for the x_j satisfy the CNF system S if and only if the corresponding $\{0, 1\}$ vector r satisfies the inequality

$$(1.4.6) \qquad A \cdot r \geq \underline{1} - q(A)$$

Polyhedral methods decide satisfiability of a CNF system by determining whether the corresponding inequality (1.4.6) has a $\{0, 1\}$ solution. We sketch the main approach.

We relax the condition that r be a $\{0, 1\}$ vector, and demand instead that r be rational and satisfy $0 \leq r \leq \underline{1}$. Let $P(A)$ be the *polyhedron* given by

$$(1.4.7) \qquad P(A) = \{r \mid A \cdot r \geq \underline{1} - q(A); \ 0 \leq r \leq \underline{1}\}$$

With these definitions, determining satisfiability for a CNF system has become equivalent to deciding whether the corresponding polyhedron $P(A)$ contains an integer vector. The latter problem is a special version of one of the core problems of *polyhedral combinatorics* where one must settle whether an arbitrary rational polyhedron, which generally is defined by a finite number of inequalities, contains an integer vector.

Polyhedral combinatorics was founded by J. Edmonds in the 1960s. Since that time, several techniques have been developed to decide whether a polyhedron contains an integer point. A good reference is Nemhauser and Wolsey (1988). For early approaches concerning the SAT problem, see Jeroslow (1989). Suffice it to say here that the presently most powerful method, called *branch and cut*, reduces a given polyhedron by adding inequalities called *cuts* to those defining the polyhedron. The cuts do not eliminate integer vectors of the polyhedron. If cuts cannot be found by a reasonable computing effort, the method switches to enumeration, where variables are fixed to integer values.

The general branch and cut method has been specialized for the polyhedron $P(A)$. Relevant references are Blair, Jeroslow, and Lowe (1986), Hooker (1988a, 1988b, 1988c, 1989, 1992, 1996), Hooker and Fedjki (1990), and Jeroslow and Wang (1989).

Besides branch and cut, other methods of polyhedral combinatorics have been used to decide whether $P(A)$ contains an integer vector. For example, see Harche, Hooker, and Thompson (1994), which is based on Harche and Thompson (1994). Additional results concerning $P(A)$ are in Wang (1993). For a complete overview of polyhedral methods for $P(A)$, see Chandru and Hooker (1997).

Encoding of Solution Space

A given CNF system S may admit a great many satisfying solutions. But that fact, by itself, does not imply that one cannot find a compact representation for these solutions. Bryant (1986) pursued that idea and produced an impressive method for encoding the solutions.

The encoding scheme accepts any Boolean formula involving the operators $\neg$, $\wedge$, and $\vee$. Together with such a formula, an ordering of the variables must be supplied. The scheme then encodes the solutions of the given formula by a directed acyclic graph. Once that graph is at hand, the satisfiability problem for the given formula is easily solved.

The size of the encoding graph generally depends not only on the formula, but also on the ordering of the variables. In fact, for the same formula, one ordering may produce a small graph, while another ordering may result in a large graph.

Bryant (1986) does not supply a method for selecting an ordering of the variables that produces a smallest encoding graph. However, Bryant

(1986) proves that the encoding scheme creates a graph that in a certain sense is smallest for a given ordering of the variables.

Implementations and Tests

A number of implementations of the above methods exist. For details and computational results, see Blair, Jeroslow, and Lowe (1986), Bryant (1986), Hooker (1988a, 1993), Gallo and Urbani (1989), Hooker and Fedjki (1990), Jeroslow and Wang (1990), Tanaka (1991), Billionnet and Sutter (1992), Larrabee (1992), Letz, Schumann, Bayerl, and Bibel (1992), Buro and Kleine Büning (1993), Pretolani (1993a, 1996), Vlach (1993), Zhang (1993), Harche, Hooker, and Thompson (1994), Mayer, Mitterreiter, and Radermacher (1995), Gallo and Pretolani (1995), Hooker and Vinay (1995), Böhm (1996), Böhm and Speckenmeyer (1996), Crawford and Auton (1996), Freeman (1996), Gent and Walsh (1996), Mitchell and Levesque (1996), Schrag and Crawford (1996), Selman and Kirkpatrick (1996), and Selman, Mitchell, and Levesque (1996).

Most test problems of the cited references are either randomly generated or represent special combinatorial problems. The results do not establish any one of the methods to be uniformly best. Instead, each one of the methods works well on some classes of problems and does not perform so well on others. From our experience, the structure of SAT instances arising from real-world problems typically is quite different from that of the cited test instances. Hence, it is not clear how the various methods perform on SAT instances of real-world problems.

So far, we have covered algorithms for the SAT problem produced by research programs of the first category. We turn to research programs of the second category. The SAT algorithms developed by these programs apply to special SAT instances having a certain structure. We sketch the cases below. Chapter 5 contains details and references.

There are three main classes of special SAT instances, plus some extensions. We describe the classes in terms of the earlier introduced representation of SAT instances by matrices A.

2SAT Matrices

Matrices of the first class have at most two nonzero entries in each row. The class of such SAT instances is called 2SAT.

Nearly Negative Matrices

Matrices of the second class have mostly -1s as nonzero entries. Specifically, each row has at most one 1. We call such matrices *nearly negative*.

The CNF systems giving rise to nearly negative matrices are usually called *Horn systems*. A generalization of nearly negative matrices permits more than one 1 in each row, but demands that a certain scaling operation be able to convert a given matrix to a nearly negative one. We call such matrices *hidden nearly negative*. The CNF systems producing such matrices typically are called *hidden* or *disguised Horn systems*.

Balanced Matrices

Matrices of the third class are called *balanced*. They are defined by the absence of certain submatrices that we shall not specify here. The reader should consult Chapter 5 for details. That chapter also describes some subclasses and extensions of the three matrix classes. Suffice it to say here that fast recognition and solution algorithms exist for most of those classes, subclasses, and extensions.

We have completed the presentation of solution approaches for the SAT problem and turn to schemes for the MINSAT problem.

MINSAT Algorithms

There are few prior results for MINSAT except for one special subclass called MIN2SAT, where each matrix has 2SAT form. That subclass is treated in Gusfield and Pitt (1992) and in Hochbaum, Megiddo, Naor, and Tamir (1993).

However, a related logic problem called MAXSAT has been investigated to quite an extent. A MAXSAT instance is a SAT instance where a positive weight has been assigned to each clause. One must produce an assignment of *True/False* values for the variables such that the total weight of the satisfied clauses is maximum.

The MAXSAT problem is difficult; see Garey, Johnson, and Stockmeyer (1976), Lieberherr and Specker (1981), Jaumard and Simeone (1987), and Arora, Lund, Motwani, Sudan, and Szegedy (1992).

Research efforts for MAXSAT have produced exact as well as approximation algorithms. Solution approaches typically rely on linear programming, polyhedral methods, and network flow techniques. Some methods have been devised for subclasses—in particular, for the subclass MAX2SAT where each matrix has 2SAT form. For details, see Johnson (1974), Lieberherr and Specker (1981), Lieberherr (1982), Poljak and Turzík (1982), Hansen and Jaumard (1990), Yannakakis (1992), Kratochvíl and Křivánek (1993), Goemans and Williamson (1994, 1995), Feige and Goemans (1995), Gallo, Gentile, Pretolani, and Rago (1997), and Cheriyan, Cunningham, Tunçel, and Wang (1996). The approach of Gallo, Gentile, Pretolani, and Rago (1997) relies on a hypergraph formulation that has been used by

Carraresi, Gallo, and Rago (1993) and Rago (1994) for constraint logic programming and first-order logic.

It is easy to show that each MAXSAT instance can be readily converted to a MINSAT instance, and vice versa. Details are included in Chapter 2. Thus, MAXSAT solution algorithms may be used to solve the MINSAT problem. We should mention, though, that approximation algorithms for MAXSAT generally are not useful for that purpose.

We have completed the overview of prior results for SAT and MINSAT. In the next section, we outline the approach taken in this book to solve these problems.

1.5 Overall Approach

Recall from Section 1.1 that real-world problems typically produce SAT or MINSAT problem classes that may be constructed as follows. For a given class, one first defines one SAT instance A or MINSAT instance (A, c). Then one declares the class to contain A or (A, c) as well as all instances that may be obtained from A or (A, c) by the deletion of rows or columns. Let $\overline{A}$ or $(\overline{A}, \overline{c})$ denote an arbitrary instance of the class.

Our approach to solving the SAT or MINSAT instances of such a class is as follows. We analyze the structure of the SAT problem A or of the MINSAT problem (A, c) defining the class. Based on that analysis, we construct a solution algorithm that can solve all instances of the class. We also compute an upper time bound for the run time of the solution algorithm. The analysis algorithm carries out these steps. The scheme contains two groups of subroutines.

The subroutines of the first group determine whether a given A or (A, c) has one of the special properties discussed in the previous section. Specifically, in the SAT case, it is tested whether the given matrix A is in the class 2SAT, is hidden nearly negative, or is balanced. In the MINSAT case, it is checked whether the matrix A of the given matrix/vector pair (A, c) is nearly negative or balanced. Note that the cited properties are inherited under submatrix taking. So if A or (A, c) has one of the properties, then this is so for all instances of the class defined by A or (A, c).

The subroutines of the second group carry out five decompositions that break down a given A or (A, c). The components of each decomposition are obtained from A or (A, c) by the deletion of some nonzero entries or by submatrix taking; the latter step may be followed by the adjoining of some rows and columns.

The analysis algorithm utilizes the subroutines of the two groups in the following manner. Given A or (A, c), the analysis algorithms first tests whether A or (A, c) has one of the special properties. If that is so, the known solution algorithm for that case solves each $\overline{A}$ or $(\overline{A}, \overline{c})$ of the class. In addition, the performance guarantee for the solution of A or (A, c) applies to each $\overline{A}$ or $(\overline{A}, \overline{c})$ as well.

If A or (A, c) does not have one of the special properties, then the analysis algorithm recursively searches for a decomposition of A or (A, c) into components, which in turn are processed analogously to A or (A, c).

Upon termination of the decomposition process, the analysis algorithm creates from the information on hand a solution algorithm that handles all instances $\overline{A}$ or $(\overline{A}, \overline{c})$ derivable from A or (A, c), and it computes an upper bound on the run time of that solution algorithm.

The above sketch of the analysis algorithm skips over a number of difficulties. For example, the decompositions must be so selected that they are in some sense mathematically compatible. Also, the decomposition process involves the solution of a number of combinatorial problems that theoretically and practically are just as difficult as SAT or MINSAT. The latter point may prompt the thought that the proposed method simply trades the difficult SAT or MINSAT problem for an equally difficult problem. But that is not so. The original task is to solve the SAT or MINSAT instances produced by a given A or (A, c). In contrast, the decomposition process demands that we find decompositions that, according to some measure, are good but not necessarily optimal. That fact allows us to use approximate or heuristic decomposition methods. We emphasize that each solution algorithm derived from the decompositions is an exact method that solves all SAT or MINSAT instances $\overline{A}$ or $(\overline{A}, \overline{c})$ arising from A or (A, c) within the time guaranteed by the performance bound.

Validity of the subroutines of the analysis algorithm is proved with a new algebra called *system* $\mathbb{B}$ that is an extension of propositional logic. The algebra uses three binary operators $\odot$, $\oplus$, and $\ominus$; they carry out $\mathbb{B}$-*multiplication*, $\mathbb{B}$-*addition*, and $\mathbb{B}$-*subtraction*, respectively.

The $\odot$ operator takes $\{0, \pm1\}$ valued α and β to $\{0, 1\}$ valued $\alpha \odot \beta$. Specifically, $\alpha \odot \beta$ is 1 if $\alpha = \beta = 1$ or $\alpha = \beta = -1$, and it is 0 otherwise. Let a value of 1 (resp. -1) for α or β represent the value *True* (resp. *False*). Declare that a value of 1 (resp. 0) for $\alpha \odot \beta$ designates the value *True* (resp. *False*). Then $\alpha \odot \beta$ is *True* if α and β are both *True* or both *False*, and it is *False* otherwise. Thus, for α and β restricted to $\{\pm1\}$ values, the $\odot$ operator acts like the "if and only if" operator $\Leftrightarrow$ of propositional logic.

The $\oplus$ operator takes $\{0, 1\}$ valued α and β to $\{0, 1\}$ valued $\alpha \oplus \beta$. That is, $\alpha \oplus \beta$ is 1 if α or β is 1, and it is 0 otherwise. Letting 1 denote *True* and 0 denote *False*, $\alpha \oplus \beta$ is *True* if α or β is *True*, and it is *False* otherwise. Thus, $\oplus$ behaves like the inclusive "or" operator $\vee$ of propositional logic.

The $\ominus$ operator takes $\{0, 1\}$ valued α and β to $\{0, 1\}$ valued $\alpha \ominus \beta$. Specifically, $\alpha \ominus \beta$ is 1 if $\alpha = 1$ and $\beta = 0$, and it is 0 otherwise. Evidently, the $\ominus$ operator is some sort of inverse of $\oplus$ and thus of the inclusive "or" operator $\vee$ of propositional logic.

As described in Section 1.2, both Leibniz and Boole defined a subtraction operator in connection with the exclusive "or." That subtraction operator was later dropped, and the exclusive "or" was replaced by the inclusive "or." Given that history, our reintroduction of a subtraction operator, this time in connection with the inclusive "or," might seem to be a backward step. Indeed, how can it be that logic subtraction, discarded long ago as not useful, suddenly has become important?

The reason is that Leibniz and Boole, and presumably subsequent researchers, evaluated the utility of logic subtraction for logic equations. In contrast, we use it in connection with inequalities involving the values 0 and 1, where, as expected, 1 is considered to be greater than 0. For example, let α, β, and γ be $\{0,1\}$ variables. Direct checking verifies that, according to the above definitions of $\oplus$ and $\ominus$, we have $\alpha \geq (\alpha \oplus \beta) \ominus \beta$, $\alpha \leq (\alpha \ominus \beta) \oplus \beta$, and $(\alpha \ominus \beta) \oplus (\beta \ominus \gamma) \geq \alpha \ominus \gamma$. These useful inequalities cannot be strengthened to become equations. For example, $\alpha \geq (\alpha \oplus \beta) \ominus \beta$ holds as a strict inequality when $\alpha = \beta = 1$.

But why are we interested in logic inequalities? The answer is that both SAT and MINSAT may be formulated using matrix inequalities of the form $A \odot s \geq b$, where A is the $\{0, \pm 1\}$ matrix representing a CNF system, b is a $\{0, 1\}$ vector, and s is a $\{0, \pm 1\}$ vector. We skip details for the time being and only mention that b defines which clauses of A must be satisfied, that s is the solution vector, and that the matrix multiplication $\odot$ is defined via the above $\odot$ and $\oplus$ in the same way that the customary matrix multiplication is defined via scalar multiplication and addition.

The representation of SAT and MINSAT by matrix inequalities over the system $\mathbb{B}$ supports the analysis of SAT and MINSAT instances with certain tools of combinatorics. The underlying notion is to analyze the matrix A using new concepts such as *Boolean independence*, *Boolean rank*, and *Boolean basis* that are adaptations of familiar concepts of linear algebra. These concepts are crucial for the approximation of features of inequalities $A \odot s \geq b$ by combinatorial structures. For that purpose, we mainly employ the ternary field GF(3), graphs, and matroids. The latter structures are a generalization of matrices over fields and graphs. The links so established between $A \odot s \geq b$ and the cited combinatorial structures are the basis for the decomposition process employed by the analysis algorithm.

The methods developed in this book for SAT and MINSAT can be extended. The system $\mathbb{B}$ can be generalized to the notion of $\mathbb{D}$-*systems*, with attendant generalization of the operators $\odot$, $\oplus$, and $\ominus$. Some combinatorial problems different from SAT and MINSAT may be formulated by $\mathbb{D}$-systems, and then they may be treated analogously to SAT and MINSAT. Chapter 4 includes details.

We conclude this section with a remark on our treatment of computational complexity.

The theory of computational complexity defines an algorithm that produces "yes" or "no" as output to be *polynomial time* if the run time can be bounded by a polynomial function of the length of the input string.

Since the subroutines employed by the analysis algorithm are often complicated, we usually describe simplified versions that, at some loss of efficiency, clearly exhibit the underlying principles and ideas. Accordingly, we typically claim those subroutines to be polynomial time and omit details about the bounding polynomial functions. It is easy to verify, though, that any reasonable implementation of the subroutines leads to computationally effective versions.

On the other hand, the solution algorithms constructed by the analysis algorithm are composed of conceptually simple subroutines. Without loss of clarity, we can describe very efficient versions of those subroutines, together with usually tight polynomial bounds on run time.

1.6 Reading Guide

We include a reading guide for the subsequent chapters. The material can be grouped into four parts.

The first part, which consists of Chapters 2–4, describes relevant background material and basic results for the system $\mathbb{B}$.

Chapter 2 covers basic definitions concerning sets, logic, graphs, matrices, and the computational complexity of algorithms.

Chapter 3 gives an introduction to a part of matroid theory.

Chapter 4 adapts concepts of linear algebra and matroid theory to the matrices over the system $\mathbb{B}$.

For a first reading, the reader may want to scan the material of Chapters 2–4 and skip all proofs.

The second part concerns special matrices over $\mathbb{B}$ and consists of Chapters 5–7.

Chapters 5 and 7 are essential for the subsequent developments.

In contrast, Chapter 6 deals with certain matrix characterizations that are not needed in later chapters. It may be skipped during a first reading.

The third part describes a number of matrix decompositions and compositions and consists of Chapters 8–12. These chapters are largely independent and may be read in any order.

The fourth part synthesizes the results of the second and third part and consists of Chapters 13 and 14.

Chapter 13 develops the analysis algorithm.

Chapter 14 determines large matrix classes that may be efficiently treated by the analysis algorithm.

The Subject Index should prove helpful as the reader navigates through the material of the various chapters.

So far, we have listed references together with the material under discussion. We change that approach in subsequent chapters, where we summarize related results and references in one section at the end of each chapter. That approach unclutters the technical presentation and makes it easier for the reader to locate references concerning a given topic.

We are ready to delve into details. In the next chapter, we introduce definitions utilized throughout the book.

Chapter 2

Basic Concepts

2.1 Overview

We cover basic definitions concerning sets, logic, graphs, matrices, and the computational complexity of algorithms. For a first pass, the reader may just scan the material.

Section 2.2 defines sets and related operations.

Section 2.3 discusses propositional logic and the problems SAT and MINSAT.

Section 2.4 defines basic concepts of first-order logic and shows that finite quantification reduces first-order logic to propositional logic.

Section 2.5 is an introduction to graph theory. Included are theorems about graph connectivity and related algorithms.

Section 2.6 deals with matrices over some fields or over an algebra called system $\mathbb{B}$.

Section 2.7 covers some elementary concepts of complexity theory—in particular, the problem class $\mathcal{NP}$ and polynomial algorithms.

The final section, 2.8, lists references.

2.2 Sets

An example of a set is $\{a, b, c\}$, the set with a, b, and c as elements. With two exceptions, all sets are assumed to be finite. The exceptions are the set of real numbers $\mathbb{R}$ and possibly the set of elements of an arbitrary field $\mathcal{F}$. Let S and T be two sets. Then $S \cup T$ is $\{z \mid z \in S \text{ or } z \in T\}$, the *union* of S and T. The set $S \cap T$ is $\{z \mid z \in S \text{ and } z \in T\}$, the *intersection* of S and T. The set $S - T$ is $\{z \mid z \in S \text{ and } z \notin T\}$, the *difference* of S and T.

Let T contain all elements of a set S. We denote this fact by $S \subseteq T$ and declare S to be a *subset* of T. We write $S \subset T$ if $S \subseteq T$ and $S \neq T$. The set S is then a *proper subset* of T. The set of all subsets of S is the *power set* of S. We denote by $|S|$ the *cardinality* of S. The set $\emptyset$ is the set without elements and is called the *empty set*.

The terms "maximal" and "minimal" are used frequently. The meaning depends on the context. When sets are involved, the interpretation is as follows. Let $\mathcal{I}$ be a collection, each of whose elements is a set. Then a set $Z \in \mathcal{I}$ is a *maximal set* of $\mathcal{I}$ if no set of $\mathcal{I}$ has Z as a proper subset. A set $Z \in \mathcal{I}$ is a *minimal set* of $\mathcal{I}$ if no proper subset of Z is in $\mathcal{I}$.

2.3 Propositional Logic

In this section and the next one, we review elementary concepts of logic. Here, we discuss propositional logic, which is concerned with results for *Boolean formulas*, also called *propositional formulas*. These formulas are constructed with *Boolean* or *propositional variables*, which are variables restricted to the values *True* or *False*.

Construction of Boolean Formula

A *Boolean variable* is a variable that may take on just two values, *True* or *False*. We construct *Boolean formulas*, which are logic expressions involving Boolean variables, as follows. First, any Boolean variable by itself is a Boolean formula. Let S and T be two Boolean formulas already constructed. Using the operators $\neg$ (denoting "not"), $\wedge$ (denoting "and"), and $\vee$ (denoting inclusive "or"), we obtain the following Boolean formulas: $\neg(S)$, $(S) \wedge (T)$, and $(S) \vee (T)$. The formula $\neg(S)$ is the *negation* of S. The formula $(S) \wedge (T)$ (resp. $(S) \vee (T)$) is the *conjunction* (resp. *disjunction*) of S and T. It is customary to omit parentheses when the intended interpretation is obvious. Thus, we usually write $\neg S$, $S \wedge T$, and $S \vee T$.

Literal

A Boolean variable may occur any number of times in a Boolean formula. Each such entry, possibly negated, is a *literal* of the formula.

Rules for ¬, ∧, and ∨

The commutative and associative laws hold for the binary operators $\land$ and $\lor$. That is, for formulas R, S, and T, we have $R \land S = S \land R$, $R \lor S = S \lor R$, $R \land (S \land T) = (R \land S) \land T$, and $R \lor (S \lor T) = (R \lor S) \lor T$. Due to these facts, we may omit parentheses when Boolean formulas are combined only by $\land$ or only by $\lor$. For example, we may write $R \land S \land T$ and $R \lor S \lor T$, or even, for $n \geq 2$ and Boolean formulas $S_1, S_2, \ldots, S_n$, $\bigvee_{i=1}^{n} S_i$ and $\bigwedge_{i=1}^{n} S_i$.

The operator $\land$ (resp. $\lor$) distributes over $\lor$ (resp. $\land$). That is, for formulas R, S, and T, we have $R \land (S \lor T) = (R \land S) \lor (R \land T)$ as well as $R \lor (S \land T) = (R \lor S) \land (R \lor T)$.

For the $\neg$ operator and two Boolean formulas S and T, we have the equations $\neg(S \lor T) = \neg S \land \neg T$, $\neg(S \land T) = \neg S \lor \neg T$, and $S = \neg(\neg S)$. Repeated use of these relationships allows us to convert any Boolean formula to one where the negation symbol is attached only to the Boolean variables, without changing the number of literals. Unless stated otherwise, we assume from now on that any given Boolean formula is of the latter form. For example, for Boolean variables s_1, s_2, s_3, s_4 we have $\neg[(s_1 \lor \neg s_2) \land (\neg s_3 \lor s_4)] = [\neg(s_1 \lor \neg s_2)] \lor [\neg(\neg s_3 \lor s_4)] = (\neg s_1 \land s_2) \lor (s_3 \land \neg s_4)$.

CNF and DNF System

A *conjunctive normal form* (CNF) *system* is, for some $n \geq 1$, the conjunction of Boolean formulas $S_1, S_2, \ldots, S_n$, that is, $\bigwedge_{i=1}^{n} S_i$, where each S_i is the disjunction of possibly negated Boolean variables. Each S_i is a CNF *clause*. For example, $(s_1 \lor \neg s_2) \land \neg s_3 \land (s_1 \lor s_4)$ is a CNF system, with CNF clauses $s_1 \lor \neg s_2$, $\neg s_3$, and $s_1 \lor s_4$. Another example is $s_1 \lor \neg s_2$, or just s_1, each consisting of just one CNF clause.

When we reverse the roles of disjunction and conjunction in a CNF system, we obtain a *disjunctive normal form* (DNF) *system*. Thus, a DNF system is a disjunction of the form $\bigvee_{i=1}^{n} S_i$, where each S_i is the conjunction of possibly negated Boolean variables. Each S_i is a DNF *clause* of the system. An example DNF system is $(s_1 \land \neg s_2) \lor \neg s_3 \lor (s_1 \land s_4)$, with DNF clauses $s_1 \land \neg s_2$, $\neg s_3$, and $s_1 \land s_4$.

In this book, we work almost exclusively with CNF systems, and thus we omit the DNF case from the definitions to follow. For the same reason, we often reduce the term CNF *clause* to just *clause* without risk of confusion.

Set Notation for CNF System

CNF systems may be nicely recorded in set notation. Each clause is then a set having the literals of the clause as elements. The fact that the clauses of the CNF system are coupled by conjunction is not made explicit. For example, the CNF system $(s_1 \vee \neg s_2) \wedge \neg s_3 \wedge (s_1 \vee s_4)$ becomes in set notation $\{s_1, \neg s_2\}, \{\neg s_3\}, \{s_1, s_4\}$.

Empty Clause

We adapt the notion of empty set to the situation at hand and declare the empty set to represent a clause without literals. Such a clause is said to be *empty*.

Trivial and Empty CNF Systems

We permit, indeed prefer, explicit declaration of the Boolean variables as part of the definition of any CNF system. An example of such a CNF system is as follows. The Boolean variables are $s_1, s_2, \ldots, s_5$. The sets representing the clauses are $\{s_1, \neg s_2\}, \{\neg s_3\}, \{s_1, s_4\}$. Note that neither s_5 nor $\neg s_5$ occurs in any of the clauses, an acceptable situation. Indeed, by the separate specification of variables and clauses, the following degenerate cases are possible. First, we may have variables but no clauses at all. Second, we may have no variables, but do have one or more clauses, all of which must be empty. Third, we may have no variables and no clauses. In the first two cases the CNF system is *trivial*, and in the third one it is *empty*.

At times, the explicit specification of Boolean variables of a CNF system would result in a cumbersome description. In such situations, we omit the explicit specification and implicitly assume that the list of Boolean variables contains precisely the variables occurring in the clauses.

CNF Subsystem

Let S be a given CNF system. We *reduce* S to a CNF *subsystem* by deleting clauses or variables. If a variable is being deleted, then all literals arising from that variable must be deleted. A subsystem of S is *proper* if at least one clause or variable has been deleted.

Value of Boolean Formula

Suppose *True/False* values have been assigned to the Boolean variables of a Boolean formula. Then we compute a value of *True* or *False* for that Boolean formula, using the steps of its construction as follows. Suppose *True/False* values have already been determined for S and T. Then $\neg S$ has the value opposite to that of S; $S \wedge T$ has the value *True* if both S and

T have the value *True*, and it has the value *False* otherwise; $S \lor T$ has the value *False* if both S and T have the value *False*, and it has the value *True* otherwise.

Two formulas are *equal* if for each possible assignment of values to the Boolean variables, the two formulas have agreeing values. For example, $S = \neg(\neg S)$.

At times, it is convenient to admit additional operators in Boolean formulas—in particular, $\Leftarrow$ (denoting "if"), $\Rightarrow$ (denoting "only if"), and $\Leftrightarrow$ (denoting "if and only if"). Examples with S and T as given Boolean formulas are $S \Leftarrow T$, $S \Rightarrow T$, and $S \Leftrightarrow T$. The values for these formulas are as follows. $S \Leftarrow T$ is *True* if S is *True* or T is *False*, and it is *False* otherwise; $S \Rightarrow T$ is *True* if S is *False* or T is *True*, and it is *False* otherwise; $S \Leftrightarrow T$ is *True* if S and T have the same value, and it is *False* otherwise.

Any Boolean formula with $\Leftarrow$, $\Rightarrow$, or $\Leftrightarrow$ is readily rewritten to an equivalent one with just $\neg$, $\land$, and $\lor$ using the relationships $S \Rightarrow T = \neg S \lor T$, $S \Leftarrow T = S \lor \neg T$, and $S \Leftrightarrow T = (\neg S \lor T) \land (S \lor \neg T)$. Hence, in subsequent sections we often confine ourselves to Boolean formulas with just the operators $\neg$, $\land$, and $\lor$, and we leave it to the reader to fill in the details for the cases of $\Leftarrow$, $\Rightarrow$, and $\Leftrightarrow$.

Tautology and Contradiction

A Boolean formula S is a *tautology* (resp. *contradiction*) if for any assignment of *True/False* values to the variables, the value of S is *True* (resp. *False*). Evidently, S is a tautology if and only if $\neg S$ is a contradiction.

Satisfiability

A Boolean formula S is *satisfiable* if there exists an assignment of *True/False* values to the Boolean variables so that the value of S is *True*. By that definition, a tautology is satisfiable, while a contradiction is unsatisfiable. Whether a given Boolean formula S is satisfiable or is a tautology or contradiction, can in principle be decided by finite enumeration of all possible *True/False* values for the Boolean variables of S and computation of the corresponding values of S.

Theorem Proving

Suppose we have expressed the facts of a situation of interest by a Boolean formula S. One sometimes says that S consists of *axioms expressing some facts* of the situation. We want to know whether the axioms imply some Boolean formula T. That is, we want to know whether $S \Rightarrow T$, which is equal to $\neg S \lor T$, is a tautology or, equivalently, whether $R = \neg(S \Rightarrow T) =$

$\neg(\neg S \vee T) = S \wedge \neg T$ is a contradiction. If this is so, we declare T to be a *theorem* of S.

To investigate the question "Is T a theorem of S?" we check whether R is unsatisfiable. If T is not a theorem, then any satisfying solution for R *certifies* that T is not a theorem of S. One might also say that any satisfying solution for R *explains* why T is not a theorem.

The reduction of the question "Is T a theorem of S?" to the satisfiability question for R implies that the former question can always be decided in a finite number of computational steps, using the earlier-mentioned enumerative method. Of course, we do not recommend that very inefficient method for practical use.

SAT Problem

Define SAT to be the following problem involving satisfiability. Any CNF system S is a problem instance. One must decide whether S is satisfiable. In the affirmative case, one must also find an assignment of *True/False* values for the Boolean variables of S so that the value of S is *True*. Since S is a conjunction of CNF clauses, say, $S = \bigwedge_{i=1}^{n} S_i$, we may restate the latter requirement as follows: We must produce an assignment of *True/False* values for the Boolean variables of S such that each CNF clause S_i has the value *True*. When this has been accomplished, we say that the clauses $S_1, S_2, \ldots, S_n$ are *satisfied* by the assigned *True/False* values, or that the values for the Boolean variables constitute a *satisfying solution* for each S_i and thus for S.

We define the empty clause to be unsatisfiable. So if S contains at least one such clause, then S is not satisfiable. In particular, if S is a trivial CNF system and has no variables but does have clauses, all of which must be empty, then S is unsatisfiable. On the other hand, if S has no clause at all, then we define S to be satisfiable. In the latter case, S must be a trivial CNF system that has variables but no clauses, or it must be the empty CNF system.

The earlier-discussed theorem-proving problem and the SAT problem are intimately connected. Recall that in the general theorem-proving problem we are given Boolean formulas S and T and must show whether T is a theorem of S or, equivalently, whether $R = S \wedge \neg T$ is unsatisfiable. We see in the next subsection that any such theorem-proving problem can be efficiently reduced to an instance of SAT. At this time, we examine a special case of S and T that turns out to be particularly important.

Specifically, we assume S to be a CNF system and suppose T to be a CNF clause, say, $T = (\bigvee_{j=1}^{m} s_j) \vee (\bigvee_{j=1}^{n} \neg t_j)$. Now $R = S \wedge \neg T = S \wedge (\bigwedge_{j=1}^{m} \neg s_j) \wedge (\bigwedge_{j=1}^{n} t_j)$ is a CNF system. Indeed, any satisfying solution for R must obey $s_1 = s_2 = \ldots = s_m = $ *False* and $t_1 = t_2 = \ldots = t_n = $ *True*.

Hence, we assign *False* to $s_1, s_2, \ldots, s_m$ and *True* to $t_1, t_2, \ldots, t_n$. This assignment allows us to reduce the satisfiability problem for R as follows.

Consider the clauses of R containing at least one literal of the form $\neg s_1, \neg s_2, \ldots, \neg s_m, t_1, t_2, \ldots,$ or t_n. Those clauses are obviously satisfied, since $s_1 = s_2 = \ldots = s_m = $ *False* and $t_1 = t_2 = \ldots = t_n = $ *True*, and we may delete them from R. From the remaining clauses of R, we delete all literals of the form $s_1, s_2, \ldots, s_m, \neg t_1, \neg t_2, \ldots,$ or $\neg t_n$, since they are irrelevant for satisfiability of R. The resulting CNF subsystem of R is a subsystem of S, and that subsystem is satisfiable if and only if this is so for R. Evidently, we have transformed the question "Is T a theorem of S?" to the satisfiability question for a subsystem of S. So if we have a fast computational method for solving the satisfiability problem for any subsystem of S, then that method may be used to answer quickly, for *any* CNF clause T, the question "Is T a theorem of S?" We record this simple but crucial observation in the following lemma.

(2.3.1) Lemma. *Let S be a CNF system, and let T be a CNF clause. Then T is a theorem of S if and only if a certain subsystem of S, which depends on T, is unsatisfiable.*

SAT Equivalence

A Boolean formula and a CNF system are SAT *equivalent* if both are satisfiable or both are unsatisfiable. One may convert any Boolean formula efficiently to a SAT equivalent CNF system that is not much larger than the given formula as follows.

We need the concept of *representation* for a Boolean formula R and a CNF system $\tilde{R}$. We say that $\tilde{R}$ *represents* R if the following holds. Let R have Boolean variables $r_1, r_2, \ldots, r_m$. The CNF system $\tilde{R}$ must contain these variables, plus *auxiliary* Boolean variables, say, $w_1, w_2, \ldots, w_k$, for some $k \geq 1$. Suppose one assigns *True/False* to the Boolean variables $r_1, r_2, \ldots, r_m$ of R and $\tilde{R}$. We then must be able to extend this partial assignment for $\tilde{R}$ to a satisfying solution for $\tilde{R}$ by a unique assignment of *True/False* to the auxiliary variables $w_1, w_2, \ldots, w_k$. In particular, the auxiliary variable w_k must receive the value *True* if and only if the *True/False* values for $r_1, r_2, \ldots, r_m$ constitute a satisfying solution for R. For this reason, we call w_k the *key* variable of $\tilde{R}$.

We next determine a representing CNF system $\tilde{R}$ when R is just r or $\neg r$. We declare $\tilde{R}$ to have the variable r and an auxiliary variable w, which also serves as the key variable. Consider the case where R is the Boolean formula r. Since $r \Leftrightarrow w = (\neg r \lor w) \land (r \lor \neg w)$, the CNF system $(\neg r \lor w) \land (r \lor \neg w)$ is readily checked to represent R, and thus it will do for $\tilde{R}$. If R is the Boolean formula $\neg r$, we take $\tilde{R}$ to be the CNF system $(r \lor w) \land (\neg r \lor \neg w)$.

We proceed inductively, assuming that we have CNF systems $\tilde{S}$ and $\tilde{T}$ representing Boolean systems S and T, respectively. We want a CNF system $\tilde{R}$ representing $R = S \wedge T$. Let $s_1, s_2, \ldots, s_m$ (resp. $t_1, t_2, \ldots, t_n$) be the Boolean variables of S (resp. T), and let $u_1, u_2, \ldots, u_k$ (resp. $v_1, v_2, \ldots, v_l$) be the auxiliary variables of $\tilde{S}$ (resp. $\tilde{T}$), with u_k (resp. v_l) serving as key variable. Evidently, $s_1, s_2, \ldots, s_m, t_1, t_2, \ldots, t_n$ are the Boolean variables of R.

We define the CNF system $\tilde{R}$ as follows. First, $\tilde{R}$ contains the Boolean variables $s_1, s_2, \ldots, s_m, t_1, t_2, \ldots, t_n$. Second, $\tilde{R}$ has $u_1, u_2, \ldots, u_k, v_1, v_2, \ldots, v_l$ plus a new variable w as auxiliary Boolean variables. The variable w is the key variable of $\tilde{R}$. The clauses of $\tilde{R}$ consist of those of $\tilde{S}$ and $\tilde{T}$, plus the clauses of a small CNF system $\overline{R}$ that is equal to $w \Leftrightarrow (u_k \wedge v_l)$. Since $w \Leftrightarrow (u_k \wedge v_l) = \left[\neg w \vee (u_k \wedge v_l)\right] \wedge \left[w \vee (\neg u_k \vee \neg v_l)\right] = (\neg w \vee u_k) \wedge (\neg w \vee v_l) \wedge (w \vee \neg u_k \vee \neg v_l)$, we take $\overline{R}$ as $(\neg w \vee u_k) \wedge (\neg w \vee v_l) \wedge (w \vee \neg u_k \vee \neg v_l)$.

We prove that $\tilde{R}$ represents R. Let arbitrary *True/False* values be assigned to the Boolean variables $s_1, s_2, \ldots, s_m, t_1, t_2, \ldots, t_n$ of R and $\tilde{R}$. Assign the same values to $s_1, s_2, \ldots, s_m$ of S and to $t_1, t_2, \ldots, t_n$ of T. By induction, there exist unique *True/False* values for the auxiliary variables $u_1, u_2, \ldots, u_k$ of S and $v_1, v_2, \ldots, v_l$ of T such that satisfying solutions are at hand for $\tilde{S}$ and $\tilde{T}$. Furthermore, the key variable u_k of S (resp. v_l of T) must have the same value as S (resp. T). Thus, $R = S \wedge T$ has the same value as $u_k \wedge v_l$.

Since $\tilde{R}$ contains all clauses of $\tilde{S}$ and $\tilde{T}$, the *True/False* values so far assigned to $s_1, s_2, \ldots, s_m, t_1, t_2, \ldots, t_n$ of $\tilde{R}$ may be extended to a satisfying solution of $\tilde{R}$ only if the auxiliary variables $u_1, u_2, \ldots, u_k, v_1, v_2, \ldots, v_l$ are fixed to the values just determined for $\tilde{S}$ and $\tilde{T}$. To satisfy the remaining clauses, which have been deduced from $w \Leftrightarrow (u_k \wedge v_l)$, we must assign the value of $u_k \wedge v_l$ to the key variable w. Since w, $u_k \wedge v_l$, and $S \wedge T$ have the same value, w has the same value as $R = S \wedge T$, and we are done.

By almost identical arguments, one may derive a representing system $\tilde{R}$ for the case $R = S \vee T$. Indeed, the only change occurs in the definition of $\overline{R}$, which now must be deduced from $w \Leftrightarrow (u_k \vee v_l)$. Since $w \Leftrightarrow (u_k \vee v_l) = \left[\neg w \vee (u_k \vee v_l)\right] \wedge \left[w \vee (\neg u_k \wedge \neg v_l)\right] = (\neg w \vee u_k \vee v_l) \wedge (w \vee \neg u_k) \wedge (w \vee \neg v_l)$, the CNF system $(\neg w \vee u_k \vee v_l) \wedge (w \vee \neg u_k) \wedge (w \vee \neg v_l)$ will do for $\overline{R}$.

We are ready to produce a SAT equivalent CNF system, say, $\hat{R}$, for a given Boolean formula R. We use the construction sequence for R to derive a representing CNF system $\tilde{R}$ for R, say, with key variable w. The details for each construction step of the sequence have just been shown. Finally, we add the clause w to $\tilde{R}$ to get $\hat{R}$. Since w is the key variable of $\tilde{R}$, the CNF system $\hat{R}$ is satisfiable if and only if this is so for R.

If R has N literals, then the above method produces a SAT equivalent CNF system $\hat{R}$ with less than $11 \cdot N$ literals. This bound can be significantly reduced when one employs a slightly more complicated method that avoids the inefficient processing of Boolean formulas with just one variable and that exploits specific structural properties of the formula R. We leave it to the reader to work out the details.

We have proved that the satisfiability problem for Boolean formulas can be efficiently transformed into one for CNF systems, that is, into the SAT problem. This fact implies that the theorem-proving problem involving Boolean formulas can also be transformed into the SAT problem.

MINSAT and MAXSAT Problems

MINSAT and MAXSAT are closely related variations of SAT. A problem instance of MINSAT (resp. MAXSAT) consists of a CNF system S, say, with Boolean variables s_1, s_2, ..., s_n and clauses S_1, S_2, ..., S_m, plus rational nonnegative *costs* c_1, c_2, ..., c_n (resp. *weights* d_1, d_2, ..., d_m) associated with the variables (resp. clauses). For the MINSAT instance, one must decide whether the clauses are satisfiable; in the affirmative case, one must produce a satisfying solution that minimizes the total cost of the variables to which the value *True* has been assigned. For the MAXSAT instance, one must determine *True/False* values for the variables so that the total weight of the satisfied clauses is maximized.

Any MINSAT instance is readily converted into a MAXSAT instance, and vice versa. We use the above notation to present details.

Let an instance of MAXSAT be given. For $i = 1, 2, \ldots, m$, we augment the clause S_i to $S_i' = (S_i \vee u_i)$, where u_i is a new Boolean variable. For $i = 1, 2, \ldots, m$, we assign a cost $c_i = d_i$ to variable u_i. The original variables s_1, s_2, ..., s_n receive a cost of 0. The resulting CNF system S' is a satisfiable instance of MINSAT. It is easily verified that the values for s_1, s_2, ..., s_n of a solution for that MINSAT instance constitute a solution for the original MAXSAT instance.

Now let an instance of MINSAT be given. The equivalent MAXSAT instance has the clauses S_1, S_2, ..., S_m, each with weight $(\sum_{j=1}^{n} c_j)+1$, plus n new clauses $\neg s_1$, $\neg s_2$, ..., $\neg s_n$ with weights c_1, c_2, ..., c_n, respectively. It is easy to see that any solution for the MAXSAT instance satisfying the clauses S_1, S_2, ..., S_m is one for the original MINSAT instance. Furthermore, if the solution of the MAXSAT instance does not satisfy one or more of the clauses S_1, S_2, ..., S_m, then the MINSAT instance is unsatisfiable.

According to the above discussion, any solution algorithm for MINSAT may be used to solve MAXSAT with essentially the same efficiency, and vice versa. Thus we may confine ourselves to the MINSAT problem without loss of generality.

MINSAT is obviously useful when a logic formulation involves costs. But there are other applications for MINSAT as well. We discuss an important case next. Let S be a CNF system, with variables s_1, s_2, ..., s_n

and $t_1, t_2, \ldots, t_k$. For some disjoint subsets J^+, J^- of $\{1, 2, \ldots, n\}$ and for $l = 1, 2, \ldots, k$, let T_l be the statement $[(\bigwedge_{j \in J^+} s_j) \wedge (\bigwedge_{j \in J^-} \neg s_j)] \Rightarrow t_l$, which is equal to the CNF clause $[(\bigvee_{j \in J^+} \neg s_j) \vee (\bigvee_{j \in J^-} s_j)] \vee t_l$. We want to prove which of the statements $T_1, T_2, \ldots, T_k$ are theorems of S. If one uses the earlier described method, k SAT instances must be solved. In another approach using MINSAT, one assigns a cost of 1 to $t_1, t_2, \ldots, t_k$ and a cost of 0 to the remaining variables $s_1, s_2, \ldots, s_n$. Finally, one fixes the s_j, $j \in J^+$, to *True* and the s_j, $j \in J^-$, to *False*. If the resulting MINSAT instance is unsatisfiable, then each T_l is a theorem of S. So assume that a MINSAT solution exists that, for some partition L^+, L^- of $\{1, 2, \ldots, k\}$, assigns *True* to t_l, $l \in L^+$, and assigns *False* to t_l, $l \in L^-$. That solution proves that the T_l, $l \in L^-$, are not theorems of S, while the T_l, $l \in L^+$, may be theorems. We decide which of the T_l, $l \in L^+$, are theorems of S by solving $|L^+|$ SAT instances as described earlier. Note that by solving one MINSAT instance we implicitly have solved $|L^-|$ SAT instances. So if the MINSAT instance is about as difficult to solve as any one of the SAT instances and if we can expect $|L^-|$ to be large, as is the case in many real-world situations, then the MINSAT approach is very effective.

2.4 First-Order Logic

First-order logic is an extension of propositional logic. It relies on predicates, also called truth functions, plus universal and existential quantification.

Predicate

Let U be some nonempty set, called the *universe*. In general, U is taken to be infinite. For any $k \geq 1$, denote the set $\{(u_1, u_2, \ldots, u_k) \mid u_i \in U, \text{ for all } i\}$, which is the *product* of k copies of U, by $\prod_{i=1}^{k} U$. When k is small, we may denote $\prod_{i=1}^{k} U$ by $U \times U \times \cdots \times U$, with k U-terms. A *predicate* or *truth function* is a function p that, for some $k \geq 1$, takes the elements of $\prod_{i=1}^{k} U$ to the set $\{True, False\}$.

Universal and Existential Quantification

Predicates are employed in connection with the *universal quantifier* $\forall$, denoting "for all" or "for each," and the *existential quantifier* $\exists$, denoting "there exists." Example statements involving these quantifiers and a predicate p that takes the pairs of $U \times U$ to $\{True, False\}$ are $\forall x (\forall y [p(x, y)])$ and $\forall x (\exists y [p(x, y)])$, meaning "for all $x \in U$ and for all $y \in U$, $p(x, y)$ holds" and "for each $x \in U$, there exists $y \in U$ such that $p(x, y)$ holds." Note the implicit use of U in each quantification.

Construction of Formula

Analogously to the case of Boolean formulas, a simple construction rule creates all possible formulas of first-order logic as follows.

First, any Boolean variable or predicate constitutes a formula. In the predicate case, each variable occurring as part of an argument of the predicate is *free*.

Second, let S and T be formulas. Then $\neg(S)$, $(S) \wedge (T)$, and $(S) \vee (T)$ are all formulas. Any free variable of S is also free in $\neg(S)$, and any free variable of S or T is also free in $(S) \wedge (T)$ and $(S) \vee (T)$.

Third, let x be a free variable of a formula S. Then $\forall x(S)$ and $\exists x(S)$ are formulas, with free variables as in S except for x.

Fourth, the construction process is allowed to stop only if the formula at hand has no free variables. Such a formula is *completed*.

As for the propositional case, we omit parentheses when the interpretation is obvious. For example, we may write $\neg S$, $S \wedge T$, $S \vee T$, $\forall x\ p(x)$, or $\exists x\ p(x)$.

We omit discussion of the conditions under which a completed formula T is considered to be a theorem of a completed formula S. We also skip computational aspects concerning verification of such conditions. Suffice it to say that one can always prove in a finite number of computational steps that T is a theorem of S if this is so. On the other hand, there is no algorithm that in finite time confirms T to be not a theorem of S, for all cases of S and T where this is so. Due to the latter result and other practical considerations that we shall not elaborate on here, we now restrict the universe U by requiring it to be always finite. Indeed, instead of assuming each quantification term to implicitly involve U, we permit specification of a particular finite set with each such term. The construction rule for completed formulas is easily amended to accommodate this change. That is, any term $\forall x$ (resp. $\exists x$) is replaced by $\forall x \in X$ (resp. $\exists x \in X$) using some subset X of the finite set U. As before, we declare the outcome of a sequence of construction steps, with finite quantification, to be a *completed formula*.

Transformation to Boolean Formula

A completed formula with finite quantification is easily translated to a Boolean formula by replacing any term of the form $\forall x \in X$ (resp. $\exists x \in X$) by $\bigwedge_{x \in X}$ (resp. $\bigvee_{x \in X}$). For example, the formula $\forall x \in X \left[\exists y \in Y\ p(x, y) \right]$ becomes the Boolean formula $\bigwedge_{x \in X} \left[\bigvee_{y \in Y} p(x, y) \right]$.

To any Boolean formula so produced, we may apply the results of the preceding section. In particular, whether a completed formula T is a theorem of a completed formula S can always be settled in finite time, in contrast to the general situation of first-order logic discussed earlier.

The switch to finite quantification permits an extension of the quantification concept. For presentation of an example, we consider the formula $\forall x \in X [\forall y \in Y \; p(x,y)]$. According to the above rule, the corresponding Boolean formula is $\bigwedge_{x \in X, y \in Y} p(x,y)$. Suppose we want to impose a restriction on x and y by demanding $(x,y) \in Z$, for some subset $Z \subseteq X \times Y$. Then the Boolean formula becomes $\bigwedge_{x \in X, y \in Y, (x,y) \in Z} p(x,y)$. The example is an instance of *restricted finite quantification*, a useful notion for compact modeling of real-world situations.

2.5 Graphs

A *graph* is given by a set of *nodes* and a set of *edges*. In principle, any edge may be *directed* or *undirected*, but here we only deal with graphs where all edges are either directed or undirected. In the former (resp. latter) case, the graph is *directed* (resp. *undirected*). For example, the graph

(2.5.1)

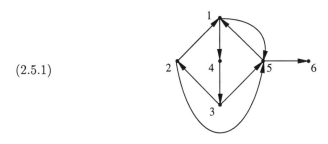

Directed graph

has nodes 1, 2..., 6 and various directed edges connecting them. The undirected version of that graph is

(2.5.2)

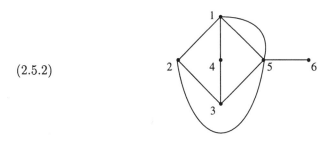

Undirected graph

A directed edge, say, going from node i to j, is specified by the ordered pair (i,j). In the undirected case, the edge is also specified by the pair (i,j), but this time the pair is considered unordered. In either case, i and j are the *endpoints* of the edge and are *adjacent*. Each one of the nodes i and j *covers* the edge (i, j). In turn, the edge (i, j) is *incident* at i and j. Nodes are also called *vertices* or *points*. Edges are also referred to as *arcs*.

We rarely consider graphs with *loops*, which are edges with just one endpoint. Loops may be directed or undirected. Unless stated otherwise, we assume that graphs do not have loops.

Suppose that several edges connect two nodes where the edges have the same direction if they are directed, or that more than one loop is incident at a node. The above notation cannot handle these situations. We then implicitly assume that a refined notation is used, say, involving an additional index, to differentiate among these edges.

Complete Graph

The undirected graph with $n \geq 2$ vertices and with every two vertices connected by an edge is denoted by K_n. It is the *complete graph* on n vertices. Small cases of K_n are as follows.

(2.5.3)

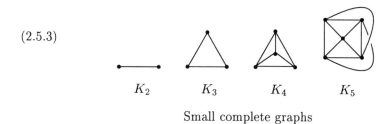

K_2 K_3 K_4 K_5

Small complete graphs

Bipartite Graph

A graph is *bipartite* if its node set can be partitioned into two nonempty subsets such that every edge connects a node of one of the sets with a node of the other set. A bipartite graph may be directed or undirected. The *complete bipartite graph* $K_{m,n}$ is undirected, has m nodes on one side and n on the other one, and has all possible edges. Small cases are as follows.

(2.5.4)

$K_{1,1}$ $K_{2,1}$ $K_{2,2}$ $K_{3,1}$ $K_{3,2}$ $K_{3,3}$

Small complete bipartite graphs

Evidently, $K_{1,1}$ is the complete graph K_2.

Bipartite graphs may be viewed as one way of encoding matrices. Details are presented in Section 2.6. For the time being, we only note that in such an encoding one of the two node subsets of the bipartite graph corresponds to the rows of the matrix, while the other node subset corresponds to the columns. Accordingly, one of the two node subsets contains the *row nodes* of the graph, and the other one contains the *column nodes*. As a matter of consistency and simplicity, we use the terminology of row/column nodes for the two node subsets of a bipartite graph even if a matrix has not been *a priori* specified.

In a *labeled*, directed, bipartite graph, each arc has been assigned 1 or 2 as label. We only consider such graphs in connection with matrices. Details are included in Section 2.6.

Deletion and Addition

The *deletion* of an edge is the removal of that edge from the graph. The *deletion* of a node involves the removal of the node and of its incident edges. The inverse of deletion is *addition*. In the case of a node addition, the edges incident at the new node must also be specified.

Contraction and Expansion

The *contraction* of an edge (i,j) is the collapsing of (i,j) to one point. The endpoints i and j of the edge become one node. If some other edge also connects i and j, then the contraction of (i,j) converts that edge to a loop. An *expansion* by an edge is the inverse of contraction. Thus, we split a node k into two new nodes, say, i and j, assign each edge previously incident at k to i or j according to some rule, and finally connect i and j by the new edge specified in the expansion step.

Our notation for nodes is not designed to accommodate complicated contraction or expansion sequences. But we rarely make use of those operations, so our admittedly cumbersome way of handling the corresponding changes of the node labels suffices for our purposes.

Graph Minor

Any sequence of deletions and contractions reduces a given graph to a *graph minor*. It is easily checked that the same minor results if the sequence is reordered, provided that one selects node labels for the new nodes produced by the contractions in a consistent manner. As a matter of convenience, we declare G also to be a minor of G. All other minors of G are *proper*.

Let G be a graph, and suppose V and W are disjoint edge subsets of G. Then $G/V\backslash W$ is the minor obtained from G by contraction of the edges of V and deletion of the edges of W.

Degree, Indegree, and Outdegree

The *degree* of a node is the number of incident edges. In the directed graph case, the *indegree* (resp. *outdegree*) of a node is the number of edges entering (resp. leaving) that node. A node is *isolated* if it has no edges incident.

Subgraph

Let G be a graph with node set V and edge set E. A *subgraph* of G is obtained by the deletion of some edges and nodes. A subgraph is *proper* if it is produced by the deletion of at least one edge or node. Let $\overline{V}$ be a subset of V, and let $\overline{E}$ be a subset of E. Suppose we delete from G all nodes of $V - \overline{V}$. The result is the subgraph *induced* by the node subset $\overline{V}$. Now suppose we delete from G all nodes having only edges of $E - \overline{E}$ incident and then delete all remaining edges of $E - \overline{E}$. The result is the subgraph *induced* by the edge subset $\overline{E}$.

Analogously to the use of "maximal" and "minimal" for sets, we use these terms in connection with graphs as follows. Suppose certain subgraphs of a given graph G have a property $\mathcal{P}$, while others do not. Then a subgraph H is a *maximal subgraph* of G with respect to $\mathcal{P}$ if no other subgraph has $\mathcal{P}$ and has H as proper subgraph. A subgraph H of G is a *minimal subgraph* of G with respect to $\mathcal{P}$ if no proper subgraph of H has $\mathcal{P}$.

Path

Suppose we walk along the edges of a graph starting at some node s, never revisit any node, and stop at a node $t \neq s$. The set P of edges we have traversed is a *path from s to t*. The nodes of the path are the nodes of the graph we encountered during the walk. The nodes s and t are the *endpoints* of P. The *length* of the path P is $|P|$. If the underlying graph is directed and if during the walk from s to t the direction of each edge agrees with the direction of the walk, then P is a *directed* path from s to t. Two paths are *node-disjoint* if they do not share any nodes. Two paths with the same endpoints are *internally node-disjoint* if they do not share any nodes except for the endpoints.

Define the empty set to represent a path that consists of just one node s. The context will make clear which node s is meant. If the underlying graph is directed (resp. undirected), the path is considered directed (resp. undirected).

Later in this chapter, the statement of Menger's theorem relies on a particular fact about the number of internally node-disjoint paths connecting two nodes. If the two nodes, say, i and j, are adjacent, then that number is unbounded, since we may declare any number of paths to consist of just the edge connecting i and j. Evidently, these paths are internally node-disjoint.

Connected Component

A graph is *connected* if, for any two vertices s and t, there is a path from s to t. The *connected components* of a graph are the maximal connected subgraphs.

A directed graph is *strongly connected* if for any two nodes i and j, there are directed paths from i to j and from j to i. Each maximal strongly connected subgraph of a directed graph is a *strong component*. Clearly, each strong component of a directed graph is induced by some node subset of the graph and has no node in common with any other strong component. A strong component of a bipartite graph consists of a row node, or of a column node, or of at least one row node and at least one column node.

A node whose deletion increases the number of connected components is an *articulation point*.

Cycle

Imagine a walk as described above for the path definition, except that we return to s. The set C of edges we have traversed is a *cycle*. The *length* of the cycle is $|C|$. Analogously to the path case, the cycle is *directed* if one can traverse it so that the direction of each edge agrees with the direction of the walk. Note that a loop is a cycle of length 1. In contrast to the path case, we do not interpret the empty set as some sort of cycle.

A directed graph is *acyclic* if it does not contain any directed cycles. Evidently, each node of an acyclic graph is a strong component of that graph. Define two nodes of an acyclic graph to be *incomparable* if they are not connected by a directed path. Declare a collection of directed paths of an acyclic graph to *cover* the nodes of the graph if each node occurs in at least one of the paths.

A *chord* of a cycle or path H is an edge incident at two nodes i and j of H that are not connected by an edge of H.

In a slight abuse of language, we say at times that a connected graph G *is* a cycle or a path, meaning that the edge set of G is a cycle or path of G. We employ terms of later defined edge subsets such as trees and cotrees similarly. We may say, for example, that a connected graph G is a tree or cotree, meaning that the edge set of G is a tree or cotree. The reader may wonder why we introduce such inaccuracies. We must describe a number of diverse graph operations that are not easily expressed with one simple set of terms. So either we tolerate a slight abuse of language, or we are

forced to introduce a number of different terms and sets. We have opted
for the former solution in the interest of simplicity and clarity.

Wheel Graph

A *wheel* is an undirected graph consisting of a *rim* and *spokes*. The rim
edges define a cycle, and the spokes are edges connecting an additional
node with each node of the rim. The wheel with n spokes is denoted by
W_n. Small wheels are as follows.

(2.5.5)

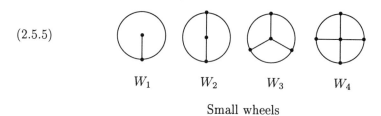

W_1 $\qquad$ W_2 $\qquad$ W_3 $\qquad$ W_4

Small wheels

Evidently, W_3 is the complete graph K_4.

Tree, Cotree, and Forest

A *tree* is the edge set of a connected graph without cycles. Note that a tree
is the empty set if and only if the graph consists of one node. A *spanning
tree* of a connected graph is a maximal tree subgraph. It is easy to show
that the cardinality of any spanning tree of a connected graph is equal to
the number of nodes of the graph minus 1. A *tip node* or *leaf node* of a tree
is a node with degree 1. The edge incident at a tip node is a *leaf edge*.

A *cotree* of a connected graph with edge set E is $E - T$ for some tree T
of the graph. A collection of trees is a *forest*. A *principal forest* of a graph
is a collection of spanning trees, one for each connected component of the
graph.

Rank

The *rank* of a graph is the cardinality of a principal forest, and thus it is the
number of nodes of the graph minus the number of connected components.

Empty Graph

As a matter of convenience, we introduce the *empty graph*. That graph
does not have any edges or nodes, and its rank is 0. We consider the empty
graph to be connected.

Coloop and Cocycle

An edge of a graph that is not in any cycle is a *coloop*. Such an edge is sometimes called a *bridge* or *isthmus*. It is easy to see that a coloop is in every principal forest of the given graph.

As one removes edges from a graph with at least one edge, eventually the number of connected components must increase. Correspondingly, the rank is reduced. A minimal set of edges whose removal reduces the rank is a *cocycle* or *minimal cutset*. Suppose one partitions the vertex set of a graph into two nonempty subsets. Then the set of edges with endpoints in both node subsets is a disjoint union of cocycles.

Recall that a coloop is contained in every principal forest. Hence, removal of a coloop leads to a drop in rank. We conclude that a set containing just a coloop is a cocycle. The definitions of principal forest and cocycle imply that a cocycle is a minimal subset of edges that intersects every principal forest.

Parallel and Series Edges

A subset of edges of a given graph G forms a *parallel class* if any two edges form a cycle and if the subset is maximal with respect to that property. We also say that the edges of the subset are *in parallel*. A subset of edges forms a *series class* (or *coparallel class*) if any two edges form a cocycle and if the subset is maximal with respect to that property. We also say that the edges of the subset are *in series* or *coparallel*. In the customary graph definition of "series," a series class of edges constitutes either a path in the graph all of whose intermediate vertices have the degree 2 or a cycle all of whose vertices, save at most one, have the degree 2. Our definition allows for these cases, but it also permits a slightly more general situation. For example, in the graph

(2.5.6)

Graph G

the edges $(2,3)$ and $(5,6)$ are in series, since $\{(2,3),(5,6)\}$ is a cocycle.

Matching

Define an edge subset of a graph to be a *matching* if no two edges of the edge subset share an endpoint.

Let X be the row or column node subset of a bipartite graph G, and let Y be the set of the remaining nodes. If a matching of G covers all nodes of X, then X has been *matched into* Y.

A matching of a graph is *perfect* if it covers all nodes. Clearly, a bipartite graph G with node subsets X and Y specified as above has a perfect matching if and only if $|X| = |Y|$ and X can be matched into Y.

Scaling

We *scale* a node of a directed graph by reversing the direction of each arc incident at that node or by leaving the graph unchanged. In the former case the *scaling factor* is -1, and in the latter case it is $+1$. If the directed graph is bipartite, then *column scaling* (resp. *row scaling*) refers to scaling of a column node (resp. row node). Node scaling of labeled, directed, bipartite graphs does not affect the arc labels.

Node Identification

Two nodes i and j of a graph are *identified* by deleting all arcs connecting i and j and then collapsing the two nodes into just one node. Thus the arcs previously incident at either i or j become incident at the new node.

Shrinking and Unshrinking

Let H be a labeled, directed, bipartite graph. Recall that the labels are 1s and 2s assigned to the arcs of H. Suppose $G_1, G_2, \ldots, G_n$ are the strong components of H. Then we *shrink* H by first collapsing, for each G_k, $k = 1$, $2, \ldots, n$, the row nodes of G_k to a new row node and collapsing the column nodes of G_k to a new column node. Of course, G_k may not have any row (resp. column) nodes. In that case, G_k has just one column (resp. row) node, and that node is not affected according to the rule for collapsing nodes. In the next step of the shrinking operation, we delete all arc labels and replace any instance of multiple arcs with the same endpoints and the same direction by just one arc each. Finally, in the reduced graph we assign to each arc the label 1 or 2, where the case of a 1 corresponds precisely to the following situation. Let the arc in question connect the row node r and the column node c of the reduced graph. Define R (resp. C) to be the set of row (resp. column) nodes of H that were collapsed to form r (resp. c). If in the reduced graph the arc in question goes from node r to node c (resp. from node c to node r), then that arc receives the label 1 if and only if in H every row node of R has exactly one arc outgoing to (resp. incoming from) the nodes of C and that arc has the label 1. The graph $\overline{H}$ resulting from these steps is the graph produced by *shrinking* from H.

We demonstrate the shrinking operation using the following graph H. Here and later we employ the convention that any arc shown without a label actually has the label 1, and that row nodes are indicated by squares.

(2.5.7)

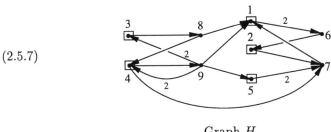

Graph H

The strong components of H containing more than one node are given by the node sets $\{1, 2, 6, 7\}$ and $\{3, 4, 8, 9\}$. Accordingly, we collapse the row node subsets $\{1, 2\}$ and $\{3, 4\}$ to one node each, say, to new nodes 10 and 11, and collapse the column node subsets $\{6, 7\}$ and $\{8, 9\}$ to new nodes 12 and 13. The reduction of arc multiplicities and the assignment of arc labels then produce the following graph $\overline{H}$.

(2.5.8)

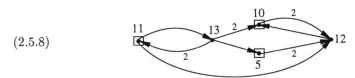

Graph $\overline{H}$ produced by shrinking of H

We verify, as examples, the labels of the three arcs connecting the nodes $r = 10$, $c_1 = 12$, and $c_2 = 13$. The nodes 1 and 2 of H are collapsed to the node $r = 10$ of $\overline{H}$, nodes 6 and 7 to $c_1 = 12$, and 8 and 9 to $c_2 = 13$. For the determination of the labels of the arcs connecting r, c_1, and c_2 in $\overline{H}$, let $R = \{1, 2\}$, $C_1 = \{6, 7\}$, and $C_2 = \{8, 9\}$. In H, each node of R has exactly one arc incoming from the nodes of C_1, and each such arc has the label 1. Thus, the label of the arc $(c_1, r) = (12, 10)$ of $\overline{H}$ is 1. On the other hand, node 1 of R has one arc with the label 2 outgoing to node 6 of C_1 and has two arcs with the label 1 incoming from nodes 8 and 9 of C_2. Thus, the arcs $(r, c_1) = (10, 12)$ and $(c_2, r) = (13, 10)$ receive the label 2.

Unshrinking is the operation inverse to shrinking. In principle, the result of unshrinking of a graph $\overline{H}$ can be any graph H that by shrinking would become $\overline{H}$.

Boolean Minor

We have already defined the term *graph minor*. Deletions and contractions in a graph produce such minors. A second type of minor called *Boolean* will also be of use. As the name suggests, Boolean minors are connected with propositional logic. The precise relationships are explained in Section 2.6. For the time being, we just introduce the definition. Let H be a labeled, directed, bipartite graph. We reduce H to a *Boolean minor* $\overline{H}$ using *column scaling*, *shrinking*, and *deletion* of column or row nodes. Any one of these operations may be omitted. But, modulo such omissions, we always consider these operations done in the specified order. The inverse operations are *addition* of nodes, *unshrinking*, and *column scaling*, always done in that order. We demand adherence to the specified order, since a resequencing of reduction steps may produce different minors or may even lead to undefined situations.

Subdivision, Isomorphism, and Homeomorphism

In a special case of expansion, we replace an edge e by a path P that contains e plus at least one more edge. We say that the edge e has been *subdivided.* The substitution process by the path is a *subdivision* of edge e.

Two graphs are *isomorphic* if they become identical upon a suitable renaming of the nodes. They are *homeomorphic* if they can be made isomorphic by repeated subdivision of certain edges in both graphs.

Terminology for Graph Minors and Boolean Minors

At times, a certain graph, say, $\overline{G}$, may be a graph minor of a graph G, or may only be isomorphic to a graph minor of G. In the first case, we say, as expected, that $\overline{G}$ is a graph minor of G or that G has $\overline{G}$ as a graph minor. For the second, more frequently occurring case, the terminology "G has a graph minor isomorphic to $\overline{G}$" is technically correct but cumbersome. So instead, we say that G has a $\overline{G}$ *graph minor.*

We employ the same terminology for Boolean minors. So we may say that a labeled, directed, bipartite graph H has a graph $\overline{H}$ as a Boolean minor or that H has an $\overline{H}$ Boolean minor. In the first case, $\overline{H}$ is a Boolean minor of H. In the second case, H has a Boolean minor that is isomorphic to $\overline{H}$.

Vertex, Cycle, and Tutte Connectivity

There are several ways to specify the *connectivity* of graphs. Two commonly used concepts of graph theory are *vertex connectivity* and *cycle connectivity.* A third connectivity concept is *Tutte connectivity*, which is important when one links graphs with matroids. For completeness and purposes of comparison, we describe all three types of graph connectivity even though later we mostly use vertex connectivity.

Let (E_1, E_2) be a pair of nonempty sets that partition the edge set E of a connected graph G. Let G_1 (resp. G_2) be the subgraph of G induced by E_1 (resp. E_2). Assume G_1 and G_2 to be connected. Suppose pairwise identification of k nodes of G_1 with k nodes of G_2 produces G. These k nodes of G_1 and G_2, as well as the k nodes of G they create, we call *connecting nodes*. Since G is connected and since both G_1 and G_2 are nonempty, we have $k \geq 1$. If $k = 1$, the single connecting node of G is an articulation point of G. For general $k \geq 1$, (E_1, E_2) is a *vertex k-separation* of G if both G_1 and G_2 have at least $k + 1$ nodes. The pair (E_1, E_2) is a *cycle k-separation* if both G_1 and G_2 contain cycles of G. Finally, (E_1, E_2) is a *Tutte k-separation* if E_1 and E_2 have at least k edges each. Correspondingly, we call G *vertex k-separable*, *cycle k-separable*, or *Tutte k-separable*. For $k \geq 2$, the graph G is *vertex k-connected* (resp. *cycle k-connected*, *Tutte k-connected*) if G does not have any vertex l-separation (resp. cycle l-separation, Tutte l-separation) for $1 \leq l < k$. Note that the empty graph is vertex, cycle, and Tutte k-connected for every $k \geq 2$. The same conclusion holds for the connected graph with just one edge. The *vertex connectivity* (resp. *cycle connectivity*, *Tutte connectivity*) of G is the largest value k for which G is vertex k-connected (resp. cycle k-connected, Tutte k-connected).

It is easy to see that any vertex l-separation or cycle l-separation is a Tutte l-separation. Thus, Tutte k-connectivity implies vertex k-connectivity and cycle k-connectivity. The converse does not hold; that is, in general, vertex k-connectivity plus cycle k-connectivity do not imply Tutte k-connectivity. A counterexample is the wheel W_3. For any $k \geq 1$, that graph is readily verified to be both vertex k-connected and cycle k-connected. But it has a Tutte 3-separation (E_1, E_2), where E_1 is one of the 3-stars and where E_2 contains the remaining three edges. There are not many other counterexamples. Indeed, it is not difficult to show that the wheels W_1 and W_2 constitute the only other counterexamples.

Some Graph Theorems

Four basic graph theorems will be of much use: Menger's theorem, which relates vertex k-separations and the existence of internally node-disjoint paths; König's theorem, which for bipartite graphs connects matchings with covering nodes; Hall's theorem, which for bipartite graphs characterizes the existence of certain matchings; and finally Dilworth's theorem, which for acyclic graphs links covering directed paths and incomparable nodes.

We list these theorems, beginning with Menger's theorem.

(2.5.9) Theorem. *For given $k \geq 1$, a graph is vertex k-connected if and* *only if for any nodes i and j there are k internally node-disjoint paths connecting i and j.*

Theorem (2.5.9) is easily seen to imply that a graph is vertex 2-connected if and only if any two edges lie on some cycle.

König's theorem relates matchings of a bipartite graph to node covers as follows.

(2.5.10) Theorem. *In a bipartite graph, the cardinality of a maximum matching is equal to the minimum number of nodes covering all edges.*

Let X be the row or column node subset of a bipartite graph G, and let Y be the set of the remaining nodes. Hall's theorem characterizes when X cannot be matched into Y.

(2.5.11) Theorem. *Let X be the row or column node subset of a bipartite graph G, and let Y be the set of the remaining nodes. Then X cannot be matched into Y if and only if there exist subsets $\overline{X} \subseteq X$ and $\overline{Y} \subseteq Y$ with $|\overline{X}| > |\overline{Y}|$ such that each arc incident at a node of $\overline{X}$ has its second endpoint in $\overline{Y}$.*

Note that Theorem (2.5.11) may be used to characterize the absence of perfect matchings.

Dilworth's theorem relates incomparable nodes to covering paths as follows.

(2.5.12) Theorem. *In an acyclic graph, the maximum number of pairwise incomparable nodes is equal to the minimum number of directed paths covering all nodes of the graph.*

Although quite distinct in appearance, Theorems (2.5.9)–(2.5.12) are nothing but manifestations of the so-called max flow min cut theorem, which we introduce in a moment. We need two optimization problems called max flow and min cut.

Let G be a directed graph. One of the edges of G, say, l, is declared to be *special*. To each edge e of G other than l, a nonnegative integer h_e is assigned and is called the *capacity* of e. Define G to have *flow value* F if there are F directed cycles, not necessarily distinct, that satisfy the following two conditions. Each cycle of the collection must contain the special edge l, and any other edge e of G is allowed to occur altogether in at most h_e of the cycles. The *max flow problem* demands that one solve the problem max F. The solution value must be accompanied by a specification of the cycles producing that value.

A companion of the max flow problem is the following *min cut problem*. For any cocycle D of G containing the special edge l, define the *capacity* of D as follows. Since D is a cocycle, $G \backslash D$ consists without loss of generality of two disjoint graphs G_1 and G_2 such that the edge l goes in G from a node of G_2 to one of G_1. Then the *capacity* of D is the sum of the capacities h_e

of the edges e of D going in G from some node of G_1 to some node of G_2. Denote the capacity of D by $h(D)$. The min cut problem asks one to solve $\min h(D)$.

The max flow min cut theorem relates $\max F$ and $\min h(D)$ as follows.

(2.5.13) Theorem. *Let G be a directed graph with a special edge l and nonnegative integral capacities h_e for all edges $e \neq l$. Then $\max F = \min h(D)$.*

Some Graph Algorithms

There are a number of practically and theoretically efficient methods for solving the max flow problem and the min cut problem. As we shall see shortly, any such method may be used to solve the optimization or decision problems implicit in Theorems (2.5.9)–(2.5.12). In each such application, the max flow problem has 1s as edge capacities. Thus, the max flow solution consists of a maximum number of directed cycles that are edge-disjoint except for the special edge l. Suppose the edge l goes from a node t to a node s. Then the max flow solution is given by a maximum number of edge-disjoint, directed paths from s to t. For this reason, s is typically called the *source node*, and t the *sink node*. Due to this designation, the special edge l is no longer needed. Furthermore, the min cut may then be specified by a partition of the node set into two sets X and $\overline{X}$ with $s \in X$, $t \in \overline{X}$ such that the number of edges going from X to $\overline{X}$ is minimum.

The Algorithm MAX FLOW given next relies on the above terminology. That algorithm is claimed to have polynomial complexity, meaning that the run time is bounded by some polynomial in the number of edges of the graph being processed. An analogous interpretation applies to other polynomial algorithms presented later for graphs or matrices. A more detailed discussion of computational complexity issues is included in Section 2.7.

The Algorithm MAX FLOW and all other algorithms of this chapter are so well known that we omit proofs of validity. Appropriate references are cited in Section 2.8.

(2.5.14) Algorithm MAX FLOW. *Solves the max flow problem and min cut problem for directed graphs with 1s as edge capacities.*

Input: Directed graph G with source node s and sink node t. The graph has no parallel edges.

Output: A maximum number of edge-disjoint, directed paths from s to t. A partition of the node set of G into X and $\overline{X}$ with $s \in X$, $t \in \overline{X}$ such that the number of edges going from X to $\overline{X}$ is minimum.

Complexity: Polynomial.

Procedure:
1. Declare all edges to be unused.
2. Label s by $(-)$.
3. Carry out the following step until t has been labeled or until no additional labeling is possible: If a labeled node i has an unused arc outgoing to or a used arc incoming from an unlabeled node j, then label j by (i).

    If t has been labeled, go to step 4. If labeling cannot continue, go to step 5.
4. Backtrack from node t to node s using the labels. That is, if backtracking has led so far to node j and the label of j is (i), then backtrack to node i. The backtracking produces a path connecting s and t. Declare each arc of the path pointing toward t (resp. s) to be used (resp. unused). Erase all labels and go to step 2.
5. The used arcs define a maximum number of edge-disjoint, directed paths going from s to t. The labeled (resp. unlabeled) nodes make up the desired node set X (resp. $\overline{X}$).

We apply Algorithm MAX FLOW (2.5.14) in the following Algorithm DISJOINT PATHS to find the matchings and node sets of Theorems (2.5.10) and (2.5.11).

(2.5.15) Algorithm DISJOINT PATHS. *Finds for undirected graphs a maximum number of node-disjoint paths connecting two disjoint node subsets X and Y, and identifies a minimum number of nodes whose removal disconnects the nodes of X from those of Y. For undirected bipartite graphs, the output may be interpreted in terms of matchings and of covers of edges by nodes.*

Input: Undirected graph G with two nonempty, disjoint node subsets X and Y.

Output: A maximum number, say, k, of node-disjoint paths connecting nodes of X with nodes of Y. A minimum cardinality node subset Z, where $|Z| = k$, so that deletion of the nodes of Z reduces G to a graph where no path connects any node of X with any node of Y.
If G is bipartite and if X is the set of row nodes or column nodes and Y is the set of the remaining nodes: The k node-disjoint paths consist of one edge each and constitute a maximum cardinality matching of G. The set Z is a minimum cardinality node subset of G covering all edges of G. If X has not been matched into Y, then $\overline{X} = X - Z$ and $\overline{Y} = Y \cap Z$ are the sets of Theorem (2.5.11); that is, all edges incident at $\overline{X}$ are also incident at $\overline{Y}$, and $|\overline{X}| > |\overline{Y}|$.

Complexity: Polynomial.

Procedure:

1. Convert G to a directed graph H as follows. Delete all edges with both endpoints in X or with both endpoints in Y. Direct all remaining edges incident at X (resp. Y) so that they point away from X (resp. into Y). Replace each edge not treated so far by two parallel, directed edges of opposite direction. Split each node $i \notin (X \cup Y)$ into two nodes i_1 and i_2, and declare each edge with direction into i (resp. away from i) to become incident at i_1 (resp. i_2). Connect each node pair i_1, i_2 by a directed edge going from i_1 to i_2. Introduce two additional nodes s and t to H, and insert directed arcs from s to each node of X and from each node of Y to t.

2. Apply Algorithm MAX FLOW (2.5.14) to H. At termination, the algorithm has declared each edge of H to be used or unused and has declared each node of H to be labeled or unlabeled.

3. For some $k \geq 0$, the used edges of H define k internally node-disjoint, directed paths from s to t and thus define k node-disjoint paths from X to Y. The corresponding node-disjoint paths in G from X to Y are the desired ones. Declare the set Z to consist of the nodes of X that are unlabeled in H, the nodes of Y that are labeled in H, and the nodes $i \notin (X \cup Y)$ of G for which in H the corresponding i_1 is labeled and i_2 is unlabeled.

A modified version of Algorithm DISJOINT PATHS determines the vertex connectivity of graphs as follows.

(2.5.16) Algorithm VERTEX CONNECTIVITY. *Determines the vertex connectivity of undirected graphs.*

Input: Undirected graph G.

Output: The vertex connectivity of G.

Complexity: Polynomial.

Procedure:
1. Repeat the steps 2 and 3 below for each pair s, t of nonadjacent nodes of G. The vertex connectivity of G is the minimum of the k-values found in step 3.

2. For the selected s and t, convert G to a directed graph H as follows. Direct all edges incident at s (resp. t) so that they point away from s (resp. into t). Replace each remaining edge by two parallel, directed edges of opposite direction. Split each node $i \neq s, t$ into two nodes i_1 and i_2, and declare each edge with direction into i (resp. away from i) to become incident at i_1 (resp. i_2). Connect each node pair i_1, i_2 by a directed edge from i_1 to i_2.

3. Apply Algorithm MAX FLOW (2.5.14) to H. At termination, the algorithm has declared each edge to be used or unused. Let k be the number of directed paths in H from s to t that are defined by the used edges.

Theorem (2.5.12) links covering paths for an acyclic graph with incomparable nodes. Algorithm PATH COVER below finds these paths and nodes.

(2.5.17) Algorithm PATH COVER. *Finds a minimum number of directed paths covering the nodes of an acyclic graph and finds a maximum number of incomparable nodes.*

Input: Directed acyclic graph G.

Output: For some $k \geq 1$, k directed paths covering all nodes, and k incomparable nodes; k is the minimum number of such directed paths and is the maximum number of incomparable nodes.

Complexity: Polynomial.

Procedure:

1. Find the *transitive closure* $\tilde{G}$ of G; that is, $\tilde{G}$ has the same node set as G and has an arc from a node i to a node j whenever G has a directed path from i to j.
2. Deduce a directed, acyclic graph H from $\tilde{G}$ as follows. For each node i of $\tilde{G}$, let i_1 and i_2 be two nodes of H. For each arc of $\tilde{G}$, say, going from i to j, insert an arc into H from i_1 to j_2. Introduce two additional nodes s and t to H, and insert directed arcs from s to each node of type i_1 and from each node of type i_2 to t.
3. Apply algorithm MAX FLOW (2.5.14) to H. At termination, that algorithm has declared each edge of H to be used or unused, and each node to be labeled or unlabeled.
4. For each used edge (i_1, j_2) of H, declare the edge (i, j) of $\tilde{G}$ to be used. The used edges of $\tilde{G}$ define a certain number of directed paths, say, k_1. The nodes of $\tilde{G}$ without any used edges incident, say, k_2 in total, define paths of length 0. Together, the $k = k_1 + k_2$ paths of $\tilde{G}$ constitute a minimum number of directed paths covering all nodes of $\tilde{G}$. The k covering paths of $\tilde{G}$ are converted to k covering paths of G by replacing any path edge of $\tilde{G}$, say, from i to j, by a directed path from i to j in G. These k directed paths of G constitute a minimum number of directed paths covering all nodes of G.

   The desired k incomparable nodes of G are determined as follows. Each one of the k covering paths of $\tilde{G}$ contains precisely one such node. That node is found by moving in $\tilde{G}$ from the source node of the path toward its destination node and stopping at the first node i for which the node i_1 of H is labeled. That node i is the desired one.

We make a first use of the above theorems and algorithms in the next section, where we discuss matrices.

2.6 Matrices

In this section, we define elementary concepts for matrices. We also link matrices to the CNF systems of Section 2.3 and to the labeled, directed, bipartite graphs of Section 2.5.

We view a matrix to be a rectangular array with entries taken from some set that typically includes 0, 1, and -1. The matrix entries are always such that they can be interpreted to be real numbers or, in a special case, to be ordered pairs of real numbers. Thus it always makes sense to consider a matrix to be over $\mathbb{R}$, the field of real numbers. But we may also consider a matrix to be over some other field—in particular, the binary field GF(2) or the ternary field GF(3). At times, we even define a matrix to be over one of two systems of axioms called $\mathbb{B}$ and BG. The system $\mathbb{B}$ is a certain extension of Boolean algebra. The system BG postulates that the nonzero matrix entries be implicitly replaced by certain real numbers; the resulting matrix is then viewed to be over $\mathbb{R}$.

Trivial and Empty Matrices

We allow a matrix to have no rows or columns. Thus, for some $k \geq 1$, a matrix A may have size $k \times 0$ or $0 \times k$. Such a matrix is *trivial*. We even permit the case 0×0, in which case A is *empty*.

When a matrix is declared to be a $\{0, \pm 1\}$, $\{0, 1\}$, or zero matrix, then we allow for the situation where the matrix is actually trivial or empty and thus has no entries.

Length, Count, and Order

The *length* of an $m \times n$ matrix A, denoted by length(A), is $m+n$. The *count* of A, denoted by count(A), is the number of nonzero entries of A. The *order* of a square matrix A is the number of rows of A. We denote any column vector containing only 1s by $\underline{1}$. Suppose a matrix A has been partitioned into two row submatrices B and C, say, $A = [\frac{B}{C}]$. For typesetting reasons, we may denote this situation by $A = [B/C]$. If A, B, and C are column vectors, say, a, b, and c, respectively, we correspondingly write $a = [b/c]$. A superscripted t denotes transpose, so A^t is the transpose of A.

Matrix Indexing

Frequently, we index the rows and columns of a matrix. We write the row indices or index subsets to the left of a given matrix and write the column indices or index subsets above the matrix. For example, we might have

(2.6.1)

$$B = \begin{array}{c@{}c} & \begin{array}{ccc} y & e & f \end{array} \\ \begin{array}{c} a \\ b \\ x \end{array} & \begin{array}{|rrr|} \hline 1 & -1 & 0 \\ 0 & 1 & -1 \\ -1 & 0 & 1 \\ \hline \end{array} \end{array}$$

Example matrix B

All row and column indices, whether directly shown or implicitly given by index sets, are considered to be distinct, except if indices or index sets are explicitly shown to apply to both rows and columns.

Appended Identity Matrix

Occasionally, we append an identity to a given matrix. In that case the index of the ith column of the identity is taken to be that of the ith row of the given matrix. From the matrix B of (2.6.1), we thus may derive the following matrix A.

(2.6.2)

$$A = \begin{array}{c@{}c} & \begin{array}{cccccc} a & b & x & y & e & f \end{array} \\ \begin{array}{c} a \\ b \\ x \end{array} & \begin{array}{|rrr|rrr|} \hline 1 & 0 & 0 & 1 & -1 & 0 \\ 0 & 1 & 0 & 0 & 1 & -1 \\ 0 & 0 & 1 & -1 & 0 & 1 \\ \hline \end{array} \end{array}$$

Matrix A produced from B of (2.6.1)

Matrix Isomorphism

We consider two matrices to be equal if up to permutation of rows and columns they are identical. Two matrices that become equal upon a suitable change of row and column indices are *isomorphic*.

Terminology for Rows and Columns

We may refer to a column directly or by its index. For example, in a given matrix B, let b be a column vector with column index y. We may refer to b as "the column vector b of B." We may also refer to b by saying "the column y of B." In the latter case, we should say more precisely "the column of B indexed by y." We have opted for the abbreviated expression "the column y of B," since references of that type occur very often in this book. We treat references to rows in an analogous manner.

Characteristic Vector and Support Matrix

Suppose a set E indexes the rows (resp. columns) of a column (resp. row) vector with $\{0,1\}$ entries. Let E' be the subset of E corresponding to the 1s of the vector. Then that vector is the *characteristic column* (resp. *row*) *vector* of E'. We abbreviate this to *characteristic vector* when it is clear from the context whether it is a row or column vector. The *support* of a matrix A is a $\{0,1\}$ matrix B of the same size as A such that the 1s of B occur in the positions of the nonzeros of A.

Monotone and Nested Matrices

A matrix is *monotone* if, when viewed over $\mathbb{R}$, all entries are nonnegative or are nonpositive. Two matrices are *nested* if, when viewed over $\mathbb{R}$, each entry of one of the two matrices is at least as large as the corresponding entry of the other matrix. A set of matrices is *nested* if any two matrices of the collection are nested.

Parallel Vectors

Two vectors are *parallel* if, when viewed over $\mathbb{R}$, they are nonzero and one of them is a scalar multiple of the other one.

Simple Matrix

A matrix is *simple* if no row or column has less than two nonzeros and if there are no parallel rows or columns.

Scaling

Let A be a matrix over a field $\mathcal{F}$. We *column scale* (resp. *row scale*) A by multiplying each column (resp. row) by some nonzero. *Scaling* refers to column or row scaling. We also use this terminology when A is over the as yet undefined systems $\mathbb{IB}$ or $\mathbb{BG}$. In that case, one temporarily considers the matrix to be over $\mathbb{R}$ to carry out the scaling operation.

Submatrix and Subregion

A *submatrix* is obtained from a given matrix by the deletion of some rows and columns. The submatrix is *proper* if at least one row or column has been deleted. A *subregion* is obtained from a given matrix by first taking a submatrix and then replacing in that submatrix some nonzero entries by zeros. The subregion is *proper* if the submatrix is proper or if at least one nonzero entry of the submatrix has been replaced by a zero.

The process of deducing a submatrix (resp. subregion) is called *submatrix taking* (resp. *subregion taking*).

Let $\mathcal{I}$ be a collection of matrices. A matrix in $\mathcal{I}$ is *maximal under submatrix taking* if it is not a proper submatrix of another matrix in $\mathcal{I}$. A matrix in $\mathcal{I}$ is *minimal under submatrix taking* if it does not have a proper submatrix that is also in $\mathcal{I}$. If $\mathcal{I}$ consists of submatrices of a given matrix, we abbreviate the above terminology to *maximal submatrix* and *minimal submatrix*.

Maximality under subregion taking and *minimality under subregion taking*, as well as *maximal subregion* and *minimal subregion*, are defined analogously.

Fields

Often, we view a matrix to be over some field $\mathcal{F}$, where $\mathcal{F}$ is almost always the binary field GF(2), the ternary field GF(3), or the field $\mathbb{R}$ of real numbers.

The binary field GF(2) has only the elements 0 and 1. Addition is given by $0 + 0 = 0$, $0 + 1 = 1$, and $1 + 1 = 0$. Multiplication is specified by $0 \cdot 0 = 0$, $0 \cdot 1 = 0$, and $1 \cdot 1 = 1$. Note that the element 1 is also the additive inverse of 1, that is, -1. Thus, we may view a $\{0, \pm 1\}$ matrix to be over GF(2). Each -1 then stands for the 1 of the field.

The ternary field GF(3) has 0, 1, and -1. Instead of the -1, we could also employ some other symbol, say, 2, but never do so. Addition is given by $0 + 0 = 0$, $0 + 1 = 1$, $0 + (-1) = -1$, $1 + 1 = -1$, $1 + (-1) = 0$, and $(-1) + (-1) = 1$. Multiplication is given by $0 \cdot 0 = 0$, $0 \cdot 1 = 0$, $0 \cdot (-1) = 0$, $1 \cdot 1 = 1$, $1 \cdot (-1) = -1$, and $(-1) \cdot (-1) = 1$.

We need a matrix terminology that indicates the underlying field. For example, consider the rank of a matrix, that is, the order of any maximal nonsingular submatrix. If the field is $\mathcal{F}$, we refer to the $\mathcal{F}$-*rank* of the matrix. For determinants we use "$\det_{\mathcal{F}}$," but in the case of GF(2) and GF(3) we simplify that notation to "$\det_2$" and "$\det_3$," respectively.

In addition, we use the terms $\mathcal{F}$-*independence*, $\mathcal{F}$-*basis*, and $\mathcal{F}$-*span* in the expected way. That is, some columns of a matrix over $\mathcal{F}$ are $\mathcal{F}$-*independent* if they are linearly independent. A maximal collection of such columns forms an $\mathcal{F}$-*basis* of the matrix. A collection of columns of a matrix $\mathcal{F}$-*spans* the remaining columns if it spans them or, equivalently, if it contains an $\mathcal{F}$-basis.

There is another reason for emphasizing the underlying field. Later in this section, we encounter the systems $\mathbb{B}$ and BG. For matrices over $\mathbb{B}$ or BG, several concepts of linear algebra—in particular, those of independence, basis, rank, and span—can be adapted. We then refer to these concepts as IB-independence, BG-independence, IB-basis, BG-basis, etc. as expected.

Pivot

Customarily, a pivot consists of the following row operations, to be performed on a given matrix A over a field $\mathcal{F}$. First, a specified row x is scaled so that a 1 is produced in a specified column y. Second, scalar multiples of the new row x are added to all other rows so that column y becomes a unit vector. In this book, the term $\mathcal{F}$-*pivot* refers to a closely related process.

Let B be a matrix over a field $\mathcal{F}$ with row index set X and column index set Y. An $\mathcal{F}$-*pivot* on a nonzero *pivot element* B_{xy} of B is carried out as follows.

(2.6.3)   We replace for every $v \in (X - \{x\})$ and every $w \in (Y - \{y\})$, B_{vw} by $B'_{vw} = B_{vw} + (B_{vy} \cdot B_{xw})/(-B_{xy})$.

(2.6.4)   We replace B_{xy} by $-B_{xy}$, and exchange the indices x and y.

We demonstrate the pivot operation using the following matrix B over GF(3).

(2.6.5)

$$B = \begin{array}{c} \\ \\ X \\ \\ \end{array} \begin{array}{c} \\ \\ a \\ b \\ x \end{array} \begin{array}{|ccc|} \multicolumn{3}{c}{Y} \\ \multicolumn{3}{c}{y \;\; e \;\; f} \\ \hline 1 & -1 & 0 \\ 0 & 1 & -1 \\ -1 & 0 & 1 \\ \hline \end{array}$$

Matrix B

A GF(3)-pivot on $B_{xy} = -1$ may be displayed as follows.

(2.6.6)

$$B = \begin{array}{c} a \\ b \\ x \end{array} \begin{array}{|ccc|} \multicolumn{3}{c}{y \;\; e \;\; f} \\ \hline 1 & -1 & 0 \\ 0 & 1 & -1 \\ ① & 0 & 1 \\ \hline \end{array} \quad \underline{\text{GF(3)-pivot}} \quad B' = \begin{array}{c} a \\ b \\ y \end{array} \begin{array}{|ccc|} \multicolumn{3}{c}{x \;\; e \;\; f} \\ \hline 1 & -1 & 1 \\ 0 & 1 & -1 \\ 1 & 0 & 1 \\ \hline \end{array}$$

Effect of GF(3)-pivot on matrix B

Here and later we use a circle to highlight the pivot element B_{xy}. To relate the above process to the row operations of the customary pivot, we append an identity matrix I to B, getting the following matrix A over GF(3).

	X			Y		
	a	b	x	y	e	f
a	1	0	0	1	-1	0
$A = X \quad b$	0	1	0	0	1	-1
x	0	0	1	-1	0	1

(2.6.7)

Matrix A

We modify A in two steps as follows.

First, we do row operations to convert column y of A to a vector containing only zeros except for the pivot element B_{xy}. In our case, we just add row x to row a to achieve this. Note that these row operations modify, for every $v \in (X - \{x\})$ and every $w \in (Y - \{y\})$, the entry B_{vw} of the submatrix B to $B'_{vw} = B_{vw} + (B_{vy} \cdot B_{xw})/(-B_{xy})$, in agreement with the above rule (2.6.3). The row operations also transform the zero in row $v \in (X - \{x\})$ and column x of the submatrix I of A to $B_{vy}/(-B_{xy})$. All other entries of A not mentioned so far, that is, row x and all columns $w \in ((X - \{x\}) \cup \{y\})$, remain unchanged.

Second, we exchange the current columns x and y and then scale column x by $-B_{xy}$ and column y by $1/B_{xy}$.

Evidently, column x of A has become column y of the original B except that B_{xy} has become $-B_{xy}$ and column y of A has become a unit vector. Accordingly, the matrix A' deduced in the above two steps from A is of the form $A' = [I \mid B']$, where B' is the matrix defined by (2.6.3) and (2.6.4) from B.

We conclude that the pivot given by (2.6.3) and (2.6.4) is an abbreviated method of displaying the effect of the row operations of the customary pivot, followed by an exchange and scaling of two columns. Below, we display A and A' for the example (2.6.7). As expected, the nonidentity portion of A' is the matrix B' of (2.6.6).

(2.6.8)

	a	b	x	y	e	f
a	1	0	0	1	-1	0
$A = b$	0	1	0	0	1	-1
x	0	0	1	①	0	1

row operations,
column exchange,
and scaling
———————→

	a	b	y	x	e	f
a	1	0	0	1	-1	1
$A' = b$	0	1	0	0	1	-1
y	0	0	1	1	0	1

Effect of row operations, column exchange,
and scaling on A of (2.6.7)

By the above discussion, every basis of A is one of A', and vice versa. We record this fact for future reference.

(2.6.9) Lemma. *Let B' be derived from B by an $\mathcal{F}$-pivot as described by (2.6.3) and (2.6.4). Append identities to both B and B' to get $A = [I \mid B]$*

and $A' = [I \mid B']$. *Declare the row index sets of B and B' to become the column index sets of the identity submatrices I of A and A', respectively. Then every column index subset of A corresponding to a basis of A also indexes a basis of A', and vice versa.*

Pivots have several important features. For the discussion below, let B, B_{xy}, and B' be the matrices just defined.

First, when we $\mathcal{F}$-pivot in B' on B'_{yx}, we obtain B again.

Second, the pivot operation is symmetric with respect to rows versus columns. Thus, the $\mathcal{F}$-pivot operation and the operation of taking the transpose commute.

Third, we may use $\mathcal{F}$-pivots to compute determinants as follows. Suppose that B is square. If we delete row y and column x from B', then the resulting matrix, say, B'', satisfies $|\det_{\mathcal{F}}(B'')| = |\det_{\mathcal{F}}(B)|/|B_{xy}|$. Thus, B is nonsingular if and only if this is so for B''. Obviously, this way of computing determinants is nothing but the well-known method based on row operations.

System BG

Let A be a matrix with row index set X and column index set Y. Then $BG(A)$ is the following undirected bipartite graph. The row index set X (resp. column index set Y) is the set of row nodes (resp. column nodes) of the graph. Each nonzero entry A_{xy} of A produces an undirected edge connecting row node x with column node y.

The system BG introduced later makes much use of the graph $BG(A)$. To begin the discussion, we assume that A is a real $k \times k$ matrix. Recall that $\mathrm{count}(A)$ is the number of nonzeros of A. Denote the nonzero entries of A by $r_1, r_2, \ldots, r_{\mathrm{count}(A)}$.

Suppose that $r_1, r_2, \ldots, r_{\mathrm{count}(A)}$ are algebraically independent over the rationals; that is, $r_1, r_2, \ldots, r_{\mathrm{count}(A)}$ cannot be the roots for any nonzero polynomial with rational coefficients and variables $x_1, x_2, \ldots, x_{\mathrm{count}(A)}$. We then have the following characterization of $\det_{\mathbb{R}}(A)$.

(2.6.10) Theorem. *Let A be a real $k \times k$ matrix whose nonzero entries $r_1, r_2, \ldots, r_{\mathrm{count}(A)}$ are algebraically independent over the rationals. Then the following statements are equivalent.*

(i) $\det_{\mathbb{R}}(A) \neq 0$.

(ii) $\det_{\mathbb{R}}(A)$ *is not the zero polynomial when $r_1, r_2, \ldots, r_{\mathrm{count}(A)}$ are viewed as variables.*

(iii) *The rows (resp. columns) of A may be permuted such that the diagonal of the resulting matrix contains only nonzeros.*

(iv) $BG(A)$ *has a perfect matching.*

Proof. According to a basic result of matrix theory, $\det_{\mathbb{R}}(A)$ is a polynomial in $r_1, r_2, \ldots, r_{\text{count}(A)}$ with integer coefficients. Indeed, each term of $\det_{\mathbb{R}}(A)$ is, for some integer α and for some indices $1 \leq i_1 < i_2 < \ldots < i_k \leq \text{count}(A)$, of the form $\alpha \cdot r_{i_1} \cdot r_{i_2} \cdot \ldots \cdot r_{i_k}$, where no two of $r_{i_1}, r_{i_2}, \ldots, r_{i_k}$ reside in the same column or row. These observations and the definition of algebraic independence imply (i)⇔(ii)⇔(iii). Finally, (iv) restates (iii) in graph language. □

Theorem (2.6.10) has the following two corollaries.

(2.6.11) Corollary. *Let A be a real $k \times k$ matrix whose nonzero entries are algebraically independent over the rationals. Then $\det_{\mathbb{R}}(A) = 0$ if and only if the row index set X and the column index set Y of A can be partitioned into X_1, X_2 and Y_1, Y_2, respectively, such that A can be depicted as*

(2.6.12)

Partition of matrix A

with $|X_1| > |Y_1|$.

Proof. By parts (i) and (iv) of Theorem (2.6.10), $\det_{\mathbb{R}}(A) = 0$ if and only BG(A) does not have a perfect matching. By Theorem (2.5.11), the latter condition holds if and only if for some subset X_1 of the row node set X and for some subset Y_1 of the column node set Y of BG(A), all arcs incident at a node of X_1 have their second endpoint in Y_1 and $|X_1| > |Y_1|$. The condition on the arcs incident at X_1 implies that the submatrix of A indexed by X_1 and $Y_2 = Y - Y_1$ is zero as shown in (2.6.12). □

(2.6.13) Corollary. *Let A be a matrix over $\mathbb{R}$ whose nonzeros are algebraically independent over the rationals. Then the row index set X and the column index set Y of A can be partitioned into X_1, X_2 and Y_1, Y_2, respectively, such that A is the matrix of (2.6.12) and $|X_2| + |Y_1| = \mathbb{R}\text{-rank}(A)$.*

Proof. Parts (i) and (iv) of Theorem (2.6.10) imply that $\mathbb{R}$-rank(A) is equal to the cardinality of a largest matching of BG(A). By Theorem (2.5.10), that cardinality is equal to the minimum number of nodes covering all edges. Accordingly, X has a subset X_2 and Y has a subset Y_1 such that deletion of the nodes of $X_2 \cup Y_1$ removes all edges from BG(A) and $\mathbb{R}$-rank(A) = $|X_2| + |Y_1|$. In matrix language, deletion of the rows indexed by X_2 and of the columns indexed by Y_1 reduces A to a zero matrix. Thus, A is given by (2.6.12). □

The expression "algebraically independent over the rationals" is rather unwieldy. Also, we want to apply that concept indirectly to matrices that are not over $\mathbb{R}$ or even over any other field. Such a situation would require an even more complex formulation unless one settles for an abbreviated terminology, as we shall do now.

Let A be a nontrivial and nonempty matrix. Derive a matrix $\tilde{A}$ from A by replacing the nonzeros of A by real numbers that are algebraically independent over the rationals. The relationship between $\mathbb{R}$-rank$(\tilde{A})$ and the above cited results for the graph $BG(\tilde{A})$, which is the same graph as $BG(A)$, then motivates the following definitions.

The $\mathbb{R}$-rank of $\tilde{A}$ is the BG-*rank* of A. If a row (resp. column) submatrix of $\tilde{A}$ has $\mathbb{R}$-independent rows (resp. columns), then the corresponding row (resp. column) submatrix of A is said to have BG-*independent* rows (resp. columns). A maximal set of BG-independent columns of A constitutes a BG-*basis* of A. Suppose certain columns of $\tilde{A}$ span the remaining columns of that matrix. Then the corresponding columns of A are said to BG-*span* the remaining columns of A.

If $\tilde{A}$ is square and $\det_{\mathbb{R}}(\tilde{A}) = 0$ (resp. $\det_{\mathbb{R}}(\tilde{A}) \neq 0$), then the BG-*determinant* of A, abbreviated $\det_{BG}(A)$, is 0 (resp. 1). In the case of $\det_{BG}(A) = 0$ (resp. $\det_{BG}(A) = 1$), A is also said to be BG-*singular* (resp. BG-*nonsingular*).

We define any trivial as well as the empty matrix to have BG-rank equal to 0, and we declare the BG-determinant of the empty matrix to be 0 as well.

Define BG to be the system of axioms defining BG-rank, BG-independence, BG-bases, and BG-determinants for arbitrary matrices A as specified above via $\tilde{A}$. When a matrix A is to be interpreted in terms of these axioms, we say that A is to be viewed as a matrix *over* BG, or that A *is over* BG. Note that we never carry out pivots in such a matrix A.

Theorem (2.6.10) implies the following useful characterizations of BG-rank, BG-independence, BG-bases, and BG-determinants for matrices over BG.

(2.6.14) Theorem. *Let A be a matrix over* BG *with row index set X and column index set Y. Then the following statements hold.*

(a) BG-rank(A) *is equal to the size of a maximum cardinality matching of* BG(A) *and is also equal to the minimum number of rows and columns whose deletion reduces A to a zero matrix.*

(b) *A has BG-independent rows (resp. columns) if and only if in* BG(A) *the node subset X (resp. Y) can be matched into the node subset Y (resp. X).*

(c) *A column submatrix $\overline{A}$ of A, say, indexed by $\overline{Y} \subseteq Y$, is a BG-basis of A if and only if some maximum cardinality matching of* BG(A) *matches $\overline{Y}$ into X.*

(d) *If A is square, then* $\det_{BG}(A) = 1$ *if and only if* $BG(A)$ *has a perfect matching.*

Proof. By Theorem (2.6.10)(i) and (iv) and the definition of the BG-determinant, a square matrix A has $\det_{BG}(A) = 1$ if and only if $BG(A)$ has a perfect matching. Thus (d) holds. That result plus Corollary (2.6.13) implies (a)–(c). ☐

We include an algorithm that for a given matrix A over BG establishes the BG-rank and related results.

(2.6.15) Algorithm BG-RANK. *Computes the BG-rank of a matrix over BG and related results.*

Input: Matrix A over BG, with row index set X and column index set Y.

Output: The BG-rank of A; whether A has BG-independent rows or columns; a column BG-basis of A; if A is square, the BG-determinant of A.

Complexity: Polynomial.

Procedure:
1. Apply Algorithm DISJOINT PATHS (2.5.15) to determine a maximum cardinality matching for the graph $BG(A)$.
2. Define BG-rank(A) to be the size of the matching. Declare the rows (resp. columns) of A to be BG-independent if the node subset X (resp. Y) of $BG(A)$ has been matched into Y (resp. X), and to be BG-dependent otherwise. Let $\overline{Y} \subseteq Y$ be the set of column nodes of $BG(A)$ having matching edges incident. Then the column submatrix $\overline{A}$ of A indexed by $\overline{Y}$ is a column BG-basis of A. If A is square, declare $\det_{BG}(A) = 1$ if the matching is perfect, and declare $\det_{BG}(A) = 0$ otherwise.

Validity of the algorithm follows directly from that of Algorithm DISJOINT PATHS and Theorem (2.6.14).

Connected Matrix

We say that a matrix A is *connected* if the graph $BG(A)$ is connected. Suppose A is trivial; that is, A is $k \times 0$ or $0 \times k$ for some $k \geq 1$. Then $BG(A)$ and hence A are connected if and only if $k = 1$. Suppose A is empty, that is, of size 0×0. Then $BG(A)$ is the empty graph. By the earlier definition, the empty graph is connected. Thus, the empty matrix is connected. A *connected block* of a matrix is a maximal connected and nonempty submatrix.

Clause/Variable Matrix

Let S be a CNF system, say, with X as the set of clauses and Y as the set of Boolean variables. Unless stated otherwise, we assume that no clause contains both a Boolean variable and the negation of that variable. Indeed, such a clause would always be satisfied for any assignment of *True/False* values to the Boolean variables and thus should be deleted. The *clause/variable matrix* of S is the $\{0, \pm 1\}$ matrix A with row index set X and column index set Y where the entry in row $x \in X$ and column $y \in Y$ is 1 if clause x contains the Boolean variable y, is -1 if clause x contains the negation of the Boolean variable y, and is 0 otherwise.

For example, let S be the CNF system with variables y_1, y_2, y_3 and clauses x_1, x_2, x_3 given in set notation by $\{y_1, \neg y_2\}$, $\{y_2, \neg y_3\}$, $\{\neg y_1, y_3\}$. The clause/variable matrix of S is then

(2.6.16)
$$
\begin{array}{c}
\quad \quad y_1\, y_2\, y_3 \\
A = \begin{array}{c} x_1 \\ x_2 \\ x_3 \end{array}
\left[\begin{array}{rrr}
1 & -1 & 0 \\
0 & 1 & -1 \\
-1 & 0 & 1
\end{array}\right]
\end{array}
$$

Clause/variable matrix A of CNF system S

Recall that a CNF system S is satisfiable if one can assign *True/False* values to the Boolean variables of S such that each clause has the value *True*. Satisfiability of S and column scaling of its clause/variable matrix are closely linked according to the following elementary result.

(2.6.17) Lemma. *Let A be the clause/variable matrix of a CNF system S. Then S is satisfiable if and only if the columns of A can be scaled by $\{\pm 1\}$ factors such that each row of the resulting matrix contains at least one $+1$.*

Proof. Suppose S is satisfiable. Thus, a certain assignment of *True/False* values to the variables of S leads to an evaluation of *True* for each clause of S. If *True* (resp. *False*) is assigned to variable y of S, then we scale column y of A by $+1$ (resp. -1). Then the evaluation of each clause to *True* manifests itself in the scaled matrix by at least one $+1$ in each row. To prove the converse part, we reverse the above arguments. $\square$

If S is satisfiable (resp. unsatisfiable), then A is also called *satisfiable* (resp. *unsatisfiable*). Let A be satisfiable. In agreement with Lemma (2.6.17), we define a *satisfying vector* for A to be a $\{\pm 1\}$ vector of column scaling factors that converts A to a matrix having at least one $+1$ in each row.

In subsequent chapters, we interpret clause/variable matrices in several ways. For example, we view such matrices to be over the field GF(3) or to be over BG. We also assume such matrices to be over a system $\mathbb{B}$ that we introduce next.

System $\mathbb{B}$

The system $\mathbb{B}$ is an extension of Boolean algebra. Its elements are 0, $+1$, and -1. Its operations are called $\mathbb{B}$-*multiplication*, $\mathbb{B}$-*addition*, and $\mathbb{B}$-*subtraction*, denoted by $\odot$, $\oplus$, and $\ominus$, respectively. We first define these operations, then interpret them in terms of the customary Boolean algebra. For $\alpha, \beta \in \{0, \pm1\}$, $\mathbb{B}$-multiplication is defined by

$$(2.6.18) \qquad \alpha \odot \beta = \begin{cases} 1 & \text{if } \alpha = \beta = 1 \text{ or } \alpha = \beta = -1 \\ 0 & \text{otherwise} \end{cases}$$

$\mathbb{B}$-addition and $\mathbb{B}$-subtraction are defined only for $\{0,1\}$ elements. For $\alpha, \beta \in \{0,1\}$, $\mathbb{B}$-addition is given by

$$(2.6.19) \qquad \alpha \oplus \beta = \begin{cases} 1 & \text{if } \alpha = 1 \text{ or } \beta = 1 \\ 0 & \text{otherwise} \end{cases}$$

$\mathbb{B}$-subtraction is some sort of inverse of $\mathbb{B}$-addition. For $\alpha, \beta \in \{0,1\}$, $\mathbb{B}$-subtraction is specified by

$$(2.6.20) \qquad \alpha \ominus \beta = \begin{cases} 1 & \text{if } \alpha = 1 \text{ and } \beta = 0 \\ 0 & \text{otherwise} \end{cases}$$

Let $P = \{0, \pm1\}$ and $R = \{0,1\}$. Interpret the elements of P as follows. Define the $+1$ to represent *True*, the -1 *False*, and the 0 "not present." For R, let the 1 stand for *True* and the 0 for *False*. View the $\mathbb{B}$-multiplication operator $\odot$ as a function from $P \times P$ to R, and view the $\mathbb{B}$-addition operator $\oplus$ as a function from $R \times R$ to R. It is then easily verified that the operator $\odot$ is an extension of the Boolean "if and only if" operator $\Leftrightarrow$, while the operator $\oplus$ is precisely the Boolean "or" operator $\vee$.

For any $\alpha, \beta, \gamma \in \{0,1\}$, it is easy to verify that $(\alpha \oplus \beta) \oplus \gamma = \alpha \oplus (\beta \oplus \gamma)$ and $\alpha \oplus \beta = \beta \oplus \alpha$, so $\mathbb{B}$-addition is both associative and commutative. Accordingly, repeated $\mathbb{B}$-additions involving, say, $\alpha_1, \alpha_2, \ldots, \alpha_n$, for some $n \geq 2$, may be carried out in any order, and a notation such as $\alpha_1 \oplus \alpha_2 \oplus \cdots \oplus \alpha_n$ or $\bigoplus_{k=1}^{n} \alpha_k$ is sufficient. It is easily checked that $\mathbb{B}$-multiplication is commutative but not associative, and that $\mathbb{B}$-subtraction is neither associative nor commutative.

$\mathbb{B}$-multiplication, $\mathbb{B}$-addition, and $\mathbb{B}$-subtraction have straightforward matrix extensions that we cover in a moment. Whenever the latter operations are applied to matrices, we say, in agreement with the terminology for fields and the system BG, that the matrices are *viewed to be over* $\mathbb{B}$, or simply that they *are over* $\mathbb{B}$.

Matrix Operations in $\mathbb{B}$

Let A and B be $\{0, \pm 1\}$ matrices over $\mathbb{B}$ of size $m \times n$ and $n \times p$, respectively. If both A and B are nontrivial and nonempty, then the matrix $C = A \odot B$ is defined to be the $m \times p$ $\{0, 1\}$ matrix whose elements C_{ij} are given by $C_{ij} = \bigoplus_{k=1}^{n} (A_{ik} \odot B_{kj})$, for $i = 1, 2, \ldots, m$ and $j = 1, 2, \ldots, p$. If at least one of A and B is trivial or empty, then $C = A \odot B$ is defined to be the $m \times p$ zero matrix. The latter convention simplifies the matrix algebra with matrices over $\mathbb{B}$, since it eliminates the otherwise necessary treatment of special cases with m, n, or p equal to 0.

Let A and B be $m \times n$ $\{0, 1\}$ matrices over $\mathbb{B}$. If A is nontrivial and nonempty, then so is B, and $C = A \oplus B$ (resp. $C = A \ominus B$) is defined to be the $m \times n$ $\{0, 1\}$ matrix whose elements C_{ij} are given by $C_{ij} = A_{ij} \oplus B_{ij}$ (resp. $C_{ij} = A_{ij} \ominus B_{ij}$), for $i = 1, 2, \ldots, m$ and $j = 1, 2, \ldots, n$. If A is trivial or empty, then B is of the same type, and both $C = A \oplus B$ and $C = A \ominus B$ are defined to be equal to A or, equivalently, B.

Satisfiability Revisited

Lemma (2.6.17) relates satisfiability of a CNF system S to the existence of certain scaling factors for the clause/variable matrix A of S. Suppose we collect the scaling factors in a vector s. Then we may rephrase Lemma (2.6.17) as follows. A CNF system S is satisfiable if and only if there exists a $\{\pm 1\}$ vector s so that column scaling of A with the entries of s results in a matrix with at least one 1 in each row.

Declare A and a scaling vector s to be over $\mathbb{B}$ and examine $A \odot s$. By the scaling condition, for each row x of A there is a column y of A such that A_{xy} is nonzero and has the same sign as s_y. Thus, $A_{xy} \odot s_y = 1$, and hence $\bigoplus_j (A_{xj} \odot s_j) = 1$. By definition of the $\odot$ operation for matrices, we then have $A \odot s = \underline{1}$ and are justified to declare s to be a *solution vector* for the latter equation.

Note that a $\{0, \pm 1\}$ vector s solving $A \odot s = \underline{1}$ becomes a $\{\pm 1\}$ solution vector when its 0s are replaced by arbitrarily selected ± 1s. Hence $A \odot s = \underline{1}$ has a $\{\pm 1\}$ solution vector if and only if it has a $\{0, \pm 1\}$ solution vector. Thus it makes sense to define a matrix A over $\mathbb{B}$ to be *satisfiable* (resp. *unsatisfiable*) if the system $A \odot s = \underline{1}$ does have (resp. does not have) a $\{0, \pm 1\}$ solution vector.

For later reference, we record the relationships between the various satisfiability definitions of this chapter in the next lemma.

(2.6.21) Lemma. *The following statements are equivalent for a CNF system S with clause/variable matrix A. The matrix A is to be viewed over $\mathbb{B}$ whenever this is appropriate.*

(i) *S is satisfiable.*

(ii) *A is satisfiable.*

(iii) *One may assign True/False values to the Boolean variables of S such that each clause has the value True.*

(iv) *There is a $\{\pm1\}$ vector s of scaling factors such that column scaling of A with these factors produces a matrix each of whose rows contains at least one 1.*

(v) *There exists a $\{\pm1\}$ solution vector for $A \odot s = \underline{1}$.*

(vi) *There exists a $\{0, \pm1\}$ solution vector for $A \odot s = \underline{1}$.*

IB-Simple Matrix

We already have introduced a definition of simple matrices. Such a matrix has no rows or columns with less than two nonzeros and has no parallel rows or columns. For matrices over IB, it is useful that we replace the exclusion of parallel rows by a weaker requirement demanding absence of duplicate rows and that we introduce two subcases. Accordingly, we define a $\{0, \pm1\}$ matrix over IB to be IB-*row simple* (resp. IB-*column simple*) if it has no rows (resp. columns) with less than two nonzeros and has no duplicate rows (resp. parallel columns). The matrix is IB-*simple* if it is both IB-row simple and IB-column simple.

Matrix Representation of Labeled, Directed, Bipartite Graph

Section 2.5 includes a definition of labeled, directed, bipartite graphs and several related operations—in particular, the taking of Boolean minors. Here, we represent these graphs by matrices and translate graph operations such as the taking of Boolean minors into matrix language.

   We start with any clause/variable matrix A and the associated bipartite graph BG(A). In the latter graph, each edge represents a $\{\pm1\}$ entry of A. The graph BG(A) does not differentiate between $+1$ and -1 entries of A. But we may encode that information for each nonzero entry A_{xy} of A by directing in BG(A) the corresponding edge, which connects row node x with column node y. Specifically, if $A_{xy} = 1$ (resp. $A_{xy} = -1$), we direct that edge from row node x to column node y (resp. column node y to row node x). We convert the resulting directed, bipartite graph to a labeled, directed, bipartite graph by assigning the label 1 to each arc. The latter graph we declare to be DBG(A), the "D" indicating "directed."

   When we reduce DBG(A) to a Boolean minor as described in Section 2.5, we might get a graph where two nodes, say, x and y, are connected by an edge from x to y and one from y to x. Also, the labels on the edges of the Boolean minor may be 1s as well as 2s, and not just 1s as for DBG(A). We would like to have a matrix representation of such Boolean minors.

There are many ways to define such a representation. For our purposes, the following approach seems advantageous.

Given the Boolean minor H or, in general, given a labeled, directed, bipartite graph H, the rows (resp. columns) of a representation matrix B of H correspond to the row nodes (resp. column nodes) of H. The entry B_{xy} in row x and column y of B is an ordered pair (α, β) where α and β are determined as follows. If H has an arc from row node x to column node y (resp. from column node y to row node x), then α (resp. β) is equal to the label of that arc. If there is no arc from x to y (resp. from y to x), then $\alpha = 0$ (resp. $\beta = 0$). We call B the *generalized clause/variable matrix* arising from H.

As an example, we encode the labeled, directed, bipartite graph H of (2.5.7), which we display again.

(2.6.22)

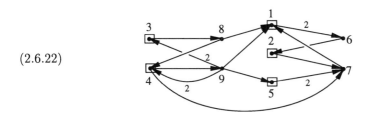

Labeled, directed, bipartite graph H

The generalized clause/variable matrix B for that graph is

(2.6.23)

$$B = \begin{array}{c} \\ 1 \\ 2 \\ 3 \\ 4 \\ 5 \end{array} \begin{array}{cccc} 6 & 7 & 8 & 9 \\ \hline (2,0) & (0,1) & (0,1) & (0,1) \\ (0,1) & (1,0) & (0,0) & (0,0) \\ (0,0) & (0,0) & (1,0) & (0,2) \\ (0,0) & (1,0) & (0,1) & (1,2) \\ (0,0) & (2,0) & (0,0) & (0,1) \end{array}$$

Generalized clause/variable matrix B of H

Given B, we obtain H again by the following rules. The row nodes (resp. column nodes) of H correspond to the rows (resp. columns) of B. If $B_{xy} = (\alpha, \beta)$ has α (resp. β) nonzero, then we introduce an arc in H from row node x to column node y (resp. column node y to row node x) with the label α (resp. β).

We know already how to encode a CNF system S by a clause/variable matrix A. We may also encode S by a generalized clause/variable matrix B, by taking B to be the matrix corresponding to the labeled, directed, bipartite graph DBG(A). We may derive B directly from S by the following rule. Each row (resp. column) of B corresponds to a clause (resp. variable) of S. The pair (α, β) in row x and column y of B is determined as follows. The entry α is 1 (resp. 0) if variable y occurs (resp. does not occur) in clause x. The entry β is 1 (resp. 0) if variable y occurs negated (resp. does not occur) in clause x. Note that the simultaneous occurrence of a variable and its negation in a given clause can be encoded in B. Such an encoding is not possible in a clause/variable matrix.

It is useful for us to extend the definition of $\mathrm{DBG}(A)$ to accommodate generalized clause/variable matrices. That is, for any such matrix B, we now declare $\mathrm{DBG}(B)$ to be the labeled, directed, bipartite graph corresponding to B. The definition is consistent with the earlier one involving clause/variable matrices, in the following sense. Let A (resp. B) be the clause/variable matrix (resp. generalized clause/variable matrix) of a CNF system S. Then $\mathrm{DBG}(A) = \mathrm{DBG}(B)$, as one would want.

Section 2.5 includes rules for deriving a Boolean minor $\overline{H}$ from a given labeled, directed, bipartite graph H. Specifically, such a minor is produced by column scaling, shrinking, and deletion of nodes, in that order. Any of these operations may be omitted. Let B be the generalized clause/variable matrix corresponding to H; that is, $H = \mathrm{DBG}(B)$.

Column scaling with a -1 factor, say, of column node y of H, corresponds to flipping of all pairs in column y of B. That is, for all x, the entry $B_{xy} = (\alpha, \beta)$ of B becomes the pair (β, α). Column scaling with a $+1$ factor leaves H and thus B unchanged.

The deletion of nodes from H becomes the deletion of rows or columns from B.

The translation of the shrinking step for H into matrix language is a bit more complicated. We first review that step for H, as described in Section 2.5.

Suppose $G_1, G_2, \ldots, G_n$ are the strong components of H. Then we shrink H by first collapsing, for each G_k, $k = 1, 2, \ldots, n$, the row nodes of G_k to a new row node and collapsing the column nodes of G_k to a new column node. In the next step of the shrinking operation, we delete all arc labels and replace any instance of multiple arcs with same endpoints and same direction by just one arc each. Finally, in the reduced graph we assign to each arc the label 1 or 2, where the case of a 1 corresponds precisely to the following situation. Let the arc in question connect the row node r and the column node c of the reduced graph. Define R (resp. C) to be the set of row (resp. column) nodes of H that were collapsed to form r (resp. c). If in the reduced graph the arc in question goes from node r to node c (resp. from node c to node r), then that arc receives the label 1 if and only if in H every row node of R has exactly one arc outgoing to (resp. incoming from) the nodes of C and that arc has the label 1.

Let $\overline{H}$ be obtained from H just by shrinking. Define $\overline{B}$ to be the generalized clause/variable matrix corresponding to $\overline{H}$; that is, $\overline{H} = \mathrm{DBG}(\overline{B})$. We may deduce $\overline{B}$ from B, without finding $\overline{H}$ first, once the strong components of H have been identified. The following two steps accomplish that task.

For the first step, let G_1, $G_2,\ldots$, G_m be the strong components of H with at least three nodes each, say, defined by row node subsets R_1, $R_2,\ldots$, R_m and column node subsets C_1, $C_2,\ldots$, C_m, respectively. For $k = 1, 2,\ldots$, m, we replace in B the columns of C_k by just one column c_k which is the sum of the columns it replaces. The summing of the columns of C_k involves the real addition of the ordered pairs residing in these columns; that is, we use $(e, f)+(g, h) = (e+g, f+h)$. In each of the resulting columns c_1, $c_2,\ldots$, c_m, we reduce any entry larger than 2 in any pair to a 2. Let B' be the resulting matrix.

For the second step, we first define a *max combination* operation for matrices having ordered pairs of real numbers as entries, as is the case for B'. The *max combination* of two pairs (e, f) and (g, h) is the pair $(\max\{e, g\}, \max\{f, h\})$. The *max combination* of several pairs is obtained by repeated application of the max combination step to two pairs at a time until just one pair is left. The *max combination* of several rows of a matrix is defined to be a row vector r where the pair in column position y of r is the max combination of the pairs in column y of the original rows.

In the second step, we replace in B', for $k = 1, 2,\ldots$, m, the rows of R_k by just one row r_k which is the max combination of the rows it replaces. The resulting matrix is, up to the indices, the matrix $\overline{B}$ corresponding to $\overline{H}$. We leave the straightforward verification of this claim to the reader.

As an example for the derivation of $\overline{B}$ directly from B or via $\overline{H}$, we consider H to be the graph of (2.6.22), which is the same graph as that of (2.5.7). According to Section 2.5, shrinking reduces H to the graph $\overline{H}$ of (2.5.8). We include the latter graph below.

(2.6.24)

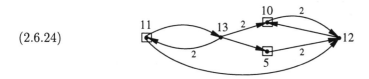

Minor $\overline{H}$ of H produced by shrinking

By direct translation of $\overline{H}$ into matrix form, or by application of the matrix shrinking operation described above, we get the following matrix $\overline{B}$ for $\overline{H}$.

(2.6.25)

$$\overline{B} = \begin{array}{c} \\ 10 \\ 11 \\ 5 \end{array} \begin{array}{cc} 12 & 13 \\ \hline (2,1) & (0,2) \\ (1,0) & (1,2) \\ (2,0) & (0,1) \end{array}$$

Matrix $\overline{B}$ for $\overline{H}$

Any matrix $\overline{B}$ obtained from a generalized clause/variable matrix B by scaling, shrinking, and column or row deletion, in that order, is a *Boolean minor* of B.

According to Lemma (2.6.21), a clause/variable matrix is satisfiable if its columns can be scaled by $\{\pm 1\}$ factors such that each row of the resulting matrix has at least one $+1$. Consistent with that result, we declare a generalized clause/variable matrix to be *satisfiable* if its columns can be scaled by $\{\pm 1\}$ factors such that each row of the resulting matrix has at least one pair (α, β) with $\alpha \geq 1$.

Extension of System IB

The extension of clause/variable matrices to generalized clause/variable matrices motivates the following extension of IB. We enlarge the set of elements $\{0, \pm 1\}$ by

$$(2.6.26) \qquad\qquad U = \{(\alpha, \beta) \mid \alpha, \beta \in \{0, 1, 2\}\}$$

and extend IB-multiplication so that for $(\alpha, \beta) \in U$ and $\gamma \in \{0, \pm 1\}$,

$$(2.6.27) \quad (\alpha, \beta) \odot \gamma = \begin{cases} 1 & \text{if } \alpha \geq 1 \text{ and } \gamma = 1, \text{ or } \beta \geq 1 \text{ and } \gamma = -1 \\ 0 & \text{otherwise} \end{cases}$$

Note that the extended IB-multiplication is no longer commutative. IB-addition and IB-subtraction need not be modified.

A matrix *over* the extension of IB has its entries in $\{0, \pm 1\} \cup U$. Matrix IB-multiplication is defined when such a matrix is postmultiplied with one having $\{0, \pm 1\}$ entries. The rules for matrix IB-addition and matrix IB-subtraction are unchanged.

Let B be a generalized clause/variable matrix. Then all entries of B are in U, and we may consider B to be over the extension of IB. According to the definition of that extension, the question whether the equation $B \odot s = \underline{1}$ has a $\{0, \pm 1\}$ vector is well-posed. If such a solution exists, we say that B is *satisfiable*; otherwise, B is *unsatisfiable*. It is easy to see that this definition is consistent with the earlier ones used in Lemma (2.6.21).

We consider any matrix over the extension of IB to be over BG by viewing any 0 or $(0, 0)$ entry of B to be zero, and we consider any other entry to be nonzero.

In the sequel, whenever we rely on the extension of IB, we explicitly say so. However, we use the terms IB-multiplication, IB-addition, and IB-subtraction for matrices over IB and for matrices over the extension of IB.

Display of Matrices

We employ a particular convention for the display of matrices. If in some portion of a matrix we explicitly list some entries but not all of them, then the omitted entries are always to be taken as zeros. This convention unclutters the appearance of matrices with complicated structure.

2.7 Complexity of Algorithms

We cover elementary notions of the computational complexity of algorithms in a summarizing discussion. Define a *problem* to be a question about binary strings that is answered each time by "yes" or "no." Any such string, say, of length s_1, represents a *problem instance*. In the setting of this book, the strings typically are graph or matrix problems encoded in binary form. The answer "yes" or "no" may be accompanied by a second binary string, say, of length s_2. We define $s_2 = 0$ if there is no such string. The *size* of each problem is then $s = s_1 + s_2$.

Suppose for a given problem we have an algorithm that determines the correct answer for each problem instance. We may imagine the algorithm to be encoded as a computer program. The algorithm is *polynomial time*, abbreviated *polynomial*, if the run time of the computer program can, for some positive integers α, β, and γ, be uniformly bounded by a polynomial of the form $\alpha \cdot s^\beta + \gamma$. We also say that the algorithm is of *order* β, and we denote this by $O(s^\beta)$. For example, the algorithm of Section 2.3 for the transformation of a general Boolean formula into a SAT equivalent CNF system is polynomial, as are Algorithms (2.5.14)–(2.5.17).

Suppose there are positive integers δ, ϵ, and ζ such that the following holds. For each problem instance of size s and with an affirmative answer, a proof of "yes" exists whose binary encoding is bounded by $\delta \cdot s^\epsilon + \zeta$. Then the problem is said to be in $\mathcal{NP}$. For example, the SAT problem is in $\mathcal{NP}$, since one may prove satisfiability of a CNF system by exhibiting *True/False* values for the variables such that each clause has the value *True*.

A problem P is *polynomially reducible* to a problem P' if there is a polynomial algorithm that transforms any instance of P into an instance of P'.

The class $\mathcal{NP}$ has a subclass of $\mathcal{NP}$-*complete* problems, which in some sense are the hardest problems of $\mathcal{NP}$. Specifically, a problem is $\mathcal{NP}$-*complete* if every problem in $\mathcal{NP}$ is polynomially reducible to it. Thus, existence of a polynomial solution algorithm for just one of the $\mathcal{NP}$-complete problems would imply existence of polynomial solution algorithms for every problem in $\mathcal{NP}$. It is an open question whether or not such polynomial algorithms exist. SAT is one of the problems of the class of $\mathcal{NP}$-complete problems; in fact, it was the first one determined to be in that class.

Let P be a given problem. If some $\mathcal{NP}$-complete problem is polynomially reducible to P, then P is $\mathcal{NP}$-*hard*.

A polynomial algorithm is not necessarily usable in practice. The constants α, β, and γ of the upper bound $\alpha \cdot s^\beta + \gamma$ on the run time may be huge, and the algorithm may require large run times even for small problem instances. The definition of "polynomial" completely ignores the magnitude of these constants.

However, almost all polynomial algorithms of this book involve constants α, β, and γ that are small enough to make the schemes practically useful.

In general, when one claims existence of an algorithm for a given class of problems, then in principle one need not exhibit an algorithm or any finite method for constructing it. Such existence claims are of little interest here. So whenever we assume or claim *existence* of an algorithm, then we mean that a complete description of the algorithm is at hand. Furthermore, if the algorithm is assumed or claimed to be polynomial, then we also mean that a polynomial function is available that bounds the worst-case run time.

2.8 References

Almost any book on logic or automated reasoning—for example, Chang and Lee (1973), Loveland (1978), Genesereth and Nilsson (1987), Hamilton (1988), Lloyd (1987), and Wos, Overbeek, Lusk, and Boyle (1992)—includes most logic definitions and results of Sections 2.3 and 2.4. For references about the MINSAT and MAXSAT problems, see Chapters 1 and 5.

The graph definitions of Section 2.5 are typically covered in the first few chapters of graph theory books—for example, in Ore (1962), Harary (1969), Wilson (1972), and Bondy and Murty (1976). The concept of Boolean minor for graphs is new. Theorems (2.5.9), (2.5.10), (2.5.11), and (2.5.12) are due to Menger (1927), König (1936), Hall (1935), and Dilworth (1950), respectively. Theorem (2.5.13) is a special case of the max flow min cut theorem for networks proved by Ford and Fulkerson (1956) and independently by Elias, Feinstein, and Shannon (1956). Algorithms (2.5.14)–(2.5.17) are simplified versions of algorithms described in the classic book on network flows by Ford and Fulkerson (1962). That book also shows Theorems (2.5.9)–(2.5.12) to be special cases of Theorem (2.5.13). The historical developments leading up to and involving Theorems (2.5.9)–(2.5.13) and Algorithms (2.5.14)–(2.5.17) are reviewed in detail in the book on matching by Lovász and Plummer (1986). Very efficient versions of Algorithms (2.5.14)–(2.5.17) are presented in Ahuja, Magnanti, and Orlin (1993).

The basic matrix definitions of Section 2.6 are included in any book on linear algebra; see, for example, Faddeev and Faddeeva (1963), Strang (1980), or Lancaster and Tismenetsky (1985). The system BG, Theorems (2.6.10) and (2.6.14), Corollaries (2.6.11) and (2.6.13), and Algorithm BG-RANK (2.6.15) are motivated by the algebra proof of Theorem (2.5.11) due to Edmonds (1967). System IB and the concept of Boolean minors for generalized clause/variable matrices are new.

Details about the computational complexity definitions may be found in Garey and Johnson (1979).

Chapter 3

Some Matroid Theory

3.1 Overview

We give a rather terse introduction to a part of matroid theory. A detailed treatment may be found in the books cited in Section 3.7.

We proceed as follows. In Section 3.2, we introduce matroids using several axiomatic definitions. We cover basic terminology and show how matroids arise from graphs and from matrices over fields.

Section 3.3 is concerned with the concept of matroid minor. We examine this idea in detail for the matroids arising from matrices over fields and compare it with those of graph minor and Boolean minor defined in Section 2.5.

In Section 3.4, we define separations and connectivity for matroids, and we relate these concepts to the graph separations and graph connectivity covered in Section 2.5. For matroids produced from matrices, we restate the separation and connectivity concepts using matrix terminology.

Sections 3.5 and 3.6 build upon Section 3.4. In Section 3.5, we show how one may efficiently locate several types of matroid and matrix separations. In Section 3.6, we introduce certain decompositions and compositions of matroids and matrices called k-sums, where k may be any positive integer. In later chapters, we employ the ideas underlying matroid and matrix separations and k-sums to obtain decompositions and compositions of logic problems.

In the final section, 3.7, we cover extensions and cite relevant references.

3.2 Definitions

We introduce the matroid concept via that of independence system. We also present several classes of matroids and define a number of matroid terms.

Independence System

Let E be a finite set. Define $\mathcal{I}$ to be a nonempty subset of the power set of E; that is, each element of $\mathcal{I}$ is a subset of E.

Suppose that for every subset $\overline{E} \subseteq E$ occurring in $\mathcal{I}$, every subset $\overline{\overline{E}} \subseteq \overline{E}$ is also in $\mathcal{I}$. We then say that the ordered pair $(E, \mathcal{I})$ is an *independence system* and that E is the *groundset* of the system. We define the subsets $\overline{E}$ of E present (resp. not present) in $\mathcal{I}$ to be *independent* (resp. *dependent*).

By these definitions, the null set must always be an element of the set $\mathcal{I}$ of an independence system, and we may replace the condition that $\mathcal{I}$ be nonempty by the demand that the null set be in $\mathcal{I}$. Thus, the following axioms define independence systems.

(3.2.1)
  (i)   The null set is in $\mathcal{I}$.
  (ii) Every subset of any set in $\mathcal{I}$ is also in $\mathcal{I}$.

Let $\overline{E}$ and $\overline{\overline{E}}$ be subsets of E such that $\overline{\overline{E}} \subseteq \overline{E}$. Suppose $\overline{\overline{E}}$ is independent. Then $\overline{\overline{E}}$ *spans* the set $\overline{E}$ if $\overline{\overline{E}}$ is a maximal independent subset of $\overline{E}$. Any independent subset of E that spans E is a *base* of $(E, \mathcal{I})$.

The rank of a subset $\overline{E} \subseteq E$, abbreviated rank$(\overline{E})$, is the cardinality of a maximum size independent subset of $\overline{E}$.

By these definitions, the independence system $(E, \mathcal{I})$ has a base $\overline{E}$ for which rank$(E) = |\overline{E}|$. But there may also be bases having fewer than rank(E) elements.

Matroid

A *matroid* is an independence system $M = (E, \mathcal{I})$ where, for any subset $\overline{E} \subseteq E$, all maximal independent subsets of $\overline{E}$ have the same cardinality. Thus, a matroid consists of a finite set E and a subset $\mathcal{I}$ of the power set of E satisfying the following axioms.

(3.2.2)
  (i)   The null set is in $\mathcal{I}$.
  (ii)   Every subset of any set in $\mathcal{I}$ is also in $\mathcal{I}$.
  (iii) For any subset $\overline{E} \subseteq E$, the maximal subsets of $\overline{E}$ that are in $\mathcal{I}$ have the same cardinality.

Let $\overline{\overline{E}}$ be an independent set that is a subset of some $\overline{E} \subseteq E$. If $\overline{\overline{E}}$ is a maximal independent subset of $\overline{E}$, then $\overline{\overline{E}}$ *spans* $\overline{E}$. By (iii) of (3.2.2), the cardinalities of any two maximal independent subsets of any $\overline{E} \subseteq E$ are the same. Thus, we may define the *rank* of $\overline{E}$, denoted by $r(\overline{E})$, to be the cardinality of any maximal independent subset of $\overline{E}$. Any independent subset of E that spans E is a *base* of M. Again by (iii) of (3.2.2), all bases of M must have the same cardinality. A *circuit* is a minimal dependent subset of E.

One can axiomatize matroids in terms of bases, circuits, and other subsets of E, or by certain functions, geometries, and operators. It is usually a simple, though at times tedious, exercise to prove equivalence of these systems. Here we just include the axioms that rely on bases, circuits, and the rank function.

Axioms Using Bases, Circuits, Rank Function

For bases, the axioms are as follows. Let $\mathcal{B}$ be a set of subsets of E. Suppose $\mathcal{B}$ observes the following axioms.

(3.2.3)
    (i) $\mathcal{B}$ is nonempty.
    (ii)   For any sets $B_1, B_2 \in \mathcal{B}$ and any $x \in (B_1 - B_2)$, there is a $y \in (B_2 - B_1)$ such that $(B_1 - \{x\}) \cup \{y\}$ is in $\mathcal{B}$.

Then $\mathcal{B}$ is the set of bases of a matroid on E.

Via circuits, we may define matroids as follows. Let $\mathcal{C}$ be the empty set, or let it be a set of nonempty subsets of E observing the following axioms.

(3.2.4)
    (i)   For any $C_1, C_2 \in \mathcal{C}$, C_1 is not a proper subset of C_2.
    (ii)   For any two $C_1, C_2 \in \mathcal{C}$ and any $z \in (C_1 \cap C_2)$, there is a set $C_3 \in \mathcal{C}$ where $C_3 \subseteq (C_1 \cup C_2) - \{z\}$.

Then $\mathcal{C}$ is the set of circuits of a matroid on E.

With the rank function, we specify a matroid as follows. Let $r(\cdot)$ be a function from the power set of E to the nonnegative integers. Assume $r(\cdot)$ satisfies the following axioms for any subsets S and T of E.

(3.2.5)
    (i) $r(S) \leq |S|$.
    (ii) $S \subseteq T$ implies $r(S) \leq r(T)$.
    (iii) $r(S) + r(T) \geq r(S \cup T) + r(S \cap T)$.

Then $r(\cdot)$ is the rank function of a matroid on E.

We omit the proofs of equivalence of the systems. It is instructive, though, to express each one of $\mathcal{I}, \mathcal{B}, \mathcal{C}$, and $r(\cdot)$ in terms of the other ones.

Suppose $\mathcal{I}$ is given. Then $\mathcal{B}$ is the set of $Z \in \mathcal{I}$ with maximum cardinality. $\mathcal{C}$ is the set of the minimal $C \subseteq E$ that are not in $\mathcal{I}$. For any $\overline{E} \subseteq E$, $r(\overline{E})$ is the cardinality of a maximal set $Z \subseteq \overline{E}$ that is in $\mathcal{I}$.

Suppose $\mathcal{B}$ is given. Then $\mathcal{I}$ is the set of all $X \in \mathcal{B}$ plus their subsets. $\mathcal{C}$ is the set of the minimal $C \subseteq E$ that are not contained in any $X \in \mathcal{B}$. For any $\overline{E} \subseteq E$, $r(\overline{E})$ is the cardinality of any maximal set $X \cap \overline{E}$ where $X \in \mathcal{B}$.

Suppose $r(\cdot)$ is given. Then $\mathcal{I}$ is the set of $\overline{E} \subseteq E$ for which $r(\overline{E}) = |\overline{E}|$. $\mathcal{B}$ is the set of $Z \subseteq E$ for which $|Z| = r(E)$. $\mathcal{C}$ is the set of the minimal $C \subseteq E$ for which $r(C) = |C| - 1$.

Matroids arise from graphs, matrices, and other settings. We present several examples.

Graphic Matroid

Let G be an undirected graph with arc set E. Define a subset $\overline{E}$ of E to be independent if deletion of the edges of $E - \overline{E}$ reduces G to a forest. Declare $\mathcal{I}$ to be the set of independent subsets of E. We prove that $M = (E, \mathcal{I})$ is a matroid. Clearly, any subset of an independent set is also independent, so M is an independence system. To establish axiom (iii) of (3.2.2), let $\overline{E}$ be a subset of E. We must show that all maximal independent subsets of $\overline{E}$ have the same cardinality. Define $\overline{G}$ to be the graph obtained from G by deletion of the edges of $E - \overline{E}$. A maximal independent subset of $\overline{E}$ is the edge set of a principal forest of $\overline{G}$. As shown in Section 2.5, all such forests have the same number of edges. Thus, M is a matroid. The matroids produced that way from graphs are called *graphic*.

Matroid Represented by Matrix

Let A be a matrix over a field $\mathcal{F}$, with columns indexed by the elements of a set E. Define a subset $\overline{E}$ of E to be independent if the columns of A indexed by $\overline{E}$ are linearly independent over $\mathcal{F}$. Collect in a set $\mathcal{I}$ the independent sets $\overline{E} \subseteq E$. Well-known arguments of linear algebra prove that $M = (E, \mathcal{I})$ satisfies the axioms of (3.2.2), so M is a matroid.

Suppose a matrix A over some field $\mathcal{F}$ and with column index set E generates a matroid $M = (E, \mathcal{I})$ as just described. In the spirit of the matroid representability definitions yet to come, we declare A to be a *nonstandard representation matrix* of M. Any elementary row operation transforms A to another nonstandard representation matrix of M. It is well known that such operations plus the deletion of zero rows can convert A to a matrix with leading identity matrix, say, $[I|B]$. As for A, the columns of $[I|B]$ are indexed by E, say, with subset $X \subseteq E$ (resp. $Y \subseteq E$) indexing the columns of I (resp. B). In conformance with the indexing convention introduced in Section 2.6, we declare the rows of $[I|B]$ and of B to be indexed by X. Clearly, the matrix B over $\mathcal{F}$, with row index set X and column index set Y, completely specifies M. Indeed, the independent sets of M are completely determined by the $\mathcal{F}$-rank of the submatrices of B. We present details of this relationship between B and M in a moment when we discuss the matroid rank function.

We call B a *standard representation matrix*, abbreviated *representation matrix*, of M. At times, we want to make the role of the underlying field $\mathcal{F}$ explicit. We then call B an $\mathcal{F}$-*representation matrix* of M.

By the discussion of the pivot operation in Section 2.6, any matrix B' obtained from B by pivots is an $\mathcal{F}$-representation matrix of M.

We discuss an instance of a nonstandard representation matrix. Consider the graphic matroid M of an undirected graph G. Define A to be the following matrix, called the *node/edge incidence matrix* of G. Each row of A corresponds to a node of G, and each column corresponds to an edge. All entries of A are zero, except that for any nonloop edge (i, j) of G, the column (i, j) of A has a 1 in row i and a second 1 in row j. Consider A to be over the binary field GF(2). It is not difficult to verify that any set of linearly independent columns of A corresponds to the edge set of a forest subgraph of G, and vice versa. Since the edge sets of forests are precisely the independent sets of M, we know that A over GF(2) is a nonstandard representation matrix of M.

$\mathcal{F}$-Matroid

For a given field $\mathcal{F}$, an $\mathcal{F}$-*matroid* is a matroid that can be represented by some matrix over $\mathcal{F}$. In particular, there are GF(2)-*matroids*, GF(3)-*matroids*, and $\mathbb{R}$-*matroids*.

BG-Matroid

Let B be a matrix over BG with row index set X and column index set Y. Derive a real matrix B' from B by replacing the nonzero entries of B by real numbers that are algebraically independent over the rationals. Recall that the BG-rank of any submatrix of B is defined to be the $\mathbb{R}$-rank of the corresponding submatrix of B'. Let M be the matroid represented by B'. Since the independent sets of M are completely determined by the $\mathbb{R}$-rank of the submatrices of B', we could also establish the independent sets of M using the BG-rank of the submatrices of B. Accordingly, we declare B to be a BG-*representation matrix* of M and call M a BG-*matroid*. Evidently, the BG-matroids are representable over $\mathbb{R}$ and thus are $\mathbb{R}$-matroids.

Matrix and Matroid Rank

Let B be a matrix over a field $\mathcal{F}$ with row index set X and column index set Y. Let M on $X \cup Y$ be the matroid $\mathcal{F}$-represented by B. We repeatedly make use of some partition (X_1, X_2) of X and some partition (Y_1, Y_2) of Y. Typically, we just specify one set of X_1, X_2 and one set of Y_1, Y_2. For any such partitions, we assume B to be partitioned as

(3.2.6)

$$B = \begin{array}{c|c|c|} & Y_1 & Y_2 \\ \hline X_1 & A^1 & D^2 \\ \hline X_2 & D^1 & A^2 \\ \hline \end{array}$$

Partitioned version of B

Let $A = [I|B]$. According to our indexing convention, A has row index set X and column index set $X \cup Y$. Let $Z \subseteq X \cup Y$ be a base of M; that is, Z is a maximal independent set. Define $X_2 = Z \cap X$ and $Y_1 = Z \cap Y$. Then the column submatrix $\overline{A}$ of A indexed by $Z = X_2 \cup Y_1$ must be an $\mathcal{F}$-basis of A of the form

(3.2.7)

$$\overline{A} = \begin{array}{c|c|c|} & X_2 & Y_1 \\ \hline X_1 & 0 & A^1 \\ \hline X_2 & \ddots_1^1 & D^1 \\ \hline \end{array}$$

Submatrix of A indexed by $Z = X_2 \cup Y_1$

The submatrix A^1 of $\overline{A}$ is square and, by cofactor expansion, $\mathcal{F}$-nonsingular. Conversely, any square and $\mathcal{F}$-nonsingular submatrix A^1 of B defined by (X_1, X_2) and (Y_1, Y_2) corresponds to a base $Z = X_2 \cup Y_1$ of M. More generally, let the submatrix A^1 of B of (3.2.7) be of arbitrary size and with $\mathcal{F}$-rank $A^1 = k$. Then the set $X_2 \cup Y_1$ of M has rank equal to $|X_2| + k$.

Let $r(\cdot)$ be the rank function of M. We have just shown that

(3.2.8) $$r(X_2 \cup Y_1) = |X_2| + \mathcal{F}\text{-rank}(A^1)$$

The above arguments are easily adjusted for the case where B is a matrix over BG and where M is the BG-matroid of B. The equation (3.2.8) then becomes

(3.2.9) $$r(X_2 \cup Y_1) = |X_2| + \text{BG-rank}(A^1)$$

When we let the submatrices D^1 and D^2 of (3.2.6) play the role of A^1, we get the respective equations

(3.2.10)
$$r(X_1 \cup Y_1) = |X_1| + \mathcal{F}\text{-rank}(D^1)$$
$$r(X_2 \cup Y_2) = |X_2| + \mathcal{F}\text{-rank}(D^2)$$

Adding these two equations and using $r(X \cup Y) = |X| = |X_1| + |X_2|$, we obtain

(3.2.11)
$$r(X_1 \cup Y_1) + r(X_2 \cup Y_2)$$
$$= r(X \cup Y) + \mathcal{F}\text{-rank}(D^1) + \mathcal{F}\text{-rank}(D^2)$$

The corresponding equation for the BG case is

(3.2.12)
$$r(X_1 \cup Y_1) + r(X_2 \cup Y_2)$$
$$= r(X \cup Y) + \text{BG-rank}(D^1) + \text{BG-rank}(D^2)$$

There are matroids that are not represented by any matrix over any field. Such matroids are *nonrepresentable*. In this book, we only use representable matroids—in particular, GF(3)-matroids and BG-matroids. Accordingly, we interpret most matroid definitions given below for an arbitrary matroid $N = (E, \mathcal{I})$ also in terms of two matrices B and $\tilde{B}$ representing two matroids M and $\tilde{M}$, respectively. The matrix B (resp. $\tilde{B}$) is assumed to be over an arbitrary field $\mathcal{F}$ (resp. over BG). Both matrices B and $\tilde{B}$ have row index set X and column index set Y. Define $A = [I|B]$ and $\tilde{A} = [I|\tilde{B}]$. We assume that $E = X \cup Y$, so the three matroids N, M, and $\tilde{M}$ have the same groundset.

Loop

An element v of N is a *loop* if the rank of v is 0. The loops of M (resp. $\tilde{M}$) correspond precisely to the zero columns of B (resp. $\tilde{B}$).

Parallel Elements and Triangle

Two nonloop elements v and w of N are *parallel* if the rank of the set $\{v, w\}$ is 1 or, equivalently, if that set is a circuit.

Two elements of M are parallel if and only if one of the following two cases applies. In the first case, one of v and w, say, v, indexes a row of B, and w indexes a column of B with exactly one nonzero, which occurs in row v. In the second case, v and w index two nonzero columns of B that form a matrix with $\mathcal{F}$-rank equal to 1.

For $\tilde{M}$ and $\tilde{B}$, two cases are also possible for parallel v and w. The first case corresponds to the first one of M and B; that is, one of v and

w, say, v, indexes a row of $\tilde{B}$, and w indexes a column of $\tilde{B}$ with exactly one nonzero, which occurs in row v. The second case for $\tilde{M}$ and $\tilde{B}$ is more restrictive than that for M and B. As before, v and w index two nonzero columns of $\tilde{B}$. But this time, the columns v and w contain just one nonzero each. The two nonzeros of the columns v and w must occur in the same row.

Evidently, "is parallel to" is an equivalence relation. The equivalence classes are the *parallel classes*.

A *triangle* is a circuit with three elements.

Fundamental Circuit

Let Z be a base of N. For any element $z \notin Z$, the set $Z \cup \{z\}$ contains exactly one circuit, which is of the form $\overline{Z} \cup \{z\}$ for some subset $\overline{Z} \subseteq Z$. The circuit $\overline{Z} \cup \{z\}$ is the *fundamental circuit* that z creates with the base Z of N.

The fundamental circuit of M that any element $y \in Y$ creates with the base X of M is of the form $\{x \in X \mid B_{xy} \neq 0\} \cup \{y\}$. The corresponding fundamental circuit of $\tilde{M}$ is $\{x \in X \mid \tilde{B}_{xy} \neq 0\} \cup \{y\}$.

Dual Matroid

The collection $\{E - Z \mid Z = \text{base of } N\}$ is the set of bases of a matroid on E, as is readily checked using the axioms of (3.2.3). We call that matroid the *dual matroid* of N. We use the asterisk to denote the dualizing operation, so N^* is the just defined dual matroid of N. Evidently, the dual matroid of the dual matroid of N is N again, so $(N^*)^* = N$.

It is not difficult to verify that the dual matroid M^* of M (resp. $\tilde{M}^*$ of $\tilde{M}$) is represented by B^t (resp. $\tilde{B}^t$), where the superscripted t denotes transpose.

If $r(\cdot)$ is the rank function of N, then the rank function $r^*(\cdot)$ of N^* is given by

$$(3.2.13) \qquad r^*(\overline{E}) = r(E - \overline{E}) + |\overline{E}| - r(E), \quad \forall\, \overline{E} \subseteq E$$

The prefix "co" dualizes a term. For example, a *cobase* (resp. *coloop*, *cocircuit*) of N is a base (resp. loop, circuit) of the dual matroid N^*. *Coparallel* or *series* elements of N are parallel elements of N^*. The equivalence classes of the equivalence relation "is coparallel to" (="is in series with") are the *coparallel* or *series classes*. A *cotriangle* or *triad* of N is a triangle of N^*.

Since B^t (resp. $\tilde{B}^t$) represents M^* (resp. $\tilde{M}^*$), it is a simple matter to describe cobases, coloops, series (=coparallel) elements, and fundamental cocircuits of M (resp. $\tilde{M}$) in terms of B (resp. $\tilde{B}$). We do this here for coloops and series elements, and we leave it to the reader to characterize the remaining items.

A coloop of M (resp. M^*) corresponds precisely to a zero row of B (resp. $\tilde{B}$).

Two elements v and w of M are in series if and only if one of the following two cases applies. In the first case, one of v and w, say, v, indexes a column of B, and w indexes a row of B with exactly one nonzero entry, which occurs in column v. In the second case, v and w index two nonzero rows of B that form a matrix with $\mathcal{F}$-rank equal to 1.

For $\tilde{M}$ and $\tilde{B}$, two cases are also possible for series elements v and w. The first case corresponds to that of M and B; that is, one of v and w, say, v, indexes a column of $\tilde{B}$, and w indexes a row of $\tilde{B}$ with exactly one nonzero entry, which occurs in column v. The second case is more restrictive than that for M and B. Here, too, v and w index two nonzero rows. But this time, the rows v and w contain just one nonzero entry each. The two nonzero entries of rows v and w must occur in the same column.

Next, we cover matroid operations producing so-called matroid minors.

3.3 Minor

We continue to refer to the three matroids N, M, and $\tilde{M}$ of Section 3.2 above, each with groundset E. The matroids M and $\tilde{M}$ are represented by B over a field $\mathcal{F}$ and by $\tilde{B}$ over BG, respectively. Both matrices have the rows indexed by X and the columns indexed by Y, so $E = X \cup Y$.

Deletion and Contraction

Let z be an element of N that is not a coloop. Define $\overline{N}$ to be the matroid on $E - \{z\}$ whose bases are the bases Z of N that do not contain z. Using the axioms of (3.2.3) for bases, one readily confirms that $\overline{N}$ is indeed a matroid. We say that $\overline{N}$ is obtained from N by the *deletion* of the element z. We denote the deletion operation by "\", so $\overline{N} = N \backslash z$.

Now let z be an element of N that is not a loop. Define $\overline{N}$ to be the matroid on $E - \{z\}$ whose bases are the sets $Z - \{z\}$, where Z ranges over the bases of N with z. Again, it is not difficult to check that $\overline{N}$ is a matroid. We say that $\overline{N}$ is obtained from N by the *contraction* of the element z. We denote the contraction operation by "/", so $\overline{N} = N/z$.

To complete the specification of the deletion and contraction operations, we declare the deletion of a coloop to be actually a contraction and declare the contraction of a loop to be actually a deletion.

A *reduction* by an element is the deletion or contraction of that element. A matroid obtained by a sequence of reductions from N is a *minor* of N. It is easy to show that the same minor results regardless of the order in which the reductions are carried out. By induction, one only needs to show that the reordering of two successive reduction steps results in the same minor.

Suppose a reduction sequence involves the deletion of elements w_1, $w_2, \ldots,$ w_n and the contraction of elements $v_1, v_2, \ldots,$ v_m. Let $V = \{v_1, v_2, \ldots, v_m\}$ and $W = \{w_1, w_2, \ldots, w_n\}$. We denote the minor of N obtained by the deletion of $w_1, w_2, \ldots,$ w_n and the contraction of $v_1, v_2, \ldots,$ v_m by $N/V\backslash W$. We also write N/V if $W = \emptyset$ and $N\backslash W$ if $V = \emptyset$. The above notation for minors is somewhat imprecise, since, for example, we write N/v when element v is contracted, and we write N/V when all elements of the set V are contracted. But our notation is the customary one and less cumbersome than, say, the alternate and formally correct notation $N/\{v\}$ when element v is contracted.

We consider N itself to be a minor of N. All other minors of N are *proper* minors of N.

Duality of Deletion and Contraction

Let z be a noncoloop element of N. The bases of the minor $N\backslash z$ are by definition the bases Z of N without z. Hence, the bases of $(N\backslash z)^*$, the dual of $N\backslash z$, are the sets $(E - Z) - \{z\}$, where Z ranges over the bases of N without z. Since the bases Z^* of N^* are the sets $E - Z$, where Z ranges over the bases of N, the bases of $(N\backslash z)^*$ are the sets $Z^* - \{z\}$, where Z^* ranges over the bases of N^* with z. By definition of the contraction operation, the just defined sets $Z^* - \{z\}$ are the bases of N^*/z. We conclude that $(N\backslash z)^* = N^*/z$. It is easy to check that the same conclusion holds when z is a coloop of N. Thus, in general, $(N\backslash z)^* = N^*/z$ and, by duality, $(N/z)^* = N^*\backslash z$.

We have shown that deletion (resp. contraction) in a matroid corresponds to contraction (resp. deletion) in the dual matroid. For the earlier defined set V and W, we have $(N/V\backslash W)^* = N^*/W\backslash V$.

Addition, Expansion, and Extension

If $\overline{N} = N\backslash z$ or $\overline{N} = N/z$, then one may obtain N from $\overline{N}$ again by inserting the element z. Specifically, if $\overline{N} = N\backslash z$ (resp. $\overline{N} = N/z$) and z is not a coloop (resp. loop) of N, then we say that N can be obtained from $\overline{N}$ by the *addition* of z (resp. *expansion* by z). We denote addition by "+" and expansion by "&". Accordingly, in the addition (resp. expansion) case, we have $N = \overline{N}+z$ (resp. $N = \overline{N}\&z$). An *extension* is an addition or expansion.

The addition and expansion operations may be extended to sets of elements in the obvious way. In particular, if N is obtained from a minor

$\overline{N}$ via the addition of the elements of a set W and via the expansion by the elements of a set V, then we write $N = \overline{N}\&V+W$.

By the duality relationship between deletion and contraction, addition (resp. expansion) in a matroid corresponds to expansion (resp. addition) in the dual matroid.

Representation Matrix of Minor

Let z be an element of M that is not a coloop. Recall that M is represented by the matrix B over $\mathcal{F}$ with row index set X and column index set Y. We know that the bases of $M \backslash z$ are the bases Z of M without z. So if z indexes a column of B, that is, if $z \in Y$, then we just delete column z from B to get a representation matrix for $M \backslash z$. Suppose z indexes a row of B; that is, $z \in X$. Since z is not a coloop of M, row z of B must be nonzero. By one pivot on any nonzero entry in row z of B, we get another representation matrix for M where z indexes a column. We conclude that, up to a pivot, deletion of a noncoloop element from M corresponds to deletion of a column from B.

Recall that the pivot operation and the taking of transpose commute, that B^t represents M^*, and that deletion and contraction are dual operations. So by duality, contraction of a nonloop element of M corresponds up to a pivot to deletion of a row from B.

By definition, the addition (resp. expansion) operation is the inverse of the deletion of a noncoloop element (resp. contraction of a nonloop element). Thus, the addition of (resp. expansion by) an element in a proper minor of M corresponds to adjoining of a column (resp. row) to any representation matrix of the minor.

Minor of Graphic Matroid

Let G be an undirected graph with edge set E. Define M to be the graphic matroid produced by G, as discussed earlier in Section 3.2. So M has E as groundset, and the edge sets of the forest subgraphs of G are the independent sets of M.

We claim that for any disjoint subset V and W of E, the matroid minor $M/V \backslash W$ is the graphic matroid of the graph minor $G/V \backslash W$. To prove this claim, one only needs to consider the deletion and contraction of just one element in M and G. We omit the elementary proof.

Matroid Minor and Boolean Minor

According to Section 2.5, labeled, directed, bipartite graphs may be reduced to Boolean minors by scaling, shrinking, and deletion of nodes. Section 2.6

contains a translation of these steps into matrix language for generalized clause/variable matrices where each entry is an ordered pair.

Suppose we exclude the shrinking step. The remaining reduction operations of column scaling and submatrix taking may also be carried out in any $\{0, \pm 1\}$ clause/variable matrix B. Indeed, any matrix $\overline{B}$ so derived from B is the clause/variable matrix of some Boolean minor of DBG(B). Suppose B has row index set X and column index set Y. Define M to be the matroid represented by B by viewing B to be over some field $\mathcal{F}$ or over the system BG.

We know that any reduction sequence for M involving contractions of some elements of X and deletion of some elements of Y reduces M to a matroid minor $\overline{M}$ that is represented by a submatrix $\overline{B}$ of B. At the same time, $\overline{B}$ is the clause/variable matrix of a Boolean minor of DBG(B). Hence, all matroid minors of M of the form $M/\overline{X}\backslash\overline{Y}$ with $\overline{X} \subseteq X$ and $\overline{Y} \subseteq Y$ correspond to Boolean minors of DBG(B).

The converse relationship also holds; that is, any Boolean minor of DBG(B) produced without shrinking corresponds to a matroid minor of the matroid M.

3.4 Connectivity

We introduce matroid separations and connectivity.

Let M be a matroid with groundset E and rank function $r(\cdot)$. Let E_1 and E_2 be two sets that partition E. For $k \geq 1$, the unordered pair (E_1, E_2) is a k-separation of M if

$$(3.4.1) \qquad \begin{aligned} |E_1|, |E_2| &\geq k \\ r(E_1) + r(E_2) &\leq r(E) + k - 1 \end{aligned}$$

The sets E_1 and E_2 are the two *sides* of the k-separation. The k-separation is *exact* if the inequality of (3.4.1) involving the rank function $r(\cdot)$ holds with equality. The matroid M is k-separable if it has a k-separation. For $k \geq 2$, M is k-connected if it does not have an l-separation for some $1 \leq l \leq k - 1$. If M is 2-connected, then it is also said to be *connected*.

Tutte graph connectivity and matroid connectivity are linked by the following theorem due to Tutte. We include that result, since it sheds light on the relationship between graphs and graphic matroids. But we make no use of the theorem and thus omit the proof.

(3.4.2) Theorem. *For any $k \geq 2$, a graph G is Tutte k-connected if and only if the graphic matroid $M(G)$ is k-connected.*

For a matroid M represented by a matrix B over a field $\mathcal{F}$, the conditions of (3.4.1) for k-separations manifest themselves in B as follows.

(3.4.3) **Lemma.** *Let M be a matroid represented by a matrix B over a field $\mathcal{F}$ or over the system BG. Let B have row index set X and column index set Y. Denote the groundset of M, which is $X \cup Y$, by E. Then (a) and (b) below hold.*

(a) *For some $k \geq 1$, let (E_1, E_2) be a k-separation of M. For $i = 1, 2$, define $X_i = E_i \cap X$ and $Y_i = E_i \cap Y$. Partition B using X_1, X_2, Y_1, and Y_2 as follows.*

(3.4.4)

$$
B = \begin{array}{c|c|c}
 & Y_1 & Y_2 \\
\hline
X_1 & A^1 & D^2 \\
\hline
X_2 & D^1 & A^2 \\
\end{array}
$$

Partitioned version of B

Then the sets X_1, X_2, Y_1, Y_2 and the submatrices D^1, D^2 of B satisfy

(3.4.5)
$$|X_1 \cup Y_1|, |X_2 \cup Y_2| \geq k$$

as well as

(3.4.6)
$$\mathcal{F}\text{-rank}(D^1) + \mathcal{F}\text{-rank}(D^2) \leq k - 1$$

for the case of the field $\mathcal{F}$, and

(3.4.7)
$$\mathrm{BG}\text{-rank}(D^1) + \mathrm{BG}\text{-rank}(D^2) \leq k - 1$$

for the case of the system BG. If the k-separation of M is exact, then the applicable inequality of (3.4.6) or (3.4.7) holds with equality.

(b) *Suppose (X_1, X_2) and (Y_1, Y_2) are partitions of X and Y, respectively. Assume that (3.4.5) and the applicable inequality of (3.4.6) or (3.4.7) hold. For $i = 1, 2$, define $E_i = X_i \cup Y_i$. Then (E_1, E_2) is a k-separation of M.*

Proof. We begin with part (a), assuming the case of field $\mathcal{F}$. The inequality $|X_1 \cup Y_1|, |X_2 \cup Y_2| \geq k$ of (3.4.5) obviously holds. To establish (3.4.6), we combine (3.2.11), which says that $r(X_1 \cup Y_1) + r(X_2 \cup Y_2) = r(X \cup Y) + \mathcal{F}\text{-rank}(D^1) + \mathcal{F}\text{-rank}(D^2)$, with (3.4.1). Thus, $r(E_1) + r(E_2) = r(X_1 \cup Y_1) + r(X_2 \cup Y_2) = r(X \cup Y) + \mathcal{F}\text{-rank}(D^1) + \mathcal{F}\text{-rank}(D^2) \leq r(E) + k - 1$. Since $E = X \cup Y$, we conclude that $\mathcal{F}\text{-rank}(D^1) + \mathcal{F}\text{-rank}(D^2) \leq k - 1$. We handle the case of (3.4.7) using (3.2.12) instead of (3.2.11). The remainder of part (a) and also part (b) are then immediate. □

We apply the terminology for matroid k-separations and k-connectivity to matrices in the expected way. So if a matrix B represents a matroid

M and if M has a k-separation $(X_1 \cup Y_1, X_2 \cup Y_2)$ (resp. has an exact k-separation, or is k-connected), then we declare B to also have a k-separation $(X_1 \cup Y_1, X_2 \cup Y_2)$ (resp. to have an exact k-separation, or to be k-connected). The matrix B may at one time be over a field $\mathcal{F}$, and at another time be over BG. To differentiate among the possible k-separations, we say, for example, that B has an $\mathcal{F}$-k-separation or has a BG-k-separation. Terms such as $\mathcal{F}$-k-connected and BG-k-connected are to be analogously interpreted.

The next lemma implies that separations of M are also separations of its dual M^*.

(3.4.8) Lemma. *A matroid M is k-separable or k-connected if and only if this is so for its dual M^*.*

Proof. The lemma follows from the equation (3.2.13) for the rank function $r^*(\cdot)$ of M^* and the k-separation conditions of (3.4.1). For representable matroids, the symmetry inherent in (3.4.5)–(3.4.7) immediately proves the result. □

3.5 Finding Separations

For the decompositions to come, we need a method that for the matroid represented by a given matrix either locates a k-separation satisfying specified conditions or determines that such a separation does not exist. If such a k-separation exists, we want one with k as small as possible. For the case of 1-separations, the specified conditions are vacuous. For k-separations with $k \geq 2$, the conditions demand that the two sides of the separation properly contain some specified sets and have at least a certain size.

We first treat the 1-separation case.

1-Separation

Define B to be a matrix over a field $\mathcal{F}$ or over BG, with row index set X and column index set Y. Let M be the matroid represented by B. We use the notation of (3.4.4). Thus, (X_1, X_2) and (Y_1, Y_2) are partitions of X and Y, respectively, and A^1, A^2, D^1, D^2 are the submatrices of B defined by these partitions.

By (3.4.5) and (3.4.6) (resp. (3.4.7)), $(X_1 \cup X_2, Y_1 \cup Y_2)$ is an $\mathcal{F}$-1-separation (resp. BG-1-separation) if and only if $|X_1 \cup Y_1|, |X_2 \cup Y_2| \geq 1$ and $\mathcal{F}$-rank(D^1)+$\mathcal{F}$-rank(D^2) (resp. BG-rank(D^1)+BG-rank(D^2)) is equal to 0. The zero matrix is the only matrix with $\mathcal{F}$-rank or BG-rank equal to 0. Thus, for any $\mathcal{F}$-1-separation or BG-1-separation, both submatrices D^1 and D^2 must be zero. The latter condition implies that the graph BG(B) is not connected. Conversely, assume that BG(B) is not connected, and let $X_1 \cup Y_1$ be the set of nodes of $BG(B)$ of some, but not all, connected components of $BG(B)$. Then $(X_1 \cup Y_1, X_2 \cup Y_2)$ is an $\mathcal{F}$-1-separation or BG-1-separation, whichever applies. We conclude that finding the $\mathcal{F}$-1-separations or BG-1-separations of B is equivalent to finding the connected components of $BG(B)$, an easy task.

We make use of these observations in the following algorithm for finding 1-separations.

(3.5.1) Algorithm 1-SEPARATION. *Finds a 1-separation of a matroid M represented by a matrix B over a field $\mathcal{F}$ or over the system BG.*

Input: Matrix B with row index set X and column index set Y.

Output: Either: A 1-separation of B and M for which the submatrix A^1 is connected. Or: "B and M do not have a 1-separation."

Complexity: Polynomial.

Procedure:
Determine the connected components of the graph $BG(B)$. If there is only one such component, declare that B does not have a 1-separation. Otherwise, output a 1-separation where the submatrix A^1 corresponds to one of the connected components.

Induced Separation

We turn to the case of k-separations with $k \geq 2$ where each side must contain specified sets and must be at least of specified size. We present a simple algorithm, called Algorithm k-SEPARATION below, for finding such separations. The algorithm relies on a subroutine where so-called induced separations are determined. The subroutine is described in this subsection, while the algorithm is covered in the next one. We should mention that Algorithm k-SEPARATION is quite satisfactory from a theoretical viewpoint, since it is polynomial whenever k is bounded from above. However, the algorithm is too inefficient to be used in actual computations. In later subsections, we address this difficulty and suggest appropriate remedies.

We still assume B to be a matrix over $\mathcal{F}$ or over BG, with row index set X and column index set Y, and assume M to be the matroid represented by B. Let $\overline{B}$ be a submatrix of B of the following form.

(3.5.2)

$$
\overline{B} = \begin{array}{c|c|c}
 & \overline{Y}_1 & \overline{Y}_2 \\
\hline
\overline{X}_1 & \overline{A}^1 & \overline{D}^2 \\
\hline
\overline{X}_2 & \overline{D}^1 & \overline{A}^2 \\
\end{array}
$$

Submatrix $\overline{B}$ of B

Define $\overline{M}$ to be the minor of M represented by $\overline{B}$. Assume that, for some $l \geq k$,

(3.5.3) $|\overline{X}_1 \cup \overline{Y}_1|, |\overline{X}_2 \cup \overline{Y}_2| \geq l$

and that

(3.5.4) $\mathcal{F}\text{-rank}(\overline{D}^1) + \mathcal{F}\text{-rank}(\overline{D}^2) = k - 1$

or

(3.5.5) $\text{BG-rank}(\overline{D}^1) + \text{BG-rank}(\overline{D}^2) = k - 1$

whichever applies. Hence, $(\overline{X}_1 \cup \overline{Y}_1, \overline{X}_2 \cup \overline{Y}_2)$ is an exact k-separation of $\overline{B}$ where each side has at least l elements.

We are to decide whether the given k-separation of $\overline{B}$ can be extended to one for B. Specifically, we must determine whether B has a k-separation $(X_1 \cup Y_1, X_2 \cup Y_2)$ where, for $i = 1, 2$, $X_i \supseteq \overline{X}_i$ and $Y_i \supseteq \overline{Y}_i$. If this is so, we say that the k-separation $(\overline{X}_1 \cup \overline{Y}_1, \overline{X}_2 \cup \overline{Y}_2)$ of $\overline{B}$ and $\overline{M}$ *induces* the k-separation $(X_1 \cup Y_1, X_2 \cup Y_2)$ of B and M.

Define $X_3 = X - (\overline{X}_1 \cup \overline{X}_2)$ and $Y_3 = Y - (\overline{Y}_1 \cup \overline{Y}_2)$. We depict B with the submatrix $\overline{B}$ and the index sets $\overline{X}_1$, $\overline{X}_2$, X_3 and $\overline{Y}_1$, $\overline{Y}_2$, Y_3 below. For reasons to become clear shortly, we have placed the submatrices $\overline{A}^1$, $\overline{A}^2$, $\overline{D}^1$, and $\overline{D}^2$ of $\overline{B}$ into the corners of B.

(3.5.6)

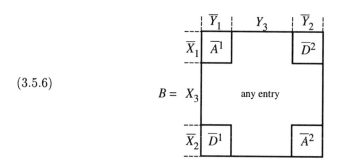

Matrix B with submatrix $\overline{B}$

By definition, an induced k-separation exists if and only if X_3 and Y_3 can be partitioned into X_{31}, X_{32} and Y_{31}, Y_{32}, respectively, such that $(\overline{X}_1 \cup \overline{Y}_1 \cup X_{31} \cup Y_{31}, \overline{X}_2 \cup \overline{Y}_2 \cup X_{32} \cup Y_{32})$ is a k-separation of B. We display B with that k-separation below.

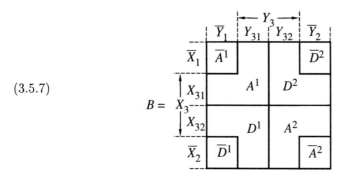

$$(3.5.7)$$

Partition of B induced by that of $\overline{B}$

By (3.5.3), we have

$$(3.5.8) \qquad |\overline{X}_i \cup \overline{Y}_i \cup X_{3i} \cup Y_{3i}| \geq l, \quad i = 1,\, 2$$

By (3.4.6), (3.4.7), (3.5.4), (3.5.5), and the fact that the matrices $\overline{D}^1$ and $\overline{D}^2$ are submatrices of D^1 and D^2, respectively, we must have

$$(3.5.9) \qquad \mathcal{F}\text{-rank}(D^1) + \mathcal{F}\text{-rank}(D^2) = k - 1$$

or

$$(3.5.10) \qquad \text{BG-rank}(D^1) + \text{BG-rank}(D^2) = k - 1$$

Hence, if an induced k-separation exists, then it must be an exact k-separation with at least l elements on each side, and

$$(3.5.11) \qquad \mathcal{F}\text{-rank}(D^i) = \mathcal{F}\text{-rank}(\overline{D}^i), \quad i = 1,\, 2$$

or

$$(3.5.12) \qquad \text{BG-rank}(D^i) = \text{BG-rank}(\overline{D}^i), \quad i = 1,\, 2$$

We utilize two different methods for deciding whether an induced k-separation exists, depending on whether the matrix B is over the field $\mathcal{F}$ or over BG.

For the case of B over $\mathcal{F}$, we employ a recursive scheme. As the measure of problem size for the recursion, we use $|X_3 \cup Y_3|$. If $|X_3 \cup Y_3| = 0$, then $(\overline{X}_1 \cup \overline{Y}_1, \overline{X}_2 \cup \overline{Y}_2)$ is the desired induced k-separation. Suppose $|X_3 \cup Y_3| > 0$. Redraw B of (3.5.6) so that an arbitrary row $x \in X_3$ and an arbitrary column $y \in Y_3$ are displayed.

(3.5.13)

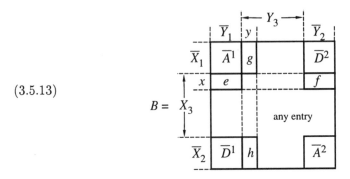

Matrix B with row $x \in X_3$ and column $y \in Y_3$

The recursive method relies on the analysis of the following three cases of B of (3.5.13). Collectively, these cases cover all situations.

In the first case, we suppose that for some row $x \in X_3$, the subvector e is not spanned by the rows of $\overline{D}^1$. We claim that, in any induced k-separation, we must have $x \in X_{31}$. For a proof, take any such separation as depicted by (3.5.7). If $x \in X_{32}$, then the subvector e of row x occurs in D^1. Since e is not spanned by the rows of $\overline{D}^1$, we have $\mathcal{F}$-rank$(D^1) >$ $\mathcal{F}$-rank$(\overline{D}^1)$, which contradicts (3.5.11). Thus, x must be in X_{31} as claimed. We examine the subvector f of row x. Suppose that subvector is not spanned by the rows of $\overline{D}^2$. Using (3.5.7) and (3.5.11) once more, we see that in any induced k-separation the subvector f forces x to be in X_{32}. But the latter requirement conflicts with the one determined earlier for x. Thus, an induced k-separation cannot exist, and we stop with that conclusion. So assume that the subvector f of row x is spanned by the rows of $\overline{D}^2$. We know already that x must be in X_{31} in any induced k-separation. Suppose in B of (3.5.13) we adjoin e to $\overline{A}^1$ and f to $\overline{D}^2$, getting a new $\overline{A}^1$ and a new $\overline{D}^2$. Evidently, the new $\overline{A}^1$, $\overline{D}^2$ plus the old $\overline{A}^2$, $\overline{D}^1$ constitutes a new matrix $\overline{B}$ for which $(\overline{X}_1 \cup \{x\} \cup \overline{Y}_1, \overline{X}_2 \cup \overline{Y}_2)$ is a k-separation, and that k-separation induces one in B if and only if this is so for the k-separation $(\overline{X}_1 \cup \overline{Y}_1, \overline{X}_2 \cup \overline{Y}_2)$ of the original $\overline{B}$. Thus, we may replace the original problem by one involving the new $\overline{B}$. By our measure of problem size, the new problem is smaller than the original one, and we may apply recursion.

In the second case, we suppose that for some column $y \in Y_3$, the subvector g is not spanned by the columns of $\overline{D}^2$. Arguing analogously to the first case via (3.5.7) and (3.5.11), we conclude that y must be in Y_{31} in any induced k-separation. Furthermore, suppose that the column subvector h of column y is not spanned by the columns of $\overline{D}^1$. Using (3.5.7) and (3.5.11) once more, we see that y must also be in Y_{32} in any induced k-separation. Thus, an induced k-separation cannot exist, and we stop with that conclusion. So suppose that h is spanned by the columns of $\overline{D}^1$. Then we adjoin g to $\overline{A}^1$, adjoin h to $\overline{D}^1$, and correspondingly redefine $\overline{B}$. The

k-separation $(\overline{X}_1 \cup \overline{Y}_1 \cup \{y\}, \overline{X}_2 \cup \overline{Y}_2)$ of the new $\overline{B}$ induces a k-separation of B if and only if this is so for the k-separation $(\overline{X}_1 \cup \overline{Y}_1, \overline{X}_2 \cup \overline{Y}_2)$ of the original $\overline{B}$. Once more, we may replace the induced k-separation problem involving the original $\overline{B}$ by one with the new $\overline{B}$. The latter problem is smaller, and we may invoke recursion.

For the discussion of the third and final case, we suppose that neither of the above cases applies. Equivalently, for all $x \in X_3$, the subvector e of row x is spanned by the rows of $\overline{D}^1$, and, for all $y \in Y_3$, the subvector g of column y is spanned by the columns of $\overline{D}^2$. By (3.5.13), $(\overline{X}_1 \cup \overline{Y}_1, \overline{X}_2 \cup X_3 \cup \overline{Y}_2 \cup Y_3)$ is a k-separation of B induced by the one of $\overline{B}$, and we stop with that conclusion.

Clearly, the above scheme has a polynomial implementation. We summarize it below.

(3.5.14) Algorithm INDUCED $\mathcal{F}$-SEPARATION. *Finds a k-separation for the matroid M represented by a matrix B over a field $\mathcal{F}$ that is induced by an exact k-separation of the minor $\overline{M}$ represented by a submatrix $\overline{B}$, or declares that such an induced separation does not exist.*

Input: Matrix B over field $\mathcal{F}$, with row index set X and column index set Y. A submatrix $\overline{B}$ of B with an exact k-separation $(\overline{X}_1 \cup \overline{Y}_1, \overline{X}_2 \cup \overline{Y}_2)$ where, for $i = 1$, 2, $\overline{X}_i \subseteq X$ and $\overline{Y}_i \subseteq Y$. The k-separation of $\overline{B}$ has at least l elements on each side.

Output: Either: A k-separation $(X_1 \cup Y_1, X_2 \cup Y_2)$ of B and M induced by the k-separation $(\overline{X}_1 \cup \overline{Y}_1, \overline{X}_2 \cup \overline{Y}_2)$ of $\overline{B}$ and $\overline{M}$; the k-separation of B is exact and has at least l elements on each side. Or: "The given k-separation of $\overline{B}$ and $\overline{M}$ does not induce a k-separation of B and M."

Complexity: Polynomial.

Procedure:

1. Consider B partitioned as in (3.5.13). Assume that B has a row $x \in X_3$ with the indicated row subvectors e and f such that $\mathcal{F}$-rank$([e/\overline{D}^1]) > \mathcal{F}$-rank$(\overline{D}^1)$. Then x must be in X_{31}. Suppose, in addition, that $\mathcal{F}$-rank$([\overline{D}^2/f]) > \mathcal{F}$-rank$(\overline{D}^2)$. Then x must also be in X_{32}; that is, B cannot be partitioned, and we stop with that declaration. On the other hand, suppose $\mathcal{F}$-rank$([\overline{D}^2/f]) = \mathcal{F}$-rank$(\overline{D}^2)$. Since x must be in X_{31}, we adjoin e to $\overline{A}^1$ and f to $\overline{D}^2$. Then we start recursively again with the new $\overline{B}$.

2. Suppose B as shown in (3.5.13) has a column $y \in Y_3$ with the indicated column subvectors g and h such that $\mathcal{F}$-rank$([g|\overline{D}^2]) > \mathcal{F}$-rank$(\overline{D}^2)$. Then y must be in Y_{31}. Suppose, in addition, $\mathcal{F}$-rank$([\overline{D}^1|h]) > \mathcal{F}$-rank$(\overline{D}^1)$. Then y must also be in Y_{32}; that is, B cannot be partitioned, and we stop with that declaration. On the other hand, suppose $\mathcal{F}$-rank$([\overline{D}^1|h]) = \mathcal{F}$-rank$(\overline{D}^1)$. Since y must be in Y_{31}, we adjoin g to $\overline{A}^1$ and h to $\overline{D}^1$. Then we start recursively again with the new $\overline{B}$.

3. Finally, suppose that, for all rows $x \in X_3$, the row subvector e satisfies $\mathcal{F}$-rank$([e/\overline{D}^1]) = \mathcal{F}$-rank$(\overline{D}^1)$, and suppose that, for all columns $y \in Y_3$, the column subvector g satisfies $\mathcal{F}$-rank$([g|\overline{D}^2]) = \mathcal{F}$-rank$(\overline{D}^2)$. Then $X_1 = \overline{X}_1$, $X_2 = \overline{X}_2 \cup X_3$, $Y_1 = \overline{Y}_1$, and $Y_2 = \overline{Y}_2 \cup Y_3$ are the sets for the desired k-separation $(X_1 \cup Y_1, X_2 \cup Y_2)$ of B.

For later use, we include two observations about the output of Algorithm INDUCED $\mathcal{F}$-SEPARATION (3.5.14).

(3.5.15) Lemma.
(a) *Any k-separation produced by Algorithm INDUCED $\mathcal{F}$-SEPARA-TION (3.5.14) has $X_1 \cup Y_1$ minimal and $X_2 \cup Y_2$ maximal, in the sense that any other k-separation $(X_1' \cup Y_1', X_2' \cup Y_2')$ of B induced by the exact k-separation $(\overline{X}_1 \cup \overline{Y}_1, \overline{X}_2 \cup \overline{Y}_2)$ of $\overline{B}$ observes $X_1 \subseteq X_1'$, $X_2 \supseteq X_2'$, $Y_1 \subseteq Y_1'$, and $Y_2 \supseteq Y_2'$.*
(b) *Let $(\overline{X}_1 \cup \overline{Y}_1, \overline{X}_2 \cup \overline{Y}_2)$ be an exact k-separation of $\overline{B}$, except that $|\overline{X}_2 \cup \overline{Y}_2|$ may be equal to $k - 1$. If B has a k-separation $(X_1' \cup Y_1', X_2' \cup Y_2')$ where, for $i = 1, 2$, $\overline{X}_i \subseteq X_i'$ and $\overline{Y}_i \subseteq Y_i'$, then one such k-separation of B is found by Algorithm INDUCED $\mathcal{F}$-SEPARATION (3.5.14).*

Proof. (a) According to Steps 1 and 2 of Algorithm INDUCED $\mathcal{F}$-SEP-ARATION (3.5.14), $\overline{X}_1$ (resp. $\overline{Y}_1$) is enlarged by $x \in X_3$ (resp. $y \in Y_3$) only if there is no induced k-separation of B with x (resp. y) on the side containing $\overline{X}_2$ (resp. $\overline{Y}_2$). Thus, any x (resp. y) added to $\overline{X}_1$ (resp. $\overline{Y}_1$) must also be in X_1' (resp. Y_1'). This implies the minimality of $X_1 \cup Y_1$ and the maximality of $X_2 \cup Y_2$.
(b) The validity of Algorithm INDUCED $\mathcal{F}$-SEPARATION (3.5.14) and the proof of part (a) are not affected if $|\overline{X}_2 \cup \overline{Y}_2|$ is equal to $k - 1$, except that possibly $(X_1 \cup Y_1, X_2 \cup Y_2)$ of the output of the algorithm has $|X_2 \cup Y_2|$ equal to $k - 1$. Assume that exceptional situation. By part (a), $X_2 \cup Y_2$ is maximal, which implies that there is no k-separation $(X_1' \cup Y_1', X_2' \cup Y_2')$ of B where, for $i = 1, 2$, $\overline{X}_i \subseteq X_i'$ and $\overline{Y}_i \subseteq Y_i'$. But this contradicts the assumptions of part (b). □

We turn to the situation where B is over the system BG. As before, B has row index set X and column index set Y, and $(\overline{X}_1 \cup \overline{Y}_1, \overline{X}_2 \cup \overline{Y}_2)$ is an exact k-separation of the given submatrix $\overline{B}$ with at least l elements on each side. First, we establish an auxiliary result that characterizes separations of $\overline{B}$ and induced separations of B by certain node subsets of the bipartite graphs BG$(\overline{B})$ and BG(B), respectively.

(3.5.16) Lemma. *Let B be a matrix over the system BG, with row index set X and column index set Y. Suppose that B has the matrix $\overline{B}$ of (3.5.2) as a submatrix. Define G to be the bipartite graph BG(B), and define $\overline{G}$ to be BG$(\overline{B})$. Then (a) and (b) below hold.*

(a) *For any $k \geq 1$, $(\overline{X}_1 \cup \overline{Y}_1, \overline{X}_2 \cup \overline{Y}_2)$ is an exact k-separation of $\overline{B}$ if and only if $|\overline{X}_1 \cup \overline{Y}_1|, |\overline{X}_2 \cup \overline{Y}_2| \geq k$ and $k - 1$ is the minimum number of* nodes of $\overline{G}$ whose removal from $\overline{G}$ disconnects the nodes of $\overline{X}_1 \cup \overline{Y}_1$ from the nodes of $\overline{X}_2 \cup \overline{Y}_2$.

(b) Suppose $(\overline{X}_1 \cup \overline{Y}_1, \overline{X}_2 \cup \overline{Y}_2)$ is an exact k-separation of $\overline{B}$. Then that k-separation induces one for B if and only if $k - 1$ is the minimum number of nodes of G whose removal from G disconnects the nodes of $\overline{X}_1 \cup \overline{Y}_1$ from the nodes of $\overline{X}_2 \cup \overline{Y}_2$.

Proof. Let Z, Z_1, and Z_2 be subsets of the node set of a graph, where Z_1 and Z_2 are disjoint. Suppose removal of the nodes of Z from the graph disconnects the nodes of Z_1 from the nodes of Z_2. We then say that Z is a *disconnecting set* of the graph for Z_1 and Z_2.

For the proof of the "if" part of (a), let Z be a minimum cardinality disconnecting set of $\overline{G}$ for $Z_1 = \overline{X}_1 \cup \overline{Y}_1$ and $Z_2 = \overline{X}_2 \cup \overline{Y}_2$. By assumption, $|Z| = k - 1$. In terms of $\overline{B}$ and its submatrices $\overline{D}^1$ and $\overline{D}^2$, the set Z is a minimum cardinality subset of the index set of $\overline{B}$ whose removal reduces $\overline{D}^1$ and $\overline{D}^2$ to zero matrices. By Theorem (2.6.14), the latter fact implies that BG-rank$(\overline{D}^1)$+BG-rank$(\overline{D}^2) = k-1$. We conclude that $(\overline{X}_1 \cup \overline{Y}_1, \overline{X}_2 \cup \overline{Y}_2)$ is an exact k-separation of $\overline{B}$.

The above arguments can be reversed to establish validity of the "only if" part of (a).

For the proof of the "if" part of (b), let Z be a minimum cardinality disconnecting set of G for $Z_1 = \overline{X}_1 \cup \overline{Y}_1$ and $Z_2 = \overline{X}_2 \cup \overline{Y}_2$. By assumption, $|Z| = k - 1$. Let G' be G minus the nodes of Z. Define X_1 (resp. Y_1) to be $\overline{X}_1$ (resp. $\overline{Y}_1$) plus the subset of nodes in X (resp. Y) that in G' may be reached from nodes in $\overline{X}_1 \cup \overline{Y}_1$. Let $X_2 = X - X_1$ and $Y_2 = Y - Y_1$. By arguments analogous to those for (a), $(X_1 \cup Y_1, X_2 \cup Y_2)$ is an exact k-separation of B induced by $(\overline{X}_1 \cup \overline{Y}_1, \overline{X}_2 \cup \overline{Y}_2)$.

For the proof of the "only if" part of (b), let $(X_1 \cup Y_1, X_2 \cup Y_2)$ be an induced k-separation of B. By a properly adapted part (a), any minimum cardinality disconnecting set of G for $X_1 \cup Y_1$ and $X_2 \cup Y_2$ contains exactly $k - 1$ nodes. Since, for $i = 1, 2$, we have $\overline{X}_i \cup \overline{Y}_i \subseteq X_i \cup Y_i$, any minimum cardinality disconnecting set of G for $\overline{X}_1 \cup \overline{Y}_1$ and $\overline{X}_2 \cup \overline{Y}_2$ has at most $k - 1$ nodes. But by (a), any disconnecting set of the subgraph $\overline{G}$ of G for $\overline{X}_1 \cup \overline{Y}_1$ and $\overline{X}_2 \cup \overline{Y}_2$ has at least $k - 1$ nodes. We conclude that any minimum cardinality disconnecting set of G for $\overline{X}_1 \cup \overline{Y}_1$ and $\overline{X}_2 \cup \overline{Y}_2$ has exactly $k - 1$ nodes.   □

The disconnecting sets specified in Lemma (3.5.16)(a) and (b) may be efficiently found by Algorithm DISJOINT PATHS (2.5.15). We use that algorithm in the following method for finding induced BG-separations.

(3.5.17) Algorithm INDUCED BG-SEPARATION. *Finds a k-separation of the matroid M represented by a matrix B over the system BG that is induced by an exact k-separation of the minor $\overline{M}$ represented by a submatrix $\overline{B}$, or declares that such an induced separation does not exist.*

Input: Matrix B over the system BG, with row index set X and column index set Y. A submatrix $\overline{B}$ of B with an exact k-separation $(\overline{X}_1 \cup \overline{Y}_1, \overline{X}_2 \cup \overline{Y}_2)$ where, for $i = 1, 2$, $\overline{X}_i \subseteq X$ and $\overline{Y}_i \subseteq Y$. The k-separation of $\overline{B}$ has at least l elements on each side.

Output: Either: A k-separation $(X_1 \cup Y_1, X_2 \cup Y_2)$ of B and M induced by the k-separation $(\overline{X}_1 \cup \overline{Y}_1, \overline{X}_2 \cup \overline{Y}_2)$ of $\overline{B}$ and $\overline{M}$; the k-separation of B and M is exact and has at least l elements on each side. Or: "The given k-separation of $\overline{B}$ and $\overline{M}$ does not induce a k-separation of B and M."

Complexity: Polynomial.

Procedure:

1. Define G to be the bipartite graph BG(B). Using Algorithm DIS-JOINT PATHS (2.5.15), find a minimum cardinality node subset Z of G so that the removal of the nodes of Z from G disconnects the nodes of $\overline{X}_1 \cup \overline{Y}_1$ from the nodes of $\overline{X}_2 \cup \overline{Y}_2$.

2. If $|Z| > k - 1$, declare that B has no induced k-separation. Otherwise, let G' be G minus the nodes of Z. Define X_1 (resp. Y_1) to be $\overline{X}_1$ (resp. $\overline{Y}_1$) plus the subset of nodes in X (resp. Y) that in G' may be reached from nodes in $\overline{X}_1 \cup \overline{Y}_1$. Let $X_2 = X - X_1$ and $Y_2 = Y - Y_1$. Then $(X_1 \cup Y_1, X_2 \cup Y_2)$ is the desired exact k-separation of B and M.

k-Separation

We present Algorithm k-SEPARATION, which for the matroid represented by a given matrix finds a certain k-separation or determines that none exists. We mentioned earlier that Algorithm k-SEPARATION is polynomial whenever k is bounded from above, but that it is not computationally usable for our purposes. In the next two subsections, we propose modified versions of Algorithm k-SEPARATION that are computationally effective for the cases of interest.

We begin with the definition of the problem solved by Algorithm k-SEPARATION. We use the same notation as before. Thus, M is a matroid represented by a matrix B over $\mathcal{F}$ or over BG. The matrix has row index set X and column index set Y. Let P_1, P_2 (resp. Q_1, Q_2) be two disjoint subsets of X (resp. Y). We want to either find partitions X_1, X_2 and Y_1, Y_2 of X and Y, respectively, such that for given integers m_1, m_2, and n

(3.5.18)
$\quad$ (i) $\quad$ $(X_1 \cup Y_1, X_2 \cup Y_2)$ is an exact k-separation of B and M with $k \leq n$.

$\quad$ (ii) $\quad$ For $i = 1, 2$, $P_i \subset X_i$ and $Q_i \subset Y_i$.

$\quad$ (iii) $\quad$ For $i = 1, 2$, $|X_i \cup Y_i| \geq |P_i \cup Q_i| + \max\{k, m_i\} + 1$.

or determine that such partitions do not exist. In the affirmative case, we want the value of k to be minimal. We assume that B does not have a separation satisfying (3.5.18) for $k = 1$. This is, for example, the case when B is connected.

We solve this problem as follows. For $k = 2, 3, 4\ldots$, we iteratively carry out the search described below until either a k-separation with the desired properties is found or k exceeds the upper bound n. In the latter case, we stop and declare that a separation of the demanded type does not exist.

Let us look at one iteration of the search, where the value of k is fixed. Inductively, we assume that for any $k' < k$, a k'-separation satisfying an appropriately adjusted (3.5.18) does not exist. For the base case $k = 2$, the induction hypothesis holds by assumption.

Let $\overline{B}$ be a submatrix of B as depicted in (3.5.2), and $\overline{M}$ be the minor of M represented by $\overline{B}$. We assume that $\overline{B}$ has the following features.

(3.5.19)
  (i) $(\overline{X}_1 \cup \overline{Y}_1, \overline{X}_2 \cup \overline{Y}_2)$ is an exact k-separation of $\overline{B}$.
  (ii) For $i = 1, 2$, $P_i \subset \overline{X}_i$ and $Q_i \subset \overline{Y}_i$.
  (iii) For $i = 1, 2$, $|\overline{X}_i \cup \overline{Y}_i| \geq |P_i \cup Q_i| + \max\{k, m_i\} + 1$.
  (iv) $\overline{B}$ is minimal with respect to (i)–(iii).

Let B and $\overline{B}$ be the input for Algorithm INDUCED $\mathcal{F}$-SEPARATION (3.5.14) or Algorithm INDUCED BG-SEPARATION (3.5.17), whichever applies.

If the algorithm finds an induced k-separation for B, then by (3.5.19) this k-separation satisfies the conditions of (3.5.18) and thus constitutes the desired separation for B.

If the algorithm does not produce an induced k-separation, we try other candidate submatrices $\overline{B}$, until we either find the desired induced k-separation for B or, having tried all possible candidates $\overline{B}$, conclude that B has no k-separation satisfying (3.5.18).

In the latter case, we increase k by 1. If the new k does not exceed the specified upper bound n, we repeat the above process. Otherwise, we stop and declare that B does not have a separation satisfying (3.5.18).

The computations involved in the derivation of the candidate submatrices $\overline{B}$ from B are rather straightforward. We remark only that the rank calculations involved in that derivation are done with $\mathcal{F}$-pivots if B is over $\mathcal{F}$, and with Algorithm DISJOINT PATHS (2.5.15) if B is over BG.

Suppose that n is bounded from above by some constant. The candidate matrices $\overline{B}$ may then be derived from B in polynomial time, as is easily checked.

We summarize the above discussion in the following algorithm.

(3.5.20) Algorithm k-SEPARATION. *Finds an exact k-separation of the matroid M represented by a matrix B over a field $\mathcal{F}$ or over the system BG, where the two sides contain specified sets and have at least a certain size, or declares that such a separation does not exist.*

Input: Matrix B over a field $\mathcal{F}$ or over the system BG, with row index set X and column index set Y. Two disjoint subsets P_1, P_2 (resp. Q_1, Q_2) of X (resp. Y). Integers m_1, m_2, and n. For $k = 1$, the matrix B does not have a separation $(X_1 \cup Y_1, X_2 \cup Y_2)$ satisfying the following conditions.

(3.5.21)

    (i) $(X_1 \cup Y_1, X_2 \cup Y_2)$ is an exact k-separation of B and M with $k \leq n$.

    (ii)   For $i = 1, 2$, $P_i \subset X_i$ and $Q_i \subset Y_i$.

    (iii) For $i = 1, 2$, $|X_i \cup Y_i| \geq |P_i \cup Q_i| + \max\{k, m_i\} + 1$.

Output: Either: A k-separation $(X_1 \cup Y_1, X_2 \cup Y_2)$ of B and M satisfying the conditions of (3.5.21) and, subject to them, with k minimal. Or: "B and M do not have an exact k-separation $(X_1 \cup Y_1, X_2 \cup Y_2)$ satisfying (3.5.21)."

Complexity: Polynomial if m_1, m_2, and n are bounded by a constant.

Procedure:

1. Initialize $k = 2$.
2. Do for each submatrix $\overline{B}$ of (3.5.2) for which the sets $\overline{X}_1$, $\overline{X}_2$, $\overline{Y}_1$, and $\overline{Y}_2$ satisfy (3.5.19):

    Let B and $\overline{B}$ be the input matrices for Algorithm INDUCED $\mathcal{F}$-SEPARATION (3.5.14) or Algorithm INDUCED BG-SEPARATION (3.5.17), whichever applies. If the algorithm finds an induced k-separation $(X_1 \cup Y_1, X_2 \cup Y_2)$, output that separation, and stop.
3. Increase k by 1. If $k \leq n$, go to Step 2. Otherwise, declare that B and M do not have a separation of the desired kind, and stop.

The matrices for which we want to locate separations typically have several hundred and sometimes several thousand rows and columns. Of specific interest are GF(3)-2-separations and BG-k-separations, where k is bounded from above by some constant. Algorithm k-SEPARATION (3.5.20) is too inefficient to find such separations with reasonable computing effort. In the next three subsections, we address this issue. Specifically, in the next two subsections we deal with GF(3)-2-separations, and in the subsequent subsection we deal with BG-k-separations.

GF(3)-2-Separation

We are given a matroid M represented by a matrix B over GF(3). The matroid M is connected and has no series or parallel elements. Correspondingly, B is connected and has no parallel vectors and no vectors with exactly one nonzero. Recall from Chapter 2 that a simple matrix has no parallel vectors and no vectors with less than two nonzeros. Since a connected matrix has no zero vectors, we may invoke the stated conditions for B by demanding that B be connected and simple.

We want to find a 2-separation of B and M or determine that none exists. Note that the sets P_1, P_2, Q_1, and Q_2, which are part of the input of Algorithm k-SEPARATION (3.5.20), are empty here.

Suppose B has a 2-separation. Thus, B can be partitioned as

(3.5.22)
$$B = \begin{array}{c|c|c|} & Y_1 & Y_2 \\ \hline X_1 & A^1 & D^2 \\ \hline X_2 & D^1 & A^2 \\ \hline \end{array}$$

Partitioned version of B

where $|X_1 \cup Y_1|$, $|X_2 \cup Y_2| \geq 2$ and where exactly one of the matrices of D^1 and D^2 has GF(3)-rank equal to 1, while the other matrix is a zero matrix. We establish some properties of B in the next two lemmas.

(3.5.23) Lemma. *Each one of the sets X_1, X_2, Y_1, and Y_2 is nonempty, and the submatrices A^1 and A^2 are nonzero.*

Proof. By the symmetry and duality, we may assume that $X_1 = \emptyset$. By $|X_1 \cup Y_1| \geq 2$, we know $|Y_1| \geq 2$. If D^1 is a zero matrix, then B has zero columns and is not connected, a contradiction. Hence GF(3)-rank$(D^1) = 1$. But then $|Y_1| \geq 2$ implies that D^1 and B have two parallel columns, another contradiction. If A^1 or A^2 is a zero matrix, then B contains a zero vector, which contradicts the connectedness of B.   □

(3.5.24) Lemma. *Matrix B has a row u with nonzero entries in some columns $v \in Y_1$ and $w \in Y_2$.*

Proof. By Lemma (3.5.23), the submatrices A^1 and A^2 are nonzero. Due to the symmetry, we may assume that D^2 is nonzero. If no row $u \in X_1$ satisfies the stated condition, then in BG(B) the nodes of Y_1 are not connected with the nodes of Y_2, which contradicts the connectedness of B.   □

Suppose that we know the row index u and the column indices $v \in Y_1$ and $w \in Y_2$ of Lemma (3.5.24). If u is in X_1, then we declare B and the submatrix $\overline{B}$ defined by $\overline{X}_1 = \{u\}$, $\overline{X}_2 = \emptyset$, $\overline{Y}_1 = \{v\}$, and $\overline{Y}_2 = \{w\}$ to be the input for Algorithm INDUCED $\mathcal{F}$-SEPARATION (3.5.14). That input and the 2-separation $(X_1 \cup Y_1, X_2 \cup Y_2)$ for B satisfy the hypotheses of Lemma (3.5.15)(b). According to that result, Algorithm INDUCED $\mathcal{F}$-SEPARATION (3.5.14) must terminate with an induced 2-separation of B. If u is in X_2, then we define $\overline{B}$ by $\overline{X}_1 = \{u\}$, $\overline{X}_2 = \emptyset$, $\overline{Y}_1 = \{w\}$, and $\overline{Y}_2 = \{v\}$. Once more, Algorithm INDUCED $\mathcal{F}$-SEPARATION (3.5.14) then produces an induced 2-separation of B.

Of course, when we are looking for a 2-separation of B, we do not know the indices u, v, or w. But we can search for them efficiently by applying Algorithm INDUCED $\mathcal{F}$-SEPARATION (3.5.14) to each submatrix $\overline{B}$ consisting of two nonzero entries of some row of B. Indeed, if the nonzero entries of each row of B are stored sequentially in a list, then according to the next lemma we can confine the search to submatrices $\overline{B}$ that contain consecutive entries in the lists.

(3.5.25) Lemma. *If a submatrix $\overline{B}$ of a row induces a 2-separation, then this must be so for some submatrix $\overline{B}$ containing two consecutive entries in the list for that row.*

Proof. Let v and w be the column indices of the entries of $\overline{B}$. Assume that $\overline{B}$ induces the 2-separation $(X_1 \cup Y_1, X_2 \cup Y_2)$ of B. Thus, $v \in Y_1$ and $w \in Y_2$, or vice versa. But then the indices of two adjacent entries of the row list, say, v' and w', satisfy $v' \in Y_1$ and $w' \in Y_2$, or vice versa. Hence, the submatrix containing these adjacent entries also induces the 2-separation $(X_1 \cup Y_1, X_2 \cup Y_2)$ of B. $\square$

According to Lemma (3.5.25), the total number of cases of $\overline{B}$ to be considered can be held in the worst case to the number of nonzeros of B minus the number of rows of B.

It is easy to see that the effort for each application of Algorithm IN-DUCED $\mathcal{F}$-SEPARATION (3.5.14) is linear in the number of nonzeros of B. Thus, total computing effort for finding a GF(3)-2-separation or determining that none exists, is in the worst case quadratic in the number of nonzeros of B. We summarize the algorithm below.

(3.5.26) Algorithm GF(3)-2-SEPARATION. *Finds an exact 2-separation of the matroid M represented by a matrix B over GF(3) or declares that such a separation does not exist.*

Input: Connected, simple matrix B over GF(3), with row index set X and column index set Y.

Output: Either: An exact 2-separation $(X_1 \cup Y_1, X_2 \cup Y_2)$ of B and M. Or: "B and M do not have a 2-separation."

Complexity: Polynomial.

Procedure:

1. Do for each row u of B:

   Do for each pair of consecutive entries in the list for row u, say, with column indices v and w:

   Let $\overline{X}_1 = \{u\}$ and $\overline{X}_2 = \emptyset$.

   Do once for $\overline{Y}_1 = \{v\}$ and $\overline{Y}_2 = \{w\}$, and a second time for $\overline{Y}_1 = \{w\}$ and $\overline{Y}_2 = \{v\}$:

Let $\overline{B}$ be defined by $\overline{X}_1$, $\overline{X}_2$, $\overline{Y}_1$, and $\overline{Y}_2$. Do Algorithm IN-DUCED $\mathcal{F}$-SEPARATION (3.5.14) with B and $\overline{B}$ as input. If a 2-separation is found, output that separation, and stop.

2. Declare that B and M do not have a 2-separation, and stop.

Algorithm GF(3)-2-SEPARATION (3.5.26) would work equally well if B was stored in column lists instead of row lists. We state and prove this fact next.

(3.5.27) Lemma. *Suppose that the matrix B is represented by column lists instead of row lists, and that in Algorithm GF(3)-2-SEPARATION (3.5.26) each candidate submatrix $\overline{B}$ consists of two consecutive nonzeros of some column. Then the modified algorithm is also valid.*

Proof. The modified algorithm is nothing but Algorithm GF(3)-2-SEPA-RATION (3.5.26) applied to B^t, which represents M^*. By Lemma (3.4.8), any separation of M is also one for M^*, and vice versa. So if M has a GF(3)-2-separation, then one such separation will be found by the modified algorithm. □

In the next subsection, we employ Algorithm GF(3)-2-SEPARATION (3.5.26) as a subroutine to decompose matroids and matrices into so-called 3-connected components.

3-Connected Components

Suppose Algorithm GF(3)-2-SEPARATION (3.5.26) has detected a 2-separation for a simple matrix B. Let that 2-separation be displayed by (3.5.22), where without loss of generality GF(3)-rank(D^1) = 1 and $D^2 = 0$. Thus, B is the matrix of (3.5.28) below, where the submatrix indexed by X_2 and Y_1 corresponds to D^1 of (3.5.22).

(3.5.28)

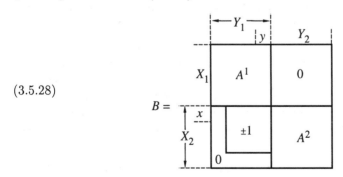

Matrix B with 2-separation

In B, we have indexed an arbitrary nonzero row (column) of D^1 by x (resp. y). Let B^1 (resp. B^2) be the submatrix of B consisting of A^1 and row x of D^1 (resp. A^2 and column y of D^1). We show B^1 and B^2 below.

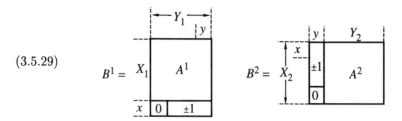

(3.5.29)

Submatrices B^1 and B^2 of B

The reader may be puzzled by the matrices B^1 and B^2. Detailed reasons for the selection of these submatrices are included in Section 3.6, where matroid sums are discussed. The reader may skip ahead to read about such sums or simply accept the selection of B^1 and B^2 for the time being.

We reduce B^1 and B^2 to simple matrices $B^{1\prime}$ and $B^{2\prime}$, respectively, by removing vectors with exactly one nonzero and by deleting vectors of parallel classes so that each class contains just one vector.

Using the bipartite graphs $BG(\cdot)$ for B, B^1, and B^2, one readily verifies that the connectedness of B implies that B^1 and B^2 as well as $B^{1\prime}$ and $B^{2\prime}$ are connected.

By the derivation, each row of $B^{1\prime}$ or $B^{2\prime}$ may be represented by a list which is obtained from the corresponding row list of B by the deletion of some entries. Note that the ordering of the entries of the row lists of $B^{1\prime}$ or $B^{2\prime}$ is consistent with that of the row lists of B.

Suppose we want to decide whether $B^{1\prime}$ or $B^{2\prime}$ has a 2-separation. We could answer this question in two applications of Algorithm GF(3)-2-SEPARATION (3.5.26). One can significantly improve upon this method if one has retained information about the processing of B by Algorithm GF(3)-2-SEPARATION (3.5.26). The key observation for the improvement is contained in Theorem (3.5.31) below. We first list a lemma that is invoked in the proof of that theorem.

(3.5.30) Lemma. *Let $\overline{B}'$ be a matrix consisting of two nonzero entries of some row of B. The two nonzero entries need not be adjacent in the row list containing them. For each one of the two entries of $\overline{B}'$, suppose there is a candidate submatrix $\overline{B}$ for which Algorithm GF(3)-2-SEPARATION (3.5.26) did not determine an induced 2-separation of B. Then that algorithm would not determine a 2-separation for B if $\overline{B}'$ were used as candidate submatrix.*

Proof. We know that the last $\overline{B}$ evaluated by Algorithm GF(3)-2-SEPARATION (3.5.26) induced the 2-separation of B given by (3.5.28). Let u be the row index of that $\overline{B}$, and let v and w be its column indices. Suppose v occurs in the list of row u ahead of w. Hence, the nonzero in row u and column w occurs in precisely one $\overline{B}$ evaluated by Algorithm GF(3)-2-SEPARATION (3.5.26). By assumption, $\overline{B}'$ cannot contain that nonzero.

All other $\overline{B}$ submatrices evaluated by Algorithm GF(3)-2-SEPARATION (3.5.26) did not result in a 2-separation of B. Then, by the contrapositive version of Lemma (3.5.25), the submatrix $\overline{B}'$ cannot induce a 2-separation of B. $\square$

(3.5.31) Theorem. For $j = 1$ or 2, let $\overline{B}'$ be a submatrix containing two consecutive nonzero entries of some row of $B^{j\prime}$. For each one of the two nonzero entries of $\overline{B}'$, assume that the entry was part of some submatrix $\overline{B}$ that did not lead to an induced 2-separation when Algorithm GF(3)-2-SEPARATION (3.5.26) was applied to B. If Algorithm GF(3)-2-SEPARATION (3.5.26) is applied to $B^{j\prime}$, then the submatrix $\overline{B}'$ of $B^{j\prime}$ does not induce a 2-separation of $B^{j\prime}$.

Proof. Suppose Algorithm GF(3)-2-SEPARATION (3.5.26) determines a 2-separation of $B^{j\prime}$ using $\overline{B}'$ as the candidate submatrix. We first convert that 2-separation of $B^{j\prime}$ to one for B^j, and later to one for B as follows.

We know that $B^{j\prime}$ can be extended to B^j by the addition of parallel vectors and of vectors with exactly one nonzero entry. If a vector v parallel to a vector w is added, then we add v to the side of the 2-separation containing w. If a vector v with exactly one nonzero is added, where the position of the nonzero is indexed by w, then once more we add v to the side containing w. It is easy to verify that this process results in a 2-separation of B^j that is induced by $\overline{B}'$.

Using (3.5.28) and (3.5.29), we may enlarge the 2-separation just determined for B^j to one for B as follows. In the case of $j = 1$ (resp. $j = 2$), we add $(X_2 - \{x\}) \cup Y_2$ (resp. $X_1 \cup (Y_1 - \{y\})$) to the side of the 2-separation of B^j containing x (resp. y). Since GF(3)-rank(D^1) = 1, one readily confirms that in both cases a 2-separation of B results that is induced by $\overline{B}'$. But Lemma (3.5.30) implies that $\overline{B}'$ cannot induce a 2-separation of B, a contradiction. $\square$

Suppose we apply Algorithm GF(3)-2-SEPARATION (3.5.26) to $B^{1\prime}$ and $B^{2\prime}$, using only those candidate submatrices $\overline{B}'$ not ruled out *a priori* by Theorem (3.5.31). If we find a 2-separation for either matrix, we repeat the process of deriving two simple submatrices analogous to (3.5.28) and (3.5.29), then search again with Algorithm GF(3)-2-SEPARATION (3.5.26) for a 2-separation of the latter submatrices, etc. The process stops when further decompositions are not possible. By the rules of the decomposition process, the matrices on hand at that time are connected and do not have a 2-separation; that is, they are 3-connected. By the reduction rules, some of the final matrices may be of size 1×0 or 0×1. We delete all such matrices and call the remaining 3-connected matrices the 3-*connected components* of B. The 3-connected components of B can be shown to be unique up to indices.

We use the same terminology for the matroid M represented by B and the minors of M represented by the 3-connected components of B.

Thus, these minors are the 3-*connected components* of M. The use of "3-connected components" for both matrices and matroids will not cause difficulties, since the meaning will always be apparent from the context.

Due to Theorem (3.5.31), total computing effort for finding the 3-connected components of B by repeated application of Algorithm GF(3)-2-SEPARATION (3.5.26) is quadratic in the number of nonzeros of B and thus is acceptable even for large matrices. Below, we summarize the entire scheme. For ease of application, we do not assume B to be simple.

(3.5.32) Algorithm 3-CONNECTED COMPONENTS. *Finds the 3-connected components of a* GF(3)*-matroid* M *by locating the 3-connected components of a representation matrix* B.

Input: A connected matrix B over GF(3), with row index set X and column index set Y. The nonzero entries of B are stored sequentially in row lists.

Output: The 3-connected components of B and M. A complete list of the separations and reductions that produce the 3-connected components of B when one starts with B.

Complexity: Polynomial.

Procedure:
1. Reduce B to a simple matrix by deleting vectors with exactly one nonzero and by reducing each class of parallel vectors to just one vector. Record these reductions. If the reduced matrix has size 1×0 or 0×1, then declare that B has no 3-connected components, and stop. Otherwise, initialize two sets C and L by defining C to contain the reduced matrix and L to be the empty set.
2. If $C = \emptyset$, stop; the set L contains the 3-connected components of B; the matrices of L represent the 3-connected components of M.
3. Remove a matrix from C and apply Algorithm GF(3)-2-SEPARATION (3.5.26) to it. During the execution of that algorithm, candidate submatrices satisfying the assumptions of Theorem (3.5.31) are to be ignored.
   If a 2-separation is not determined by Algorithm GF(3)-2-SEPARATION (3.5.26), add the matrix to L and go to Step 2.
   If a 2-separation is found, decompose the matrix into two matrices using (3.5.28) and (3.5.29). Reduce the latter matrices to simple matrices analogously to the reduction of B in Step 1. Add the simple matrices that are not of size 1×0 or 0×1 to C. Record the 2-separation and the reductions, and go to Step 2.

We turn to the problem of finding BG-k-separations.

BG-k-Separation

We want to locate BG-k-separations where k is bounded by some constant. We already have a polynomial method for this task, Algorithm k-SEPARATION (3.5.20). But that scheme is too slow for practical use. In this subsection, we develop a heuristic algorithm that on one hand is fast, but that on the other hand may fail to locate a separation even though one exists.

The heuristic algorithm is very similar to Algorithm k-SEPARATION (3.5.20) in that it selects candidate submatrices $\overline{B}$ and attempts to extend separations of these submatrices to separations of B. However, the selection of the submatrices $\overline{B}$ is handled by some heuristic method that limits the number of $\overline{B}$ instances. Furthermore, any $\overline{B}$ that for some $l \leq k$ has an exact l-separation with the required number of elements on each side is deemed acceptable. This contrasts with the selection rule of Algorithm k-SEPARATION (3.5.20), where each $\overline{B}$ must have an exact k-separation.

Since l may be less than k, it does not make much sense to demand that the l-separation of $\overline{B}$ induce an l-separation of B. Instead, the algorithm searches for an l'-separation of B, where $l \leq l' \leq k$ and where the two sides of the l'-separation of B contain the two sides of the l-separation of $\overline{B}$.

According to the next lemma, the search for such an l'-separation of B is easy. We omit the proof, since it is very similar to that of Lemma (3.5.16).

(3.5.33) Lemma. *Let B be a matrix over the system BG, with row index set X and column index set Y. Suppose that B has the matrix $\overline{B}$ of (3.5.2) as submatrix. Define G to be the bipartite graph BG(B).*

For $l \leq k$, let $(\overline{X}_1 \cup \overline{Y}_1, \overline{X}_2 \cup \overline{Y}_2)$ be an l-separation of $\overline{B}$ with at least $k+1$ elements on each side. Then B has an l'-separation $(X_1 \cup Y_1, X_2 \cup Y_2)$, where $l \leq l' \leq k$, where, for $i = 1, 2$, $\overline{X}_i \subseteq X_i$ and $\overline{Y}_i \subseteq Y_i$, and where subject to these conditions l' is minimal, if and only if $l' - 1$ is the minimum number of nodes of G whose removal from G disconnects the nodes of $\overline{X}_1 \cup \overline{Y}_1$ from the nodes of $\overline{X}_2 \cup \overline{Y}_2$.

Lemma (3.5.33) validates the following heuristic algorithm.

(3.5.34) Heuristic BG-k-SEPARATION. *Finds an exact k-separation of the matroid M represented by a matrix B over the system BG, where the two sides contain specified sets and have at least a certain size, or declares that the method cannot find such a separation.*

Input: Matrix B over the system BG, with row index set X and column index set Y. Two disjoint subsets P_1, P_2 (resp. Q_1, Q_2) of X (resp. Y). Integers m_1, m_2, and n. For $k = 1$, the matrix B does not have a separation $(X_1 \cup Y_1, X_2 \cup Y_2)$ satisfying the following conditions.

(3.5.35)
    (i) $(X_1 \cup Y_1, X_2 \cup Y_2)$ is an exact k-separation of B and M with $k \leq n$.

    (ii)   For $i = 1, 2$, $P_i \subset X_i$ and $Q_i \subset Y_i$.

    (iii) For $i = 1, 2$, $|X_i \cup Y_i| \geq |P_i \cup Q_i| + \max\{k, m_i\} + 1$.

Output: Either: A k-separation $(X_1 \cup Y_1, X_2 \cup Y_2)$ of B and M satisfying the conditions of (3.5.35). Or: "The heuristic algorithm cannot locate an exact k-separation $(X_1 \cup Y_1, X_2 \cup Y_2)$ for B and M satisfying (3.5.35)."

Complexity: Polynomial if m_1, m_2, and n are bounded by a constant.

Procedure:

1. Define G to be the bipartite graph $BG(B)$. Initialize $k = 2$.
2. Use any convenient polynomial heuristic to find a collection of sub-matrices $\overline{B}$ of the form (3.5.2) whose index sets $\overline{X}_1$, $\overline{X}_2$, $\overline{Y}_1$, and $\overline{Y}_2$ satisfy the following conditions.

$$
\begin{array}{ll}
& \text{(i)} \quad (\overline{X}_1 \cup \overline{Y}_1, \overline{X}_2 \cup \overline{Y}_2) \text{ is an exact } l\text{-separation of } \overline{B}, \\
& \quad\quad \text{for some } l \leq k. \\
\text{(3.5.36)} & \text{(ii)} \quad \text{For } i = 1, 2, P_i \subset \overline{X}_i \text{ and } Q_i \subset \overline{Y}_i. \\
& \text{(iii)} \quad \text{For } i = 1, 2, |\overline{X}_1 \cup \overline{Y}_1| \geq |P_i \cup Q_i| + \max\{k, m_i\} + 1. \\
& \text{(iv)} \quad \overline{B} \text{ is minimal with respect to (i)–(iii).}
\end{array}
$$

3. Do for each submatrix $\overline{B}$ determined in Step 2:
   Using Algorithm DISJOINT PATHS (2.5.15), find a minimum cardi-nality node subset Z of G so that the removal of the nodes of Z from G disconnects the nodes of $\overline{X}_1 \cup \overline{Y}_1$ from the nodes of $\overline{X}_2 \cup \overline{Y}_2$. If $|Z| \leq k - 1$, go to Step 5.
4. Increase k by 1. If $k \leq n$, go to Step 2. Otherwise, declare that the algorithm cannot locate the desired separation of B and M, and stop.
5. Let G' be G minus the nodes of Z. Define X_1 (resp. Y_1) to be $\overline{X}_1$ (resp. $\overline{Y}_1$) plus the subset of nodes in X (resp. Y) that in G' may be reached from nodes in $\overline{X}_1 \cup \overline{Y}_1$. Let $X_2 = X - X_1$ and $Y_2 = Y - Y_1$. Redefine k to be $|Z| + 1$. Then declare $(X_1 \cup Y_1, X_2 \cup Y_2)$ to be the desired exact k-separation of B and M.

We briefly discuss a method for the selection of the candidate submatrices $\overline{B}$ in Step 2 of Heuristic BG-k-SEPARATION (3.5.34). Depending on which of the sets $P_1 \cup Q_1$ and $P_2 \cup Q_2$ are empty, the method proceeds as follows. Let k be given.

First, suppose that $P_1 \cup Q_1$ and $P_2 \cup Q_2$ are nonempty. For $i = 1, 2$, we try to enlarge the node subsets P_i and Q_i of G by neighboring nodes of $P_i \cup Q_i$ until the resulting sets $\overline{X}_i$ and $\overline{Y}_i$ are larger than the original P_i and Q_i and have together at least $|P_i \cup Q_i| + \max\{k, m_i\} + 1$ elements. We also enforce the following condition: $\overline{X}_1 \cup \overline{Y}_1$ and $\overline{X}_2 \cup \overline{Y}_2$ must be disjoint node subsets of G that, for some $l \leq k$, define an exact l-separation for the subgraph $\overline{G}$ of G induced by $\overline{X}_1 \cup \overline{Y}_1 \cup \overline{X}_2 \cup \overline{Y}_2$. If we can identify such sets, we have obtained a candidate submatrix $\overline{B}$. By varying the selection of $\overline{X}_i$ and $\overline{Y}_i$ while enforcing the above conditions, one obtains a suitable number of candidate matrices $\overline{B}$.

Second, assume that both $P_1 \cup Q_1$ and $P_2 \cup Q_2$ are empty. Using any convenient shortest route method, we select two nodes i and j of G so that the distance of any shortest path connecting these two nodes is large. Temporarily, we place node i into $P_1 \cup Q_1$, place node j into $P_2 \cup Q_2$, and carry out the above method for nonempty $P_1 \cup Q_1$ and $P_2 \cup Q_2$. If desired, one may repeat this process using alternate pairs of nodes i and j for which the distance of any shortest path connecting them is large.

Third, assume without loss of generality that $P_1 \cup Q_1$ is empty and $P_2 \cup Q_2$ is nonempty. Temporarily, declare $P_1 \cup Q_1$ to consist of a node i whose shortest distance to the nodes of $P_2 \cup Q_2$ is large, then apply the method for the case of nonempty $P_1 \cup Q_1$ and $P_2 \cup Q_2$. Similarly to the second case, one may select several nodes i and thus select several nonempty $P_1 \cup Q_1$. Each such $P_1 \cup Q_1$ constitutes, together with $P_2 \cup Q_2$, an instance to which the method of the first case can be applied.

3.6 Sums

We describe ways of decomposing or composing representable matroids, using a class of constructs called k-sums, where k ranges over the positive integers. Thus, there are 1-sums, 2-sums, 3-sums, etc. We describe these sums here, since they are related to, and indeed are the motivation for, the IB-k-sums introduced in Chapter 4.

To simplify the discussion, we confine ourselves to k-sums involving GF(3)-matroids.

1- and 2-Sums

We start with the 1-sum case. Let M be a matroid represented by a matrix B over GF(3). According to the discussion immediately preceding Algorithm 1-SEPARATION (3.5.1), the matroid M has a 1-separation if and only if the graph BG(B) is not connected. Assume the latter case. Clearly, B can be partitioned as shown in (3.6.1) below, with $|X_1 \cup Y_1|, |X_2 \cup Y_2| \geq 1$. The latter inequality is equivalent to demanding that the submatrices A^1 and A^2 of B are nonempty; that is, they are not 0×0 matrices.

(3.6.1)

$$B = \begin{array}{c|c|c} & Y_1 & Y_2 \\ \hline X_1 & A^1 & 0 \\ \hline X_2 & 0 & A^2 \\ \hline \end{array}$$

Matrix B with 1-separation

We declare that the matroids represented by A^1 and A^2, say, M_1 and M_2, are the two *components* of a 1-*sum decomposition* of M. The decomposition is reversed in the obvious way, giving a 1-*sum composition* of M_1 and M_2 to M. We mean either process when we say that M is a 1-*sum* of M_1 and M_2. We apply the same terminology to B, B^1, and B^2. For example, B is a 1-*sum* of B^1 and B^2.

We move to the more interesting case of 2-sums. We assume that the given GF(3)-matroid M is connected and has a GF(3)-2-separation. Since M is connected, that separation must be exact. According to (3.5.28), the GF(3)-2-separation corresponds to the following partition of B.

(3.6.2)

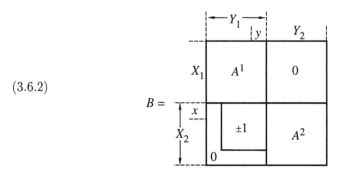

Matrix B with 2-separation

We refer to the submatrix of B indexed by X_2 and Y_1 as D^1, in agreement with (3.4.4). We know that GF(3)-rank$(D^1) = 1$. We want to extract from B two submatrices B^1 and B^2 that contain A^1 and A^2, respectively, and that also contain enough information to reconstruct B. Evidently, the latter requirement is equivalent to the condition that we must be able to compute D^1 from B^1 and B^2. Since GF(3)-rank$(D^1) = 1$, knowledge of one nonzero row of D^1 and one nonzero column of D^1 suffices for the computation of D^1.

With this insight, we choose B^1 to be A^1 plus one nonzero row of D^1, say, row x, and B^2 to be A^2 plus one nonzero column of D^1, say, column y. The two indices x and y are shown in (3.6.2). We display B^1 and B^2 below.

(3.6.3)

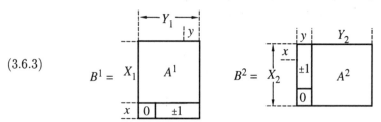

Matrices B^1 and B^2 of 2-sum

We reconstruct D^1, and thus implicitly B, from B^1 and B^2 by computing

$$(3.6.4) \qquad D^1 = (\text{column } y \text{ of } B^2) \cdot (\text{row } x \text{ of } B^1)$$

Let M_1 and M_2 be the minors of M represented by B^1 and B^2. We call these minors the *components* of a 2-*sum decomposition* of M. The reverse process, which corresponds to a reconstruction of B from B^1 and B^2, is a 2-*sum composition* of M_1 and M_2 to M. Both cases are handled by saying that M is a 2-*sum* of M_1 and M_2. We use the same terminology for B, B^1, and B^2. For example, B is a 2-*sum* of B^1 and B^2.

We saw the matrices B^1 and B^2 earlier, under (3.5.29) and in connection with Algorithm 3-CONNECTED COMPONENTS (3.5.32). That algorithm locates the 3-connected components of a given matrix B using GF(3)-2-separations and certain reductions involving vectors with one nonzero entry and parallel vectors. In the terminology of this section, the algorithm processes each case of GF(3)-2-separation as a 2-sum decomposition. We now show that the reductions made by the algorithm may also be viewed as 2-sum decomposition steps, provided that the length of the matrix being processed, say, B with row index set X and column index set Y, is at least 4.

For example, suppose that B has two parallel columns y and z. Then $X_1 = \emptyset$, $X_2 = X$, $Y_1 = \{y, z\}$, and $Y_2 = Y - \{y, z\}$ define a 2-separation $(X_1 \cup Y_1, X_2 \cup Y_2)$ of B. Using the above rule for 2-sum decomposition, B^1 is a 1×2 matrix with two nonzero entries, and B^2 is B minus column z. The matrix B^1 has length 3 and can be reduced to a 1×0 or 0×1 matrix. Effectively, we are thus left with B^2. On the other hand, one reduction step in Algorithm 3-CONNECTED COMPONENTS (3.5.32) converts B up to indices to B^2 as well. We conclude that reduction of a parallel class by one vector is equivalent to a 2-sum decomposition. By similar arguments, the other reductions can also be shown to be equivalent to 2-sum decompositions. Thus, the 3-connected components of B are obtainable, up to indices, by repeated 2-sum decompositions plus reductions involving matrices with length at most 3.

We turn to the more complex case of k-sums with $k \geq 3$.

k-Sums

We still assume that M is a matroid represented by a matrix B over GF(3), with row index set X and column index set Y. For some $k \geq 3$, we also know an exact k-separation $(X_1 \cup Y_1, X_2 \cup Y_2)$. We want to decompose M in some useful way. We investigate this problem using the partitioned version of B of (3.5.22). Slightly enlarged, we repeat that matrix below.

(3.6.5)

$$B = $$

	Y_1	Y_2
X_1	A^1	D^2
X_2	D^1	A^2

Matrix B with exact k-separation

Recall that by (3.5.9)

(3.6.6) $\text{GF}(3)\text{-rank}(D^1) + \text{GF}(3)\text{-rank}(D^2) = k - 1$

We want to decompose M into two matroids M_1 and M_2 that correspond to two submatrices B^1 and B^2 of B. As in the 2-sum case, we postulate that B^1 and B^2 include A^1 and A^2, respectively. Furthermore, B^1 and B^2 must permit a reconstruction of B. The latter requirement can be satisfied by including in B^1 (resp. B^2) a row (resp. column) submatrix of D^1 with the same rank as D^1 and a column (resp. row) submatrix of D^2 with the same rank as D^2. Indeed, the submatrices D^1 and D^2 of B can be computed from these row and column submatrices. We provide the relevant formulas in a moment. Last but not least, we want B^1 and B^2 to be proper submatrices of B.

There are numerous ways to satisfy these requirements. In the most general case, both B^1 and B^2 intersect all four submatrices A^1, A^2, D^1, and D^2 of B and thus induce the following rather complicated looking partition of B.

(3.6.7)

$$B = $$

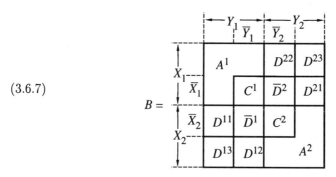

Partition of B displaying k-sum

In the notation of (3.6.7), the submatrix B^1 of B, which is not explicitly indicated, is indexed by $X_1 \cup \overline{X}_2$ and $Y_1 \cup \overline{Y}_2$. Furthermore, the submatrix

B^2 is indexed by $\overline{X}_1 \cup X_2$ and $\overline{Y}_1 \cup Y_2$. Hence, B^1 contains A^1 and intersects A^2 in C^2, D^1 in $[D^{11}|\overline{D}^1]$, and D^2 in $[D^{22}/\overline{D}^2]$. The submatrix B^2 contains A^2 and intersects A^1 in C^1, D^1 in $[\overline{D}^1/D^{12}]$, and D^2 in $[\overline{D}^2|D^{21}]$. We assume that C^1 (resp. C^2) is a proper submatrix of A^1 (resp. A^2). This implies that both B^1 and B^2 are proper submatrices of B. Observe that $\overline{D}^1$ (resp. $\overline{D}^2$) is the submatrix of D^1 (resp. D^2) contained in both B^1 and B^2.

(3.6.8)

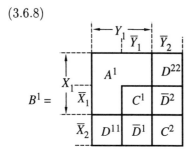

Matrices B^1 and B^2 of k-sum

We assume that both submatrices $[D^{11}|\overline{D}^1]$ and $[\overline{D}^1/D^{12}]$ of D^1 (resp. $[\overline{D}^2|D^{21}]$ and $[D^{22}/\overline{D}^2]$ of D^2) have the same GF(3)-rank as D^1 (resp. D^2). By an elementary argument of linear algebra, this implies that

(3.6.9) $\text{GF(3)-rank}(D^i) = \text{GF(3)-rank}(\overline{D}^i), \quad i = 1, 2$

By (3.6.6) and (3.6.9), we conclude that

(3.6.10) $\text{GF(3)-rank}(\overline{D}^1) + \text{GF(3)-rank}(\overline{D}^2) = k - 1$

The decomposition of B into B^1 and B^2 corresponds to a decomposition of M into two matroids, say, M_1 and M_2, that are represented by B^1 and B^2. We call that decomposition of M a k-sum decomposition and declare M_1 and M_2 to be the components of the k-sum.

The decomposition process is readily reversed. All submatrices of B except for the submatrices D^{13} and D^{23} are present in B^1 and B^2 and thus are already known.

For the computation of D^{13} from B^1 and B^2, we first depict D^1 and its submatrices.

(3.6.11)

$$D^1 = \begin{array}{c|c|c} & Y_1 & \\ & \overline{Y}_1 & \overline{Y}_1 \\ \hline X_2 & D^{11} & \overline{D}^1 \\ \hline & D^{13} & D^{12} \end{array}$$

Partitioned version of D^1

By (3.6.9), GF(3)-rank(D^1) = GF(3)-rank($\overline{D}^1$), so there is a matrix F^1 solving the equation $[D^{13}|D^{12}] = F^1 \cdot [D^{11}|\overline{D}^1]$. Thus,

$$(3.6.12) \qquad\qquad D^{12} = F^1 \cdot \overline{D}^1$$

and

$$(3.6.13) \qquad\qquad D^{13} = F^1 \cdot D^{11}$$

We first solve (3.6.12) for F^1. The solution may not be unique, but any solution is acceptable. Then we use F^1 in (3.6.13) to obtain D^{13}.

If $\overline{D}^1$ is square and nonsingular, then $F^1 = D^{12} \cdot (\overline{D}^1)^{-1}$. With that solution, we compute D^{13} according to (3.6.13) as

$$(3.6.14) \qquad\qquad D^{13} = D^{12} \cdot (\overline{D}^1)^{-1} \cdot D^{11}$$

The computation of D^{23} proceeds analogously. We first solve

$$(3.6.15) \qquad\qquad D^{22} = F^2 \cdot \overline{D}^2$$

for F^2, then use that solution in

$$(3.6.16) \qquad\qquad D^{23} = F^2 \cdot D^{21}$$

If $\overline{D}^1$ is square and nonsingular, we have

$$(3.6.17) \qquad\qquad D^{23} = D^{22} \cdot (\overline{D}^2)^{-1} \cdot D^{21}$$

The reconstruction of B from B^1 and B^2 corresponds to a k-sum composition of M_1 and M_2 to M. We use the term k-sum to refer to both the k-sum decomposition of M and the k-sum composition of M_1 and M_2 to M. We apply the same terminology to B, B^1, and B^2. For example, B is a k-sum of B^1 and B^2.

It is a simple exercise to prove that the 1- and 2-sums discussed earlier are special instances of the above situation. We leave the details to the reader.

The above discussion of matroid sums, though brief, suffices for the purposes of this book.

3.7 Extensions and References

Matroids were formulated by Whitney (1935) as an abstraction of matrices and graphs. That pioneering paper already contains almost all matroid concepts of Sections 3.2 and 3.3. The material may also be found in many books on matroid theory—for example, in Tutte (1971), Lawler (1976), Aigner (1979), Recski (1989), Oxley (1992), and Truemper (1992).

The definitions of separation and connectivity of Section 3.4 are due to Tutte (1966).

A detailed discussion of the historical developments and earlier work on matroid connectivity and matroid sums is included in Truemper (1992). Here, we mention only that a fundamental investigation of 2-separations for matroids and other combinatorial structures is made in Cunningham and Edmonds (1980), and that the first profound use of matroid k-sums with $k = 3$ is due to Seymour (1980). Actually, Seymour (1980) uses a k-sum definition that is different from that of Section 3.6, which is taken from Truemper (1992). But for $k = 3$ and for the class of matroids considered in Seymour (1980), the two definitions lead to very similar constructs.

Most algorithms of Sections 3.5 and 3.6 rely on results of Truemper (1992). There is one difference, though. In Truemper (1992), pivots are freely employed to reduce the number of cases that need to be considered. But here pivots have been avoided as much as possible to simplify the adaptation of results to matrices over the system $\mathbb{B}$.

Truemper (1992) contains a generalization of matrices over fields to so-called abstract matrices. The latter matrices are nothing but an encoding of matroids. Several algorithms of Sections 3.5 and 3.6 for GF(3)-matroids can be extended to general matroids by a switch from matrices over GF(3) to abstract matrices. We omit details here, but mention that the adaptation of the algorithms is not difficult.

The reader interested in a survey of the historical developments of matroid theory should turn to Kung (1986).

Chapter 4

System IB, Linear Algebra, and Matroids

4.1 Overview

We adapt concepts of linear algebra and matroid theory to the matrices over IB. The chapter contains a number of definitions, algorithms, and theorems, so for a first pass the reader may want to skip all proofs and may just scan any material concerning conjectures and counterexamples.

We first motivate the ideas and notions that have led us to link the system IB to linear algebra and matroid theory. According to Lemma (2.6.21), deciding satisfiability of a matrix A over IB is equivalent to solving matrix equations of the form $A \odot s = \underline{1}$ over IB. Such equations are reminiscent of matrix equations of linear algebra, and one may be tempted to modify relevant solution techniques of linear algebra to the case at hand. That idea has produced attractive results—for example, the polyhedral method summarized in Chapter 1. In this book, we use that idea as well, but proceed in a different fashion. The main notion is to first analyze the structure of a given system $A \odot s = \underline{1}$ using concepts and algorithms of linear algebra and matroid theory, and then to build a solution algorithm with the insight so gained. The indirect use of linear algebra and matroid theory forces us to deal with a number of new constructs for the matrices over IB, a definite drawback. On the other hand, the method leads to polynomial and efficient solution algorithms for large classes of logic problems.

In Section 4.2, we begin the investigation into the structure of $A \odot s = \underline{1}$ by deducing some basic equations and inequalities for vectors over IB.

In Section 4.3, we adapt concepts of linear algebra and matroid theory such as independence and rank to the matrices over IB.

In Section 4.4, we use the independence concept for $\mathbb{B}$ to derive for each matrix over $\mathbb{B}$ a so-called $\mathbb{B}$-independence system. That system is related in certain ways to the BG-matroid and the GF(3)-matroid of A, which are the matroids represented by A when that matrix is taken to be over BG or GF(3), respectively.

In Sections 4.5–4.7, we adapt the matroid concepts and methods of Chapter 3 to $\mathbb{B}$-independence.

In Section 4.5, we translate matroid separations and connectivity to so-called $\mathbb{B}$-separations and $\mathbb{B}$-connectivity for $\mathbb{B}$-independence systems.

In Section 4.6, we develop algorithms for finding particular $\mathbb{B}$-separations.

In Section 4.7, we deduce so-called $\mathbb{B}$-sum decompositions and compositions from the related matroid concepts.

In Section 4.8, we summarize how the concepts of $\mathbb{B}$-independence, $\mathbb{B}$-separation, $\mathbb{B}$-sum, etc. are put to use in later chapters.

The underlying ideas of this chapter are applicable to settings other than logic once one extends the system $\mathbb{B}$ to a more general system. In Section 4.9, we introduce the axioms of that extension and sketch how it may be used to solve certain combinatorial problems.

In the final section, 4.10, we discuss extensions and list references.

4.2 Basic Equations and Inequalities

We introduce basic equations and inequalities for the system $\mathbb{B}$. These results will be used later for several decompositions and compositions of matrices over $\mathbb{B}$.

According to (2.6.18)–(2.6.20), the system $\mathbb{B}$ has 0, +1, and −1 as elements, and the operations of $\mathbb{B}$-multiplication, $\mathbb{B}$-addition, and $\mathbb{B}$-subtraction, denoted by $\odot$, $\oplus$, and $\ominus$, respectively, are as follows.

For $\alpha, \beta \in \{0, \pm 1\}$, $\mathbb{B}$-multiplication is defined by

$$(4.2.1) \qquad \alpha \odot \beta = \begin{cases} 1 & \text{if } \alpha = \beta = 1 \text{ or } \alpha = \beta = -1 \\ 0 & \text{otherwise} \end{cases}$$

$\mathbb{B}$-addition and $\mathbb{B}$-subtraction are defined only for $\{0, 1\}$ elements. For $\alpha, \beta \in \{0, 1\}$, $\mathbb{B}$-addition is given by

$$(4.2.2) \qquad \alpha \oplus \beta = \begin{cases} 1 & \text{if } \alpha = 1 \text{ or } \beta = 1 \\ 0 & \text{otherwise} \end{cases}$$

For $\alpha, \beta \in \{0, 1\}$, $\mathbb{B}$-subtraction is specified by

$$(4.2.3) \qquad \alpha \ominus \beta = \begin{cases} 1 & \text{if } \alpha = 1 \text{ and } \beta = 0 \\ 0 & \text{otherwise} \end{cases}$$

The extension of these operations to matrices over IB is accomplished as follows.

Let A and B be $\{0, \pm 1\}$ matrices over IB of size $m \times n$ and $n \times p$, respectively. If both A and B are nontrivial and nonempty, then the matrix $C = A \odot B$ is defined to be the $m \times p$ $\{0, 1\}$ matrix whose elements C_{ij} are given by $C_{ij} = \bigoplus_{k=1}^{n} (A_{ik} \odot B_{kj})$, for $i = 1, 2, \ldots, m$ and $j = 1, 2, \ldots, p$. If at least one of A and B is trivial or empty, then $C = A \odot B$ is defined to be the $m \times p$ zero matrix.

Let A and B be $m \times n$ $\{0, 1\}$ matrices over IB. If A is nontrivial and nonempty, then so is B, and $C = A \oplus B$ (resp. $C = A \ominus B$) is defined to be the $m \times n$ $\{0, 1\}$ matrix whose elements C_{ij} are given by $C_{ij} = A_{ij} \oplus B_{ij}$ (resp. $C_{ij} = A_{ij} \ominus B_{ij}$), for $i = 1, 2, \ldots, m$ and $j = 1, 2, \ldots, n$. If A is trivial or empty, then B is of the same type, and both $C = A \oplus B$ and $C = A \ominus B$ are defined to be equal to A or, equivalently, B.

The next three lemmas summarize elementary results for system IB.

(4.2.4) Lemma. *Let a be a $\{0, 1\}$ vector, and let e be a $\{0, \pm 1\}$ vector. View a and e to be over IB. Then the following relationships hold.*

$$(4.2.5) \qquad\qquad e \odot 0 = 0$$

$$(4.2.6) \qquad\qquad a \oplus 0 = a \oplus a = a \ominus 0 = a$$

$$(4.2.7) \qquad\qquad a \ominus a = 0$$

Proof. We may confine ourselves to the case where a and e are scalars. The equations then follow directly from the definitions (4.2.1)–(4.2.3). $\qquad\square$

(4.2.8) Lemma. *Let a, b, and c be $\{0, 1\}$ vectors over IB of the same dimension. Then the following relationships hold.*

$$(4.2.9) \qquad a \oplus b = b \oplus a \quad (\oplus \text{ is commutative})$$

$$(4.2.10) \qquad (a \oplus b) \oplus c = a \oplus (b \oplus c) \quad (\oplus \text{ is associative})$$

$$(4.2.11) \qquad (a \ominus b) \ominus c = a \ominus (b \oplus c)$$

$$(4.2.12) \qquad (a \ominus b) \oplus (b \ominus c) \geq a \ominus c$$

$$(4.2.13) \qquad a \oplus b \geq c \text{ if and only if } a \geq c \ominus b$$

Proof. As for Lemma (4.2.4), we may restrict ourselves to the case where a, b, and c are scalars.

(4.2.9), (4.2.10): These equations hold by the definition of $\oplus$.

(4.2.11): Both sides of the equation are equal to 1 if and only if $a = 1$ and $b = c = 0$.

(4.2.12): If $a = 0$ or $c = 1$, then the right-hand side of the inequality is equal to 0, and the inequality holds trivially. For the remaining case of

$a = 1$ and $c = 0$, the left-hand side is equal to 1 regardless of the value of b.

(4.2.13): The claim is obviously correct if $a = 1$ or $c = 0$. For the remaining case of $a = 0$ and $c = 1$, the claim is verified by checking its validity for $b = 0$ and $b = 1$.   □

In the next lemma, we deduce several elementary relationships from (4.2.9)–(4.2.13). For reasons to become clear shortly, we avoid any reference to system $\mathbb{B}$.

(4.2.14) Lemma. *Let a partial order $\geq$ and binary operators $\oplus$ and $\ominus$ be defined for a given set. Suppose that, for any elements a, b, and c of the set, (4.2.9)–(4.2.13) are satisfied. Then, for any elements a, b, c, and d of the set, the following relationships hold.*

$$(4.2.15) \qquad (a \ominus b) \ominus c = a \ominus (b \oplus c) = (a \ominus c) \ominus b$$

$$(4.2.16) \qquad a \geq (a \oplus b) \ominus b$$

$$(4.2.17) \qquad a \leq (a \ominus b) \oplus b$$

$$(4.2.18) \qquad a \geq b \text{ implies } a \ominus c \geq b \ominus c$$

$$(4.2.19) \qquad a \geq b \text{ implies } a \oplus c \geq b \oplus c$$

$$(4.2.20) \qquad a \geq b \text{ and } c \geq d \text{ imply } a \oplus c \geq b \oplus d$$

Proof. (4.2.15): The equation follows directly from (4.2.9) and (4.2.11).
(4.2.16): (4.2.13) and $a \oplus b \geq a \oplus b$ imply $a \geq (a \oplus b) \ominus b$.
(4.2.17): (4.2.13) and $a \ominus b \geq a \ominus b$ imply $a \leq (a \ominus b) \oplus b$.
(4.2.18): By (4.2.17) and $a \geq b$ we have $(a \ominus c) \oplus c \geq a \geq b$, and thus by (4.2.13) we have $a \ominus c \geq b \ominus c$.
(4.2.19): By (4.2.16) and $a \geq b$ we have $a \geq b \geq (b \oplus c) \ominus c$, and thus by (4.2.13) we have $a \oplus c \geq b \oplus c$.
(4.2.20): We use $a \geq b$, $c \geq d$, (4.2.9), and two applications of (4.2.19) to deduce $a \oplus c \geq b \oplus c \geq b \oplus d$.   □

The equations $e \odot 0 = 0$ of (4.2.5) and $a \oplus 0 = a$ of (4.2.6), plus the commutativity and associativity of $\oplus$ established by (4.2.9) and (4.2.10), are essential for the constructs, algorithms, and theorems of the subsequent sections of this chapter. In fact, those results are valid for matrices over systems other than $\mathbb{B}$ if matrix multiplication and addition, say, denoted by $\odot$ and $\oplus$, have the properties just mentioned, and if these operations are defined in terms of scalar multiplication and addition analogously to the case of $\mathbb{B}$. Thus, we may view these properties as *axioms* that make the results of this chapter possible.

In a similar vein, the equations and inequalities of Lemmas (4.2.8) and (4.2.14) are essential for the validity of several decompositions, compositions, and solution algorithms of later chapters. Since Lemma (4.2.14) follows from (4.2.9)–(4.2.13) of Lemma (4.2.8), only (4.2.9)–(4.2.13) are needed to establish these decompositions, compositions, and algorithms. Accordingly, we also view (4.2.9)–(4.2.13) as *axioms*.

Altogether, we thus have the following axioms.

(4.2.21) $$e \odot 0 = 0$$

(4.2.22) $$a \oplus 0 = a$$

(4.2.23) $$a \oplus b = b \oplus a \quad (\oplus \text{ is commutative})$$

(4.2.24) $$(a \oplus b) \oplus c = a \oplus (b \oplus c) \quad (\oplus \text{ is associative})$$

(4.2.25) $$(a \ominus b) \ominus c = a \ominus (b \oplus c)$$

(4.2.26) $$(a \ominus b) \oplus (b \ominus c) \geq a \ominus c$$

(4.2.27) $$a \oplus b \geq c \text{ if and only if } a \geq c \ominus b$$

In Section 4.9, we introduce important combinatorial problems that may be expressed by matrices over certain systems and by operations $\oplus$, $\ominus$, and $\odot$ to which (4.2.21)–(4.2.27) apply. According to the preceding discussion, the constructs of this chapter and several decompositions, compositions, and solution algorithms of subsequent chapters are valid for these problems.

4.3 System IB and Linear Algebra

We adapt concepts of linear algebra to the matrices over IB. We first introduce two auxiliary notions called range and subrange.

Range and Subrange

Let A be an $m \times n$ $\{0, \pm 1\}$ matrix over IB. For the moment, we view A to be a function that takes a given $\{0, \pm 1\}$ vector s of size n to the $\{0, 1\}$ vector $b = A \odot s$ of size m. Thus, the *domain* of the function A is the set of $\{0, \pm 1\}$ vectors of size n, and the *range* of A is the following set of $\{0, 1\}$ vectors of size m.

(4.3.1) $$\text{range}(A) = \{b \mid b = A \odot s; \ s_j \in \{0, \pm 1\}, \ \forall \, j\}$$

When we restrict the vector s of (4.3.1) to $\{\pm 1\}$ entries, we get an important subset of range(A) that we call *subrange*. Thus,

(4.3.2) $$\text{subrange}(A) = \{b \mid b = A \odot s; \ s_j \in \{\pm 1\}, \ \forall \, j\}$$

Occasionally, we require sets that lie between range(A) and subrange(A). The vectors of such a set are of the form $b = A \odot s$ as in the definition (4.3.1) of range(A). But this time, only the entries s_j indexed by a given set J may take on $\{0, \pm 1\}$ values, while the remaining s_j are confined to $\{\pm 1\}$ values. The set of such vectors s, say, S, is therefore

$$(4.3.3) \qquad S = \{s \mid s_j \in \{0, \pm 1\}, \ \forall\, j \in J; \ s_j \in \{\pm 1\}, \ \forall\, j \notin J\}$$

In a slight abuse of notation, we denote the subset of range(A) corresponding to the vectors $s \in S$ by range(A, J). Thus,

$$(4.3.4) \qquad \text{range}(A, J) = \{b \mid b = A \odot s; \ s \in S\}$$

Let Y be the column index set of A. Evidently,

$$(4.3.5) \qquad \begin{aligned} \text{range}(A) &= \text{range}(A, Y) \\ \text{subrange}(A) &= \text{range}(A, \emptyset) \end{aligned}$$

and, for any $J \subseteq Y$,

$$(4.3.6) \qquad \text{subrange}(A) \subseteq \text{range}(A, J) \subseteq \text{range}(A)$$

Recall the following convention for matrix multiplication for IB introduced in Section 2.6: If in a matrix product $A \odot B$ with $m \times n$ A and $n \times p$ B at least one of A and B is trivial or empty, then $A \odot B$ is the $m \times p$ zero matrix. This convention plus the above definitions of range(A), subrange(A), and range(A, J) implies that these sets are always nonempty. In particular, let the $m \times n$ matrix A be trivial or empty; that is, $m = 0$ or $n = 0$. Then, regardless of the choice of the $n \times 1$ vector s, the vector $b = A \odot s$ is the $m \times 1$ zero vector. Thus, each one of range(A), subrange(A), and range(A, J) contains just that zero vector.

Computation of Range

Computation of range(A), subrange(A), and range(A, J) is not difficult. We first treat the case of range(A), assuming the nontrivial situation where $m, n \geq 1$. Inductively, we assume that for some column submatrix B of A we already have range(B). Let j index a column vector a of A that is not in B. We compute range($[B|a]$) as follows. Define R_α to be the subset of vectors $c = [B|a] \odot s$ of range($[B|a]$) for which $s_j = \alpha$. Clearly, any vector of R_α is the sum of a vector of range(B) and the vector $a \odot \alpha$; that is,

$$(4.3.7) \qquad R_\alpha = \{c \mid c = b \oplus (a \odot \alpha); \ b \in \text{range}(B)\}$$

Since $R_0 = \text{range}(B)$, we have

(4.3.8) $\text{range}([B|a]) = R_0 \cup R_{+1} \cup R_{-1} = \text{range}(B) \cup R_{+1} \cup R_{-1}$

If $A = [B|a]$, we stop with $\text{range}(A) = \text{range}([B|a])$. Otherwise, we redefine B to be $[B|a]$ and repeat the above process with the new B and a new column a of A.

At times, we are given the range of a matrix $[B/D]$ and desire the range of a second matrix $A = [B|E]$. Note that B is a submatrix of both matrices. We allow for the situation where $[B/D]$ has no columns, in which case $\text{range}([B/D])$ contains only a zero vector of appropriate dimension. We compute $\text{range}([B|E])$ in two steps.

First, we delete from each vector of $\text{range}([B/D])$ all entries corresponding to the rows of D. The resulting set is $\text{range}(B)$. We call this step the *projection* of $\text{range}([B/D])$ onto $\text{range}(B)$.

Second, we extend $\text{range}(B)$ to $\text{range}([B|E])$ by processing each column of E as described above for the computation of $\text{range}(A)$.

The sets $\text{subrange}(A)$ and $\text{range}(A, J)$ are found by computations that are almost identical to those for $\text{range}(A)$. It suffices that we discuss the modifications for the case of $\text{range}(A, J)$. Define J' (resp. J'') to be the restriction of J to the column index of B (resp. $[B|a]$). In the modified procedure, the projection step takes $\text{range}([B/D], J')$ to $\text{range}(B, J')$, the sets R_α of (4.3.7) are replaced by

(4.3.9) $R_\alpha = \{c \mid c = b \oplus (a \odot \alpha); \ b \in \text{range}(B, J')\}$

and the formula of (4.3.8) becomes

(4.3.10) $\text{range}([B|a], J'') = \begin{cases} \text{range}(B, J') \cup R_{+1} \cup R_{-1} & \text{if } j \in J \\ R_{+1} \cup R_{-1} & \text{otherwise} \end{cases}$

We summarize the procedure below.

(4.3.11) Algorithm RANGE. *Finds* $\text{range}(A, J)$. *If J is equal to the column index set of A, then* $\text{range}(A, J)$ *is equal to* $\text{range}(A)$. *If J is empty, then* $\text{range}(A, J)$ *is equal to* $\text{subrange}(A)$.

Input: Matrix $A = [B|E]$ over IB, of size $m \times n$; a subset J of the column index set of A; a second matrix $[B/D]$ over IB and $\text{range}([B/D], J')$, where J' is the restriction of J to the column index set of B.

Output: $\text{range}(A, J)$. If J is equal to the column index set of A (resp. is empty), then $\text{range}(A, J)$ is also $\text{range}(A)$ (resp. $\text{subrange}(A)$).

Complexity: Polynomial in terms of the maximum of m and n and in terms of the cardinalities of $\text{range}(A)$ and $\text{range}([B/D], J')$. Theorem (4.4.19) given in the next section provides complexity formulas for two special cases.

Procedure:

1. If $m = 0$ or $n = 0$, declare range(A, J) to just contain the $m \times 1$ zero vector, and stop.

2. Delete from each vector of range$([B/D], J')$ the entries indexed by the row index set of D. The resulting set is range(B, J'). Set R equal to that set.

3. Do for each column vector a of E, say, with index j:
   For $\alpha = +1, -1$, compute $R_\alpha = \{c \mid c = b \oplus (a \odot \alpha); \ b \in R\}$. If $j \in J$, update R to $R \cup R_{+1} \cup R_{-1}$; otherwise, update R to $R_{+1} \cup R_{-1}$.

4. Output range$(A, J) = R$, and stop.

Define a column j of E to be *redundant* if during the processing of that column in Step 3 of Algorithm RANGE (4.3.11) the set R is not changed. Note that the order in which the columns of E are processed may influence whether a given column j of E is redundant. Of course, the term is well-defined when the order of processing the columns is fixed. For such an order, let F be E minus the redundant columns determined in Step 3. Let K be the restriction of J to the index set of F. By definition of redundancy, we have range$([B|F], K) = $ range$([B|E], J)$, a simple but useful fact.

By (4.3.6), range(A) (resp. subrange(A)) is the unique maximal (resp. minimal) set among the possible range(A, J) sets. So if we have range$(A) = $ subrange(A), then, for all J, range$(A) = $ range(A, J) as well. This happens, for example, when A is a $\{0, 1\}$ matrix. The sets R_0 and R_{-1} defined by (4.3.9) are then equal, and Step 3 of Algorithm RANGE (4.3.11) effectively updates R to $R \cup R_{+1}$ for any J.

There are situations where the cardinality of range(A) is much larger than that of subrange(A). For example, let A be a square matrix of order m having -1s on its diagonal and 1s in all off-diagonal positions. Then it is not difficult to verify that range(A) contains 2^m vectors, while subrange(A) has $m+1$ vectors, each of which contains at most one 0 entry. In Chapter 7, we define a class of matrices A over $\mathbb{B}$ called closed where subrange(A) is bounded by a linear function of n.

Range and Satisfiability

We show later, in Section 4.10, that deciding whether a given $\{0, 1\}$ vector b is in range(A, J) can be reduced to determining whether a certain submatrix of A is satisfiable. When $b = \underline{1}$, the statement of that result and its proof become very simple. We include details next.

(4.3.12) Lemma. *The following statements are equivalent for a matrix A over $\mathbb{B}$.*

(i) *A is satisfiable.*

(ii) $|\text{range}(A)| = |\text{range}([A|\underline{1}])|$.

(iii) $\underline{1} \in \text{range}(A)$.

Statements (ii) *and* (iii) *remain equivalent to* (i) *when* range(A) *is replaced by* subrange(A) *or* range(A, J), *for some J.*

Proof. By Lemma (2.6.21), A is satisfiable if and only if $A \odot s = \underline{1}$ has a $\{\pm 1\}$ solution s. The latter statement is equivalent to (ii) and (iii), even when range(A) in (ii) or (iii) is replaced by subrange(A) or range(A, J), for some J.   □

According to the discussion of Section 2.7, deciding satisfiability of a matrix A over IB is difficult in general. By Lemma (4.3.12), determining the cardinality of range(A), subrange(A), or range(A, J), or deciding whether the vector $\underline{1}$ is in range(A), subrange(A), or range(A, J), is difficult as well.

On the other hand, the polynomial bound on Algorithm RANGE (4.3.11) implies that the latter problem is easy if the cardinality of any one of the sets range(A), subrange(A), or range(A, J) is small. In the next subsection, we look at simple cases of such matrices A.

Matrices with Small Range

The following theorem establishes the structure of matrices A over IB with small range(A). Unless noted otherwise, the display of matrices of the theorem relies on the following special convention. If the columns (resp. rows) of a submatrix are explicitly indexed, then at least one column (resp. row) of that submatrix must be present. Columns or rows not explicitly indexed may be absent.

(4.3.13) Theorem. *Let A be a matrix over IB with $|\text{range}(A)| \leq 4$. Then up to scaling of columns by $\{\pm 1\}$ factors and change of index sets, A is one of the matrices of the applicable case of* (a)–(d) *below.*

(a) $|\text{range}(A)| = 1$: *A is a zero matrix.*

(b) $|\text{range}(A)| = 2$: *A is the following matrix.*

(4.3.14)

$$A = \begin{array}{c|c|} & Y_1 \\ \hline X_1 & 1s \\ \hline & 0 \\ \end{array}$$

Matrix A with $|\text{range}(A)| = 2$

(c) $|\text{range}(A)| = 3$: *A is the matrix of* (4.3.15) *or* (4.3.16) *below.*

(4.3.15)

$$A = \begin{array}{c|c} X_1 & \underline{1} \\ \hline X_2 & \underline{1} \end{array} \quad 0$$

Matrix A with $|\text{range}(A)| = 3$, case 1

(4.3.16)

$$A = \begin{array}{c|cc} & Y_1 & Y_2 \\ \hline X_1 & \text{1s} & \\ X_2 & & 0 \end{array}$$

Matrix A with $|\text{range}(A)| = 3$, case 2

(d) $|\text{range}(A)| = 4$: A is one of the matrices of (4.3.17)–(4.3.19) below.

(4.3.17)

$$A = \begin{array}{c|c} & Y_1 \\ \hline X_1 & \\ X_2 & -\text{1s} \quad \text{1s} \quad \text{1s} \\ & 0 \end{array}$$

Matrix A with $|\text{range}(A)| = 4$, case 1

where $|Y_1| \geq 2$.

(4.3.18)

$$A = \begin{array}{c|ccc} & y & Y_2 & Y_3 \\ \hline X_1 & \underline{1} & \text{1s} & \\ X_2 & -\underline{1} & & \text{1s} \\ X_3 & & & 0 \end{array}$$

Matrix A with $|\text{range}(A)| = 4$, case 2

where X_3, Y_2, or Y_3 may be empty, but $Y_2 \cup Y_3$ is nonempty, and where Y_3 must be empty if X_3 is nonempty.

(4.3.19)

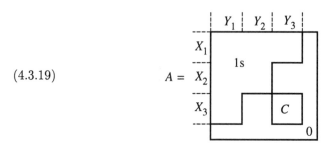

$$A = \begin{array}{c|ccc} & Y_1 & Y_2 & Y_3 \\ \hline X_1 & & & \\ X_2 & \text{1s} & & \\ X_3 & & & C \\ & & & 0 \end{array}$$

Matrix A with $|\text{range}(A)| = 4$, case 3

where the submatrix C is a zero matrix or contains only 1s. In the latter case, X_1 or Y_1 may be empty.

We need some definitions and a lemma for the proof of Theorem (4.3.13).

A matrix A is *solid triangular* if for all $i < j$, $A_{ij} = 0$, and for all $i \geq j$, $A_{ij} = 1$. When we add parallel or zero vectors any number of times to a solid triangular matrix, we get a *solid staircase* matrix. A typical example of such a matrix is given below.

(4.3.20)

$$\begin{array}{c} 0 \\ \text{1s} \end{array}$$

Solid staircase matrix

The next lemma characterizes solid staircase matrices.

(4.3.21) Lemma. *A* $\{0, 1\}$ *matrix is a solid staircase matrix if and only if it has no* 2×2 *identity submatrix.*

Proof. The "only if" part is elementary. The "if" part is proved by a straightforward inductive argument. One removes a row with maximum number of 1s, invokes induction, then adds that row again for the desired conclusion. □

Proof of Theorem (4.3.13). It is easy to check that the range of each matrix A of (4.3.14)–(4.3.19) has the claimed cardinality. Hence, we assume for (a)–(d) that range(A) has the appropriate cardinality, and we deduce that A is one of the claimed matrices.

$|\text{range}(A)| = 1$: Then range(A) contains just the zero vector, and A must be a zero matrix.

$|\text{range}(A)| = 2$: If A has a column with a $+1$ and a -1, or if A has a 2×2 submatrix that either is an identity or contains exactly three 1s, then it is easily seen that $|\text{range}(A)| \geq 3$, a contradiction. By Lemma (4.3.21), we may therefore assume that A is a $\{0, 1\}$ solid staircase matrix. Since 2×2 submatrices with exactly three 1s are ruled out, (4.3.14) is the only case.

$|\text{range}(A)| = 3$: If A has a column with a $+1$ and a -1, then it is easily checked that A has no other nonzero column and thus is the matrix of (4.3.15). Hence, we may assume that A is a $\{0, 1\}$ matrix. If A has a 2×2 identity submatrix, then $|\text{range}(A)| \geq 4$, which is not possible. By Lemma (4.3.21), matrix A is a solid staircase matrix. Since $|\text{range}(A)| = 3$, A must be the matrix of (4.3.16).

$|\text{range}(A)| = 4$: Given the detailed discussion of the previous cases, we just summarize the proof here and leave it for the reader to fill in the details. Let Y_1 be the index set of the columns with both a $+1$ and a -1. If $|Y_1| \geq 2$ (resp. $Y_1 = \{y\}$), then A is the matrix of (4.3.17) (resp. (4.3.18)). If $Y_1 = \emptyset$, then A is given by (4.3.19).   □

We are ready to adapt several concepts of linear algebra to the system ℝ.

Independence

In linear algebra, a collection of vectors is declared to be (linearly) independent if any linear combination of the vectors involving at least one nonzero coefficient is not zero. There is an equivalent definition that is more useful for our purposes: A collection of vectors is independent if the vector space generated by the vectors is reduced when one removes any one of the vectors from the collection. We utilize the latter idea as follows. The columns (resp. rows) of a matrix A over ℝ are ℝ-*independent* if, for any column (resp. row) submatrix B of A, we have $|\text{range}(B)| < |\text{range}(A)|$. If any columns or rows are not ℝ-independent, then they are ℝ-*dependent*. We may abbreviate the terms ℝ-*independence* and ℝ-*dependence* to *independence* and *dependence* whenever confusion with other independence concepts is unlikely.

Note that any $m \times 0$ (resp. $0 \times n$) matrix over ℝ has independent columns (resp. rows).

In linear algebra, subsets of independent columns or rows are also independent. The next theorem establishes that result for system ℝ.

(4.3.22) Theorem. *Let A be a matrix over ℝ.*

(a) *If A has independent columns (resp. rows), then every column (resp. row) submatrix of A has independent columns (resp. rows) as well.*

(b) *If A has independent columns (resp. rows), then addition of any number of rows (resp. columns) preserves independence of columns (resp. rows).*

Proof. Let A have size $m \times n$.

We prove (a) by induction. Hence, we only need to show that independence of columns (resp. rows) is preserved under deletion of column n (resp. row m). In either case, denote the reduced matrix by B.

Assume the column case. If B has dependent columns, then deletion of some column j reduces B to a matrix B' for which $|\text{range}(B')| = |\text{range}(B)|$. Hence Algorithm RANGE (4.3.11) can skip column j when computing range(B) or range(A). But then the columns of A are dependent, a contradiction.

Assume the row case. Let i be any row index of B. Define A' (resp. B') to be A (resp. B) minus row i. Since the rows of A are independent, $|\text{range}(A')| < |\text{range}(A)|$. Now range($A'$) is obtained from range(A) by projection, so $|\text{range}(A')| < |\text{range}(A)|$ implies that range(A) contains two vectors a and b that differ only in element i. Define a' (resp. b') to be a (resp. b) minus the element m. The vectors a' and b' occur in range(B). They differ only in element i and thus establish $|\text{range}(B')| < |\text{range}(B)|$. Hence, the rows of B are independent.

The proof of (b) is also handled by induction. Hence, we only need to show that independence of columns (resp. rows) is preserved under addition of a single row (resp. column). In either case, denote the enlarged matrix by C.

Assume the case of independent columns. If C has dependent columns, then deletion of some column j reduces C to a matrix, say, C', for which $|\text{range}(C')| = |\text{range}(C)|$. By projection, deletion of column j produces a matrix A' for which $|\text{range}(A')| = |\text{range}(A)|$. But then the columns of A are dependent, a contradiction.

Assume the case of independent rows. Let j be the index of the column added to A. If C has dependent rows, then deletion of some row i reduces C to a matrix C' for which $|\text{range}(C')| = |\text{range}(C)|$. Define A' to be A minus row i. Since $|\text{range}(C')| = |\text{range}(C)|$, every vector of range($C'$) has a unique extension to a vector of C. This is particularly so for the vectors $C' \odot s$ of range(C') and $C \odot s$ of range(C) where the entry of s corresponding to column j is fixed to 0. But the latter vectors are those of range(A') and range(A), so $|\text{range}(A')| = |\text{range}(A)|$. Hence, the rows of A are dependent, a contradiction.   □

Basis

We translate the notion of basis from linear algebra as follows. A *column* (resp. *row*) IB-*basis* of a matrix A over IB is a column (resp. row) submatrix B of A whose columns (resp. rows) are independent and for which $|\text{range}(B)| = |\text{range}(A)|$. If confusion with other basis concepts is unlikely, we may simply use *column* (resp. *row*) *basis*. These definitions imply that, for any m, $n \geq 0$, the $m \times n$ zero matrix has the $m \times 0$ matrix as unique column basis and the $0 \times n$ matrix as unique row basis.

Many fundamental results of linear algebra do not apply to the case at hand. We cite two instances.

In linear algebra, the number of columns in any column basis is equal to the number of rows in any row basis. We demonstrate with two matrices that this important result does not hold here. The first example matrix is

(4.3.23)

$$A = \begin{array}{c} \begin{array}{cccccc} a & b & c & d & e & f \end{array} \\ \left[\begin{array}{cc|cccc} 1 & -1 & 1 & 0 & 1 & 0 \\ -1 & 1 & 0 & 1 & 0 & 1 \\ 1 & 1 & 0 & 0 & 1 & 1 \end{array} \right] \end{array}$$

Matrix A, example 1

It is easy to check that columns a and b form a column basis of A, as do columns c, d, e, and f. Also, the three rows of A are independent, so A itself is a row basis. Thus, A has a column basis with two columns, has another column basis with four columns, and is itself a row basis with three rows.

The second example matrix is

(4.3.24)

$$A = \begin{bmatrix} 0 & 1 & 1 & 0 \\ 0 & 0 & 1 & 1 \\ 1 & 1 & 0 & 1 \end{bmatrix}$$

Matrix A, example 2

It is easily verified that the four columns as well as the three rows of A are independent. Hence, A itself is a column basis with four columns and is a row basis with three rows.

We turn to the second instance. In linear algebra, a maximal collection of independent columns always constitutes a column basis. In contrast, consider the matrix A over IB given by

(4.3.25)

$$A = \begin{array}{c} \begin{array}{cccccc} a & b & c & d & e & f \end{array} \\ \begin{bmatrix} 1 & 0 & 0 & 1 & 1 & 0 \\ 0 & 1 & 0 & 0 & 1 & 1 \\ 0 & 0 & 1 & 1 & 0 & 1 \end{bmatrix} \end{array}$$

Matrix A, example 3

Simple checking confirms that columns c, d, e, and f of A make up a maximal independent column submatrix B with $|\text{range}(B)| < |\text{range}(A)|$. Thus, B is not a column basis of A.

The above negative conclusions notwithstanding, some important results of linear algebra do carry over to IB. For example, in linear algebra any column basis of a matrix A intersects any row basis of A in a matrix B with independent rows and columns. Indeed, such B is a minimal submatrix of A generating the essentially same vector space as A. For this reason, one could call B a *basis* of A. Analogously, we define a *basis* of a matrix A over IB to be any submatrix B with independent columns and rows satisfying $|\text{range}(B)| = |\text{range}(A)|$. Note that if A is any zero matrix, then A has just one basis, which is the 0×0 empty matrix.

The next theorem shows that the above linear algebra result for bases carries over to IB.

(4.3.26) Theorem. *Let A be a matrix over IB of the form*

(4.3.27)
$$A = \begin{array}{|c|c|} \hline B & E \\ \hline D & F \\ \hline \end{array}$$

Partitioned matrix A

Then the following statements are equivalent.

(i) *The column submatrix $[B/D]$ is a column basis of A, and the row submatrix $[B|E]$ is a row basis of A.*
(ii) *The submatrix B is a basis of A.*
(iii) *The submatrix B satisfies $|\text{range}(B)| = |\text{range}(A)|$ and is minimal subject to that condition.*

Proof. We prove (i)$\Rightarrow$(ii) by inducting on the number of rows of D plus the number of columns of E. If that number is 0, then $A = B$, and we are done. So assume that number to be positive.

If D has at least one row, delete one such row i from A, say, changing A, D, and F to A', D', and F', respectively. We are done by induction once we show that $[B/D']$ is a column basis of A' and $[B|E]$ is a row basis of A'.

We begin with $[B|E]$. We know that $|\text{range}([B|E])| \leq |\text{range}(A')| \leq |\text{range}(A)| = |\text{range}([B|E])|$. Evidently, all inequalities must be tight, so $|\text{range}([B|E])| = |\text{range}(A')| = |\text{range}(A)|$. Since the rows of $[B|E]$ are independent, we conclude that $[B|E]$ is a row basis of A'.

We turn to $[B/D']$. Since $|\text{range}([B/D])| = |\text{range}(A)|$, we have by projection $|\text{range}([B/D'])| = |\text{range}(A')|$. Since $|\text{range}(A')| = |\text{range}(A)|$, we also have $|\text{range}([B/D'])| = |\text{range}(A)|$. By assumption, deletion of any column j from $[B/D]$ causes a reduction of range. By projection and the fact that $|\text{range}([B/D'])| = |\text{range}(A)| = |\text{range}([B/D])|$, a reduction of range must also occur when column j is deleted from $[B/D']$. Thus, the columns of $[B/D']$ are independent, and by $|\text{range}([B/D'])| = |\text{range}(A')|$ they constitute a column basis of A'.

We are done with the case where D has a row, and we begin the case where E has a column. Delete one such column j from A, say, changing A, E, and F to A', E', and F', respectively. We are done by induction once we show that $[B/D]$ is a column basis of A' and $[B|E']$ is a row basis of A'.

We start with $[B/D]$. We know that $|\text{range}([B/D])| \leq |\text{range}(A')| \leq |\text{range}(A)| = |\text{range}([B/D])|$. Evidently, all inequalities must be tight, so $|\text{range}([B/D])| = |\text{range}(A')| = |\text{range}(A)|$. Since the columns of $[B/D]$ are independent, $[B/D]$ is a column basis of A'.

We turn to $[B|E']$. Since $|\text{range}([B|E])| = |\text{range}(A)|$, each vector of range($[B|E]$) has a unique extension to a vector of range(A). This is particularly so for the vectors $[B|E] \odot s$ of range($[B|E]$) and $A \odot s$ of range(A) where the entry of s corresponding to column j is fixed to 0. But such vectors are precisely those of range($[B|E']$) and range(A'), so $|\text{range}([B|E'])| = |\text{range}(A')|$. It remains to be shown that $[B|E']$ has independent rows. Suppose we delete any row i from $[B|E]$, say, getting $[B''|E'']$. By assumption, $|\text{range}([B''|E''])| < |\text{range}([B|E])|$. Since $|\text{range}([B|E])| = |\text{range}(A)| = |\text{range}(A')| = |\text{range}([B|E'])|$, we have $|\text{range}([B''|E''])| < |\text{range}([B|E'])|$. Now deletion of row i from $[B|E']$ results in a matrix whose range contains at most $|\text{range}([B''|E''])|$ vectors and thus strictly less than $|\text{range}([B|E'])|$ vectors. Thus, the rows of $[B|E']$ are independent, and by $|\text{range}([B|E'])| = |\text{range}(A')|$ they constitute a row basis of A'.

Next, we show (i)⇐(ii). Thus we assume B to be a basis; that is, B has independent columns and rows, and $|\text{range}(B)| = |\text{range}(A)|$. The latter condition implies that the following obvious inequalities $|\text{range}(B)| \leq |\text{range}([B/D])| \leq |\text{range}(A)|$ as well as $|\text{range}(B)| \leq |\text{range}([B|E])| \leq |\text{range}(A)|$ must all be tight. Hence, $|\text{range}([B/D])| = |\text{range}([B|E])| = |\text{range}(A)|$. By Theorem (4.3.22)(b), any addition of rows (resp. columns) to a matrix preserves independence of columns (resp. rows). Thus, independence of the columns (resp. rows) of B implies independence of columns of $[B/D]$ (resp. rows of $[B|E]$). We conclude that $[B/D]$ is a column basis of A, and $[B|E]$ is a row basis.

Finally, the proof of the implication (ii)⇔(iii) follows from the fact that when $|\text{range}(B)| = |\text{range}(A)|$, independence of columns and rows of B is equivalent to minimality of B.   □

We remark that the submatrix B of A of Theorem (4.3.26) need not be square, in contrast to the linear algebra case. An example of a nonsquare B is supplied by the matrix A of (4.3.23) when columns a and b are taken as column basis and A itself as row basis.

Finding a Basis

We present an algorithm for finding a basis for a given matrix A over IB. The method is polynomial if $|\text{range}(A)|$ is bounded by some constant. When that assumption is removed, the task may become difficult as follows.

Let B be a basis of a matrix A over IB. By Theorem (4.3.26), the column (resp. row) index set of B defines a column (resp. row) basis of A. Thus, any algorithm for finding a basis may be employed to locate a column or row basis. In the next subsection, we reduce the generally difficult satisfiability problem for the matrices over IB to finding a column basis for such matrices. We conclude that finding a column basis or a basis is a generally difficult task.

(4.3.28) Algorithm BASIS. *Finds a basis B for a matrix A over IB.*

Input: Matrix A over IB, of size $m \times n$.

Output: A partition of A of the form

(4.3.29)
$$A = \begin{array}{|c|c|} \hline B & E \\ \hline D & F \\ \hline \end{array}$$

Partitioned version of A

such that B is a basis of A. Furthermore, the submatrix $[B/D]$ (resp. $[B|E]$) is a column (resp. row) basis of A.

Complexity: Polynomial if $|\text{range}(A)|$ is bounded by some constant.

Procedure:

1. Compute $\text{range}(A)$ with Algorithm RANGE (4.3.11). Define A^0 to be A minus the columns of A found to be redundant in Step 3 of Algorithm RANGE. Thus $\text{range}(A^0) = \text{range}(A)$. Define n_0 to be the number of columns of A^0. Let the rows (resp. columns) of A^0 be indexed by $x_1, x_2 \ldots, x_m$ (resp. $y_1, y_2 \ldots, y_{n_0}$).

2. Do for $j = 1, 2, \ldots, n_0$:
   Let C be A^{j-1} minus column y_j. Compute $\text{range}(C)$ with Algorithm RANGE. Define

   (4.3.30)
   $$A^j = \begin{cases} A^{j-1} & \text{if } |\text{range}(C)| < |\text{range}(A)| \\ C & \text{otherwise} \end{cases}$$

3. Output the column index set of A^{n_0} as the column index set of B of (4.3.29). Redefine A^0 to be equal to A^{n_0}.

4. Do for $i = 1, 2, \ldots, m$:
   Let C be A^{i-1} minus row x_i. Compute $\text{range}(C)$ with Algorithm RANGE. Define

   (4.3.31)
   $$A^i = \begin{cases} A^{i-1} & \text{if } |\text{range}(C)| < |\text{range}(A)| \\ C & \text{otherwise} \end{cases}$$

5. Output the row index set of A^m as the row index set of B of (4.3.29), and stop.

Proof of Validity. By Steps 1 and 2 and induction, the matrix A^{no} of Step 3 satisfies $|\text{range}(A^{no})| = |\text{range}(A)|$. Let A' be A^{no} minus an arbitrary column y_j, and take C to be the matrix of Step 2 when column y_j was processed. Note that C has A' as column submatrix.

Since column y_j occurs in A^{no}, according to (4.3.30) we must have $|\text{range}(A')| \leq |\text{range}(C)| < |\text{range}(A)| = |\text{range}(A^{no})|$. Hence, A^{no} has independent columns, and by $|\text{range}(A^{no})| = |\text{range}(A)|$ it is a column basis of A.

The case for A^m of Step 5 follows in the same manner once one observes that a subset of rows of A is independent if and only if this is so for the corresponding subset of A^0 defined in Step 3.

By the definition of the index sets of B in Steps 3 and 5, the column basis A^{no} of Step 3 is the submatrix $[B/D]$ of A, while the row basis A^m of Step 5 is the submatrix $[B|E]$. By Theorem (4.3.26), the intersection of these two matrices, which is B, is a basis of A. ◻

The subrange of any column basis, row basis, or basis of A is related to subrange(A) as follows.

(4.3.32) Theorem. *Let A be a matrix over $\mathbb{B}$ of the form*

(4.3.33)
$$A = \begin{array}{|c|c|} \hline B & E \\ \hline D & F \\ \hline \end{array}$$

Partitioned matrix A

where B is a basis of A. Then

(4.3.34)
$$|\text{subrange}(B)| = |\text{subrange}([B/D])| \geq |\text{subrange}(A)|$$
$$|\text{subrange}(B)| \geq |\text{subrange}([B|E])| = |\text{subrange}(A)|$$

Proof. Since B is a basis of A, the obvious inequalities $|\text{range}(B)| \leq |\text{range}([B/D])| \leq |\text{range}(A)|$ must all be tight. Thus, every vector of range(B) has a unique extension to a vector of range($[B/D]$). In particular, every vector of subrange(B) has a unique extension to a vector of subrange($[B/D]$), so $|\text{subrange}(B)| = |\text{subrange}([B/D])|$.

To show $|\text{subrange}([B/D])| \geq |\text{subrange}(A)|$, let b be any vector in subrange(A). Thus, $b = ([B/D] \odot s) \oplus ([E/F] \odot t)$ for some $\{\pm 1\}$ vectors s and t, which implies $[B/D] \odot s \leq b$.

Since b is also in range(A), and since range($[B/D]$) = range(A), there is a $\{0, \pm 1\}$ vector v for which $b = [B/D] \odot v$. Define a $\{\pm 1\}$ vector w

from s and v by setting $w_j = v_j$ if $v_j \neq 0$, and $w_j = s_j$ otherwise. Since $[B/D] \odot s \leq b$ and $[B/D] \odot v = b$, we have $[B/D] \odot w = b$, which implies $b \in$ subrange$([B/D])$. Hence, $|$subrange$([B/D])| \geq |$subrange$(A)|$.

Finally, we prove $|$subrange$(B)| \geq |$subrange$([B|E])| = |$subrange$(A)|$ by adapting the above arguments and using the fact that, since B is a basis, the inequalities $|$range$(B)| \leq |$range$([B|E])| \leq |$range$(A)|$ must all be tight.   □

The case $|$subrange$(B)| > |$subrange$(A)|$ permitted by (4.3.34) is indeed possible. For example, let A be the matrix

(4.3.35)

$$A = \begin{array}{c} \begin{array}{ccccc} a & b & c & d & e \end{array} \\ \begin{bmatrix} 1 & 0 & 0 & 0 & 1 \\ 0 & 1 & 0 & 0 & 1 \\ 0 & 0 & 1 & 0 & -1 \\ 0 & 0 & 0 & 1 & -1 \end{bmatrix} \end{array}$$

Matrix A

Define B to consist of columns a, b, c, and d. Let D be the 0×4 trivial matrix, let E be column e, and let F be the 0×1 trivial matrix. The column unit vectors of B are independent and $|$range$(B)| = |$range$(A)|$. Due to the presence of column e in A, the four column unit vectors occur in subrange(B), but not in subrange(A). Hence, $|$subrange$(B)| > |$subrange$(A)|$.

Theorem (4.3.32) supports the following bounds on the cardinality of range(A) and subrange(A).

(4.3.36) Corollary. *Let A be a matrix over IB of the form (4.3.33) such that B, say, of size $p \times q$, is a basis of A. Then*

(4.3.37)
$$|\text{range}(A)| \leq 3^{\min\{p,q\}}$$
$$|\text{subrange}(A)| \leq 2^{\min\{p,q\}}$$

Proof. By the definition of range in (4.3.1), $|$range$([B/D])| \leq 3^q$. Since $[B|E]$ has p rows, $|$range$([B|E])| \leq 2^p$. By Theorem (4.3.26), $[B/D]$ is a column basis of A, and $[B|E]$ is a row basis. Thus, $|$range$(A)| = |$range$([B/D])| = |$range$([B|E])|$, so $|$range$(A)| \leq 3^{\min\{p,q\}}$.

By the definition of subrange in (4.3.2), $|$subrange$([B/D])| \leq 2^q$. Since $[B|E]$ has p rows, $|$subrange$([B|E])| \leq 2^p$. By Theorem (4.3.32), $|$subrange$(A)| = |$subrange$([B|E])| \leq |$subrange$([B/D])|$. We conclude that $|$subrange$(A)| \leq 2^{\min\{p,q\}}$.   □

Satisfiability and Column Basis

Any algorithm for identifying a column basis for any matrix over IB may be used to decide satisfiability of the matrices over IB as follows. For a given matrix A over $\mathbb{B}$, we find a column basis B for A, then determine a column basis C for $[B|\underline{1}]$. Since the columns of B are independent, C cannot be a proper submatrix of B. Thus, $C = [B|\underline{1}]$, or $C = B$, or C includes $\underline{1}$ and a proper column submatrix, say, B', of B.

In the first case, we have by the independence of the columns of C, $\underline{1} \notin$ range(B) and thus $\underline{1} \notin$ range(A). By Lemma (4.3.12), A is not satisfiable.

In the second case, $C = B$ implies range(C) = range(B) = range(A) and $\underline{1} \in$ range(C). Thus, $\underline{1} \in$ range(A), and A is satisfiable.

In the third case, range(C) = range($[B|\underline{1}]$) and the given independence of the columns of B imply $\underline{1} \cup$ range(B') = range($[B'|\underline{1}]$) = range(C) $\supseteq$ range(B) $\supset$ range(B'). Thus, range(B) and range(B') differ exactly by the vector $\underline{1}$. Accordingly, $\underline{1}$ is in range(B) = range(A), and A is satisfiable.

Rank

There are several ways to adapt the rank concept of linear algebra to the system $\mathbb{B}$. Each of the possible definitions seems advantageous for some settings and not so useful for others. For our purposes, the following definition seems suitable. Let A be a matrix over $\mathbb{B}$. Then the $\mathbb{B}$-*rank* of A, denoted by $\mathbb{B}$-rank(A), is min$\{p,q \mid A$ has a basis B of size $p \times q\}$. Since the 0×0 matrix is the unique basis of any zero matrix, the $\mathbb{B}$-rank of any such matrix is 0.

An alternate formula for $\mathbb{B}$-rank(A) is given in the next lemma.

(4.3.38) Lemma. *Let q (resp. p) be the smallest integer such that a matrix A over $\mathbb{B}$ has a column (resp. row) basis with q columns (resp. p rows). Then*

$$(4.3.39) \qquad \mathbb{B}\text{-rank}(A) = \min\{p,q\}$$

Proof. By Theorem (4.3.26), any column basis with q columns intersects any row basis with p rows in a $p \times q$ basis. Thus, $\mathbb{B}$-rank(A) = min$\{p,q\}$ where the minimization is over p and q for which there is a $p \times q$ basis or, equivalently, for which there are a column basis with q columns and a row basis with p rows. $\qquad \square$

There is a polynomial algorithm for finding $\mathbb{B}$-rank(A) if $|$range(A)$|$ is bounded by a constant, but the scheme we have in mind is not so nice. We omit details, since in the sequel we do not require an algorithm for $\mathbb{B}$-rank(A).

Using (4.3.39), one may express the bounds for range and subrange of Corollary (4.3.36) in terms of $\mathbb{B}$-rank as follows.

(4.3.40) Corollary.

$$(4.3.41) \qquad \begin{aligned} |\text{range}(A)| &\leq 3^{\mathbb{B}\text{-rank}(A)} \\ |\text{subrange}(A)| &\leq 2^{\mathbb{B}\text{-rank}(A)} \end{aligned}$$

We may simply use the terms *rank* and rank(A) when confusion with other concepts of rank is unlikely.

In linear algebra, the rank of a matrix cannot decrease as one adds columns or rows. Unfortunately, this is not so here, as the next lemma shows.

(4.3.42) Lemma. *The rank of a matrix B over IB may drop if one adds columns; this may be so even if B has independent rows.*

Proof. We use the matrix A of (4.3.23). We define B to consist of columns c, d, e, and f of A. The four columns of B are independent, as are the three rows. Thus, rank(B) = 3. The matrix A is obtained from B by addition of the columns a and b. The latter two columns are a column basis of A and lead to rank(A) = 2 < rank(B). ☐

Span

In linear algebra, a matrix B spans the columns of a matrix C if both B and $[B|C]$ produce the same vector space. The matrix B spans the rows of a matrix C if the vector space produced by $[B/C]$ may be viewed as a higher-dimensional embedding of the vector space generated by B.

We replace the notion of vector space by that of range to arrive at analogous definitions for IB. Let B and C be matrices over IB. Then B *IB-spans* the columns of C if range(B) = range($[B|C]$), and B *IB-spans* the rows of C if range(B) = range($[B/C]$). When it is clear from the context whether the columns or rows of C are meant, we may simply say that B *IB-spans* C.

If confusion with other concepts of span is unlikely, we may abbreviate *IB-span* to *span*.

According to the next lemma, one may test whether B spans the columns or rows of a matrix by testing one column or row at a time.

(4.3.43) Lemma. *Let B and C be matrices over IB. If B spans each column (resp. row) of C, taken one at a time, then B spans the columns (resp. rows) of C.*

Proof. For the column case, we apply Algorithm RANGE (4.3.11) to the matrix $[B|C]$. Since B spans each column of C, that algorithm determines range(B) as range for $[B|C]$.

For the row case, consider an arbitrary vector $b \in$ range(B). Let $S(b) = \{s \mid b = B \odot s; \ s_j \in \{0, \pm 1\}, \ \forall \ j\}$. By assumption, for any row c of C, $|\text{range}(B)| = |\text{range}([B/c])|$. Thus, the set $\{d \mid d = [B/c] \odot s; \ s \in S(b)\}$ contains just one vector. But then the set $\{e \mid e = [B/C] \odot s; \ s \in S(b)\}$ contains just one vector as well. We conclude that $|\text{range}(B)| = |\text{range}([B/C])|$. ☐

We rely on Lemma (4.3.43) in the next algorithm, which determines the columns or rows spanned by a matrix B.

(4.3.44) Algorithm SPAN. *Finds the columns and rows of a matrix A over IB that are spanned by a submatrix B of A.*

Input: Matrix A over IB, partitioned as follows.

(4.3.45)
$$A = \begin{array}{|c|c|} \hline B & E \\ \hline D & F \\ \hline \end{array}$$

Partitioned matrix A

Output: The unique maximal column (resp. row) submatrix of E (resp. D) spanned by B.

Complexity: Polynomial if $|\text{range}(B)|$ is bounded by some constant.

Procedure:

1. Use Algorithm RANGE (4.3.11) as if the range of $A' = [B|E]$ were to be found. However, once processing of the columns of the submatrix B of A' in Step 3 of Algorithm RANGE has been completed, modify Step 3 for the processing of the columns of E as follows.

   Suppose column j of E is being considered in the current iteration of Step 3 of Algorithm RANGE. If the set R on hand at the beginning of that iteration is not equal to the set $R \cup R_{+1} \cup R_{-1}$ of that iteration, then label column j of E as not spanned by B. Otherwise, label column j as spanned by B. In both cases, do not update R.

   When Algorithm RANGE stops, let E' consist of the columns of E that are labeled as spanned by B. Output E' as the maximal spanned column submatrix of E.

2. Tentatively label each row of D as spanned by B. Use Algorithm RANGE as if the range of $A'' = [B/D]$ were to be found. However, modify Step 3 of that algorithm as follows.

   Let R be the set on hand at the end of an arbitrary iteration of Step 3, that is, after the updating. Consider each vector of R to be of the form $[b/c]$ where the partition agrees with that of $[B/D]$. If R contains two vectors $[b/c]$ and $[b/c']$ where $c \neq c'$, then relabel the rows i of D for which $c_i \neq c_i'$ as not spanned by B, and delete the vector $[b/c']$ from R. Repeat the reduction process until no pair of vectors of R satisfies the above condition. At that point, the iteration has been completed. When Algorithm RANGE stops, let D' consist of the rows of D that are labeled as spanned by B. Output D' as the maximal spanned row submatrix of D, and stop.

Proof of Validity. Lemma (4.3.43) directly proves validity of Step 1. The lemma also implies that the rows of D labeled in Step 2 as not spanned by B are correctly classified. Suppose that at least one of the remaining rows of D is, contrary to its label, not spanned by B. Then range($[B/D']$) contains two vectors $[b/e]$ and $[b/e']$ where $e \neq e'$. But then in Step 2 at least one additional row of D would have been relabeled as not spanned by B, a contradiction.   □

We extend the definition of span to submatrices B of a given matrix A in the following manner. Suppose A is a matrix over IB of the form

(4.3.46)
$$A = \begin{array}{|c|c|} \hline B & E \\ \hline D & F \\ \hline \end{array}$$

Partitioned matrix A

Then the submatrix B IB-*spans* A if $|\mathrm{range}(B)| = |\mathrm{range}(A)|$.

Basic results about A spanned by B are summarized in the next theorem.

(4.3.47) Theorem. *Let A be a matrix over IB that is partitioned as given by (4.3.46). Then the following statements are equivalent.*

(i) *B spans A.*
(ii) *B spans the rows of D, and $[B/D]$ spans the columns of $[E/F]$.*
(iii) *B spans the columns of E, and $[B|E]$ spans the rows of $[D|F]$.*
(iv) *$[B/D]$ spans the columns of $[E/F]$, and $[B|E]$ spans the rows of $[D|F]$.*

Furthermore, if B spans A, then D spans the columns of F, and E spans the rows of F.

Proof. We show that (i) implies (ii), (iii), and (iv). Since B spans A, we have $|\mathrm{range}(B)| = |\mathrm{range}(A)|$. The latter equation implies that the obvious inequalities $|\mathrm{range}(B)| \leq |\mathrm{range}([B/D])| \leq |\mathrm{range}(A)|$ and $|\mathrm{range}(B)| \leq |\mathrm{range}([B|E])| \leq |\mathrm{range}(A)|$ must all be tight. Thus, we have $|\mathrm{range}(B)| = |\mathrm{range}([B/D])| = |\mathrm{range}(A)|$, which implies that B spans the rows of D and $[B/D]$ spans the columns of $[E/F]$. Also, $|\mathrm{range}(B)| = |\mathrm{range}([B|E])| = |\mathrm{range}(A)|$, which implies that B spans the columns of E and $[B|E]$ spans the rows of $[D|F]$.

For the proof that (ii) or (iii) implies (i), one only needs to reverse the above arguments.

We prove (iv)$\Rightarrow$(i). If B does not span the rows of D, then $[B|E]$ cannot span the rows of $[D|F]$, a contradiction. Thus, B spans the rows of D, and we may invoke the already proved implication (ii)$\Rightarrow$(i) to conclude (i).

We establish the additional statement of the theorem using (i)⇒(iv). By (iv), $[B/D]$ spans the columns of $[E/F]$, so by projection D spans the columns of F. Also by (iv), $[B|E]$ spans the rows of $[D|F]$, so clearly E spans the rows of F.   ☐

Basis and Span

The definitions of basis and span directly imply the following link between these concepts.

(4.3.48) Lemma. *Let A be a matrix over IB, and let B be a submatrix of A.*
(a) *B is a column (resp. row) basis of A if and only if B is a column (resp. row) submatrix with independent columns (resp. rows) that spans the remaining columns (resp. rows) of A.*
(b) *B is a basis of A if and only if B spans A and has independent columns and rows.*

Theorem (4.3.32) relies on bases to link the range and subrange of a matrix A to those of certain submatrices. We draw analogous conclusions using span.

(4.3.49) Theorem. *Let A be a matrix over IB of the form*

(4.3.50)
$$A = \begin{array}{|c|c|} \hline B & E \\ \hline D & F \\ \hline \end{array}$$

Partitioned matrix A

where the submatrix B spans A. Then

(4.3.51)
$$|\text{subrange}(B)| = |\text{subrange}([B/D])| \geq |\text{subrange}(A)|$$
$$|\text{subrange}(B)| \geq |\text{subrange}([B|E])| = |\text{subrange}(A)|$$

Proof. By Theorem (4.3.47), if B spans A, then B spans the rows of D and the columns of E, $[B/D]$ spans the columns of $[E/F]$, and $[B|E]$ spans the rows of $[D|F]$. These relationships are precisely the facts used in the proof of Theorem (4.3.32) to show (4.3.34), which is (4.3.51) here.   ☐

Rank and Span

In linear algebra, if a submatrix B of a matrix A spans A, then the two matrices have the same rank. For system IB, one can only prove that rank(A) cannot exceed rank(B), as shown in the next theorem.

(4.3.52) Theorem. *Let B be a submatrix of a matrix A over IB. If B spans A, then*

$$(4.3.53) \qquad\qquad \mathrm{rank}(A) \leq \mathrm{rank}(B)$$

Proof. We may assume that A is the matrix of (4.3.50). Recall that the rank of a matrix is defined by $\min\{p, q\}$, where the minimum is taken over p and q such that the matrix has a basis of size $p \times q$. Thus, $\mathrm{rank}(A) \leq \mathrm{rank}(B)$ holds if each basis B' of B is also a basis of A. We prove the latter assumption as follows.

By Theorem (4.3.26), the basis B' is a minimal submatrix of B satisfying $|\mathrm{range}(B')| = |\mathrm{range}(B)|$. Since B spans A, $|\mathrm{range}(B)| = |\mathrm{range}(A)|$. Hence, $|\mathrm{range}(B')| = |\mathrm{range}(A)|$, and B' is a basis of A. □

(4.3.54) Corollary. *Let B be a submatrix of a matrix A over IB. If B spans A, then*

$$(4.3.55) \qquad\qquad |\mathrm{range}(A)| \leq 3^{\mathrm{IB\text{-}rank}(B)}$$
$$|\mathrm{subrange}(A)| \leq 2^{\mathrm{IB\text{-}rank}(B)}$$

Proof. The two inequalities $|\mathrm{range}(A)| \leq 3^{\mathrm{IB\text{-}rank}(A)}$ and $|\mathrm{subrange}(A)| \leq 2^{\mathrm{IB\text{-}rank}(A)}$ of (4.3.41) plus $\mathrm{rank}(A) \leq \mathrm{rank}(B)$ of (4.3.53) yield the inequalities of (4.3.55). □

Satisfiability and Span

Lemma (4.3.12) plus the definition of span implies the following result.

(4.3.56) Lemma. *A matrix A over IB is satisfiable if and only if A spans $\underline{1}$.*

Proof. By Lemma (4.3.12) and the definition of span, the following statements are equivalent: matrix A is satisfiable; $\underline{1} \in \mathrm{range}(A)$; $\mathrm{range}(A) = \mathrm{range}([A|\underline{1}])$; A spans $\underline{1}$. □

(4.3.57) Corollary. *The following statements are equivalent for a matrix A over IB.*

(i) *A is satisfiable.*
(ii) *Every column basis of A spans $\underline{1}$.*
(iii) *Some column submatrix of A with independent columns spans $\underline{1}$.*

Proof. The equivalence follows from Lemma (4.3.56) and the definitions of independence, column basis, and span. □

Extension of System IB

According to (2.6.26) and (2.6.27), the system IB may be extended as follows. The set of elements $\{0, \pm 1\}$ is enlarged by

$$(4.3.58) \qquad U = \{(\alpha, \beta) \mid \alpha, \beta \in \{0, 1, 2\}\}$$

IB-multiplication is extended so that for $(\alpha, \beta) \in U$ and $\gamma \in \{0, \pm 1\}$,

$$(4.3.59) \quad (\alpha, \beta) \odot \gamma = \begin{cases} 1 & \text{if } \alpha \geq 1 \text{ and } \gamma = 1, \text{ or } \beta \geq 1 \text{ and } \gamma = -1 \\ 0 & \text{otherwise} \end{cases}$$

IB-addition and IB-subtraction are not affected.

A matrix over the extension of IB has its entries in $\{0, \pm 1\} \cup U$. Matrix IB-multiplication is defined when such a matrix is postmultiplied with one having $\{0, \pm 1\}$ entries. The rules for matrix IB-addition and matrix IB-subtraction are unchanged.

Satisfiability of a matrix A is defined as before, via an equation of the form $A \odot s = \underline{1}$.

All definitions and results of this section for matrices A over IB that do not specifically require A to be a $\{0, \pm 1\}$ matrix apply to the extended setting. In particular, the definitions of range(A), subrange(A), and range(A, J) by (4.3.1)–(4.3.4) are appropriate, and these sets are correctly computed by Algorithm RANGE (4.3.11). Every one of the lemmas, theorems, and corollaries following that algorithm remains valid except for Theorem (4.3.13) and Lemma (4.3.21), which deal with $\{0, \pm 1\}$ and $\{0, 1\}$ matrices, respectively.

In the next section, we derive from the matrices over IB so-called IB-independence systems. Subsequently, we adapt a number of matroid concepts to such systems.

4.4 IB-Independence System

We deduce from the matrices over IB so-called IB-independence systems, then link those systems to certain matroids of Chapter 3. We first review relevant portions of Section 3.2.

Let E be a finite set. Define $\mathcal{I}$ to be a nonempty subset of the power set of E; that is, each element of $\mathcal{I}$ is a subset of E. The pair $(E, \mathcal{I})$ is an independence system if the following axioms are satisfied.

(4.4.1)
    (i)   The null set is in $\mathcal{I}$.
    (ii) Every subset of any set in $\mathcal{I}$ is also in $\mathcal{I}$.

The set E is the groundset of the system, and $\mathcal{I}$ is the set of independent subsets of E.

A matroid is an independence system $M = (E, \mathcal{I})$ where, for any subset $\overline{E} \subseteq E$, all maximal independent subsets of $\overline{E}$ have the same cardinality. Thus, a matroid consists of a finite set E and a subset $\mathcal{I}$ of the power set of E satisfying the following axioms.

(4.4.2)

	(i)	The null set is in $\mathcal{I}$.
	(ii)	Every subset of any set in $\mathcal{I}$ is also in $\mathcal{I}$.
	(iii)	For any subset $\overline{E} \subseteq E$, the maximal subsets of $\overline{E}$ that are in $\mathcal{I}$ have the same cardinality.

Let B be a matrix over a field $\mathcal{F}$, say, with row index set X and column index set Y. Append an identity matrix I to B to get a matrix $A = [I|B]$ over $\mathcal{F}$. Consistent with the indexing convention of Section 2.6, the rows of A are indexed by X, and the columns of the submatrices I and B of A are indexed by X and Y, respectively. Declare $\mathcal{I}$ to consist of the index sets of the $\mathcal{F}$-independent column submatrices of A. Then $M = (X \cup Y, \mathcal{I})$ is a matroid called the $\mathcal{F}$-matroid represented by B over $\mathcal{F}$. The matrix B is an $\mathcal{F}$-representation matrix of M.

Of particular interest is the case of $\mathcal{F} = \mathrm{GF}(3)$, where M is the GF(3)-matroid represented by B over GF(3).

The same construction applies when B is over the system BG. In that case, M is the BG-matroid represented by B.

We have completed the review and now motivate in an informal discussion the definition of IB-independence systems yet to come.

Let B be a matrix over IB with row index set X and column index set Y. By Corollary (4.3.57), B is satisfiable if and only if some column submatrix B' of B with independent columns spans $\underline{1}$. Hence, if B is satisfiable, we can demonstrate this by exhibiting such a B'. But how would we exhibit unsatisfiability of B?

An easy way out is to append an identity matrix I to B and to declare the resulting matrix $A = [I|B]$ to be over IB. The matrix I spans $\underline{1}$, so A is satisfiable. Suppose we ask for a column submatrix A' of A with independent columns that spans $\underline{1}$ and that avoids as many columns of I as possible. Since I is satisfiable, such an A' must exist. Furthermore, B is satisfiable if and only if A' does not contain any column of I. Thus, A' may be used to demonstrate satisfiability or unsatisfiability of B, whichever applies.

When we originally followed this simple line of reasoning, we were reminded of the construction of the representable matroids reviewed above, and we wondered whether some matroid concepts or results are relevant for the satisfiability problem. As we explored that question, it became evident that matroids are indeed useful for the solution of that problem. In the remainder of this chapter, we establish that connection with matroids and prove basic results.

We are ready to define the IB-independence system for B. We let $X \cup Y$ be the groundset E and let $\mathcal{I}$ contain the subsets of $X \cup Y$ that index column submatrices of $A = [I|B]$ over IB with IB-independent columns. By Theorem (4.3.22), any subset of a set of IB-independent columns is also IB-independent. Thus, $\mathcal{I}$ is maintained under subset taking. We conclude that $(X \cup Y, \mathcal{I})$ satisfies the axioms of (4.4.1) and thus is an independence system. We call it the IB-*independence system represented* by B.

One might hope that the IB-independence system $(X \cup Y, \mathcal{I})$ of B is a matroid. But this is not necessarily so. A counterexample is given by

(4.4.3)

$$B = \; X \begin{array}{c} \\ \end{array} \begin{array}{|ccccccc|} \multicolumn{6}{c}{\overset{\longleftarrow \; Y \; \longrightarrow}{a \; b \; c \; d \; e \; f}} \\ \hline 1\text{-}1\; 1\; 0\; 1\; 0 \\ \text{-}1\; 1\; 0\; 1\; 0\; 1 \\ 1\; 1\; 0\; 0\; 1\; 1 \\ \hline \end{array}$$

Matrix B over IB

which essentially is the matrix of (4.3.23). As shown following (4.3.23), the column submatrices indexed by $\{a, b\}$ and $\{c, d, e, f\}$ are two column bases. Accordingly, these submatrices are maximal IB-independent column submatrices of the submatrix B of $A = [I|B]$. In the terminology of independence systems, the sets $\{a, b\}$ and $\{c, d, e, f\}$ are maximal independent subsets of the set $Y = \{a, b, c, d, e, f\}$ of the IB-independence system $(X \cup Y, \mathcal{I})$ of B. If $(X \cup Y, \mathcal{I})$ is a matroid, then by (4.4.2) the sets $\{a, b\}$ and $\{c, d, e, f\}$ must have the same cardinality, which clearly is not the case.

Despite the nonmatroidal nature of IB-independence systems, we venture to adapt matroid concepts to these systems and attempt to build algorithms for the satisfiability problem for B based on the insight so gained. Also, if computations for IB-independence systems become too complex or cumbersome, we approximate these systems by matroids for which the corresponding computations are manageable. Of particular use are GF(3)-matroids and BG-matroids.

For the approximations, we establish some basic inequalities linking the system IB, the field GF(3), and the system BG.

We need an auxiliary result for connecting IB with GF(3).

(4.4.4) Lemma. *Let E be a matrix over IB. Assume that for each column vector of E, the matrix E also contains the negative of that column. Define A to be any matrix over IB derived from E by first adding duplicate or zero columns or rows and then taking a submatrix. Then*

(4.4.5)
$$|\text{range}(A)| \leq |\text{range}(E)|$$
$$|\text{subrange}(A)| \leq \max\{|\text{subrange}(E')|\}$$

where the maximum in the second inequality is taken over all column sub-matrices E' of E.

Proof. For any matrix A, $|\text{range}(A)|$ and $|\text{subrange}(A)|$ cannot decrease when rows are added to A. Thus, we may assume that the row index set of A contains the row index set of E. The addition of duplicate or zero rows does not change $|\text{range}(A)|$ or $|\text{subrange}(A)|$, so we may suppose that the row index set of A is equal to the row index set of E.

For any matrix A, $\text{range}(A)$ and $\text{subrange}(A)$ do not change if one adds a column vector e for which both e and $-e$ are already present. Thus, we may suppose that A is a column submatrix E' of E.

These observations prove the inequality about $\text{subrange}(A)$. Since the addition of columns at most enlarges $\text{range}(A)$, the inequality for $\text{range}(A)$ follows as well. □

(4.4.6) Theorem. *Let A be a $\{0, \pm 1\}$ matrix, to be viewed over $\mathbb{B}$ or GF(3) as appropriate. Then*

$$(4.4.7) \qquad |\text{range}(A)| \leq \begin{cases} 4 & \text{if GF(3)-rank}(A) = 1 \\ 69 & \text{if GF(3)-rank}(A) = 2 \end{cases}$$

$$(4.4.8) \qquad |\text{subrange}(A)| \leq \begin{cases} 3 & \text{if GF(3)-rank}(A) = 1 \\ 15 & \text{if GF(3)-rank}(A) = 2 \end{cases}$$

Proof. If GF(3)-rank$(A) = 1$, then up to column scaling all nonzero columns are identical, and clearly $|\text{range}(A)| \leq 4$ and $|\text{subrange}(A)| \leq 3$.

For the case of GF(3)-rank$(A) = 2$, let C be the matrix

$$(4.4.9) \qquad C = \begin{array}{c c} & \begin{array}{cccc} a & b & c & d \end{array} \\ \begin{array}{c} e \\ f \\ g \\ h \end{array} & \left| \begin{array}{rrrr} 1 & 0 & 1 & 1 \\ 0 & 1 & 1 & -1 \\ 1 & 1 & -1 & 0 \\ 1 & -1 & 0 & -1 \end{array} \right| \end{array}$$

Matrix C

Due to the 2×2 identity matrix in the top left corner of C, the first two columns (resp. rows) of C are GF(3)-independent. The following claims about the remaining columns and rows are easily checked. Column c (resp. d) of C is the sum (resp. difference) of columns a and b. Row g (resp. h) of C is the sum (resp. difference) of rows e and f. Thus, GF(3)-rank$(C) = 2$.

In two steps, we construct a larger matrix E with GF(3)-rank$(E) = 2$ from C. First, a matrix D is obtained by appending to C the negative of each row of C; that is, $D = [C/(-C)]$. Second, the matrix E is obtained from D by appending to D the negative of each column of D; that is, $E = [D|(-D)]$.

It is easy to see that any matrix A over GF(3) with GF(3)-rank$(A) = 2$ may be derived from E by repeatedly adding duplicate or zero rows or columns and then taking a submatrix. By Lemma (4.4.4), $|\text{range}(A)| \leq |\text{range}(E)|$ and $|\text{subrange}(A)| \leq \max\{|\text{subrange}(E')|\}$, where the maximum is taken over all column submatrices E' of E.

Direct but tedious calculations verify that $|\text{range}(E)| = 69$ and that $\max\{|\text{subrange}(E')|\} = 15$. These equations plus the cited inequalities of Lemma (4.4.4) prove (4.4.7) and (4.4.8). $\qquad \square$

We link IB with BG. Recall from Section 2.6 that a subregion is obtained from a given matrix by first taking a submatrix and then replacing some nonzero entries in that submatrix by zeros. Let A be a matrix over IB. A *subregion cover* of A is a finite collection of subregions of A, say, A^1, $A^2, \ldots,$ A^k, having the same size as A and observing the following condition. For each nonzero entry A_{ij} of A, there is at least one matrix A^l containing that entry. Any such matrix A^l *covers* the entry A_{ij}.

The following lemma relates the range and subrange of a matrix to the range and subrange of the matrices of a subregion cover.

(4.4.10) Lemma. *Let A be a matrix over IB. Define A^1, $A^2, \ldots,$ A^k to be the matrices of a subregion cover of A. Then*

(4.4.11)
$$|\text{range}(A)| \leq \prod_{l=1}^{k} |\text{range}(A^l)|$$
$$|\text{subrange}(A)| \leq \prod_{l=1}^{k} |\text{subrange}(A^l)|$$

Proof. We establish the inequality for $|\text{range}(A)|$. Let b be any vector of range(A). Thus, for some $\{0, \pm 1\}$ vector s, $b = A \odot s$. For $l = 1, 2, \ldots, k$, the vector $b^l = A^l \odot s$ is in range(A^l).

Since each nonzero entry of A is covered by some A^l, one readily verifies that $b = \bigoplus_{l=1}^{k}(A^l \odot s) = \bigoplus_{l=1}^{k} b^l$. Accordingly, each $b \in \text{range}(A)$ may be constructed by selecting for each l some $b^l \in \text{range}(A^l)$ and adding up the vectors so chosen. There are $\prod_{l=1}^{k} |\text{range}(A^l)|$ different ways of selecting vectors b^l, so $|\text{range}(A)|$ is bounded from above by $\prod_{l=1}^{k} |\text{range}(A^l)|$.

The inequality for $|\text{subrange}(A)|$ is handled by the above arguments once we consider the vector b to be in subrange(A), the vector s to be a $\{\pm 1\}$ vector, and each vector b^l to be in subrange(A^l). $\qquad \square$

(4.4.12) Theorem. *Let A be a $\{0, \pm 1\}$ matrix, viewed to be over IB or*

BG *as appropriate. Then*

(4.4.13)
$$|\text{range}(A)| \le 3^{\text{BG-rank}(A)}$$
$$|\text{subrange}(A)| \le 2^{\text{BG-rank}(A)}$$

Proof. By Theorem (2.6.14), BG-rank(A) is equal to the minimum number of rows and columns that must be deleted to reduce A to a zero matrix. Hence, A has a subregion cover with two matrices A^1 and A^2 where A^1 has p nonzero columns, A^2 has q nonzero rows, and $p + q = \text{BG-rank}(A)$. Clearly, IB-rank($A^1$) $\le p$ and IB-rank(A^2) $\le q$.

For $l = 1$, 2, we have according to Corollary (4.3.40) $|\text{range}(A^l)| \le 3^{\text{IB-rank}(A^l)}$ and $|\text{subrange}(A^l)| \le 2^{\text{IB-rank}(A^l)}$. Using IB-rank($A^1$) $\le p$ and IB-rank(A^2) $\le q$, we conclude that $|\text{range}(A^1)| \le 3^p$, $|\text{range}(A^2)| \le 3^q$, $|\text{subrange}(A^1)| \le 2^p$, and $|\text{subrange}(A^2)| \le 2^q$.

According to Lemma (4.4.10), the range and subrange of A and its submatrices A^1 and A^2 are linked by the inequalities $|\text{range}(A)| \le |\text{range}(A^1)| \cdot |\text{range}(A^2)|$ and $|\text{subrange}(A)| \le |\text{subrange}(A^1)| \cdot |\text{subrange}(A^2)|$. We combine these inequalities with the ones above to get $|\text{range}(A)| \le 3^{p+q} = 3^{\text{BG-rank}(A)}$ and $|\text{subrange}(A)| \le 2^{p+q} = 2^{\text{BG-rank}(A)}$.   □

The similarity between the inequalities $|\text{range}(A)| \le 3^{\text{IB-rank}(A)}$ and $|\text{subrange}(A)| \le 2^{\text{IB-rank}(A)}$ of Corollary (4.3.40) on one hand and the inequalities $|\text{range}(A)| \le 3^{\text{BG-rank}(A)}$ and $|\text{subrange}(A)| \le 2^{\text{BG-rank}(A)}$ on the other hand might induce one to conjecture that IB-rank and BG-rank, or IB-rank and GF(3)-rank, are related by some simple inequality. For example, one might conjecture that IB-rank(A) $\le$ BG-rank(A) for all matrices A.

We list four matrices A below that disprove all such conjectures involving IB-rank and either GF(3)-rank or BG-rank. The first matrix A is

(4.4.14)
$$A = \begin{bmatrix} 1 & 1 \\ -1 & -1 \end{bmatrix}$$

Matrix A, example 1

and has IB-rank(A) = 2 > GF(3)-rank(A) = 1. The second matrix A is

(4.4.15)
$$A = \begin{bmatrix} 1 & -1 & 1 \\ -1 & 1 & 0 \\ 1 & 1 & 0 \end{bmatrix}$$

Matrix A, example 2

and has IB-rank(A) = 2 < GF(3)-rank(A) = 3. We construct the third matrix A using the following matrix C.

(4.4.16)

$$C = \begin{bmatrix} 0 & 1 & 1 & 1 \\ 1 & 0 & 1 & 1 \\ 1 & 1 & 0 & 1 \\ 1 & 1 & 1 & 0 \\ -1 & 0 & 0 & 0 \end{bmatrix}$$

Matrix C

Direct checking establishes that both the 5×4 matrix C and its transpose, C^t, have IB-independent columns and rows. Also, BG-rank$(C) =$ BG-rank$(C^t) = 4$. Then the 9×9 matrix A

(4.4.17)

$$A = \begin{array}{|c|c|} \hline C & 0 \\ \hline 0 & C^t \\ \hline \end{array}$$

Matrix A, example 3

has IB-independent columns and rows. Accordingly, IB-rank$(A) = 9 >$ BG-rank$(A) = 8$. The fourth matrix A is

(4.4.18)

$$A = \begin{bmatrix} 1 & 1 \\ 1 & 1 \end{bmatrix}$$

Matrix A, example 4

and has IB-rank$(A) = 1 <$ BG-rank$(A) = 2$.

Complexity of Algorithm RANGE (4.3.11)

We establish complexity formulas for Algorithm RANGE (4.3.11) using the dimensions of the given matrix, as well as its BG-rank and IB-rank.

(4.4.19) Theorem. Let A be an $m \times n$ matrix over IB. Define $k =$ BG-rank(A) and $l =$ IB-rank(A). The computational effort of Algorithm RANGE (4.3.11) for finding range(A) of A is then $O(2^{\alpha} \cdot m \cdot n)$, where $\alpha = \min\{m, 1.6k, 1.6l\}$. The effort for finding subrange(A) is $O(2^{\beta} \cdot m \cdot n)$, where $\beta = \min\{k, 1.6l\}$.

Proof. We assume the situation where just the input matrix A is given. Thus, in the notation of Algorithm RANGE (4.3.11), we take the second input matrix $[B/D]$ to be trivial and take the input set J to be equal to the column index set of A or to be empty, depending on whether range(A) or subrange(A) is to be found. Below, we repeatedly use the fact that $2^{1.6}$ is a bit larger than 3.

Algorithm RANGE (4.3.11) processes the matrix A column by column, each time producing the range or subrange of a larger column submatrix of A. Let α' (resp. β') be real numbers such that $2^{\alpha'}$ (resp. $2^{\beta'}$) is equal to the cardinality of the largest range (resp. subrange) set on hand during these iterations. It is easily checked that, in the range (resp. subrange) case, the effort for processing one column is then $O(2^{\alpha'} \cdot m)$ (resp. $O(2^{\beta'} \cdot m)$), and that the effort for processing all columns of A is $O(2^{\alpha'} \cdot m \cdot n)$ (resp. $O(2^{\beta'} \cdot m \cdot n)$). Hence, it suffices to show that α and β of the theorem are upper bounds for α' and β', respectively.

Since the range or subrange sets on hand during any one iteration consist of $\{0,1\}$ vectors with m entries, any such set cannot contain more than 2^m vectors. Accordingly, we know $\alpha' \leq m$ and $\beta' \leq m$.

By Theorem (4.4.12), $|\text{range}(A)| \leq 3^{\text{BG-rank}(A)}$ and $|\text{subrange}(A)| \leq 2^{\text{BG-rank}(A)}$. Using $k = \text{BG-rank}(A) \leq m$ and the fact that the BG-rank of a matrix cannot decrease as columns are added, we deduce that, for any column submatrix B of A, we have $|\text{range}(B)| \leq 3^k < 2^{1.6k}$ and $|\text{subrange}(B)| \leq 2^k$. Hence, $\alpha' \leq \min\{m, 1.6k\}$ and $\beta' \leq \min\{m, k\} = k$.

Finally, we make use of $l = \text{IB-rank}(A)$. Lemma (4.3.42) states that the IB-rank of a matrix may drop as one adds columns. Hence, we cannot argue as in the BG-rank case. However, range(A) does contain as subsets the range and subrange of any column submatrix of A. Using that fact and Corollary (4.3.40), which supplies $|\text{range}(A)| \leq 3^{\text{IB-rank}(A)} < 2^{1.6l}$, we improve the bounds obtained so far for α' and β' to $\alpha' \leq \min\{m, 1.6k, 1.6l\} = \alpha$ and $\beta' \leq \min\{k, 1.6l\} = \beta$, as desired. $\qquad\square$

One might conjecture that the term $1.6l$ in the exponent β of the complexity formula of Theorem (4.4.19) for subrange(A) is just an artifact of the proof, and that that term can be reduced to l. It can be shown that this is not possible, using arbitrarily large block diagonal matrices where each block is a copy of the matrix of (4.3.23). We omit the straightforward arguments and only mention that one needs to arrange that, for any such block diagonal matrix, Algorithm RANGE (4.3.11) first processes all columns containing the block vectors labeled c, d, e, and f in (4.3.23).

Extension of System IB

As defined in Section 2.6 and again at the end of Section 4.3, the system IB may be extended by enlarging the set of elements $\{0, \pm1\}$ by

$$(4.4.20) \qquad U = \{(\alpha, \beta) \mid \alpha, \beta \in \{0, 1, 2\}\}$$

and by defining IB-multiplication of $(\alpha, \beta) \in U$ and $\gamma \in \{0, \pm1\}$ by

$$(4.4.21) \quad (\alpha, \beta) \odot \gamma = \begin{cases} 1 & \text{if } \alpha \geq 1 \text{ and } \gamma = 1, \text{ or } \beta \geq 1 \text{ and } \gamma = -1 \\ 0 & \text{otherwise} \end{cases}$$

A matrix over the extension of $\mathbb{B}$ has its entries in $\{0, \pm 1\} \cup U$. When such a matrix is declared to be over BG, then any 0 or $(0,0)$ entry is considered to be zero and any other entry is considered to be nonzero.

It is readily seen that the definition of $\mathbb{B}$-independence system introduced earlier in this section also accommodates the extension of $\mathbb{B}$ and that all results except Theorem (4.4.6), which makes sense only for $\{0, \pm 1\}$ matrices, remain valid.

In the next three sections, we adapt the matroid concepts and methods of Chapter 3 to $\mathbb{B}$-independence systems.

4.5 Connectivity

We derive from the notion of matroid separations and connectivity the concept of $\mathbb{B}$-separation and $\mathbb{B}$-connectivity for $\mathbb{B}$-independence systems.

We review relevant material of Section 3.4. In that section, matroid separations and connectivity are defined for general matroids using the matroid rank function. In Lemma (3.4.3), these definitions are shown for representable matroids to be equivalent to certain conditions on representation matrices. To achieve a short review, we restate here just the conditions of Lemma (3.4.3) as if they were the defining conditions for matroid separations of representable matroids. This is valid due to the equivalency established in that lemma.

Let M be a matroid represented by a matrix B over a field $\mathcal{F}$ or over the system BG. Let B have row index set X and column index set Y. Suppose B is partitioned as follows.

(4.5.1)

$$
B = \begin{array}{c|c|c|}
 & Y_1 & Y_2 \\
\hline
X_1 & A^1 & D^2 \\
\hline
X_2 & D^1 & A^2 \\
\hline
\end{array}
$$

Partitioned version of B

Suppose the index sets X_1, X_2, Y_1, Y_2 and the submatrices D^1, D^2 of B satisfy, for some $k \geq 1$,

(4.5.2)
$$|X_1 \cup Y_1|, |X_2 \cup Y_2| \geq k$$

as well as

(4.5.3)
$$\mathcal{F}\text{-rank}(D^1) + \mathcal{F}\text{-rank}(D^2) \leq k - 1$$

for the case of the field $\mathcal{F}$, and

(4.5.4) $\text{BG-rank}(D^1) + \text{BG-rank}(D^2) \leq k - 1$

for the case of the system BG. Then $(X_1 \cup Y_1, X_2 \cup Y_2)$ is a k-separation of M. The k-separation is exact if the applicable inequality of (4.5.3) or (4.5.4) holds with equality. For $k \geq 2$, M is k-connected if it does not have an l-separation for some $1 \leq l \leq k - 1$. If M is 2-connected, then it is also said to be connected.

The above terminology for matroid k-separations and k-connectivity is applied to their representation matrices as follows. If a matrix B represents a matroid M and if M has a k-separation $(X_1 \cup Y_1, X_2 \cup Y_2)$ (resp. has an exact k-separation, is k-connected, or is connected), then we declare B to also have a k-separation $(X_1 \cup Y_1, X_2 \cup Y_2)$ (resp. to have an exact k-separation, to be k-connected, or to be connected). The matrix B may at one time be over a field $\mathcal{F}$ and at another time be over BG. To differentiate among the possible k-separations, we may say, for example, that B has an $\mathcal{F}$-k-separation or has a BG-k-separation. Terms such as $\mathcal{F}$-k-connected and BG-k-connected are to be analogously interpreted. Since $\mathcal{F}$-rank$(D^1)+$ $\mathcal{F}$-rank$(D^2) = 0$ or BG-rank$(D^1) + $BG-rank$(D^2) = 0$ if and only if both D^1 and D^2 are zero matrices, the specification of $\mathcal{F}$ or BG is not needed for the terms 1-separation, 2-connected, or connected.

We translate the above concepts to matrices over IB in the expected way. Let B be such a matrix, with row index set X and column index set Y. Assume B is partitioned as follows.

(4.5.5)

$$B = \begin{array}{c|c|c} & Y_1 & Y_2 \\ \hline X_1 & A^1 & D^2 \\ \hline X_2 & D^1 & A^2 \end{array}$$

Partitioned version of B

Suppose the index sets X_1, X_2, Y_1, Y_2 and the submatrices D^1, D^2 of B satisfy, for some $k \geq 1$,

(4.5.6) $|X_1 \cup Y_1|, |X_2 \cup Y_2| \geq k$

and

(4.5.7) $\text{IB-rank}(D^1) + \text{IB-rank}(D^2) \leq k - 1$

Then $(X_1 \cup Y_1, X_2 \cup Y_2)$ is a k-separation of the IB-independence system $(X \cup Y, \mathcal{I})$ represented by B, and of the matrix B itself. The k-separation is *exact* if the inequality of (4.5.7) holds with equality. For $k \geq 2$, $(X \cup Y, \mathcal{I})$ and B are k-*connected* if they do not have an l-separation for some $1 \leq l \leq k-1$. If $(X \cup Y, \mathcal{I})$ or B is 2-connected, then it is also said to be *connected*. Since $\text{IB-rank}(D^1) + \text{IB-rank}(D^2) = 0$ if and only if both D^1 and D^2 are zero matrices, the specification of IB is not needed for the terms 1-*separation*, 2-*connected*, or *connected*. In fact, the usage of these terms for matrices over IB fully agrees with that for matrices over $\mathcal{F}$ or BG.

At times, we view a matrix B to be over IB, over GF(3), or over BG. To differentiate among the possible cases, we then call a k-separation of B over IB a *IB-k-separation*. Terms like *IB-k-connected* are to be analogously interpreted.

Since the IB-rank function is not easy to compute, we desire alternate, more easily checked sufficient conditions that guarantee that a given partition of a matrix B over IB corresponds to a IB-separation. The next lemma provides such conditions.

(4.5.8) Lemma. *Let B be a matrix over IB that is partitioned as in (4.5.5). Suppose that the submatrices D^1 and D^2 of B satisfy the following conditions.*

(4.5.9)

    (i)   *The columns of D^1 or the rows of D^2 are IB-dependent.*

    (ii)   *The rows of D^1 or the columns of D^2 are IB-dependent.*

Then for

$$(4.5.10) \qquad k = \min\{|X_1 \cup Y_1|, |X_2 \cup Y_2|\}$$

$(X_1 \cup Y_1, X_2 \cup Y_2)$ *is a IB-k-separation of B.*

Proof. For $i = 1, 2$, let D^i be of size $p_i \times q_i$. Then (4.5.9) and (4.5.10) imply that $k \leq |X_1 \cup Y_1| = p_2 + q_1 > \text{IB-rank}(D^1) + \text{IB-rank}(D^2)$, $k \leq |X_2 \cup Y_2| = p_1 + q_2 > \text{IB-rank}(D^1) + \text{IB-rank}(D^2)$, and $k = |X_1 \cup Y_1|$ or $|X_2 \cup Y_2|$. Thus, $|X_1 \cup Y_1|, |X_2 \cup Y_2| \geq k$ and $\text{IB-rank}(D^1) + \text{IB-rank}(D^2) < k$, and both (4.5.6) and (4.5.7) are satisfied. $\square$

IB-separations are the central ingredients for the decompositions of later chapters. We include an overview in Section 4.8. At this point, we only mention that the range and certain structural properties of the submatrices D^1 and D^2 of B of (4.5.5) are important for these decompositions. In particular, upper bounds on the product of the cardinalities of the range of D^1 and D^2; that is, upper bounds on $|\text{range}(D^1)| \cdot |\text{range}(D^2)|$, are needed. The next theorem supplies such bounds in terms of the IB-rank, BG-rank, and GF(3)-rank of D^1 and D^2.

(4.5.11) Theorem. *Let B be a matrix over IB that is partitioned as follows.*

(4.5.12)

$$B = \begin{array}{c|c|c} & Y_1 & Y_2 \\ \hline X_1 & A^1 & D^2 \\ \hline X_2 & D^1 & A^2 \end{array}$$

Partitioned version of B

(a) *If $(X_1 \cup Y_1, X_2 \cup Y_2)$ is a IB-k-separation or a BG-k-separation, then*

(4.5.13)
$$|\text{range}(D^1)| \cdot |\text{range}(D^2)| \le 3^{k-1}$$
$$|\text{subrange}(D^1)| \cdot |\text{subrange}(D^2)| \le 2^{k-1}$$

(b) *If $(X_1 \cup Y_1, X_2 \cup Y_2)$ is a GF(3)-k-separation with $1 \le k \le 3$, then*

(4.5.14)
$$|\text{range}(D^1)| \cdot |\text{range}(D^2)| \le \begin{cases} 1 & \text{if } k = 1 \\ 4 & \text{if } k = 2 \\ 69 & \text{if } k = 3 \end{cases}$$

and

(4.5.15)
$$|\text{subrange}(D^1)| \cdot |\text{subrange}(D^2)| \le \begin{cases} 1 & \text{if } k = 1 \\ 3 & \text{if } k = 2 \\ 15 & \text{if } k = 3 \end{cases}$$

Proof. By (4.3.41), we have, for $i = 1$ and 2, $|\text{range}(D^i)| \le 3^{\text{IB-rank}(B)}$ and $|\text{subrange}(D^i)| \le 2^{\text{IB-rank}(B)}$. By condition (4.5.7) of IB-k-separations, IB-rank(D^1) + IB-rank$(D^2) \le k - 1$. Hence, $|\text{range}(D^1)| \cdot |\text{range}(D^2)| \le 3^{\text{IB-rank}(D^1)+\text{IB-rank}(D^2)} \le 3^{k-1}$ as well as $|\text{subrange}(D^1)| \cdot |\text{subrange}(D^2)| \le 2^{\text{IB-rank}(D^1)+\text{IB-rank}(D^2)} \le 2^{k-1}$. Thus, (4.5.13) holds for B over IB.

The case of (4.5.13) for B over BG is handled analogously. Specifically, (4.4.13) supplies $|\text{range}(A)| \le 3^{\text{BG-rank}(A)}$ and $|\text{subrange}(A)| \le 2^{\text{BG-rank}(A)}$, and once more we conclude $|\text{range}(D^1)| \cdot |\text{range}(D^2)| \le 3^{k-1}$ and $|\text{subrange}(D^1)| \cdot |\text{subrange}(D^2)| \le 2^{k-1}$.

We turn to the claim about B over GF(3). By (4.4.7) and (4.4.8), for any matrix A over GF(3),

(4.5.16)
$$|\text{range}(A)| \le \begin{cases} 4 & \text{if GF(3)-rank}(A) = 1 \\ 69 & \text{if GF(3)-rank}(A) = 2 \end{cases}$$

(4.5.17)
$$|\text{subrange}(A)| \le \begin{cases} 3 & \text{if GF(3)-rank}(A) = 1 \\ 15 & \text{if GF(3)-rank}(A) = 2 \end{cases}$$

Since $(X_1 \cup Y_1, X_2 \cup Y_2)$ is a GF(3)-k-separation for $1 \le k \le 3$, we know by (4.5.3) that GF(3)-rank(D^1) + GF(3)-rank$(D^2) \le k - 1$. Below, we argue each one of the cases $k = 1$, 2, and 3. It suffices that we consider exact GF(3)-k-separations, where GF(3)-rank(D^1) + GF(3)-rank$(D^2) = k - 1$.

If $k = 1$, then both D^1 and D^2 must be zero matrices, and both their range and subrange sets contain just the zero vector. Hence, |range$(D^1)| \cdot$ |range$(D^2)|$ = |subrange$(D^1)| \cdot$ |subrange$(D^2)|$ = 1.

If $k = 2$, then one of D^1 and D^2 must have GF(3)-rank equal to 1, while the other one has GF(3)-rank equal to 0. Using (4.5.16) and (4.5.17), we verify the inequalities |range$(D^1)| \cdot$ |range$(D^2)| \le 4$ and |subrange$(D^1)| \cdot$ |subrange$(D^2)| \le 3$.

If $k = 3$, then one of D^1 and D^2 must have GF(3)-rank equal to 2 and the other one has GF(3)-rank equal to 0, or both D^1 and D^2 have GF(3)-rank equal to 1. Once more using (4.5.16) and (4.5.17), we see that |range$(D^1)| \cdot$ |range$(D^2)| \le 69$ and |subrange$(D^1)| \cdot$ |subrange$(D^2)| \le 15$. □

Extension of System IB

As defined in Section 2.6 and repeated in Sections 4.3 and 4.4, the system IB may be extended by enlarging the set of elements $\{0, \pm1\}$ by

$$(4.5.18) \qquad U = \{(\alpha, \beta) \mid \alpha, \beta \in \{0, 1, 2\}\}$$

and by defining IB-multiplication of $(\alpha, \beta) \in U$ and $\gamma \in \{0, \pm1\}$ by

$$(4.5.19) \quad (\alpha, \beta) \odot \gamma = \begin{cases} 1 & \text{if } \alpha \ge 1 \text{ and } \gamma = 1, \text{ or } \beta \ge 1 \text{ and } \gamma = -1 \\ 0 & \text{otherwise} \end{cases}$$

A matrix over the extension of IB has its entries in $\{0, \pm1\} \cup U$. When such a matrix is declared to be over BG, then any 0 or (0,0) entry is considered to be zero and any other entry is considered to be nonzero.

The results of this section apply to the extension of IB except for the inequalities of (4.5.14) and (4.5.15), which make sense only for $\{0, \pm1\}$ matrices. The proof of this claim follows directly from the fact that the results of earlier sections invoked in this section, except those concerning GF(3), also hold for the extension of IB.

4.6 Finding Separations

Chapter 3 includes several algorithms that find certain separations for matrices over GF(3) or BG. In this section, we add to this arsenal an algorithm that analogously to Algorithm INDUCED-$\mathcal{F}$-SEPARATION (3.5.14) finds so-called induced IB-separations.

Induced Separation

We review relevant material of Section 3.5. For current purposes, it suffices that we consider matrices over $\mathcal{F}$ and ignore matrices over BG. Thus, we are given a matrix B over $\mathcal{F}$, with row index set X and column index set Y. Let $\overline{B}$ be a submatrix of B of the following form.

(4.6.1)

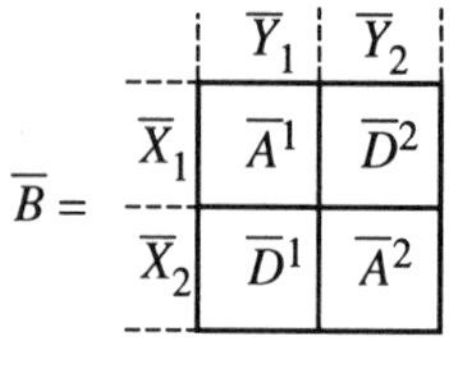

$$\overline{B} = \begin{array}{c|c|c} & \overline{Y}_1 & \overline{Y}_2 \\ \hline \overline{X}_1 & \overline{A}^1 & \overline{D}^2 \\ \hline \overline{X}_2 & \overline{D}^1 & \overline{A}^2 \end{array}$$

Submatrix $\overline{B}$ of B

Assume that, for some $l \geq k$,

(4.6.2)
$$|\overline{X}_1 \cup \overline{Y}_1|, |\overline{X}_2 \cup \overline{Y}_2| \geq l$$

and that

(4.6.3)
$$\mathcal{F}\text{-rank}(\overline{D}^1) + \mathcal{F}\text{-rank}(\overline{D}^2) = k - 1$$

Hence, $(\overline{X}_1 \cup \overline{Y}_1, \overline{X}_2 \cup \overline{Y}_2)$ is an exact k-separation of $\overline{B}$ where each side has at least l elements.

If B has a k-separation $(X_1 \cup Y_1, X_2 \cup Y_2)$ where, for $i = 1, 2$, $X_i \supseteq \overline{X}_i$ and $Y_i \supseteq \overline{Y}_i$, then the k-separation $(\overline{X}_1 \cup \overline{Y}_1, \overline{X}_2 \cup \overline{Y}_2)$ of $\overline{B}$ is said to induce the k-separation $(X_1 \cup Y_1, X_2 \cup Y_2)$ of B.

Define $X_3 = X - (\overline{X}_1 \cup \overline{X}_2)$ and $Y_3 = Y - (\overline{Y}_1 \cup \overline{Y}_2)$. We depict B with the submatrix $\overline{B}$ and the index sets $\overline{X}_1$, $\overline{X}_2$, X_3 and $\overline{Y}_1$, $\overline{Y}_2$, Y_3 below.

(4.6.4)

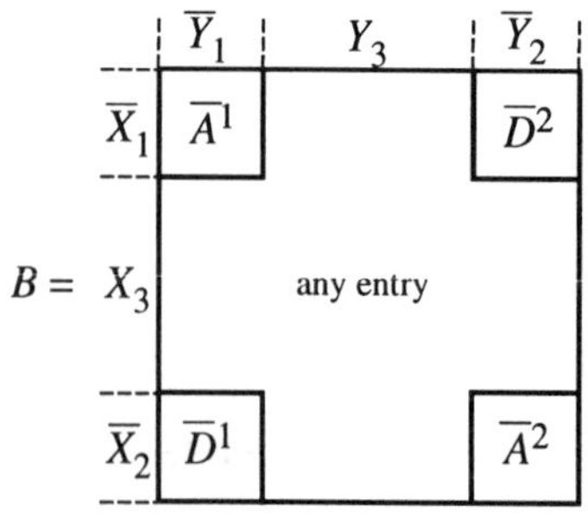

$$B = \begin{array}{c|c|c|c} & \overline{Y}_1 & Y_3 & \overline{Y}_2 \\ \hline \overline{X}_1 & \overline{A}^1 & & \overline{D}^2 \\ \hline X_3 & & \text{any entry} & \\ \hline \overline{X}_2 & \overline{D}^1 & & \overline{A}^2 \end{array}$$

Matrix B with submatrix $\overline{B}$

By definition, an induced k-separation exists if and only if X_3 and Y_3 can be partitioned into X_{31}, X_{32} and Y_{31}, Y_{32}, respectively, such that

$(\overline{X}_1 \cup \overline{Y}_1 \cup X_{31} \cup Y_{31}, \overline{X}_2 \cup \overline{Y}_2 \cup X_{32} \cup Y_{32})$ is a k-separation of B. We display B with that k-separation below.

(4.6.5)

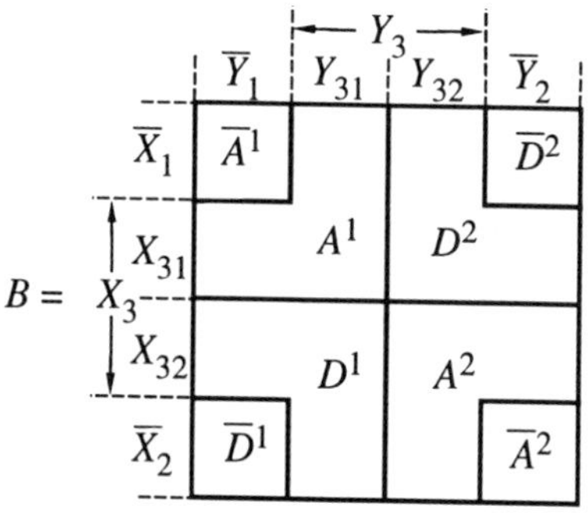

$$B = \text{(matrix diagram)}$$

Partition of B induced by that of $\overline{B}$

As argued in Section 3.5, an induced k-separation exists if and only if

$$(4.6.6) \qquad \mathcal{F}\text{-rank}(D^i) = \mathcal{F}\text{-rank}(\overline{D}^i), \quad i = 1, \ 2$$

Define a submatrix $\overline{A}$ of a matrix A over $\mathcal{F}$ to $\mathcal{F}$-*span* A if $\overline{A}$ and A have the same $\mathcal{F}$-rank. The condition (4.6.6) can then be restated as

$$(4.6.7) \qquad \overline{D}^i \ \mathcal{F}\text{-spans } D^i, \quad i = 1, \ 2$$

We are ready to adapt the induced separation concept to matrices over $\mathbb{B}$. Let $\overline{B}$ be a matrix over $\mathbb{B}$ that is partitioned as in (4.6.1). Assume that the partition of $\overline{B}$ corresponds to a $\mathbb{B}$-separation; that is, for some $l \geq k$

$$(4.6.8) \qquad |\overline{X}_1 \cup \overline{Y}_1|, |\overline{X}_2 \cup \overline{Y}_2| \geq l$$

and

$$(4.6.9) \qquad \mathbb{B}\text{-rank}(\overline{D}^1) + \mathbb{B}\text{-rank}(\overline{D}^2) \leq k - 1$$

Let B over $\mathbb{B}$ have $\overline{B}$ as a submatrix and be the matrix of (4.6.4). Then the $\mathbb{B}$-separation $(\overline{X}_1 \cup \overline{Y}_1, \overline{X}_2 \cup \overline{Y}_2)$ of $\overline{B}$ *induces* a $\mathbb{B}$-separation $(X_1 \cup Y_1, X_2 \cup Y_2)$ if, for $i = 1, 2$, we have $X_i \supseteq \overline{X}_i$ and $Y_i \supseteq \overline{Y}_i$ and if for the corresponding partition of B as given by (4.6.5) we have

$$(4.6.10) \qquad \overline{D}^i \ \mathbb{B}\text{-spans } D^i, \quad i = 1, \ 2$$

or, equivalently,

$$(4.6.11) \qquad |\text{range}(\overline{D}^i)| = |\text{range}(D^i)|, \quad i = 1, \ 2$$

The reader may wonder why we have not used the condition IB-rank(D^i) = IB-rank($\overline{D}^i$) analogously to (4.6.6), and why we have not enforced equality in (4.6.9) analogously to (4.6.3). The changes are mainly due to Lemma (4.3.42), according to which the IB-rank of a matrix may drop as columns are added. That fact rules out a direct translation of (4.6.6) where $\mathcal{F}$-rank is replaced by IB-rank. In view of (4.6.7), which expresses the condition for an induced $\mathcal{F}$-separation in terms of $\mathcal{F}$-span, the switch to IB-span seems appropriate. Due to that change, it suffices that $\overline{B}$ has just a IB-k-separation instead of an exact IB-k-separation. Indeed, for the discussion of induced IB-separations, the particular value of k is irrelevant, and we may omit its specification and simply say that the partition of $\overline{B}$ given by (4.6.1) depicts a IB-separation $(\overline{X}_1 \cup \overline{Y}_1, \overline{X}_2 \cup \overline{Y}_2)$.

We derive an algorithm for finding induced IB-separations from Algorithm INDUCED $\mathcal{F}$-SEPARATION (3.5.17) by almost trivial changes. For a review of the latter method, we redraw B of (4.6.4) so that an arbitrary row $x \in X_3$ and an arbitrary column $y \in Y_3$ are displayed.

(4.6.12)

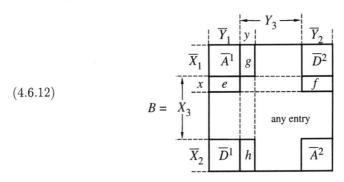

Matrix B with row $x \in X_3$ and column $y \in Y_3$

The recursive steps of Algorithm INDUCED $\mathcal{F}$-SEPARATION (3.5.17) are based on the following three observations.

First, suppose that the subvector e of row x is not $\mathcal{F}$-spanned by the rows of $\overline{D}^1$. If the subvector f of row x is $\mathcal{F}$-spanned by the rows of $\overline{D}^2$, we adjoin e to $\overline{A}^1$ and f to $\overline{D}^2$ and invoke recursion; otherwise, we declare that an induced $\mathcal{F}$-separation does not exist and stop.

Second, suppose that the subvector g of column y is not $\mathcal{F}$-spanned by the columns of $\overline{D}^2$. If the subvector h of column y is $\mathcal{F}$-spanned by the columns of $\overline{D}^1$, we adjoin g to $\overline{A}^1$ and h to $\overline{D}^1$ and invoke recursion; otherwise, we declare that an induced $\mathcal{F}$-separation does not exist and stop.

Third, suppose that, for all $x \in X_3$ and all $y \in Y_3$, neither of the above two cases applies. Then $(\overline{X}_1 \cup \overline{Y}_1, \overline{X}_2 \cup X_3 \cup \overline{Y}_2 \cup Y_3)$ is an induced $\mathcal{F}$-separation of B, and we stop with that conclusion.

By Lemma (4.3.43) and Theorem (4.3.47), we may test whether a submatrix IB-spans a matrix containing it by iteratively adding one column or row at a time, each time testing with Algorithm SPAN (4.3.44) whether that additional column or row leaves the cardinality of range unchanged. This nice behavior of IB-span parallels that of $\mathcal{F}$-span and thus permits a straightforward adaptation of Algorithm INDUCED $\mathcal{F}$-SEPARATION (3.5.14) to matrices over IB as follows.

(4.6.13) Algorithm INDUCED IB-SEPARATION. *Finds a IB-separation for a matrix B over IB that is induced by a IB-separation of a submatrix $\overline{B}$, or declares that such an induced separation does not exist.*

Input: Matrix B over IB, with row index set X and column index set Y. A submatrix $\overline{B}$ of B with a IB-separation $(\overline{X}_1 \cup \overline{Y}_1, \overline{X}_2 \cup \overline{Y}_2)$ where, for $i = 1$, 2, $\overline{X}_i \subseteq X$ and $\overline{Y}_i \subseteq Y$. The IB-separation of $\overline{B}$ has at least l elements on each side.

Output: Either: A IB-separation $(X_1 \cup Y_1, X_2 \cup Y_2)$ of B induced by the IB-separation $(\overline{X}_1 \cup \overline{Y}_1, \overline{X}_2 \cup \overline{Y}_2)$ of $\overline{B}$; the IB-separation of B has at least l elements on each side. Or: "The given IB-separation of $\overline{B}$ does not induce a IB-separation of B."

Complexity: Polynomial if for the submatrices $\overline{D}^1$ and $\overline{D}^2$ of $\overline{B}$, both $|\text{range}(\overline{D}^1)|$ and $|\text{range}(\overline{D}^2)|$ are bounded by some constant.

Procedure:

1. Consider B partitioned as in (4.6.12). Assume that B has a row $x \in X_3$ with the indicated row subvectors e and f such that $|\text{range}([e/\overline{D}^1])| > |\text{range}(\overline{D}^1)|$. Then x must be in X_{31}. Suppose, in addition, that $|\text{range}([\overline{D}^2/f])| > |\text{range}(\overline{D}^2)|$. Then x must also be in X_{32}; that is, B cannot be partitioned, and we stop with that declaration. On the other hand, suppose $|\text{range}([\overline{D}^2/f])| = |\text{range}(\overline{D}^2)|$. Since x must be in X_{31}, we adjoin e to $\overline{A}^1$ and f to $\overline{D}^2$. Then we start recursively again with the new $\overline{B}$.

2. Suppose B as shown in (4.6.12) has a column $y \in Y_3$ with the indicated column subvectors g and h such that $|\text{range}([g|\overline{D}^2])| > |\text{range}(\overline{D}^2)|$. Then y must be in Y_{31}. Suppose, in addition, $|\text{range}([\overline{D}^1|h])| > |\text{range}(\overline{D}^1)|$. Then y must also be in Y_{32}; that is, B cannot be partitioned, and we stop with that declaration. On the other hand, suppose $|\text{range}([\overline{D}^1|h])| = |\text{range}(\overline{D}^1)|$. Since y must be in Y_{31}, we adjoin g to $\overline{A}^1$ and h to $\overline{D}^1$. Then we start recursively again with the new $\overline{B}$.

3. Finally, suppose that, for all rows $x \in X_3$, the row subvector e satisfies $|\text{range}([e/\overline{D}^1])| = |\text{range}(\overline{D}^1)|$, and suppose that, for all columns $y \in Y_3$, the column subvector g satisfies $|\text{range}([g|\overline{D}^2])| = |\text{range}(\overline{D}^2)|$. Then $X_1 = \overline{X}_1$, $X_2 = \overline{X}_2 \cup X_3$, $Y_1 = \overline{Y}_1$, and $Y_2 = \overline{Y}_2 \cup Y_3$ are the sets for the desired IB-separation $(X_1 \cup Y_1, X_2 \cup Y_2)$ of B.

Analogously to part (a) of Lemma (3.5.15), we have the following conclusion about the output of Algorithm INDUCED IB-SEPARATION (4.6.13).

(4.6.14) Lemma. *Any IB-separation produced by Algorithm INDUCED IB-SEPARATION (4.6.13) has $X_1 \cup Y_1$ minimal and $X_2 \cup Y_2$ maximal, in the sense that any other IB-separation $(X_1' \cup Y_1', X_2' \cup Y_2')$ of B induced by the IB-separation $(\overline{X}_1 \cup \overline{Y}_1, \overline{X}_2 \cup \overline{Y}_2)$ of $\overline{B}$ observes $X_1 \subseteq X_1'$, $X_2 \supseteq X_2'$, $Y_1 \subseteq Y_1'$, and $Y_2 \supseteq Y_2'$.*

Proof. According to Steps 1 and 2 of Algorithm INDUCED IB-SEPARATION (4.6.13), $\overline{X}_1$ (resp. $\overline{Y}_1$) is enlarged by $x \in X_3$ (resp. $y \in Y_3$) only if there is no induced IB-separation of B with x (resp. y) on the side containing $\overline{X}_2$ (resp. $\overline{Y}_2$). Thus, any x (resp. y) added to $\overline{X}_1$ (resp. $\overline{Y}_1$) must also be in X_1' (resp. Y_1'). This implies the minimality of $X_1 \cup Y_1$ and the maximality of $X_2 \cup Y_2$. □

We discuss an application of Algorithm INDUCED IB-SEPARATION (4.6.13). Recall from Section 3.5 that Algorithm k-SEPARATION (3.5.20) finds k-separations with minimal k for matrices over $\mathcal{F}$ or BG as follows. From a given input matrix B, all minimal submatrices $\overline{B}$ with certain separations are selected. The conditions imposed on the submatrices $\overline{B}$ are given by (3.5.19). Then, for each of these submatrices $\overline{B}$, Algorithm INDUCED $\mathcal{F}$-SEPARATION (3.5.14) or Algorithm INDUCED BG-SEPARATION (3.5.17) is used to check whether the separation of $\overline{B}$ induces a separation of B. The process stops as soon as a separation of B has been found or when all selected submatrices $\overline{B}$ have been tried.

We convert Algorithm k-SEPARATION (3.5.20) to a method for finding IB-separations by invoking Algorithm INDUCED IB-SEPARATION (4.6.13) instead of Algorithm INDUCED $\mathcal{F}$-SEPARATION (3.5.14) or Algorithm INDUCED BG-SEPARATION (3.5.17) and by making some notational adjustments. We present details next.

The conditions (3.5.19) imposed on the submatrices $\overline{B}$ become those of (4.6.15) below.

(4.6.15)
- (i) $(\overline{X}_1 \cup \overline{Y}_1, \overline{X}_2 \cup \overline{Y}_2)$ is an exact IB-k-separation of $\overline{B}$.
- (ii) For $i = 1, 2$, $P_i \subset \overline{X}_i$ and $Q_i \subset \overline{Y}_i$.
- (iii) For $i = 1, 2$, $|\overline{X}_i \cup \overline{Y}_i| \geq |P_i \cup Q_i| + \max\{k, m_i\} + 1$.
- (iv) $\overline{B}$ is minimal with respect to (i)–(iii).

Here is the algorithm derived from Algorithm k-SEPARATION (3.5.20).

(4.6.16) Algorithm IB-k-SEPARATION. *Finds an exact IB-k-separation of a matrix B over IB where the two sides contain specified sets and have at least a certain size, or declares that such a separation does not exist.*

Input: Matrix B over IB, with row index set X and column index set Y. Two disjoint subsets P_1, P_2 (resp. Q_1, Q_2) of X (resp. Y). Integers m_1, m_2, and n. For $k = 1$, the matrix B does not have a separation $(X_1 \cup Y_1, X_2 \cup Y_2)$ satisfying the following conditions.

$$(4.6.17) \quad \begin{array}{ll} \text{(i)} & (X_1 \cup Y_1, X_2 \cup Y_2) \text{ is an exact IB-}k\text{-separation of} \\ & B \text{ with } k \le n. \\ \text{(ii)} & \text{For } i = 1, 2, P_i \subset X_i \text{ and } Q_i \subset Y_i. \\ \text{(iii)} & \text{For } i = 1, 2, |X_i \cup Y_i| \ge |P_i \cup Q_i| + \max\{k, m_i\} + 1. \end{array}$$

Output: Either: An exact IB-k-separation $(X_1 \cup Y_1, X_2 \cup Y_2)$ of B satisfying the conditions of (4.6.17) and, subject to them, with k minimal. Or: "B does not have an exact IB-k-separation $(X_1 \cup Y_1, X_2 \cup Y_2)$ satisfying (4.6.17)."

Complexity: Polynomial if m_1, m_2, and n are bounded by a constant.

Procedure:
1. Initialize $k = 2$.
2. Do for each submatrix $\overline{B}$ of (4.6.1) for which the sets $\overline{X}_1$, $\overline{X}_2$, $\overline{Y}_1$, and $\overline{Y}_2$ satisfy (4.6.15):
   Let B and $\overline{B}$ be the input matrices for Algorithm INDUCED IB-SEPARATION (4.6.13). If the algorithm finds an induced IB-separation $(X_1 \cup Y_1, X_2 \cup Y_2)$, output that separation, and stop.
3. Increase k by 1. If $k \le n$, go to Step 2. Otherwise, declare that B and M do not have a separation of the desired kind, and stop.

Proof of Validity. The arguments are almost identical to those validating Algorithm k-SEPARATION (3.5.20). We omit details except for the discussion of one aspect. Suppose Algorithm INDUCED IB-SEPARATION (4.6.13) finds an induced IB-separation for B in Step 2, using a submatrix $\overline{B}$ with an exact IB-k-separation satisfying (4.6.15). The latter conditions, Lemma (4.3.42), and the arguments validating Algorithm INDUCED IB-SEPARATION (4.6.13) imply that the IB-separation for B so found satisfies (4.6.17) except that it possibly is not exact. In the exceptional situation, we have, for some $l \le k - 1$, an exact IB-l-separation for B that satisfies a modified (4.6.17) where k has been replaced by l. But then an exact IB-l-separation of B would have been found in an earlier iteration through Step 2, a contradiction. $\qquad\Box$

Extension of System IB

The definition of induced IB-separation and the related Algorithm INDUCED IB-SEPARATION (4.6.13) and Algorithm IB-k-SEPARATION (4.6.16) fully apply to the extension of IB where the set $\{0, \pm 1\}$ of matrix elements is enlarged by

$$(4.6.18) \qquad\qquad U = \{(\alpha, \beta) \mid \alpha, \beta \in \{0, 1, 2\}\}$$

and where IB-multiplication is extended so that for $(\alpha, \beta) \in U$ and $\gamma \in \{0, \pm 1\}$,

$$(4.6.19) \quad (\alpha, \beta) \odot \gamma = \begin{cases} 1 & \text{if } \alpha \geq 1 \text{ and } \gamma = 1, \text{ or } \beta \geq 1 \text{ and } \gamma = -1 \\ 0 & \text{otherwise} \end{cases}$$

Recall from Section 2.6 that Boolean minors of $\{0, \pm 1\}$ clause/variable matrices B over IB are generalized clause/variable matrices $\overline{B}$ that one may view to be over the above extension of IB. Any such minor is produced by column scaling, shrinking, and column or row deletion, in that order. Details about these operations are included in Sections 2.5 and 2.6. Suffice it to say here that each column (resp. row) of $\overline{B}$ corresponds to a subset of the columns (resp. rows) of B. Furthermore, the subsets of columns (resp. rows) of B corresponding to any two distinct columns (resp. rows) of $\overline{B}$ are disjoint. This implies that any submatrix of $\overline{B}$—in particular, $\overline{B}$ itself—uniquely corresponds to some submatrix of B.

Suppose that a Boolean minor $\overline{B}$ of a clause/variable matrix B has a IB-separation $(\overline{X}_1 \cup \overline{Y}_1, \overline{X}_2 \cup \overline{Y}_2)$ as displayed by (4.6.1). Let $\tilde{B}$ be the submatrix of B corresponding to $\overline{B}$, and, for $i = 1$, 2, let $\tilde{X}_i$ and $\tilde{Y}_i$ be the index sets of $\tilde{B}$ corresponding to $\overline{X}_i$ and $\overline{Y}_i$ of $\overline{B}$. Consistent with the notation of (4.6.1), define $\tilde{D}^1$ (resp. $\tilde{D}^2$) to be the submatrix of $\tilde{B}$ indexed by $\tilde{X}_2$ and $\tilde{Y}_1$ (resp. $\tilde{X}_1$ and $\tilde{Y}_2$). We say that the given IB-separation $(\overline{X}_1 \cup \overline{Y}_1, \overline{X}_2 \cup \overline{Y}_2)$ of $\overline{B}$ induces a IB-separation $(X_1 \cup Y_1, X_2 \cup Y_2)$ of B if the following two conditions are satisfied.

First, $(\tilde{X}_1 \cup \tilde{Y}_1, \tilde{X}_2 \cup \tilde{Y}_2)$ must be a IB-separation of $\tilde{B}$ for which

$$(4.6.20) \qquad |\text{range}(\overline{D}^i)| = |\text{range}(\tilde{D}^i)|, \quad i = 1, 2$$

Second, the IB-separation $(\tilde{X}_1 \cup \tilde{Y}_1, \tilde{X}_2 \cup \tilde{Y}_2)$ of $\tilde{B}$ must induce the IB-separation $(X_1 \cup Y_1, X_2 \cup Y_2)$ of B. Since the latter condition implies

$$(4.6.21) \qquad |\text{range}(\tilde{D}^i)| = |\text{range}(D^i)|, \quad i = 1, 2$$

we thus have

$$(4.6.22) \qquad |\text{range}(\overline{D}^i)| = |\text{range}(\tilde{D}^i| = |\text{range}(D^i)|, \quad i = 1, 2$$

In the next section, we decompose and compose the matrices over IB in several ways.

4.7 Sums

In Section 3.6, a k-sum decomposition and composition is described for matrices over $GF(3)$ and the matroids represented by them. The k-sum decomposition of a matrix B over GF(3) produces two submatrices B^1 and B^2 according to (3.6.7) and (3.6.8). The inverse composition process is given by (3.6.11)–(3.6.17). In this section, we adapt the main ideas of that k-sum to the matrices over $\mathbb{B}$ and thus obtain several sum decompositions and compositions. We cover these sums in detail in subsequent chapters. Here, we give an overview that should assist the reader to place the results of those chapters in an overall context.

Each sum decomposition requires a particular separation of a given matrix B over $\mathbb{B}$ and, with one exception that we ignore here, results in two matrices B^1 and B^2. The inverse sum composition of B^1 and B^2 creates B again. We say that B^1 and B^2 are the *components* of a *sum decomposition* of B, and that B is obtained by a *sum composition* of B^1 and B^2. For an abbreviated terminology, we simply say that B is a *sum* of B^1 and B^2, meaning both the sum decomposition and sum composition.

The separations involved in the sums are found by the separation algorithms of Chapter 3 and of this chapter, as well as by special methods. Details about the separations and the algorithms locating them are provided in later chapters.

We need some auxiliary definitions for a summarizing description of the separations and sums. According to Lemma (2.6.21), solving the satisfiability problem for a matrix A over $\mathbb{B}$ is equivalent to finding a $\{0, \pm 1\}$ vector s solving the equation $A \odot s = \underline{1}$. We extend the latter problem by considering for an arbitrary $\{0, 1\}$ vector a the inequality $A \odot s \geq a$. If that inequality has a $\{0, \pm 1\}$ solution vector s, we declare the matrix A to be *a-satisfiable*.

Let S be the CNF system producing A. We say that S is *a-satisfiable* if an assignment of *True/False* values exists for the variables of S such that at least the clauses i of S with $a_i = 1$ evaluate to *True*. By these definitions, $\underline{1}$-satisfiability of A or S is the same as satisfiability of A or S.

Analogously to Lemma (2.6.21), we have the following link between a-satisfiability of A and S. We omit the elementary proof.

(4.7.1) Lemma. *The following statements are equivalent for a CNF system S with clause/variable matrix A, and a $\{0, 1\}$ vector a. The matrix A is to be viewed over $\mathbb{B}$ whenever this is appropriate.*

(i) S *is a-satisfiable.*

(ii) A *is a-satisfiable.*

(iii) *One may assign True/False values to the Boolean variables of S such that each clause i with $a_i = 1$ has value True.*

(iv) *There is a $\{\pm 1\}$ vector s of scaling factors such that column scaling of A with these factors produces a matrix where each row i with $a_i = 1$ contains at least one 1.*

(v) *There exists a $\{\pm 1\}$ solution vector for $A \odot s \geq a$.*

(vi) *There exists a $\{0, \pm 1\}$ solution vector for $A \odot s \geq a$.*

The reader may wonder why we do not consider the equation $A \odot s = a$ instead of the inequality $A \odot s \geq a$. It turns out that the inequality $A \odot s \geq a$ is important for the sum decompositions, while the equation $A \odot s = a$ is not. We remark, though, that the equation $A \odot s = a$ comes up when one wants to decide membership of a in any one of the sets range(A, J). Details of that membership test are included in Section 4.10.

In subsequent chapters, it is shown that for each sum decomposition of a matrix B over IB into B^1 and B^2 and for each $\{0,1\}$ vector b, the matrix B is b-satisfiable if and only if, for $i = 1, 2$, there exists a $\{0,1\}$ vector b^i such that a certain column submatrix $\overline{B}^i$ of B^i is b^i-satisfiable. Assume that result. Furthermore, assume that we are given B and b, as well as B^1 and B^2. We want to decide whether B is b-satisfiable. If we knew b^1 and b^2, then we could reduce the b-satisfiability problem for B to the b^1-satisfiability problem for $\overline{B}^1$ and the b^2-satisfiability problem for $\overline{B}^2$. Unfortunately, b^1 or b^2 are not always easily determined. But it turns out that we can always carry out the following alternate process.

First, we determine certain vectors b^1 and solve for these vectors the b^1-satisfiability problem for $\overline{B}^1$.

Second, given the results of those computations, we construct certain vectors b^2 and solve for these vectors the b^2-satisfiability problem for $\overline{B}^2$. At that point, we can decide whether B is b-satisfiable.

Finally, if B is found to be b-satisfiable, we combine the solution of one of the b^2-satisfiability problems for $\overline{B}^2$ in a backtracking step with the solution of one of the b^1-satisfiability problems for $\overline{B}^1$ to a solution for the b-satisfiability problem for B.

We classify each sum B according to worst-case upper bounds on the number of b^1- and b^2-satisfiability problems for $\overline{B}^1$ and $\overline{B}^2$ that may have to be solved by the SAT algorithm we have developed for that sum. If that upper bound is 1 for both $\overline{B}^1$ and $\overline{B}^2$, the sum is said to be of *type* I. If the upper bound is at least 2 for $\overline{B}^1$ and is 1 for $\overline{B}^2$, then the sum is of *type* II. In the remaining case, where both upper bounds are at least 2, the sum is of *type* III.

There are a total of five sums, called 1-sum, monotone sum, closed sum, augmented sum, and linear sum. We sketch them shortly in the indicated order. The respective separations are the 1-separation of Section 3.5 and yet to be defined separations called monotone, closed, augmentable, and linear. Below, we describe the separations and sums using a matrix B with row index X and column index set Y. Each case involves a partition of X (resp. Y) into X_1, X_2 (resp. Y_1, Y_2). We begin with the 1-sum.

1-Sum

If the matrix B has a 1-separation of the form

(4.7.2)

$$B = \begin{array}{c|c|c} & Y_1 & Y_2 \\ \hline X_1 & A^1 & 0 \\ \hline X_2 & 0 & A^2 \\ \hline \end{array}$$

Matrix B with 1-separation

then B is a 1-*sum* of $B^1 = A^1$ and $B^2 = A^2$, denoted by $B = B^1 \boxplus_1 B^2$. Let a given $\{0,1\}$ vector b be partitioned into b^1 and b^2 in agreement with the partition of the rows of B. Evidently, $B \odot s \geq b$ has a solution if and only if, for $i = 1, 2$, $B^i \odot s \geq b^i$. We conclude that the 1-sum is of type I.

Monotone Sum

Suppose that B has a partition of the form

(4.7.3)

$$B = \begin{array}{c|c|c} & Y_1 & Y_2 \\ \hline X_1 & A^1 & 0 \\ \hline X_2 & D & A^2 \\ \hline \end{array}$$

Matrix B with monotone separation

where each row of A^1 has at most one $+1$ and where $D \leq 0$. Then $(X_1 \cup Y_1, X_2 \cup Y_2)$ is a *monotone separation* of B.

We decompose B by declaring B^1 (resp. B^2) to be equal to the submatrix A^1 (resp. $[D \mid A^2]$) of the matrix of (4.7.3). Thus,

(4.7.4)

$$B^1 = \begin{array}{c|c} & Y_1 \\ \hline X_1 & A^1 \\ \hline \end{array} \qquad B^2 = \begin{array}{c|c|c} & Y_1 & Y_2 \\ \hline X_2 & D & A^2 \\ \hline \end{array}$$

Components B^1 and B^2 of monotone sum B

The matrix B is the *monotone sum* of B^1 and B^2, denoted by $B = B^1 \boxplus_m B^2$. Details of the monotone sum, including an explanation for the probably puzzling conditions on A^1 and D, are provided in Chapter 9. There it is also shown that the monotone sum is of type I.

Closed Sum

Let B have a partition of the form

(4.7.5)

$$B = \begin{array}{c|c|c} & Y_1 & Y_2 \\ \hline X_1 & A^1 & 0 \\ \hline X_2 & D & A^2 \end{array}$$

Matrix B with closed separation

where the submatrix D has a property called *Boolean closedness*. The latter property is defined in Chapter 7. We declare $(X_1 \cup Y_1, X_2 \cup Y_2)$ to be a *closed separation* of B.

There are two ways to decompose B. In the first case, we take B^1 (resp. B^2) to be the column (resp. row) submatrix of B indexed by Y_1 (resp. X_2). Thus,

(4.7.6)

$$B^1 = \begin{array}{c|c} & Y_1 \\ \hline X_1 & A^1 \\ \hline X_2 & D \end{array} \qquad B^2 = \begin{array}{c|c|c} & Y_1 & Y_2 \\ \hline X_2 & D & A^2 \end{array}$$

Components B^1 and B^2 of closed sum B, first case

In the second case, the roles of B^1 and B^2 of (4.7.6) are reversed. That is, B^1 (resp. B^2) is the row (resp. column) submatrix of B indexed by X_2 (resp. Y_1). Thus,

(4.7.7)

$$B^1 = \begin{array}{c|c|c} & Y_1 & Y_2 \\ \hline X_2 & D & A^2 \end{array} \qquad B^2 = \begin{array}{c|c} & Y_1 \\ \hline X_1 & A^1 \\ \hline X_2 & D \end{array}$$

Components B^1 and B^2 of closed sum B, second case

In both cases, B is the *closed sum* of B^1 and B^2, denoted by $B = B^1 \boxplus_c B^2$. Details about the closed sum are presented in Chapter 10. There it is proved that the closed sum is of type II.

Augmented Sum

Let B be partitioned according to (4.7.8) below. Assume that the submatrices A^1 and A^2 are nonempty. We then declare $(X_1 \cup Y_1, X_2 \cup Y_2)$ to be an *augmented separation* of B.

(4.7.8)

$$B = \begin{array}{c|c|c} & Y_1 & Y_2 \\ \hline X_1 & A^1 & E \\ \hline X_2 & D & A^2 \end{array}$$

Matrix B with augmented separation

Define D^1 to be the row submatrix of D that contains all nonzero rows of D, say, indexed by $X_{21} \subseteq X_2$. Analogously, define E^1 to contain all nonzero columns of E, say, indexed by $Y_{21} \subseteq Y_2$. We deduce B^1 from B by replacing D by D^1, E by E^1, and A^2 by a zero matrix of suitable dimension.

To obtain B^2, we replace in B the submatrices A^1, D, and E by certain new matrices $\tilde{F}$, $\tilde{D}$, and $\tilde{E}$, respectively. The new $\tilde{D}$ (resp. $\tilde{E}$) has the same number of rows (resp. columns) as D (resp. E), but the number of columns (resp. rows) may be different. Accordingly, the submatrix $\tilde{F}$ may not be of the same size as A^1. Altogether, we have

(4.7.9)

$$B^1 = \begin{array}{c|c|c} & Y_1 & Y_{21} \\ \hline X_1 & A^1 & E^1 \\ \hline X_{21} & D^1 & 0 \end{array} \qquad B^2 = \begin{array}{c|c|c} & \tilde{Y}_1 & Y_2 \\ \hline \tilde{X}_1 & \tilde{F} & \tilde{E} \\ \hline X_2 & \tilde{D} & A^2 \end{array}$$

Components B^1 and B^2 of augmented sum B

Then B is the *augmented sum* of B^1 and B^2, denoted by $B = B^1 \boxplus_a B^2$. The terminology is motivated by the fact that the submatrices $\tilde{F}$, $\tilde{D}$, and $\tilde{E}$ of B^2 need not be submatrices of B. Chapter 11 covers details of the augmented sum, including a proof that it is of type II.

Linear Sum

Linear sums may involve any number of components. We summarize the case with two components. Let B be partitioned as for the augmentable separation. That is,

(4.7.10)

$$B = \begin{array}{c|c|c} & Y_1 & Y_2 \\ \hline X_1 & A^1 & E \\ \hline X_2 & D & A^2 \end{array}$$

Matrix B with linear separation

Assume that both submatrices A^1 and A^2 are nonempty. We then call $(X_1 \cup Y_1, X_2 \cup Y_2)$ a *linear separation* of B. Note that the submatrices D and E may have any form.

We deduce B^1 (resp. B^2) from B by replacing the submatrix A^2 (resp. A^1) by a zero matrix. Thus,

(4.7.11)

$$
B^1 = \begin{array}{c|c|c|}
 & Y_1 & Y_2 \\
\hline
X_1 & A^1 & E \\
\hline
X_2 & D & 0 \\
\hline
\end{array}
\qquad
B^2 = \begin{array}{c|c|c|}
 & Y_1 & Y_2 \\
\hline
X_1 & 0 & E \\
\hline
X_2 & D & A^2 \\
\hline
\end{array}
$$

Components B^1 and B^2 of linear sum B

The matrix B is the *linear sum* of B^1 and B^2, denoted by $B = B^1 \boxplus_l B^2$. Details are covered in Chapter 12. There it is shown that the linear sum with two components is of type III.

In the next section, we summarize how the above sums are used in Chapter 13 to obtain solution algorithms for the problems SAT and MINSAT.

4.8 A Glimpse Ahead

We project how the material developed so far plus the results of Chapters 5–12 are employed in Chapter 13 to construct solution algorithms for SAT and MINSAT. Since the two types of problems are handled by similar methods, we focus here on the SAT case. For the purposes of this section, we define the *satisfiability problems* of a matrix A over IB to be the a-satisfiability problems for all column submatrices $\overline{A}$ of A arising from all possible $\{0, 1\}$ vectors a. We want an algorithm for solving the satisfiability problems of A. Any such algorithm is a *solution algorithm* for A. We construct solution algorithms with an *analysis algorithm* that generally proceeds as follows.

First, we explore whether some fast algorithm can directly decide the satisfiability problems for A. Several such algorithms are presented in Chapter 5. If one of those algorithms is applicable, we are done. Otherwise, we check whether an extension of those algorithms, described in Chapter 8, is capable of solving the satisfiability problems. If the answer is affirmative, then once more we are done. Otherwise, we repeatedly carry out the sum decompositions of Chapters 9–12, which are summarized in Section 4.7 above, until A has been decomposed into sufficiently simple component matrices such that each satisfiability problem of A can be reformulated in terms of certain satisfiability problems of the component matrices. The latter satisfiability problems are solved by the algorithms of Chapter 5 or 8.

Let us look at the analysis algorithm in more detail. Inductively, we assume that for given A over $\mathbb{B}$ we have constructed, for some small $n \geq 1$, matrices B^1, B^2, ..., B^n and have selected certain algorithms. Together, the matrices and algorithms constitute a solution algorithm that can solve any satisfiability problem of A. We sketch that algorithm. The input consists of a column submatrix $\overline{A}$ of A and a $\{0,1\}$ vector a.

Define $\overline{B}^1$ to be a certain column submatrix of B^1, and derive certain vectors b^{1j} from the given vector a. For $i = 1, 2, \ldots, n$, carry out the following steps. In iteration i, first solve for each vector b^{ij} on hand the b^{ij}-satisfiability problem of $\overline{B}^i$; second, if $i < n$, derive from these solutions a column submatrix $\overline{B}^{i+1}$ from B^{i+1} and certain vectors $b^{i+1,j}$ for the next iteration.

Stop if during any iteration i it is detected that $\overline{A}$ is not a-satisfiable. Otherwise, upon solution of the b^{nj}-satisfiability problems for $\overline{B}^n$, backtrack through the problems in the order $i = n, n-1, \ldots, 1$, and assemble a solution for the a-satisfiability problem of $\overline{A}$.

We make three assumptions about the solution algorithm. First, we suppose to have, for $i = 1, 2, \ldots, n$, a worst-case bound α_i on the number of b^{ij} vectors that might ever be produced by the solution algorithm. Second, each bound α_i is assumed not to exceed some given small constant. Third, for $i = 1, 2, \ldots, n$, a rational number β_i is supposed to be known that bounds the time for solving any one b^{ij}-satisfiability problem for any column submatrix $\overline{B}^i$ of B^i. In contrast to the bounds α_i, the β_i are allowed to be large.

The base case of the inductive assumption, where $n = 1$, involves $B^1 = A$ and a single vector $b^{11} = a$. The solution algorithm is based on one of the algorithms of Chapter 5 or 8. Clearly, the three assumptions above are satisfied.

By these definitions and assumptions, the total run time of the solution algorithm for answering any satisfiability problem of A is bounded by $\sum_{i=1}^{n} \alpha_i \cdot \beta_i$. Suppose that time bound is unacceptably large. Since n and the α_i are small, at least one of the β_i must be large. Let l be the smallest index for which this is so. We then look for a sum decomposition of B^l into B^{l1} and B^{l2} that allows us to replace B^l in the sequence B^1, B^2, ..., B^l, ..., B^n by B^{l1} and B^{l2}. Correspondingly, we strive for an extension of the solution algorithm so that the inductive assumptions are again satisfied. Chapter 13 contains the details of the search for such a decomposition and the corresponding adjustment of the solution algorithm. Here, we just mention that an alternate process is sometimes used where we first compose B^l and B^{l+1} and then carry out a decomposition of the resulting matrix. The latter procedure may not make much sense at this point, but it turns out to be useful.

Suppose that the analysis algorithm eventually produces a solution algorithm with a small time bound. Evidently, the latter algorithm constitutes an efficient method for any satisfiability problem of A. It is shown in Chapter 14 that this desirable situation prevails for large classes of matrices A arising from real-world applications. On the other hand, if the analysis algorithm does not produce a small time bound, we still have a potentially useful solution algorithm for the satisfiability problems of A, except that we do not have a tight bound on its performance.

In typical applications—for example, in expert systems—we are interested in solving numerous satisfiability problems for a given matrix A. Thus, it is appropriate that the solution algorithm be efficiently implemented. Chapter 13 discusses this aspect in detail.

In the next section, we introduce the concept of ID-system, which generalizes the system IB and its extension.

4.9 ID-System

The concepts and algorithms of this book can be generalized so that problems other than the logic problems SAT and MINSAT are handled. The generalization builds on a relaxation of the axioms of the system IB that replaces the sets $\{0, \pm1\}$ and $\{0, 1\}$ of IB by more general sets, among them an index set X. At the same time, the operations $\odot$, $\oplus$, and $\ominus$ of the system IB are generalized to families of operations whose members are indexed by the elements x of X and denoted by $\odot_x$, $\oplus_x$, and $\ominus_x$. We call any system observing the new axioms a ID-*system*.

ID-systems may be employed in the following three-step solution process for certain combinatorial problems. Given a problem instance, we first define a particular ID-system by specifying the underlying sets and the operations $\odot_x$, $\oplus_x$, and $\ominus_x$. Second, we formulate the problem instance as a matrix inequality over that ID-system. Third, we solve the matrix inequality, and thus the problem instance, by adapting the solution approach of the preceding section.

In this section, we define the axioms of ID-systems and discuss the three-step solution process. We begin with the axioms.

Axioms

Let P, Q, R, and X be four nonempty sets. We suppose that the sets Q and R contain an element called zero and denoted by 0, and that R is totally ordered. We summarize these conditions below.

(4.9.1)
	(i)	P, Q, R, and X are nonempty.
	(ii)	$0 \in Q \cap R$.
	(iii)	R is totally ordered.

Analogously to the usual ordering of the reals, we call the elements of R that are greater (resp. less) than 0 *positive* (resp. *negative*).

We need families of multiplication, addition, and subtraction operations, with members indexed by the elements of X and denoted by $\odot_x$, $\oplus_x$, and $\ominus_x$, respectively. The domain and range for each $\odot_x$, $\oplus_x$, and $\ominus_x$ are defined as follows.

$$\odot_x : P \times Q \rightarrow R$$
(4.9.2)
$$\oplus_x : R \times R \rightarrow R$$
$$\ominus_x : R \times R \rightarrow R$$

In a moment, we will impose several axioms on these operations. For the time being, it suffices that we demand the $\oplus_x$ operation to be associative and commutative.

We extend the above operations to matrix multiplication, addition, and subtraction analogously to the case of system $\mathbb{B}$. For this, we need a convenient way of specifying that the elements of a given matrix or vector are taken from one of the sets P, Q, or R. We accomplish this by the prefix P, Q, or R. For example, we say *P-matrix* or *P-vector* when P is the set in question.

Here are the definitions of the matrix operations. Let A be a P-matrix of size $m \times n$, with rows indexed by a given $\overline{X} \subseteq X$. Define B to be a Q-matrix of size $n \times p$. If both A and B are nontrivial and nonempty, then the matrix $C = A \odot B$ is defined to be the $m \times p$ R-matrix whose elements C_{xj} are given by $C_{xj} = (A_{x1} \odot_x B_{1j}) \oplus_x (A_{x2} \odot_x B_{2j}) \oplus_x \ldots \oplus_x (A_{xn} \odot_x B_{nj})$, for $x \in \overline{X}$ and $j = 1, 2, \ldots, p$. The definition relies on the fact that $\oplus_x$ is associative and commutative. If at least one of A and B is trivial or empty, then $C = A \odot B$ is defined to be the $m \times p$ zero matrix.

Let A and B be $m \times n$ R-matrices, each having its rows indexed by a given $\overline{X} \subseteq X$. If A is nontrivial and nonempty, then so is B, and $C = A \oplus B$ (resp. $C = A \ominus B$) is defined to be the $m \times n$ R-matrix whose elements C_{xj} are given by $C_{xj} = A_{xj} \oplus_x B_{xj}$ (resp. $C_{xj} = A_{xj} \ominus_x B_{xj}$), for $x \in \overline{X}$ and $j = 1, 2, \ldots, n$. If A is trivial or empty, then B is of the same type, and both $C = A \oplus B$ and $C = A \ominus B$ are defined to be equal to A or, equivalently, B.

We need one additional set Z, which is a set of R-vectors satisfying the following conditions.

(4.9.3)

(i)   Each vector of Z is indexed by some subset $\overline{X} \subseteq X$.

(ii) For all $\overline{X} \subseteq X$, the set Z contains the 0 vector indexed by $\overline{X}$.

(iii) For any P-vector e indexed by any $\overline{X} \subseteq X$, and for any element α of Q, the vector $e \odot \alpha$ is in Z.

(iv) For any two vectors a and b of Z indexed by any $\overline{X} \subseteq X$, both $a \oplus b$ and $a \ominus b$ are in Z.

Let e be a P-vector, and let a, b, and c be vectors of Z. Suppose each vector has its elements indexed by a given $\overline{X} \subseteq X$. We impose the following axioms on $\odot$, $\oplus$, and $\ominus$. The axioms are nothing but the equations and inequalities of (4.2.21)–(4.2.27), which, according to Section 4.2, hold for the system IB.

(4.9.4) $$e \odot 0 = 0$$

(4.9.5) $$a \oplus 0 = a$$

(4.9.6) $$a \oplus b = b \oplus a \quad (\oplus \text{ is commutative})$$

(4.9.7) $$(a \oplus b) \oplus c = a \oplus (b \oplus c) \quad (\oplus \text{ is associative})$$

(4.9.8) $$(a \ominus b) \ominus c = a \ominus (b \oplus c)$$

(4.9.9) $$(a \ominus b) \oplus (b \ominus c) \geq a \ominus c$$

(4.9.10) $$a \oplus b \geq c \text{ if and only if } a \geq c \ominus b$$

For any P-matrix A, say, with row index set $\overline{X} \subseteq X$ and column index set Y, and for any subset $J \subseteq Y$, we define sets $\mathbb{D}$-range(A), $\mathbb{D}$-subrange(A), and $\mathbb{D}$-range(A, J) analogously to the specification of the sets range(A), subrange(A), and range(A, J) by (4.3.1)–(4.3.4). In the definitions below, s is assumed to be a Q-vector.

$$\mathbb{D}\text{-range}(A) = \{b \mid b = A \odot s\}$$
(4.9.11) $$\mathbb{D}\text{-subrange}(A) = \{b \mid b = A \odot s;\ s_j \neq 0,\ \forall\, j\}$$
$$\mathbb{D}\text{-range}(A, J) = \{b \mid b = A \odot s;\ s_j \neq 0,\ \forall\, j \notin J\}$$

By the conditions of (4.9.3) for Z and by the above definition (4.9.11), each one of the sets $\mathbb{D}$-range(A), $\mathbb{D}$-subrange(A), and $\mathbb{D}$-range(A, J) is a subset of Z.

We need one additional axiom. It demands that, for any P-matrix A with rows indexed by any $\overline{X} \subseteq X$,

(4.9.12) $$\mathbb{D}\text{-range}(A) \text{ is finite.}$$

The definitions of (4.9.11) imply that $|\mathbb{D}\text{-range}(A)| \geq |\mathbb{D}\text{-range}(A, J)| \geq |\mathbb{D}\text{-subrange}(A)|$. Hence, the finiteness condition (4.9.12) for $\mathbb{D}$-range(A) implies that $\mathbb{D}$-range(A, J) and $\mathbb{D}$-subrange(A) are finite as well.

We call any system observing the above definitions and axioms (4.9.1)–(4.9.12) a $\mathbb{D}$-*system*.

The system IB is a particular $\mathbb{D}$-system as follows. We take $P = Q = \{0, \pm 1\}$ and $R = \{0, 1\}$, omit the set X, since it is not needed, and declare Z to be the set of all $\{0, 1\}$ vectors. The element $1 \in R$ is defined to be positive. Finally, we specify $\odot$, $\oplus$, and $\ominus$ by (4.2.1)–(4.2.3). As argued in Section 4.2, the operations $\odot$, $\oplus$, and $\ominus$ satisfy (4.2.21)–(4.2.27), which are (4.9.4)–(4.9.10) here. The finiteness condition (4.9.12) on ID-range(A) holds trivially, since R is finite. Thus, modulo a trivial adjustment of the axioms (4.9.3) for Z that accounts for the absence of X, the system $\mathbb{B}$ is indeed a ID-system.

With similar ease one verifies that the extension of $\mathbb{B}$, listed several times in this chapter, most recently under (4.6.18) and (4.6.19), is another ID-system.

Solution Process

A number of combinatorial problems can be formulated and solved using ID-systems with certain sets P, Q, R, X, and Z and with certain operations $\odot_x$, $\oplus_x$, and $\ominus_x$. The general approach is as follows.

For a given problem instance, one defines a ID-system and encodes the instance by a P-matrix A, with rows indexed by some $\overline{X} \subseteq X$, and by an R-vector d. The ID-system and the arrays A and d are so selected that solving the problem instance is equivalent to finding a Q-vector s satisfying the inequality $A \odot s \geq d$. Frequently, a cost is associated with each Q-vector s. In that case, one wants a solution s for $A \odot s \geq d$ with least cost. The SAT and MINSAT problems of logic are examples, since each instance can be expressed in the above way using the ID-system $\mathbb{B}$.

Most of the definitions, ideas, and solution techniques described in this book for SAT and MINSAT and the underlying system $\mathbb{B}$ apply directly or after some minor modification to combinatorial problems that may be formulated via ID-systems. Thus, this book implicitly contains ideas and constructs that are useful for the solution of those combinatorial problems. In fact, the book could have been written using the more general framework of ID-systems instead of that of system $\mathbb{B}$. We did not take that route because it would have resulted in a rather abstract exposition that would have masked the central ideas. Instead, we decided on a direct treatment of SAT and MINSAT using the system $\mathbb{B}$, along with a summarizing discussion about combinatorial problems involving ID-systems. That discussion follows next.

We first adapt Lemma (4.2.14) to ID-systems. According to that lemma, (4.2.9)–(4.2.13) imply (4.2.15)–(4.2.20). Since (4.2.9)–(4.2.13) are identical to the axioms (4.9.6)–(4.9.10) for ID-systems, we have the following result for such systems.

(4.9.13) Lemma. *For a given ID-system, let a, b, c, and d be vectors of the set Z indexed by some $\overline{X} \subseteq X$. Then the following relationships hold.*

(4.9.14) $$(a \ominus b) \ominus c = a \ominus (b \oplus c) = (a \ominus c) \ominus b$$

(4.9.15) $$a \geq (a \oplus b) \ominus b$$

(4.9.16) $$a \leq (a \ominus b) \oplus b$$

(4.9.17) $$a \geq b \text{ implies } a \ominus c \geq b \ominus c$$

(4.9.18) $$a \geq b \text{ implies } a \oplus c \geq b \oplus c$$

(4.9.19) $$a \geq b \text{ and } c \geq d \text{ imply } a \oplus c \geq b \oplus d$$

Proof. For any ID-system, (4.9.6)–(4.9.10) hold, which are the equations and inequalities (4.2.9)–(4.2.13) assumed in Lemma (4.2.14). That lemma concludes (4.2.15)–(4.2.20), which are (4.9.14)–(4.9.19) here.   □

The definitions of (4.9.11) for the sets ID-range(A), ID-subrange(A), and ID-range(A, J) of a ID-system correspond to the definitions of (4.3.1)–(4.3.4) for the sets range(A), subrange(A), and range(A, J) of the system IB. This fact plus the finiteness condition (4.9.12) on ID-range(A) allows a direct translation of the results of Section 4.3 from the system IB to arbitrary ID-systems. That way, we acquire notions of ID-independence, ID-basis, ID-rank, and ID-span, as well as related algorithms, lemmas, and theorems.

The link between ID-systems and matroids, which for the system IB is established and utilized in Sections 4.4–4.6, is more complicated. Few general results can be claimed here, and one must rely on particular features of a given ID-system to prove useful relationships. This is particularly so for results concerning separations and connectivity. The notion of sum also requires specialization to particular ID-systems, except for the linear sum. According to Chapter 12, the latter sum requires just the axioms for ID-systems plus (4.9.14)–(4.9.19) of Lemma (4.9.13) and thus is always applicable.

For the sake of discussion, suppose we have modified or extended the notions of separation, connectivity, and sum so that they apply to a class of ID-systems. Then the results of Section 4.8 can be readily translated to an overall approach for solving the combinatorial problems formulated with those ID-systems. In particular, the concept of algorithm construction, which involves special algorithms, sum decomposition, and related bounds, is fully applicable and leads to efficient algorithms for many problem instances formulated with those ID-systems. In the next two subsections, we present representative problems that can be so handled.

Covering Problem

As first example, we discuss the so-called *covering problem*. An instance is expressed by integer arrays and an inequality using integer arithmetic. The arrays are all nonnegative and consist of an $m \times n$ matrix A, an $m \times 1$ vector d, and an $n \times 1$ vector c. We must find an $n \times 1$ $\{0, 1\}$ vector s that satisfies $A \cdot s \geq d$ and that, subject to that condition, minimizes $\sum_{j=1}^{n} c_j s_j$. To rule out trivial cases of infeasibility, we assume that $A \cdot \underline{1} \geq d$.

We define the following ID-system for a given instance. The sets P and R are the nonnegative integers, and $Q = \{0, 1\}$. The set X is defined to be the index set of the rows of A. We assume that X also indexes the elements of d.

The matrix operations $\odot$, $\oplus$, and $\ominus$ are defined by scalar operations $\odot_x$, $\oplus_x$, and $\ominus_x$ as follows. For $\alpha \in P$ and $\beta \in Q$, multiplication is specified by

$$(4.9.20) \qquad \alpha \odot_x \beta = \min\{\alpha \cdot \beta, d_x\}$$

For $\alpha, \beta \in R$, addition is given by

$$(4.9.21) \qquad \alpha \oplus_x \beta = \min\{\alpha + \beta, d_x\}$$

and subtraction is defined by

$$(4.9.22) \qquad \alpha \ominus_x \beta = \max\{\alpha - \beta, 0\}$$

The set Z is the set of nonnegative integer vectors a indexed by any $\overline{X} \subseteq X$ and satisfying, for all $x \in \overline{X}$, $a_x \leq d_x$. It is readily checked that Z obeys the axioms of (4.9.3), that $\odot$, $\oplus$, and $\ominus$ satisfy the axioms of (4.9.4)–(4.9.10), and that the finiteness condition (4.9.12) holds. Thus, the sets P, Q, R, X, and Z plus the operations $\odot$, $\oplus$, and $\ominus$ define a ID-system. We encode a problem instance by the given P-matrix A, the R-vector d, and the integer vector c. To solve the given instance, we must find a Q-vector s that satisfies $A \odot s \geq d$, and that, subject to that condition, minimizes $\sum_{j=1}^{n} c_j s_j$.

The covering problem where A is restricted to be a $\{0, 1\}$ matrix and where $d = \underline{1}$ is called the *set covering problem*. Each column of A may be viewed as the incidence vector of a subset of X, and the covering problem asks that a least cost collection of such subsets be determined whose union is equal to and hence *covers* X. According to the above discussion, the set covering problem may be expressed as the problem of finding a least cost solution of $A \odot s \geq \underline{1}$. Note that in this case we can reduce both P and R to the set $\{0, 1\}$. Since $P = Q = R = \{0, 1\}$ and $d = \underline{1}$, the definitions of $\odot$, $\oplus$, and $\ominus$ via (4.9.20)–(4.9.22) become a particular case of (4.2.1)–(4.2.3). We conclude that the set covering problem can be handled by the ID-system IB. In fact, it is nothing but a special case of MINSAT.

Packing Problem

Another example is the so-called *packing problem*. As for the covering problem, an instance is given by nonnegative integer arrays A, d, and c of size $m \times n$, $m \times 1$, and $n \times 1$, respectively. But this time, we must find a $\{0,1\}$ vector s that satisfies $A \cdot s \leq d$ and that, subject to that condition, maximizes $\sum_{j=1}^{n} c_j s_j$. To rule out trivial inequalities in $A \cdot s \leq d$, we assume that $A \cdot \underline{1} > d$.

We transform any instance of the packing problem to one of the covering problem by the variable transformation $s = \underline{1} - s'$. Evidently, s' is a $\{0,1\}$ vector. The inequality $A \cdot s \leq d$ becomes $A \cdot (\underline{1} - s') \leq d$, which may be rewritten as $A \cdot s' \geq (A \cdot \underline{1}) - d$. Using $d' = (A \cdot \underline{1}) - d$, we have the inequality $A \cdot s' \geq d'$ of the covering problem. The objective function $\sum_{j=1}^{n} c_j s_j$ becomes $\sum_{j=1}^{n} c_j (1 - s'_j)$. Maximizing the latter function is equivalent to minimizing $\sum_{j=1}^{n} c_j s'_j$. Thus, A, d', and c define an instance of the covering problem, as desired.

The special packing problem where A is a $\{0,1\}$ matrix and where $d = \underline{1}$ is called the *set packing problem*. We interpret each column j of A as the incidence vector of some $\overline{X} \subseteq X$ and assign c_j as its value. The problem demands that we find a disjoint union of these subsets with maximum total value. Note that the transformation of an instance of the set packing problem to one of the covering problem need not result in an instance of the set covering problem. But there is a special situation, treated next, where this is so.

We restrict A of the set packing problem by requiring exactly two 1s in each row. Thus, $A \cdot \underline{1} = 2 \cdot \underline{1}$. According to Section 3.2, the matrix A may be viewed to be the transpose of the node/edge incidence matrix of some undirected graph G. Declare a note subset J of G to be *independent* if no two nodes of J are joined by an edge. The set packing problem for A effectively demands that we find an independent node subset J of G that maximizes $\sum_{j \in J} c_j$. The latter problem is usually called the *independent node set problem*. Since $A \cdot \underline{1} = 2 \cdot \underline{1}$ and $d = \underline{1}$, the above transformation to an instance of the covering problem results in $d' = (A \cdot \underline{1}) - d = 2 \cdot \underline{1} - \underline{1} = \underline{1}$, so we obtain an instance of the set covering problem. We have already observed that the latter problem is a particular case of MINSAT.

According to the above definitions and transformations, the independent node set problem essentially is also a covering problem, a set covering problem, a packing problem, and a set packing problem. The independent node set problem is known to be difficult in general, so the same conclusion applies to each one of the other problems.

We stop the discussion of examples and uses of ID-systems and go on to the final section, where we provide extensions and list references.

4.10   Extensions and References

It would be redundant to once more list the references of Chapter 3 concerning the linear algebra and matroid material of Sections 4.3–4.7. But we do emphasize that the original approach of Whitney (1935) to matroids plays the central role in the formulation of the basic definitions.

Indeed, Whitney begins with matrices over fields, expresses notions such as independence, rank, etc. of linear algebra in terms of the column index sets of such matrices, and finally defines matroids by postulating certain axioms concerning independence, rank, etc. for the index sets.

We mimic Whitney's approach in Sections 4.3–4.4. We begin with the matrices over $\mathbb{B}$, adapt the notions of independence, rank, etc. from linear algebra, and finally define $\mathbb{B}$-independence systems.

In Section 4.3, it is shown that several fundamental problems arising from the independence concept for matrices over $\mathbb{B}$ are difficult in general, but become easy when range(A) is a small set. We show here that this is also the case for the problem of deciding membership in range(A) or, more generally, in range(A, J). By (4.3.3) and (4.3.4), the latter set may be defined as

$$(4.10.1) \qquad \text{range}(A, J) = \{b \mid b = A \odot s; \ s_j \neq 0, \ \forall \ j \notin J\}$$

If range(A) is small, then Algorithm RANGE (4.3.11) efficiently determines range(A, J), and deciding if a given vector b is in range(A, J) is easy.

Now suppose that range(A, J) is large. We show that determining membership of b in range(A, J) is equivalent to solving a certain SAT instance. To answer the membership question, we must decide whether there exists a vector $\{0, \pm 1\}$ s satisfying $A \odot s = b$ and $s_j = \pm 1$, for all $j \notin J$. To solve the latter problem, we first examine each row i of A for which $b_i = 0$. If that row of A has a 1 (resp. -1) in a column $j \notin J$, then $b \in$ range(A, J) implies that there is a solution s for $A \odot s = b$ with $s_j = -1$ (resp. $s_j = 1$); if $j \in J$, the conclusion becomes $s_j = 0$ or -1 (resp. $s_j = 0$ or 1). Thus, we impose these conditions on s. When the conditions resulting from all rows i with $b_i = 0$ are simultaneously imposed on s, we get, for each column j, a restriction of the values for s_j to a set S_j that is empty or equal to $\{1\}$, $\{-1\}$, $\{0, 1\}$, $\{0, -1\}$, or $\{0, \pm 1\}$.

If any S_j is empty, we know that $b \notin$ range(A, J). Assume otherwise. We delete from the matrix equation $A \odot s = b$ all rows i for which $b_i = 0$. The resulting system, say, $\overline{A} \odot s = \overline{b} = \underline{1}$, has a solution s where each s_j is in S_j if and only if $b \in$ range(A, J). Since the equation $\overline{A} \odot s = \underline{1}$ is equivalent to $\overline{A} \odot s \geq \underline{1}$, we may ignore, for each column j with S_j equal to $\{0, 1\}$, $\{0, -1\}$, or $\{0, \pm 1\}$, the possibility $s_j = 0$. Thus, the problem of determining whether $b \in$ range(A, J) is a satisfiability problem for $\overline{A}$ where some variables are fixed to the value 1 or -1. The latter problem is an instance of SAT, which is difficult in general.

The material of Section 4.8 is discussed in detail in Chapter 13. Relevant references are included at that time.

In Chapter 2, it is indicated that the system IB is a generalization of Boolean algebra. Here, we examine the relationship between ID-systems and general algebraic systems. For details about the latter systems, the reader may consult any text on algebra—for example, Cohn (1982), Lang (1984), Jacobson (1985), or MacLane and Birkhoff (1988).

A *monoid* is an algebraic system given by a set S and an associative operation $\Box : S \times S \rightarrow S$. The set S has an identity $\alpha \in S$; that is, for all $\beta \in S$, $\alpha \Box \beta = \beta \Box \alpha = \beta$.

By (4.9.5)–(4.9.7), the addition of ID-systems is associative and commutative and has the identity 0. Thus, the addition defines a commutative monoid.

On the other hand, the multiplication of ID-systems is arbitrary except for the simple condition of (4.9.4) concerning multiplication by 0, and the subtraction must obey the equation and inequalities of (4.9.8)–(4.9.10). One might reasonably say that, in terms of standard concepts of algebra, the axioms for ID-systems are rather rudimentary. Nevertheless, there is an interesting connection to the more structured *abelian groups* of algebra. Such a group is given by a set S and an associative and commutative operation $\Box : S \times S \rightarrow S$. The set S has an identity α and, for each $\beta \in S$, has an inverse element β^{-1}; that is, $\beta \Box \beta^{-1} = \beta^{-1} \Box \beta = \alpha$. We need a few definitions to describe the relationship.

An *integer program*, abbreviated IP, is specified by integer arrays A, c, and d of size $m \times n$, $n \times 1$, and $m \times 1$, respectively. One must find an $n \times 1$ solution vector s^* for the following minimization problem. Below, s.t. stands for "subject to."

$$
\begin{aligned}
\min \quad & c^t \cdot s \\
\text{s. t.} \quad & A \cdot s = d \\
& s \geq 0 \\
& s \text{ integer}
\end{aligned}
$$

(4.10.2)

Without loss of generality, we may assume that the rows of A are linearly independent.

The generally difficult example problems of Section 4.9 can be, and indeed frequently are, formulated as IPs. Thus, IPs are generally difficult as well.

A large body of theory exists for the solution of IPs. The so-called *group-theoretic approach* proposed by Gomory (1965, 1967, 1969) represents a particular way of attack. It is based on a relaxation of (4.10.2) where the nonnegativity condition for the variables corresponding to a certain basis B of A are omitted. We shall not attempt to even summarize the known results for the group-theoretic approach. Details may be found in standard texts on integer programming—for example, in Hu (1969), Salkin (1975), Garfinkel and Nemhauser (1972), Schrijver (1986), or Nemhauser and Wolsey (1988).

Suffice it to say here that the relaxed version of (4.10.2) is formulated as a minimization problem involving the following abelian group. The operation $\square$ of the group is the addition of rational vectors modulo 1; that is, for any rational vectors a and b, the sum $a \square b$ is the fractional part of the rational sum of a and b. The set S of the group consists of the vectors derived from the matrix A by the following process. Partition A as $A = [B|E]$, where B is the basis of A mentioned above. Then the elements of S are the columns of the matrix $B^{-1} \cdot E$ plus all vectors obtainable from these columns by repeated use of the $\square$ operation.

The minimization problem involving the abelian group is called the *group problem*. In general, (4.10.2) is solved by solving a sequence of group problems.

The definition of the abelian group in terms of the columns of the matrix $B^{-1} \cdot E$ is reminiscent of the range definition for ID-systems. Indeed, the basic idea leading to the abelian group on one hand, and to the range set of ID-systems on the other hand, is the same. That is, both systems are used to represent the behavior of a given matrix. However, the goals are quite different. In the group-theoretic approach, the abelian group leads to the group problem, which is an approximation of the given IP instance. In the ID-system case, one moves from the range set via an independence system to matroids, which in turn are used to analyze the structure of the given problem instance. That analysis and additional considerations support the construction of a solution algorithm.

Chapter 5

Special Matrix Classes

5.1 Overview

So far, we have introduced basic concepts, results, and algorithms and are now ready for the main developments. To set the stage, let us review our goals and overall approach.

Assume the following situation. We want to construct an effective expert system for a given problem domain. The axioms of the expert system have been encoded as a CNF system S. Suppose that each query posed to the expert system is equivalent to a theorem-proving problem where some CNF clause is proved or disproved to be a theorem of S. According to Section 2.2, the latter theorem-proving problem is nothing but a SAT instance defined by some subsystem of S. Hence, if S is represented by a matrix A over $\mathbb{B}$, then each theorem-proving problem is a SAT instance involving some submatrix of A.

It seems appropriate that during the construction process of the expert system we analyze A rather carefully, with the aim of identifying a fast solution algorithm for the queries, that is, for the SAT instances of the submatrices of A. The algorithm used for the investigation of A is the *analysis algorithm* introduced in Section 1.5. Any algorithm produced by the analysis algorithm for the solution of the SAT instances is a *solution algorithm* for A.

From a theoretical viewpoint, we want the analysis algorithm to be polynomial for the matrices over $\mathbb{B}$, and we want the solution algorithms to be polynomial for large subclasses of the class of matrices over $\mathbb{B}$. For practical use of the methods, we impose the additional requirement that the coefficients and exponents of the bounding polynomials be reasonable for the analysis algorithm and be small for the solution algorithms.

We subdivide the construction of a solution algorithm into two parts.

In the first part, we look for special matrix classes for which the SAT instances are very easy and for which the recognition problem can be efficiently answered. This chapter and Chapters 6 and 7 are concerned with such matrix classes.

In the second part, we apply efficient decomposition methods that repeatedly break down a given matrix until each component matrix is in one of the classes identified in the first part. Chapters 8–12 contain these methods and their use for solving SAT instances.

Taken together, the recognition algorithms for special matrix classes plus the decomposition algorithms constitute the analysis algorithm, while the solution algorithms for the special matrix classes plus certain methods utilizing the various decompositions are the components of the solution algorithms. Details of the analysis algorithm are given in Chapter 13.

The above discussion applies to MINSAT and not just SAT, once we consider matrix/vector pairs instead of matrices as representing the axioms. The vectors are rational and nonnegative, and they represent the costs encountered when variables are given the value *True*. The cost for assigning *False* is assumed to be 0. The nonnegativity condition imposed on the cost of *True* and the assumed zero cost for *False* do not impose a restriction in view of the following *reduction* of the general logic cost minimization problem to MINSAT. We omit the elementary proof of validity of the reduction.

If a variable has been assigned some cost value α for *True* and a nonzero cost value β for *False*, then we declare the difference $\alpha - \beta$ to be the cost for *True* and declare 0 to be the cost for *False*.

If the cost for assigning *True* to a variable is $\alpha < 0$ and the cost for assigning *False* is $\beta = 0$, then we replace that variable by its complement, replace the cost for *True* by $-\alpha$, and retain $\beta = 0$ as the cost for *False*. The change essentially involves scaling of the corresponding column of the clause/variable matrix and of the cost value α by -1.

According to the above reduction, nonnegativity of costs for the assignment of *True* and zero costs for the assignment of *False* can always be achieved. We have imposed these conditions, since they simplify the characterization of certain easily solved minimization problems as well as the description of related solution algorithms.

We are ready for an overview of the remaining sections of this chapter.

In Section 5.2, we formalize the above discussion and define a class of matrices over $\mathbb{B}$ to be *central for* SAT, abbreviated SAT *central*, if the class is maintained under submatrix taking and if there are polynomial algorithms for recognition as well as for solution of the SAT instances. A similar definition, involving matrix/vector pairs, applies to MINSAT.

In Section 5.3, we establish some properties of centrality and derive elementary reductions for SAT and MINSAT that produce so-called SAT simple and MINSAT simple matrices.

In Sections 5.4–5.7, we establish several classes of matrices that are central for SAT or MINSAT.

We begin in Section 5.4 with the SAT central class of 2SAT matrices, which have at most two nonzero entries in each row. When 2SAT matrices are paired with rational nonnegative cost vectors, then the resulting MINSAT subclass becomes $\mathcal{NP}$-hard. Accordingly, MINSAT centrality is unlikely to hold for any class containing all such pairs.

In Section 5.5, we discuss the SAT central class of nearly negative matrices, which have at most one +1 in each row. When we pair nearly negative matrices with rational nonnegative cost vectors, we obtain a central class for MINSAT.

In Section 5.6, we treat the SAT central class of hidden nearly negative matrices, which become nearly negative upon an appropriate column scaling by $\{\pm 1\}$ factors. A pairing of the hidden nearly negative matrices with rational nonnegative cost vectors that remain nonnegative under one such scaling produces a central class for MINSAT.

In Section 5.7, we discuss the SAT central class of balanced matrices, which do not contain certain circuit-type submatrices. Balanced matrices may be paired with rational nonnegative cost vectors to yield a central class for MINSAT.

In Section 5.8, we compare the above matrix classes and see that the 2SAT matrices and the nearly negative matrices essentially are subsumed by the hidden nearly negative matrices.

The final section, 5.9, includes additional central classes, extensions, and references.

5.2 Centrality

In this section, we make precise the intuitive notion of matrix classes for which the SAT or MINSAT instances can be efficiently solved. We begin with the SAT case.

SAT Centrality

Let C be a class of matrices over $\mathbb{B}$. Then C is *central for* SAT, abbreviated

SAT *central*, if the following conditions are satisfied.

(5.2.1)

(i)   If $A \in C$, then any submatrix of A is also in C.
(ii) There is a polynomial algorithm for solving the SAT instances given by the matrices of C.
(iii) There is a polynomial algorithm for recognizing the matrices of C.

The class C is *semicentral for* SAT, abbreviated SAT *semicentral*, if it observes (5.2.1)(i) and (ii). Thus, semicentrality is like centrality, except that we do not demand a polynomial algorithm for the recognition problem.

Some SAT semicentral classes C of later chapters are produced by certain compositions. In each case, it is assumed that the polynomial solution algorithm for C receives the applicable composition information as part of the input. If C is SAT central, then the polynomial recognition algorithm derives the desired composition information as part of the membership test for C.

As an example, we show that the class C of matrices over $\mathbb{B}$ for which the cardinality of the range is bounded by some constant k is SAT central. Define A to be an arbitrary matrix in C.

According to Section 4.3, the range of any submatrix of A is obtained from range(A) by projection and subset taking. Thus, the range of any submatrix of A has cardinality of at most k, and (5.2.1)(i) holds.

By Lemma (4.3.12), A is satisfiable if and only if $\underline{1} \in$ range(A). Thus, Algorithm RANGE (4.3.11), which determines the range of matrices, may be used to determine satisfiability of A. The algorithm is polynomial, since |range(A)| is bounded by k. If in Algorithm RANGE (4.3.11) we record for each range vector b a solution vector for $A \odot s = b$, then in the case of satisfiability of A we also obtain a satisfying solution. We conclude that C satisfies (5.2.1)(ii).

Upon the following modification, Algorithm RANGE (4.3.11) recognizes the matrices of C in polynomial time and thus establishes (5.2.1)(iii). During each iteration through Step 3 of Algorithm RANGE (4.3.11), it is checked whether the set R of that step contains more than k vectors. If this is so, the algorithm declares that the matrix in question is not in C and stops. If no such termination occurs, then the given matrix has been proved to be in C.

MINSAT Centrality

We turn to the centrality definitions for MINSAT. This time, we let C be a set of matrix/vector pairs (A, c), where A is a matrix over $\mathbb{B}$ and c is a rational nonnegative vector. We assume that the column index set of A is also the index set of the elements of c. We interpret each entry of c as the cost of assigning *True* to the variable associated with the corresponding column of A. The cost for assigning *False* is assumed to be 0.

We define a *submatrix pair* of (A, c) to be any pair obtained from (A, c) by deletion of some columns from A and of the corresponding elements from c, and by deletion of some rows from A. We declare C to be *central for* MINSAT, abbreviated MINSAT *central*, if the following conditions are satisfied.

(5.2.2)

(i)   If $(A, c) \in C$, then any submatrix pair of (A, c) is also in C.

(ii)   There is a polynomial algorithm for solving the MINSAT instances given by the matrix/vector pairs of C.

(iii) There is a polynomial algorithm for recognizing the matrix/vector pairs of C.

The class C is *semicentral for* MINSAT, abbreviated MINSAT *semicentral*, if it observes (5.2.2)(i) and (ii). Hence, semicentrality for MINSAT is like centrality, except that we do not demand a polynomial membership test for C.

Analogously to the SAT case, some MINSAT semicentral classes C of later chapters are produced by certain compositions. In each case, it is assumed that the polynomial solution algorithm for C receives the applicable composition information as part of the input. If C is MINSAT central, then the polynomial recognition algorithm derives the desired composition information as part of the membership test for C.

We include an example of a MINSAT central class, using again the matrices A over $\mathbb{B}$ for which the cardinality of the range is bounded by some constant k. Let C be the class of matrix/vector pairs (A, c) with such matrices and with rational nonnegative vectors c.

We have already proved (5.2.2)(i) and (iii) for C in the context of the SAT centrality example.

A polynomial solution algorithm establishing (5.2.2)(ii) may be deduced from Algorithm RANGE (4.3.11) as follows. Suppose we are at the beginning of an arbitrary iteration through Step 3 of the algorithm. Define B to be the submatrix of A containing the columns of A processed so far. Thus, the set R at hand is equal to range(B). Define d to be the subvector of c corresponding to B.

Inductively, assume that, for each $b \in R$, we have a solution vector $s = s^b$ for the following problem.

(5.2.3)
$$\begin{aligned} \min \quad & d^t \cdot s \\ \text{s. t.} \quad & B \odot s = b \end{aligned}$$

Let $z_b = d^t \cdot s^b$. We say that s^b *produces* b from B with minimal cost z_b. Let a be the column of A processed in the given iteration through

Step 3. According to that step, each vector $b' \in \text{range}([B|a])$ is of the form $b' = b \oplus (a \odot \alpha)$, where $b \in R$ and $\alpha \in \{0, \pm 1\}$. Using s^b and z_b for the possible vectors b, plus the entry in c corresponding to a, we obtain a solution vector that produces b' from $[B|a]$ with minimal cost. By induction, we have at termination of the modified Algorithm RANGE (4.3.11) solution vectors that produce the vectors $b \in \text{range}(A)$ from A with minimal cost. If A is satisfiable, the solution vector for $b = \underline{1}$ solves the MINSAT problem for (A, c). Clearly, the entire procedure is polynomial, so (5.2.3)(ii) holds and C is MINSAT central.

The solution algorithm described above for MINSAT is based on the so-called *principle of optimality* of *dynamic programming*. Suppose a given optimization problem has a certain decomposition into subproblems. Then the principle of optimality roughly says that an optimal solution for the problem can be built up from optimal solutions of the subproblems. For details, the reader should consult the references given in Section 5.9.

5.3 Properties of Centrality

We establish basic properties of centrality and introduce some reductions for SAT and MINSAT as well as an extension of central classes.

Basic Properties

The next two lemmas say that each type of centrality is preserved under appropriate subclass taking and finite union and intersection of classes. We omit the elementary proofs.

(5.3.1) Lemma. *Let C be a class of matrices or matrix/vector pairs. Define $\overline{C}$ to be a subclass of C that is maintained under submatrix taking. Then (a) and (b) below hold.*

(a) *If C is SAT or MINSAT semicentral, then $\overline{C}$ also has that property.*
(b) *Suppose membership in $\overline{C}$ can be tested in polynomial time provided that membership in C is known. If C is SAT or MINSAT central, then $\overline{C}$ also has that property.*

(5.3.2) Lemma. *For given $n \geq 2$, let $C_1, C_2, \ldots, C_n$ be classes of matrices or matrix/vector pairs. Assume that the classes have a given centrality property, that is, SAT or MINSAT centrality or semicentrality. Then the union and the intersection of these classes also have that property.*

Reductions

Recall from Section 2.6 that an array over $\mathbb{B}$ is monotone if all entries are nonnegative or nonpositive, and that two columns over $\mathbb{B}$ are parallel if they are identical up to scaling. Define a matrix A over $\mathbb{B}$ to be SAT *simple* if A has no rows with less than two nonzeros, has no duplicate rows, and has no parallel or monotone columns. The matrix A is MINSAT *simple* if A has no rows with less than two nonzeros, has no duplicate rows, and has no nonpositive columns. Note that A is SAT simple if it is $\mathbb{B}$-simple, as defined in Section 2.6, and has no monotone columns.

The following algorithm deduces from a given matrix A over $\mathbb{B}$ a SAT or MINSAT simple submatrix $\overline{A}$ with attractive properties.

(5.3.3) Algorithm SIMPLE SUBMATRIX. *Derives from a given matrix A over $\mathbb{B}$ a SAT or MINSAT simple submatrix $\overline{A}$ that essentially contains all SAT or MINSAT simple submatrices of A.*

Input: Matrix A over $\mathbb{B}$. Specification whether a SAT or MINSAT simple submatrix $\overline{A}$ is desired.

Output: A SAT or MINSAT simple submatrix $\overline{A}$ of A, as demanded by the input, and the sequence of row/column deletions of Step 2 below that reduce A to $\overline{A}$. The matrix $\overline{A}$ is maximum and unique in the following sense. In the case of a SAT (resp. MINSAT) simple $\overline{A}$, each SAT (resp. MINSAT) simple submatrix of A is, up to column scaling and up to row and column indices (resp. up to row indices), a submatrix of $\overline{A}$.

Complexity: Polynomial.

Procedure:
1. Initialize $\overline{A} = A$.
2. SAT case: Iteratively delete from $\overline{A}$ any row with at most one nonzero entry, any monotone column, and, in the case of two duplicate rows or of two parallel columns, one of the two rows or columns.
   MINSAT case: Iteratively delete from $\overline{A}$ any row with at most one nonzero entry, any nonpositive column, and, in the case of two duplicate rows, one of the two rows.
3. Output the final matrix $\overline{A}$ and the sequence of row/column deletions that were carried out in Step 2, and stop.

Proof of Validity. We confine ourselves to the SAT case, since the arguments for the MINSAT case are similar.

By Step 2, the output matrix $\overline{A}$ is SAT simple. We show by induction on the length of A that every SAT simple submatrix B of A is a submatrix of $\overline{A}$, up to column scaling and up to row and column indices. If A itself is SAT simple—in particular, if A has length 0 and hence is the empty matrix—then we have $\overline{A} = A$, and the conclusion follows trivially. Otherwise, examine the first reduction of A in Step 2. Several cases must be considered.

Assume that the reduction involves a row with at most one nonzero, or a monotone column. Clearly, $\overline{A}$ as well as B does not contain that row or column. We delete that row or column from A and apply induction.

Assume that the reduction removes a column z from A that is parallel to a column y. If B contains columns indexed by y and z, then, regardless of the form of B, those two columns are parallel or are zero vectors. Either case contradicts the fact that B is SAT simple. If B contains a column indexed by z, and hence no column y, then we relabel that column z as y and scale it if necessary so that it becomes a subvector of column y of A. Thus, we may assume that B does not have z as column index, and we may delete column z from A and apply induction.

The final case, where the reduction removes one of two duplicate rows, is argued analogously to that for two parallel columns.   □

Since the SAT (resp. MINSAT) simple matrix $\overline{A}$ produced by Algorithm SIMPLE SUBMATRIX (5.3.3) from a given matrix A contains, up to column scaling and indices, each SAT (resp. MINSAT) simple submatrix of A, we refer to the matrix $\overline{A}$, in a slight abuse of terms, as *the maximum* SAT (resp. MINSAT) *simple submatrix* of A.

One may reduce any SAT or MINSAT instance involving a submatrix of A to an instance involving a submatrix of the maximum SAT or MINSAT simple submatrix $\overline{A}$ of A. Details are given next.

Reduction for SAT

The reduction for the SAT case is accomplished by the following algorithm. The complexity claim of the algorithm refers to the function $count(A)$, which according to Section 2.6 is the number of nonzero entries in a matrix A.

(5.3.4) Algorithm REDUCE SAT INSTANCE. *Reduces the SAT instance of a submatrix of a given matrix A over* $\mathbb{B}$ *to one involving a submatrix of the maximum SAT simple matrix $\overline{A}$ of A.*

Input: Matrix A over $\mathbb{B}$, of size $m \times n$. The maximum SAT simple submatrix $\overline{A}$ of A, plus the sequence of row/column deletions that transform A to $\overline{A}$. A submatrix A' of A for which the SAT problem is to be solved.

Output: Either: A' is not satisfiable. Or: A submatrix A'' of $\overline{A}$ such that A' is satisfiable if and only if A'' is satisfiable. A method for deriving a satisfying solution for A' from one for A'' is given following the algorithm.

Complexity: $O(m + n + count(A))$.

Procedure:
  1. Initialize $A'' = A'$.

2. Process the reductions that transform A to $\overline{A}$ one by one. If the reduction specifies

   - deletion of a zero row x: If row x occurs in A'', then declare A' to be unsatisfiable, and stop.

   - deletion of a unit row vector x: If row x occurs in A'' and is zero, then declare A' to be unsatisfiable, and stop. If row x occurs in A'' and is a unit vector with a $+1$ (resp. -1) entry in a column y, then assign the value *True* (resp. *False*) to column y, and reduce A'' by deleting all rows now satisfied and column y.

   - deletion of a monotone column y: If column y occurs in A'' and is zero, arbitrarily assign *True* or *False* to column y, and update A'' by deleting column y. If column y occurs in A'' and is nonnegative (resp. nonpositive) and nonzero, then assign the value *True* (resp. *False*) to column y, and update A'' by deleting all rows now satisfied and column y.

   - deletion of a column z parallel to a column y: If both columns y and z occur in A'' and are zero, arbitrarily assign *True* or *False* to columns y and z of A'', then update A'' by deleting columns y and z. If both columns y and z occur in A'' and are nonzero, then assign *True/False* values to columns y and z such that all rows of A'' with a nonzero entry in columns y and z become satisfied— for example, if column y is a duplicate of column z, then assign *True* to column y and *False* to column z—and finally reduce A'' by deleting the now satisfied rows as well as columns y and z. If column z occurs in A'' but not column y, then change the label of column z in A'' to y; if in addition $\overline{A}$ has a column y, then scale the column now labeled y in A'' such that the resulting column becomes a subvector of column y of $\overline{A}$.

   - deletion of a row z that is a duplicate of a row x: If both rows x and z occur in A'', then update A'' by deleting row z. If row z occurs in A'' but not row x, then relabel row z of A'' as row x.

Proof of Validity. The arguments validating Algorithm SIMPLE SUB-MATRIX (5.3.3) may be used to show that the final A'' is a submatrix of $\overline{A}$. Hence, we omit a detailed proof of that part.

During each iteration in Step 2, the specified *True/False* values, if any, plus the reduction of A'' correspond to an obvious problem reduction of the SAT instance A'. Hence, A' is satisfiable if and only if this is so for A''. The procedure clearly has the claimed complexity.   □

We sketch a method that converts a satisfying solution for A'' to one for A'. For this, we keep track of the following relationships during execution of Algorithm REDUCE SAT INSTANCE (5.3.4). Let a be a column vector that occurs in the final A'' or that is deleted during some iteration of the algorithm. We need to record to which column of A' the vector a corresponds, and how often the algorithm scales that column with factor -1. Assume this information as well as a satisfying solution for the final A'' is at hand. If the algorithm has scaled the column an even (resp. odd) number of times with factor -1, then we assign to the column of A' corresponding to vector a the *True/False* value that agrees with (resp. is the opposite of) the value given to vector a in Algorithm REDUCE SAT INSTANCE (5.3.4) or in the satisfying solution for A'', whichever case applies.

Reduction for MINSAT

The MINSAT case is handled by a very similar algorithm. We omit the proof of validity, since it is almost identical to that for the SAT case.

(5.3.5) Algorithm REDUCE MINSAT INSTANCE. *Reduces the MINSAT instance of a submatrix pair of a given matrix/vector pair (A, c) to a pair involving a submatrix of the maximum MINSAT simple matrix $\overline{A}$ of A.*

Input: Pair (A, c), where A is an $m \times n$ matrix over $\mathbb{B}$ and where c is a rational nonnegative cost vector with elements indexed by Y. The maximum MINSAT simple submatrix $\overline{A}$ of A, plus the sequence of row/column deletions that transform A to $\overline{A}$. A submatrix pair (A', c') of (A, c) for which the MINSAT problem is to be solved.

Output: Either: A' is not satisfiable. Or: A submatrix A'' of $\overline{A}$ and a vector c'' such that A' is satisfiable if and only if A'' is satisfiable, and such that a MINSAT solution for (A', c') may be readily derived from any MINSAT solution of (A'', c''). The method for deriving such a solution is a simplified version of the one already given for the SAT case.

Complexity: $O(m + n + \text{count}(A))$.

Procedure:

1. Initialize $A'' = A'$ and $c'' = c'$.
2. Process the reductions that transform A to $\overline{A}$ one by one. If the reduction specifies
   - deletion of a zero row x: If row x occurs in A'', then declare A' to be unsatisfiable, and stop.
   - deletion of a nonpositive column y: If column y occurs in A'', then assign *False* to column y, reduce A'' by deleting all rows now satisfied and column y, and delete element y from c''.
   - deletion of a unit row vector x: If row x occurs in A'' and is zero, then declare A' to be unsatisfiable, and stop. If row x occurs in A'' and is a unit vector with a $+1$ (resp. -1) entry in a column y, then assign the value *True* (resp. *False*) to column y, reduce A''

by deleting column y and all rows now satisfied, and reduce c'' by deleting element y.

- deletion of a row z that is a duplicate of a row x: If both rows x and z occur in A'', then update A'' by deleting row z. If row z occurs in A'' but not row x, then relabel row z of A'' as row x. Vector c'' is not changed.

Extension of Central Classes

We use Algorithms SIMPLE SUBMATRIX (5.3.3), REDUCE SAT IN-STANCE (5.3.4), and REDUCE MINSAT INSTANCE (5.3.5) in the proof of the following extension result for central or semicentral classes.

(5.3.6) Theorem. *Let C be a class of matrices that is maintained under submatrix taking, and let C' be a subclass of C. If C' is SAT central (resp. semicentral) and if C consists precisely of the matrices whose maximum SAT simple matrix is in C', then C is SAT central (resp. semicentral) as well. The MINSAT version of the above statements also holds provided that C and C' are classes of matrix/vector pairs.*

Proof. We show validity for the case of SAT centrality of C'. The case of SAT semicentrality of C' and the MINSAT version are handled similarly.

By assumption, C is maintained under submatrix taking, so (5.2.1)(i) holds for C.

To solve the SAT problem for a given matrix A in C, we first identify with Algorithm SIMPLE SUBMATRIX (5.3.3) the maximum SAT simple submatrix $\overline{A}$ of A. By assumption, $\overline{A}$ and its submatrices are in C'. Thus, we may use Algorithm REDUCE SAT INSTANCE (5.3.4) plus the assumed polynomial solution algorithm for C' to solve the SAT problem for A. This proves (5.2.1)(ii) for C.

Finally, let A be a matrix for which membership in C is to be determined. We compute with Algorithm SIMPLE SUBMATRIX (5.3.3) the maximum SAT simple submatrix $\overline{A}$ of A. Using the assumed polynomial membership test for C', we decide whether $\overline{A}$ is in C'. By assumption, this is so if and only if A is in C, so (5.2.1)(iii) holds. $\Box$

(5.3.7) Theorem. *Given a class C' with a certain centrality property, there exists a unique maximum class C that, together with C', satisfies the assumptions of Theorem (5.3.6).*

Proof. There exists a maximal class C that, together with C', satisfies the assumptions of Theorem (5.3.6), since $C = C'$ is a candidate. If there are two maximal classes that satisfy these assumptions, then the union of those two classes does so as well. Hence, there is a unique maximum C. $\Box$

Given C', one typically can describe an appealing construction of the maximum C. As an example, we treat the case of a SAT central class C' of matrices that is maintained under scaling of columns by $\{\pm 1\}$ factors and changes of row/column indices. We construct the matrices A of the maximum C for such C' as follows. For each simple matrix A' of C', we derive, in all possible ways, matrices A, each of which is obtained from A' by addition of any number of vectors. Each such vector is a row with at most one nonzero, a monotone column, a parallel column, or a duplicate row. We leave it to the reader to verify that C is the desired maximum class.

The construction of the maximum C from given C' usually is easy. Hence, in subsequent sections or chapters we shall not discuss that extension when covering central or semicentral classes. An exception is Chapter 14, where we explicitly treat such extensions.

In the remainder of this chapter, we introduce several SAT or MIN-SAT central classes of matrices or matrix/vector pairs as well as related recognition and solution algorithms. We begin in the next section with a class whose matrices have very simple structure. Nevertheless, the matrices may be used to formulate a number of important combinatorial problems.

5.4 2SAT Matrices

Define 2SAT to be the subclass of SAT where each matrix instance has at most two nonzeros in each row. In this section, we show that the class 2SAT is SAT central. We also prove that the MINSAT problem involving 2SAT matrices is generally difficult.

Evidently, the class 2SAT is maintained under submatrix taking, and testing for membership in 2SAT is trivial. Thus, the conditions (5.2.1)(i) and (iii) for SAT centrality are satisfied. The remaining condition (5.2.1)(ii) is established by the following algorithm.

Solution Algorithm for 2SAT

Let A be a given 2SAT matrix for which the SAT problem is to be solved. If A has no rows, then A is satisfiable, and any *True/False* values for the columns constitute a satisfying solution. If A has a zero row, then A is unsatisfiable. If A has a row with just one nonzero, then we carry out the obvious reduction of A.

In the remaining case, each row of A has two nonzero entries. We arbitrarily select a column of A, say, y, and fix its value to *False*. That choice may satisfy any number of rows. We delete those rows and column y. In turn, that reduction may produce rows with just one nonzero, which permit further reductions. We carry out all such reductions and finally terminate with one of the following three situations. (1) The reduced matrix has a zero row and thus is unsatisfiable. We then know that column y must receive the value *True* in any satisfying solution. We repeat the above process with column y fixed to *True* instead of *False*. Then either we detect unsatisfiability again, in which case the original matrix is unsatisfiable, or we terminate in one of the cases described next. (2) The reduced matrix has no rows. We arbitrarily assign *True/False* to the columns. These values plus those assigned earlier constitute a satisfying solution for the original matrix. (3) In the remaining case, the reduced matrix has at least one row, and each row has two nonzero entries. We claim that the reduced matrix is satisfiable if and only if the original matrix is satisfiable. The "only if" part follows from the derivation of the reduced matrix. The "if" part is trivially true, since addition of zero columns to the reduced matrix results in a row submatrix of the original matrix. We conclude that we may apply the above process to the reduced matrix.

A formal description of the algorithm follows.

(5.4.1) Algorithm SOLVE 2SAT. *Solves the* SAT *problem for a given* 2SAT *matrix A over* $\mathbb{B}$.

Input: Matrix A over $\mathbb{B}$, of size $m \times n$, with at most two nonzero entries in each row.

Output: Either: A satisfying solution for A. Or: "A is unsatisfiable."

Complexity: $O(m + n)$ when properly implemented.

Procedure:

1. If A has a zero row: Declare A to be unsatisfiable, and stop.

   If A has no rows: Assign arbitrary *True/False* values to the columns of A. These values plus the earlier assigned *True/False* values, if any, constitute a satisfying solution for the input matrix. Stop with that solution.

   If A has a row with just one nonzero, say, in column z: If that nonzero is $+1$ (resp. -1), assign *True* (resp. *False*) to column z, delete from A all rows now satisfied and column z, and repeat Step 1.

2. (A has two nonzeros in each row; try assignment of *False* to a column y.) Retain A as a matrix A'. Arbitrarily select a column y of A. Assign *False* to column y of A, and delete from A all rows now satisfied and column y.

3. Apply the process of Step 1 to A, with two exceptions: If A becomes a matrix with a zero row, go to Step 4; if A becomes a matrix with two nonzeros in each row, go to Step 2.

4. (Restore A, and try alternate assignment of *True* to column y.) Redefine A to be A'. Assign *True* to column y of A, and delete from A all rows now satisfied and column y. Go to Step 1.

Proof of Validity. Given the previous discussion, we only need to establish the claimed complexity of $O(m+n)$. When we assign the value *False* to a column y in Step 2, we also assign, in a parallel execution, the value *True* as specified in Step 4. We evaluate the consequences of the two choices in parallel until in both cases we conclude unsatisfiability of A or until one case would proceed to Step 2. In the first instance, the input matrix has been proved to be unsatisfiable. In the second situation, we stop the parallel evaluation and go with the applicable matrix to Step 2.   □

We record the SAT centrality of the class of 2SAT matrices for future reference.

(5.4.2) Theorem. *The class of* 2SAT *matrices is* SAT *central.*

Applications of 2SAT

A number of interesting problems may be reduced to 2SAT. We present three cases below and discuss others in subsequent sections.

The first such problem, which we call SELECT SET, is as follows. Given are $n \geq 1$ nonempty and disjoint sets $R_1, R_2, \ldots, R_n$, with two elements each, and two binary relations α and β on $R = \cup_{i=1}^{n} R_i$. One must decide whether there is a subset $S \subseteq R$ such that, for all i, $R_i \cap S$ is nonempty, and such that, for each $r \in R$ and $s \in S$, $r \ \alpha \ s$ implies $r \in S$ and $r \ \beta \ s$ implies $r \notin S$.

We reduce a given instance of SELECT SET to one of 2SAT as follows. We consider each element $r \in R$ to be a Boolean variable. For each i, the set R_i, say, $R_i = \{r_{i1}, r_{i2}\}$, produces the clause $r_{i1} \vee r_{i2}$. Each case of $r \ \alpha \ s$ (resp. $r \ \beta \ s$) produces a clause $s \Rightarrow r = \neg s \vee r$ (resp. $s \Rightarrow \neg r = \neg s \vee \neg r$). We link any satisfying solution for the 2SAT instance with selection of a set S for the SELECT SET instance by considering the assignment of *True* to variable $r \in R$ to be equivalent to placing r into S. The reader should have no difficulty confirming that the 2SAT instance correctly represents the SELECT SET instance.

The second example concerns $n \geq 1$ unordered pairs of sets, say, $(R_{11}, R_{12}), (R_{21}, R_{22}), \ldots, (R_{n1}, R_{n2})$. From each pair, one must select at least one set such that altogether the selected sets are pairwise disjoint. We call this problem DISJOINT SETS.

We reduce an instance of DISJOINT SETS to one of 2SAT. Each R_{ij} is considered to be a Boolean variable. For each i, we require $R_{i1} \vee R_{i2}$. Furthermore, for any two sets R_{ij} and R_{kl} with nonempty intersection, we enforce $\neg R_{ij} \vee \neg R_{kl}$. It is easily established that the assignment of *True* to R_{ij} in the 2SAT instance is equivalent to the selection of the set R_{ij} in the DISJOINT SETS instance.

For the third example, we define a subset of the nodes of an undirected graph G to be *independent* if no edge of G connects any two nodes of the subset. The problem of deciding whether G has, for given $k \geq 1$, at least k independent nodes is called INDEPENDENT SET and is known to be $\mathcal{NP}$-complete in general. We treat a polynomially solvable case. Recall from Section 2.5 that an edge incident at a node is said to cover that node, and that a matching of G is a subset of the edge set such that any node of G is covered by at most one edge of the subset.

Let F be a matching of G with maximum cardinality, and define l to be the number of nodes of G not covered by F. Since at most one of the endpoints of any matching edge can be in any independent set, G has at most $k = |F| + l$ independent nodes. Thus, it makes sense to ask whether G has $k = |F| + l$ independent nodes. We reduce this special instance of INDEPENDENT SET to an instance of 2SAT.

First, we note that the l nodes not covered by any edge of F are independent and must be part of the solution set if it exists. Indeed, if an edge connects two such nodes, then F is not maximal, a contradiction, and if one of the l nodes is not selected, then due to F one cannot possibly obtain a set of $|F| + l$ independent nodes.

Second, for each edge of F, at least one of the two endpoints must be selected if a total of $|F| + l$ independent nodes is to be found.

These observations support the following 2SAT formulation. For each node r of G, we define a Boolean variable r. The value *True* for that variable implies that node r is selected as part of a set of independent nodes.

For each of the l nodes r not covered by an edge of F, we demand $r = True$. For each edge of G, say, connecting nodes r and s, we introduce the clause $\neg r \lor \neg s$. If that edge is in F, we also add the clause $r \lor s$.

It is easy to check that the resulting 2SAT instance is satisfiable if and only if G has at least $|F| + l$ independent nodes.

Resolution and 2SAT

It generally is interesting to check whether the resolution procedure introduced in Section 1.4 maintains a given structural property of SAT instances. For the case at hand, we show that the resolution procedure maintains the property of being a 2SAT matrix. This fact is useful for the comparison of matrix classes in Section 5.8.

We first review the resolution procedure. When applied to a CNF system, the procedure eliminates one variable in each iteration, each time obtaining an equivalent CNF system. The following algorithm carries out one such iteration.

(5.4.3) Algorithm RESOLUTION FOR CNF SYSTEM. *Converts a given CNF system S to an equivalent CNF system S' with one less variable.*

Input: CNF system S with a set Y of variables. A variable $y \in Y$.

Output: A CNF system S' that has $Y' = Y - \{y\}$ as set of variables and that is equivalent to S in the following sense. Each satisfying solution for S becomes upon deletion of the *True/False* value for variable y a satisfying solution for S'. Conversely, for each satisfying solution for S', there exists a *True/False* value for variable y such that a satisfying solution for S results.

Complexity: Polynomial.

Procedure:

1. Delete from S any clause containing both y and $\neg y$. Initialize S' as the CNF system with variable set $Y' = Y - \{y\}$ and without any clauses.
2. Add each clause of S that does not contain y or $\neg y$, to S'.
3. For each clause C_i of S containing y and for each clause C_j of S containing $\neg y$, define $D_i = C_i - \{y\}$ and $D_j = C_j - \{\neg y\}$, and add the clause $D_i \cup D_j$ to S'.

Proof of Validity. Let a satisfying solution for S be given. By definition, that solution satisfies the clauses added to S' in Step 2. It also satisfies the clauses C_i and C_j of S referenced in Step 3 and thus, regardless of the *True/False* value assigned to y, at least one of $D_i = C_i - \{y\}$ and $D_j = C_j - \{\neg y\}$. Accordingly, the solution for S also satisfies the clause $D_i \cup D_j$ added to S' in Step 3.

Conversely, let a solution for S' be given. Since that solution satisfies, in the notation of Step 3, all clauses $D_i \cup D_j$, it also satisfies all D_i or all D_j. If both cases apply, we arbitrarily assign *True* or *False* to y. Otherwise, if all D_i (resp. D_j) are satisfied, we assign *False* (resp. *True*) to y. No matter which case is at hand, we then have a satisfying solution for each $C_i = D_i \cup \{y\}$, for each $C_j = D_j \cup \{\neg y\}$, and for each clause of S that does not contain y or $\neg y$. $\qquad\square$

In Step 3 of Algorithm RESOLUTION FOR CNF SYSTEM (5.4.3), we may omit from S' any redundant clause $D_i \cup D_j$ that, for some variable z, contains both z and $\neg z$. When this is done, the algorithm can be rephrased for matrices of SAT instances as follows.

(5.4.4) Algorithm RESOLUTION FOR MATRIX. *Converts a matrix A over $\mathbb{B}$ to an equivalent matrix A' with one less column.*

Input: Matrix A over $\mathbb{B}$ with a column index set Y. A column index $y \in Y$.

Output: A matrix A' that has $Y' = Y - \{y\}$ as the set of variables, and that is equivalent to A in the following sense. Each satisfying solution for A becomes upon deletion of the *True/False* value for column y a satisfying solution for A'. Conversely, for each satisfying solution for A', there exists a *True/False* value for column y such that a satisfying solution for A results.

Complexity: Polynomial.

Procedure:
1. Initialize A' as the trivial matrix with column index set $Y' = Y - \{y\}$ and without any rows.
2. For each row of A with a zero entry in column y: Remove that zero entry from the row, and adjoin the reduced row to A'.
3. For any two rows i and j for which column y is the unique column having nonzeros with opposite sign in rows i and j: Add a row to A' where, for each $z \in Y'$, the entry in column z is $+1$ (resp. -1, 0) if the rows i and j of A contain at least one $+1$ (resp. at least one -1, two 0s) in column z.

We have the following result for the processing of 2SAT matrices by Algorithm RESOLUTION FOR MATRIX (5.4.4).

(5.4.5) Theorem. *Algorithm* RESOLUTION FOR MATRIX (5.4.4) *reduces each* 2SAT *instance to another* 2SAT *instance.*

Proof. By Steps 2 and 3 of Algorithm RESOLUTION FOR MATRIX (5.4.4), each row of the output matrix A' derived from a given 2SAT matrix A is up to one zero entry a row of A, or is obtained from two rows of A as follows. A $+1$ is deleted from one of the rows, a -1 is deleted from the other row, and the remaining nonzeros of the two rows are combined into one or two nonzeros of the row for A'. Thus, A' is a 2SAT matrix as claimed. □

MIN2SAT Problem

Define MIN2SAT to be the subclass of MINSAT where the matrix A of each matrix/vector pair has at most two nonzero entries in each row. In contrast to the well-solved 2SAT subclass of SAT, the subclass MIN2SAT of MINSAT is $\mathcal{NP}$-hard. The easy proof relies on the following $\mathcal{NP}$-hard problem, which is known as VERTEX COVER. Given is an undirected graph G. Let a *vertex cover* of G be a subset of the vertex set of G such that each edge of G is covered by at least one node of the subset. For given $k \geq 1$, one must decide whether G has a vertex cover of cardinality at most k.

We reduce an instance of VERTEX COVER to one of MIN2SAT as follows. Let A be the transpose of the node/edge incidence matrix of the given graph G. Thus, A has exactly two 1s in each row. View A to be over $\mathbb{B}$, and declare it to be the matrix of a MIN2SAT instance where the cost vector contains only 1s. Evidently, that instance is trivially satisfiable. A least cost satisfying solution where at most k variables receive the value *True* exists if and only if G has a vertex cover of cardinality at most k.

In Section 5.7, we establish a subclass of MIN2SAT that is MINSAT central. For details, see Theorem (5.7.29).

We turn to the next matrix class.

5.5 Nearly Negative Matrices

Define a matrix A over $\mathbb{B}$ to be *nearly negative* if each row of A contains at most one $+1$. In this section, we prove that the class of nearly negative matrices is SAT central and that the corresponding class of matrix/vector pairs is MINSAT central. Clearly, the two classes are maintained under submatrix taking, and testing for membership is elementary. Thus, the conditions (5.2.1)(i) and (iii) for SAT centrality and (5.2.2)(i) and (iii) for MINSAT centrality are satisfied by the respective classes. Below, we provide an efficient algorithm that solves any instance of SAT or MINSAT involving a nearly negative matrix and thus prove the remaining conditions (5.2.1)(ii) and (5.2.2)(ii).

Solution Algorithm for Nearly Negative Matrices

We sketch the algorithm. Let A be the given matrix over $\mathbb{B}$. For the MINSAT case, define c to be the rational nonnegative cost vector.

If A has a zero row, then the given instance of SAT or MINSAT is not satisfiable. If A has a row with just one nonzero entry and if that entry is a $+1$, say, in column y, then in any satisfying solution column y must receive the value *True*. Hence, we assign that value to column y, delete from A all rows now satisfied and column y, and apply the algorithm to the reduced matrix.

Assume that neither of the two situations above applies. Since A is nearly negative, each row of A must have at least one -1. Hence, we may assign *False* to each column of A to obtain a satisfying solution.

We list the algorithm next and then prove that it solves not just the SAT case, but also the MINSAT case.

(5.5.1) Algorithm SOLVE NEARLY NEGATIVE SAT OR MINSAT. *Solves the* SAT *or* MINSAT *problem involving a given nearly negative matrix A over $\mathbb{B}$. In the* MINSAT *case, the cost vector is a given rational nonnegative vector c.*

Input: Matrix A over $\mathbb{B}$, of size $m \times n$. In the MINSAT case, rational nonnegative vector c.

Output: Either: A solution for the SAT instance A or the MINSAT instance (A, c), whichever applies. Or: "A is unsatisfiable."

Complexity: $O(m + n + \text{count}(A))$.

Procedure:
1. If A has a zero row: Declare A to be unsatisfiable, and stop.
   If A has no rows: Assign *False* to each column of A. These values plus the earlier assigned *True/False* values, if any, constitute a solution for the SAT or MINSAT instance, whichever applies.

If A has a row with a $+1$, say, in column y, and without -1s: Assign *True* to column y, delete from A all rows now satisfied and column y, and repeat Step 1.

2. (A has at least one -1 in each row.) Assign *False* to each column of A. These values plus the earlier assigned *True/False* values, if any, constitute a solution for the SAT or MINSAT instance, whichever applies.

The algorithm is clearly valid for the SAT case. For discussion of the MINSAT case, define a satisfying solution for A to be *minimum with respect to True* if that solution has the value *True* for a column of A only if every satisfying solution for A must have *True* for that column. The key observation for the MINSAT case is given in the following result.

(5.5.2) Theorem. *Let A be a nearly negative matrix that is satisfiable. Then A has a satisfying solution that is minimum with respect to True, and this solution is found by Algorithm SOLVE NEARLY NEGATIVE SAT OR MINSAT (5.5.1).*

Proof. The algorithm assigns *True* to a column y of A only if A has a row with a $+1$ in column y and without -1s. Since A is nearly negative, this $+1$ entry is the only nonzero entry of the row in question. Accordingly, the value *True* must be assigned to column y in every satisfying solution. The theorem follows from this fact and induction on the length of A.   □

(5.5.3) Corollary. *Algorithm SOLVE NEARLY NEGATIVE SAT OR MINSAT (5.5.1) is valid for the MINSAT case.*

Proof. Assume that the matrix A of a MINSAT instance (A, c) is satisfiable. By Theorem (5.5.2), Algorithm SOLVE NEARLY NEGATIVE SAT OR MINSAT (5.5.1) finds the satisfying solution for A that is minimum with respect to *True*. Since the cost vector c for the assignment of *True* is nonnegative while the cost for *False* is zero, this satisfying solution for A must have least cost.   □

We record the SAT and MINSAT centrality results just proved.

(5.5.4) Theorem.
(a) *The class of nearly negative matrices is SAT central.*
(b) *The class of matrix/vector pairs (A, c) where A is nearly negative and c is a rational nonnegative vector is MINSAT central.*

Note that Theorem (5.5.4) becomes invalid if the nonnegativity condition on c is dropped. Indeed, the $\mathcal{NP}$-hard problem VERTEX COVER discussed in Section 5.4 can be formulated by pairs (A, c) where A is nonpositive and thus nearly negative as follows. We take A to be the transpose of the negated node/edge incidence matrix of the given graph G and define c to be the vector containing only -1s. That formulation is equivalent to the one given in Section 5.4, where A is the transpose of the node/edge incidence matrix of G and where c contains only $+1$s.

Applications of Nearly Negative Matrices

Nearly negative matrices typically arise from logic problems where each clause can be stated in one of the following two ways, using some variables $x_1, x_2, \ldots, x_n$, and y: "If x_1 and x_2 and ... and x_n, then y" or "If x_1 and x_2 and ... and x_n, then $\neg y$." Indeed, the first statement is equivalent to $\neg x_1 \vee \neg x_2 \ldots \vee \neg x_n \vee y$, and the second one to $\neg x_1 \vee \neg x_2 \ldots \vee \neg x_n \vee \neg y$. In either case, the corresponding row of the clause/variable matrix contains at most one $+1$. Thus, that matrix is nearly negative.

Section 2.3 discusses the following important application of MINSAT. Let S be a CNF system with variables $s_1, s_2, \ldots, s_n$ and $t_1, t_2, \ldots, t_k$. For some disjoint subsets J^+, J^- of $\{1, 2, \ldots, n\}$ and for $l = 1, 2, \ldots, k$, let T_l be the statement $[(\bigwedge_{j \in J^+} s_j) \wedge (\bigwedge_{j \in J^-} \neg s_j)] \Rightarrow t_l$, which is equal to the CNF clause $[(\bigvee_{j \in J^+} \neg s_j) \vee (\bigvee_{j \in J^-} s_j)] \vee t_l$. We want to prove which of the statements $T_1, T_2, \ldots, T_k$ are theorems of S. If one relies on the commonly used method, k SAT instances must be solved. In another approach using MINSAT, one assigns a cost of 1 to $t_1, t_2, \ldots, t_k$ and a cost of 0 to the remaining variables $s_1, s_2, \ldots, s_n$. Finally, one fixes the s_j, $j \in J^+$, to *True* and the s_j, $j \in J^-$, to *False*. If the resulting MINSAT instance is unsatisfiable, then each T_l is a theorem of S. So assume that a MINSAT solution exists that, for some partition L^+, L^- of $\{1, 2, \ldots, k\}$, assigns *True* to t_l, $l \in L^+$, and *False* to t_l, $l \in L^-$. That solution proves that the T_l, $l \in L^-$, are not theorems of S, while the T_l, $l \in L^+$, may be theorems.

If the underlying matrix A of S is nearly negative, then we may assume that the MINSAT solution has been determined by Algorithm SOLVE NEARLY NEGATIVE SAT OR MINSAT (5.5.1). By Theorem (5.5.2), this solution is minimum with respect to *True*. In particular, the columns t_l, $l \in L^+$, which have the value *True* in that solution, must have the value *True* in every satisfying solution of A. Hence, each T_l, $l \in L^+$, has been proved to be a theorem. Evidently, we have reduced the k SAT instances to *one* MINSAT instance that can be solved as efficiently as any one of the SAT instances. This conclusion is substantially stronger than the one drawn in Section 2.3 for the general case, where one MINSAT instance and $|L^+|$ SAT instances are solved to decide which of the T_l are theorems of S.

Resolution and Nearly Negative Matrices

It turns out that the resolution procedure maintains near negativity of matrices. We record and prove this result next.

(5.5.5) Theorem. *Algorithm* RESOLUTION FOR MATRIX (5.4.4) *reduces each nearly negative matrix to another nearly negative matrix.*

Proof. Recall that by Steps 2 and 3 of Algorithm RESOLUTION FOR MATRIX (5.4.4), each row of the output matrix A' derived from a given input matrix A is up to a zero entry a row of A, or is obtained from two rows i and j of A as follows.

Without loss of generality, row i (resp. j) has a $+1$ (resp. -1) in the column y that is to be eliminated, and no other column has nonzeros of opposite sign in rows i and j. Then, for each column $z \neq y$, the resulting row of A' has a $+1$ (resp. $-1, 0$) in column z if the rows i and j of A contain at least one $+1$ (resp. at least one -1, two 0s) in column z. Since A is nearly negative, all entries of row i in columns other than y are nonpositive, while row j contains at most one $+1$. Thus, the row of A' resulting from rows i and j has at most one $+1$.

We conclude that A' is nearly negative.   □

In the next section, we generalize the notion of nearly negative matrices to that of hidden nearly negative matrices.

5.6 Hidden Nearly Negative Matrices

Define a matrix A over $\mathbb{B}$ to be *hidden nearly negative* if the matrix becomes nearly negative upon an appropriate scaling of its columns by $\{\pm 1\}$ factors. In this section, we show that the class C of hidden nearly negative matrices A is SAT central and that a certain class C' of pairs (A, c) with hidden nearly negative matrix A and rational nonnegative vector c is MINSAT central. Specifically, the matrix A of each pair (A, c) must become nearly negative by a column scaling that is restricted to the columns of A for which the corresponding entries of c are zero.

Evidently, the classes C and C' are maintained under submatrix taking, so condition (5.2.1)(i) for SAT centrality and condition (5.2.2)(i) for MINSAT centrality are satisfied. Below, we establish the remaining conditions (ii) and (iii) of (5.2.1) and (5.2.2) by constructing polynomial recognition and solution algorithms.

Recognition Algorithm for Hidden Nearly Negative Matrices

The recognition problem for the classes C and C' may be formulated as the following scaling problem. Given is a partitioned matrix $A = [D|E]$ over $\mathbb{B}$. One must either scale the columns of the submatrix E with $\{\pm 1\}$

factors such that A becomes a nearly negative matrix or conclude that such scaling is not possible. When A can be so scaled, we say that $A = [D|E]$ is *hidden nearly negative relative to* E. When $A = E$, we simply say, in agreement with the earlier definition, that A is *hidden nearly negative*.

We test membership of a matrix A in the class C by defining E to be A itself. To decide membership of a pair (A, c) in the class C', we let E consist of the columns of A that correspond to the zero entries of c.

We first show that the scaling problem can be formulated as an instance of 2SAT. Let $A = [D|E]$ be given.

We declare each column index y of A to be a Boolean variable y of the 2SAT instance. A value of *True* (resp. *False*) for y means that column y of A is to be scaled by the factor $+1$ (resp. -1).

The clauses of the 2SAT instance are as follows. For each column y of D, we demand $y = $ *True* and thus enforce $+1$ as scaling factor. For each row x of A and for each pair of columns y and z for which the entries A_{xy} and A_{xz} are nonzero, we introduce one clause with two literals using the variables y and z. Specifically, if $A_{xy} = +1$ (resp. $A_{xy} = -1$), then the literal arising from variable y is $\neg y$ (resp. y). Correspondingly, if $A_{xz} = +1$ (resp. $A_{xz} = -1$), then the literal arising from variable z is $\neg z$ (resp. z).

We claim that the 2SAT instance correctly represents the scaling problem. The proof is as follows.

Suppose we have scaling factors that transform A to a nearly negative matrix. We claim that the corresponding *True/False* values satisfy all clauses of the 2SAT instance. Take an arbitrary clause, say, defined from nonzero A_{xy} and A_{xz}. Consider the case $A_{xy} = A_{xz} = +1$. According to the above rules, the clause is $\neg y \vee \neg z$. Since the scaling factors achieve a nearly negative matrix, $A_{xy} = A_{xz} = +1$ implies that at least one of the factors for columns y and z is -1. Correspondingly, at least one of the variables y and z of the 2SAT instance has been assigned the value *False*, and thus the clause $\neg y \vee \neg z$ is satisfied. The remaining three cases for A_{xy} and A_{xz} are argued analogously.

Conversely, suppose that the 2SAT instance has a satisfying solution and that we scale A with the corresponding scaling factors. If the resulting matrix A' is not nearly negative, then in some row x of A' there are two $+1$ entries, say, A'_{xy} and A'_{xz}. But then it is easily checked that the assumed solution for the 2SAT instance does not satisfy the clause arising from A_{xy} and A_{xz}, a contradiction.

The 2SAT formulation of the scaling problem may seem attractive, but it nevertheless is an inefficient way to formulate and solve the scaling problem. By the definition of the 2SAT instance, the number of clauses may be quadratic in count(A), the number of nonzeros of A. Thus, just the formulation of the 2SAT instance may involve a computing effort that is quadratic in count(A).

One can significantly improve on that performance by a direct attack on the scaling problem. We describe the main ideas next. As before, let $A = [D|E]$ be given.

If D is not nearly negative, then due to the scaling restriction A cannot be scaled to become nearly negative, and we stop with that conclusion. So suppose that D is nearly negative.

If A has no rows, then A is hidden nearly negative. Formally, we assign arbitrary $\{\pm 1\}$ scaling factors to the columns of E and stop. So suppose that A has at least one row.

If a column of D is nonpositive, then that column does not influence the scaling of E, and we delete it from A and D.

If a row x of D is nonpositive and row x of E has at most one nonzero entry or if a row x of D contains exactly one $+1$ and row x of E is zero, then row x of A does not influence the choice of scaling factors, and we delete it from A, D, and E.

If a row x of D contains exactly one $+1$ and row x of E has a nonzero entry, say, in column z, then the latter entry E_{xz} forces the scale factor for column z as follows. If $E_{xz} = +1$ (resp. $E_{xz} = -1$), then column z of E must be scaled with -1 (resp. $+1$) if a nearly negative matrix is to result. Hence, we scale column z of E with that factor, then move the scaled column from E to D.

We recursively apply the above rules until (1) the current D is found to be not nearly negative, or (2) the current A has no rows, or (3) none of the rules applies.

In case (1), the original matrix A has been demonstrated to be not hidden nearly negative relative to the original submatrix E, and we stop with that conclusion.

In case (2), we have determined scaling factors that turn the original matrix A into a nearly negative matrix, and thus we may stop.

In case (3), additional processing is needed. We claim that the current D has no columns and that each row of the current E has at least two nonzero entries. The proof is as follows.

According to the reduction rules, the nearly negative D cannot have a row with a $+1$, since otherwise that row is deleted or some column of E is scaled and transferred to D. Hence, D is nonpositive. Since all nonpositive columns must be deleted from D, that matrix cannot have any column at all. Finally, since D has no columns, E cannot have a row with at most one nonzero entry, since any such row must be deleted.

We continue with case (3), again using recursion. Note that the original $A = [D|E]$ has been reduced to a current $A = [D|E]$ for which D has no columns and E has at least two nonzero entries in each row. We store a copy of A as a matrix A'.

We scale an arbitrarily selected column y of E by $+1$ (this has no effect, of course), move the scaled column to D, and apply the earlier process. When that process stops, we again have one of the cases (1)–(3).

As shown above, in case (2) we have the desired scaling factors, while in case (3) the current A is smaller than A' and we may invoke recursion.

In case (1), the current D is not nearly negative. We conclude that the scaling factor $+1$ for column y cannot lead to a nearly negative matrix. We restore A to the saved matrix A'. Then we scale column y of E by -1, move the scaled column to D, and once more apply the earlier process. We either terminate in case (1) again, in which situation the original matrix A is not hidden nearly negative relative to the original submatrix E, or we have one of the cases (2) or (3), which have already been treated.

We summarize the algorithm below.

(5.6.1) Algorithm TEST HIDDEN NEAR NEGATIVITY. *Tests whether a given matrix $A = [D|E]$ over $\mathbb{B}$ can be column scaled with $\{\pm 1\}$ factors so that a nearly negative matrix results. The scaling is restricted to the columns of E.*

Input: Matrix $A = [D|E]$ over $\mathbb{B}$, of size $m \times n$.

Output: Either: Scaling factors for the columns of E that convert A to a nearly negative matrix. Or: "A is not hidden nearly negative relative to E."

Complexity: $O(m + n + \text{count}(A))$ when properly implemented.

Procedure:

1. If D is not nearly negative: Declare that the input matrix A is not hidden nearly negative relative to the input submatrix E, and stop.

   If A has no rows: Assign arbitrary $\{\pm 1\}$ scaling factors to the columns of E. These values plus the earlier assigned scaling factors, if any, constitute the desired scaling factors for the input matrix A. Output those factors, and stop.

   If a column z of D is nonpositive: Delete column z from A and D, and repeat Step 1.

   If a row x of D is nonpositive and row x of E has at most one nonzero entry or if a row x of D contains exactly one $+1$ and row x of E is zero: Delete row x from A, D, and E, and repeat Step 1.

   If a row x of D contains exactly one $+1$ and row x of E has a nonzero entry, say, in column z: Scale column z of E by -1 (resp. $+1$) if $E_{xz} = +1$ (resp. $E_{xz} = -1$), retain that scaling factor for possible output, move the scaled column from E to D, and repeat Step 1.

2. ($A = [D|E]$ where D has no columns and where each row of E has at least two nonzero entries; try a scaling factor of $+1$ for a column y of E.) Retain A as a matrix A'. Arbitrarily select a column y of E. Move column y from E to D, and record $+1$ as scaling factor for column y for possible output.

3. Apply the process of Step 1 to $A = [D|E]$, with two exceptions: If D becomes a matrix that is not nearly negative, go to Step 4; if D

becomes a matrix without columns and E becomes a matrix where each row has at least two nonzero entries, go to Step 2.

4. (Restore A, and try alternate scaling factor -1 for column y.) Redefine A to be A'. Scale column y by -1, and move the scaled column from E to D. Update the recorded scaling factor for column y to -1. Go to Step 1.

Proof of Validity. We only need to establish the claimed complexity of $O(m + n + \text{count}(A))$. The proof mimics that for Algorithm SOLVE 2SAT (5.4.1). When we use the scaling factor $+1$ for column y in Step 2, we also assign, in a parallel execution, the scaling factor -1 as specified in Step 4. We evaluate the consequences of the two choices in parallel until in both cases a D is encountered that is not nearly negative or until one case would proceed to Step 2. In the first instance, the input matrix A has been proved to be not hidden nearly negative relative to the input submatrix E. In the second situation, we stop the parallel evaluation and go with the applicable matrix to Step 2.   □

We have already seen that the scaling problem for hidden nearly negative matrices can be formulated as an instance of 2SAT. There also is the following, closely related result.

(5.6.2) Theorem. *Let A be a matrix over $\mathbb{B}$.*

(a) *If A has at most two nonzero entries in each row and is satisfiable, then A is hidden nearly negative.*

(b) *If A has exactly two nonzero entries in each row and is hidden nearly negative, then it is satisfiable.*

Proof.

To show part (a), we suppose that the matrix A is satisfiable. By Lemma (2.6.21), there are $\{\pm 1\}$ scaling factors for the columns of A that convert A to a matrix A' where each row has at least one $+1$. Since A has at most two nonzero entries in each row, we may equivalently say that each row of A' has at most one -1. When we scale each column of A' by -1, we obtain a matrix A'' that contains in each row at most one $+1$. Thus, A'' is nearly negative, and A has been shown to be hidden nearly negative.

An almost trivial reversal of the above arguments establishes part (b).   □

By Theorem (5.6.2), testing satisfiability of a 2SAT matrix A is essentially equivalent to deciding whether A is hidden nearly negative. Accordingly, one might expect a close relationship between Algorithm SOLVE 2SAT (5.4.1) and Algorithm TEST HIDDEN NEAR NEGATIVITY (5.6.1) when the latter scheme processes a 2SAT instance. Indeed, the latter algorithm essentially becomes the former one when we declare a scaling factor of $+1$ (resp. -1) in Algorithm TEST HIDDEN NEAR NEGATIVITY (5.6.1) to correspond to the assignment of *False* (resp. *True*) in Algorithm SOLVE 2SAT (5.4.1). We leave the easy verification of this claim to the reader.

Solution Algorithm for Hidden Nearly Negative Matrices

Construction of a solution algorithm is now simple. For the SAT case, we assume that the given matrix A is hidden nearly negative. We use Algorithm TEST HIDDEN NEAR NEGATIVITY (5.6.1) to find $\{\pm 1\}$ scaling factors that convert A to a nearly negative matrix A', then use Algorithm SOLVE NEARLY NEGATIVE SAT OR MINSAT (5.5.1) to find a satisfying solution for A' or to determine that no such solution exists.

In the first case, we use the scaling factors to convert the satisfying solution for A' to one for A, by switching a value of *True* (resp. *False*) for a column y of A' to a value of *False* (resp. *True*) for column y of A whenever the scaling factor for that column is -1.

In the second case, unsatisfiability of A' implies the same conclusion for A.

The MINSAT case is handled almost identically. We assume that the matrix A over $\mathbb{B}$ of a given pair (A, c), with rational nonnegative c, is hidden nearly negative relative to the column submatrix E of A whose columns correspond to the zero entries of c.

We apply Algorithm TEST HIDDEN NEAR NEGATIVITY (5.6.1) to determine $\{\pm 1\}$ scaling factors for E that convert A to a nearly negative matrix A'. Since any column of A scaled by -1 corresponds to a zero entry of c, the MINSAT instances (A, c) and (A', c) are equivalent.

We use Algorithm SOLVE NEARLY NEGATIVE SAT OR MINSAT (5.5.1) to solve the instance (A', c). If (A', c) is satisfiable, then we transform the optimal solution for (A', c) to one for (A, c) as in the SAT case. If (A', c) is unsatisfiable, then (A, c) is unsatisfiable as well.

The above discussion establishes the following algorithm.

(5.6.3) Algorithm SOLVE HIDDEN NEARLY NEGATIVE SAT OR MINSAT. *Solves the* SAT *or* MINSAT *problem involving a given hidden nearly negative matrix A over $\mathbb{B}$. In the* MINSAT *case, the cost vector is a given rational nonnegative vector c, and A is hidden nearly negative relative to the column submatrix E of A whose columns correspond to the zero entries of c.*

Input: Hidden nearly negative matrix A over $\mathbb{B}$, of size $m \times n$. In the MINSAT case, a rational nonnegative vector c is also given, and A is hidden nearly negative relative to the column submatrix E of A whose columns correspond to the zero entries of c.

Output: Either: A solution for the SAT instance A or the MINSAT instance (A, c), whichever applies. Or: "A is unsatisfiable."

Complexity: $O(m + n + \text{count}(A))$.

Procedure:

1. Partition A as $A = [D|E]$ where in the SAT case D has no columns and thus $A = E$, and where in the MINSAT case the columns of E correspond to the zero entries of c.

2. Use Algorithm TEST HIDDEN NEAR NEGATIVITY (5.6.1) to obtain $\{\pm 1\}$ scaling factors for E that scale A to a nearly negative matrix A'.

3. Apply Algorithm SOLVE NEARLY NEGATIVE SAT OR MINSAT (5.5.1) to solve the SAT instance A' or the MINSAT instance (A', c). If that instance is satisfiable, convert a solution for A' or (A', c) to one for A or (A, c), by switching a value of *True* (resp. *False*) for a column y of A' to a value of *False* (resp. *True*) for column y of A whenever column y occurs in E and the scaling factor is -1.

   If A' or (A', c) is unsatisfiable, declare A or (A, c) to be unsatisfiable as well.

We summarize the SAT and MINSAT centrality established by the above algorithms.

(5.6.4) Theorem.

(a) *The class of hidden nearly negative matrices A is SAT central.*

(b) *The class of the following matrix/vector pairs (A, c) is MINSAT central. The vector c is rational nonnegative, and the matrix A is hidden nearly negative relative to the column submatrix E of A whose columns correspond to the zero entries of c.*

Applications of Hidden Nearly Negative Matrices

The notion of hidden near negativity is obviously useful when satisfiability or logic minimization problems are to be solved. It also is attractive in the context of the theorem-proving problem posed and solved in Section 5.5 for nearly negative matrices. We summarize the discussion of that section.

For some disjoint index sets J^+ and J^- and for $l = 1, 2, \ldots, k$, statements $T_l = [(\bigwedge_{j \in J^+} s_j) \wedge (\bigwedge_{j \in J^-} \neg s_j)] \Rightarrow t_l$ are given. We want to decide which of these statements are theorems of a given CNF system S. Denote by A the matrix over $\mathbb{B}$ corresponding to S.

Let L be any nonempty subset of $\{t_1, t_2, \ldots, t_k\}$. Define c to be the rational $\{0, 1\}$ cost vector that assigns a cost of 1 to the columns $t_l \in L$ of A and assigns a cost of 0 to all remaining columns. Partition A as $A = [D|E]$ where the columns of D are indexed by L.

The discussion in Section 5.5 implies the following. If A is hidden nearly negative relative to E, then by solving the MINSAT instance (A, c) we can ascertain which of the statements T_l, $l \in L$, are theorems of S. Indeed, such a T_l is a theorem if and only if the MINSAT solution assigns the value *True* to the column t_l of A.

In the ideal situation, the above process works for $L = \{t_1, t_2, \ldots, t_k\}$. We may not be as lucky, but the process might work for some subsets L of $\{t_1, t_2, \ldots, t_k\}$ whose union is $\{t_1, t_2, \ldots, t_k\}$. It is easy to see that such subsets, say, L_1, L_2, $\ldots$, L_m, may be considered to be disjoint. Indeed, if A is hidden nearly negative relative to E defined via L_i, then this is so for any subset of L_i.

With disjoint L_1, L_2, $\ldots$, L_m at hand, one must solve m MINSAT instances to settle which statements T_l are theorems. Accordingly, one would like to choose a collection with small m. Let us call the problem of finding a collection with minimum m PARTITION FOR MINSAT. Below, we show that this problem is generally difficult.

Recall from Section 5.4 that a node subset of an undirected graph is independent if no edge of G connects any two nodes of the subset. Define PARTITION INTO INDEPENDENT NODE SUBSETS to be the problem where, for a given undirected graph G, one must partition the node set of G into a minimum number of independent subsets. This problem is $\mathcal{NP}$-hard, since it is the optimization version of the so-called GRAPH K-COLORABILITY problem, which is known to be $\mathcal{NP}$-complete.

We reduce an instance G of PARTITION INTO INDEPENDENT NODE SETS to an instance of PARTITION FOR MINSAT.

We let A be the transpose of the node/edge incidence matrix of G. Thus, each node (resp. edge) of G produces a column (resp. row) of A; if an edge of G connects nodes i and j, then the corresponding row of A has one 1 in column i and a second 1 in column j.

Let L be a node subset of G. Partition $A = [D|E]$ so that the columns of D correspond to the nodes in L. Since A is a $\{0, 1\}$ matrix, the following four statements are equivalent: A is hidden nearly negative relative to E; A becomes nearly negative when each column of E is scaled with -1; D has at most one $+1$ in each row; the node subset L of G is independent.

Thus, finding a minimum number of disjoint independent node subsets of G whose union is equal to the node set of G is equivalent to finding a minimum number of disjoint subsets of the column index set of A, say, L_1, $L_2, \ldots$, L_m, with the following property. For each L_i, the matrix A must be hidden nearly negative relative to the column submatrix whose indices do not occur in L_i, and the union of the L_i must be the column index set of A. The latter problem is an instance of PARTITION FOR MINSAT.

The $\mathcal{NP}$-hardness of PARTITION FOR MINSAT should not deter us from the use of the above ideas. For example, one might employ simple heuristics to search for an attractive collection L_1, $L_2, \ldots$, L_m.

Resolution and Hidden Nearly Negative Matrices

Resolution maintains hidden near negativity according to the following theorem.

(5.6.5) Theorem. *Let* $A = [D|E]$ *be a matrix over* $\mathbb{B}$ *that is hidden nearly negative relative to* E. *If Algorithm* RESOLUTION FOR MATRIX *(5.4.4) reduces* A *to a matrix* $\overline{A} = [\overline{D}|\overline{E}]$, *where the columns of* $\overline{D}$ *(resp.* $\overline{E}$) *correspond to those of* D *(resp.* E), *then* $\overline{A}$ *is hidden nearly negative relative to* $\overline{E}$.

Proof. By assumption, we may scale the columns of the submatrix E of A such that A becomes a nearly negative matrix $A' = [D|E']$. Suppose we apply Algorithm RESOLUTION FOR MATRIX (5.4.4) to A as well as A', getting $\overline{A} = [\overline{D}|\overline{E}]$ and $\overline{A'} = [\overline{D'}|\overline{E'}]$, respectively.

By Theorem (5.5.5), the algorithm maintains near negativity, so $\overline{A'}$ is nearly negative. Elementary case checking verifies that $\overline{D'}$ is equal to $\overline{D}$ and that $\overline{E'}$ is up to column scaling equal to $\overline{E}$. Hence, $\overline{A}$ is hidden nearly negative relative to $\overline{E}$. □

We turn to the fourth class of matrices covered in this chapter.

5.7 Balanced Matrices

For $k \geq 2$, declare a $k \times k$ $\{0, \pm 1\}$ matrix to be a *cycle matrix* if it is connected and has exactly two nonzeros in each row and column. Thus, a cycle matrix has the following form.

(5.7.1)

$$\begin{bmatrix} \pm 1 & & & \pm 1 \\ \pm 1 & \pm 1 & & 0 \\ & \pm 1 & \ddots & \\ & & \ddots & \pm 1 \\ 0 & & & \pm 1 & \pm 1 \end{bmatrix}$$

Cycle matrix

Define a cycle matrix to be *balanced* if the integer sum of its entries is divisible by 4, that is, if that sum is equal to 0(mod 4). Define a $\{0, \pm 1\}$ matrix A to be *balanced* if every cycle submatrix of A is balanced.

In this section, we prove that the class of balanced matrices over $\mathbb{B}$ is SAT central and that the class of matrix/vector pairs (A, c) with balanced A and rational nonnegative c is MINSAT central.

We claim that the nonnegativity condition on the vector c of a MINSAT instance (A, c) effectively does not pose a restriction, in contrast to the situation in Sections 5.5 or 5.6. The proof relies on the following lemma.

(5.7.2) Lemma. *Any matrix derived from a $\{0, \pm 1\}$ balanced matrix by scaling of rows and columns with $\{\pm 1\}$ factors and submatrix taking is balanced as well.*

Proof. Scaling of a row or column of a cycle matrix evidently changes the integer sum of its entries by a multiple of 4. Thus, balancedness of a cycle matrix is preserved under scaling. Since balancedness of a general $\{0, \pm 1\}$ matrix is defined via balancedness of its cycle submatrices, the lemma follows. □

Let a MINSAT instance (A, c) with balanced A be given. If the rational vector c contains negative entries, then we scale these entries and the corresponding columns of A by -1 and get an equivalent MINSAT instance (A', c') where A' is balanced by Lemma (5.7.2) and where c' is nonnegative.

By Lemma (5.7.2), the SAT and MINSAT classes with balanced matrices are maintained under submatrix taking, so condition (5.2.1)(i) for SAT centrality and condition (5.2.2)(i) for MINSAT centrality are satisfied. Below, we establish the remaining conditions (ii) and (iii) of (5.2.1) and (5.2.2) by polynomial recognition and solution algorithms. Some of these algorithms are quite complicated. We also discuss special subclasses for which somewhat simpler or even elementary recognition and solution algorithms exist.

Recognition Algorithm for Balanced Matrices

Let A be a matrix over $\mathbb{B}$. There is a complicated but polynomial algorithm by Conforti, Cornuéjols, Kapoor, and Vušković (1994b) for testing balancedness of A. The algorithm is too long and complicated to be included here in its entirety, so we settle for a summarizing description.

(5.7.3) Algorithm TEST BALANCEDNESS. *Tests balancedness of a given $\{0, \pm 1\}$ matrix A.*

Input: $\{0, \pm 1\}$ matrix A.

Output: Either: "A is balanced." Or: "A is not balanced."

Complexity: Polynomial.

Procedure: (See Conforti, Cornuéjols, Kapoor, and Vušković (1994b) for details.)
1. Initialize a candidate list L of matrices to $L = \{A\}$.
2. Remove a matrix, say, B, from L. Check if B contains as a submatrix one of several nonbalanced matrices. If this is so, declare the input matrix A to be nonbalanced, and stop.
3. Using several subroutines, verify that B is balanced, or decompose B into several matrices, which are then added to L.

4. If L is nonempty: Go to Step 2.

5. Declare the input matrix A to be balanced, and stop.

There are two important nested subclasses of balanced matrices. The larger one of the two subclasses consists of the *totally unimodular matrices*. Each square submatrix of such a matrix, when viewed to be over the rationals, has determinant 0, $+1$, or -1.

The second subclass consists of the *network matrices*, which are the totally unimodular matrices with at most two nonzero entries in each row or in each column.

For the two subclasses, the following analogue of Lemma (5.7.2) holds.

(5.7.4) Lemma. *Any matrix derived from a $\{0, \pm1\}$ totally unimodular (resp. network) matrix by scaling of rows and columns with $\{\pm1\}$ factors and submatrix taking is also a totally unimodular (resp. network) matrix.*

Proof. Scaling of a row or column changes at most the sign of the determinants of square submatrices. Submatrix taking at most reduces column counts. Thus, total unimodularity and the network property are preserved by scaling and submatrix taking.   □

The next lemma establishes that the totally unimodular matrices, and hence the network matrices, are balanced.

(5.7.5) Lemma. *Any $\{0, \pm1\}$ totally unimodular matrix—in particular, any network matrix—is balanced.*

Proof. Let D be a cycle submatrix of a totally unimodular matrix. It is easy to see from (5.7.1) that we can scale the rows and columns of D with $\{\pm1\}$ factors so that a matrix D' of the form

(5.7.6)
$$
\begin{bmatrix}
1 & & & & \alpha \\
-1 & 1 & & 0 & \\
 & -1 & \cdot & & \\
 & & \cdot & \cdot & 1 \\
 & 0 & & \cdot & \\
 & & & -1 & 1
\end{bmatrix}
$$

Scaled cycle matrix

results where α is $+1$ or -1. By Lemma (5.7.4), total unimodularity is maintained by scaling with $\{\pm1\}$ factors and submatrix taking, so D' has determinant 0, $+1$, or -1.

By cofactor expansion and counting, we confirm that the determinant of D' is 2 (resp. 0) if and only if $z = 1$ (resp. $z = -1$), which holds if and only if the entries of D' sum to 2 (resp. 0).

Thus, the determinant of D' must be 0, and its entries must sum to 0. Hence, D' is balanced, and by Lemma (5.7.1), D is balanced as well. We conclude that A is balanced.   □

We demonstrate that the class of balanced matrices properly contains the class of totally unimodular matrices and that the latter class properly contains the class of network matrices. We use the matrix

(5.7.7)
$$\begin{array}{|cccc|} 1 & 1 & 1 & 1 \\ 1 & 1 & 0 & 0 \\ 1 & 0 & 1 & 0 \\ 1 & 0 & 0 & 1 \end{array}$$

Balanced matrix

for this purpose. Straightforward checking confirms that the matrix is balanced, has determinant equal to -2, and hence is not totally unimodular. When the last row and the last column are deleted from the matrix, a totally unimodular matrix results that is not a network matrix.

Recognition Algorithm for Totally Unimodular Matrices

Algorithm TEST BALANCEDNESS (5.7.3) as described in Conforti, Cornuéjols, Kapoor, and Vušković (1994b) is polynomial, but also contains enumerative subroutines requiring substantial computational effort. Therefore, that algorithm may require considerable time for processing large matrices.

We next include a polynomial algorithm described in Truemper (1990) for testing total unimodularity. That algorithm is also quite complicated, so we just summarize the main steps. However, the algorithm is computationally quite efficient and handles large matrices with reasonable computational effort.

(5.7.8) Algorithm TEST TOTAL UNIMODULARITY. *Tests total unimodularity of a given* $\{0, \pm 1\}$ *matrix* A.

Input: $\{0, \pm 1\}$ matrix A, of size $m \times n$.

Output: Either: "A is totally unimodular." Or: "A is not totally unimodular."

Complexity: $O((m+n)^3)$.

Procedure: (See Truemper (1990) for details.)

1. Derive a $\{0, 1\}$ matrix A' from A by replacing each -1 of A by $+1$. View A' to be over GF(2). Initialize a candidate list L of matrices to $L = \{A'\}$.

2. Remove a matrix, say, B, from L.
   If B can be decomposed in one of three ways into two component matrices: Carry out one such decomposition, add the component matrices to L, and repeat Step 2.

3. Test if B is one of two special 5×5 matrices, or if one can derive from B a network matrix by certain operations. If neither case applies, declare the input matrix A to be not totally unimodular, and stop.

4. If L is nonempty: Go to Step 2.

5. Sign the 1s of A' so that a totally unimodular matrix A'' results. Test if A'' can be transformed by row and column scaling with $\{\pm 1\}$ factors to the input matrix A. The matrix A is totally unimodular if and only if such scaling factors exist. Output the appropriate conclusion, and stop.

Recognition Algorithm for Network Matrices

Testing for the network property turns out to be simple, as we see next.

(5.7.9) Algorithm TEST NETWORK PROPERTY. *Tests whether a given $\{0, \pm 1\}$ matrix A has the network property.*

Input: $\{0, \pm 1\}$ matrix A, of size $m \times n$.

Output: Either: "A is a network matrix." Or: "A is not a network matrix."

Complexity: $O(m + n)$.

Procedure:

1. If A has at least three nonzeros in some row and at least three nonzeros in some column, declare that A is not a network matrix, and stop.
   If A has at least three nonzeros in some row, replace A by its transpose.

2. (A has at most two nonzeros in each row.) Derive from A a $\{0, 1\}$ matrix A' by first deleting any row with at most one nonzero, and then replacing in the remaining rows each -1 by $+1$.
   If A' has no rows, declare A to be a network matrix, and stop.

3. (A' has exactly two nonzeros in each row.) Let G be the graph that has the transpose of A' as node/edge incidence matrix. Declare any edge of G to be special if the corresponding row of A contains two $+1$s or two -1s.

4. Contract each nonspecial edge of G, getting a minor $\overline{G}$ of G.

5. If $\overline{G}$ contains a loop, declare that A does not have the network property, and stop. Otherwise, use breadth-first-search to test whether $\overline{G}$ is bipartite. The matrix A is a network matrix if and only if $\overline{G}$ is bipartite. Output the appropriate conclusion, and stop.

Proof of Validity. We may assume the case where A has exactly two nonzeros in each row. We analyze two mutually exclusive cases, depending on whether A is balanced.

Suppose A is not balanced and hence contains a cycle submatrix whose entries sum to $2 \pmod 4$. Thus, the number of rows of the submatrix containing two $+1$s or two -1s is odd. Correspondingly, the graph G defined in Step 3 has a cycle with an odd number of special edges, and the minor $\overline{G}$ defined in Step 4 has a loop or a cycle with an odd number of edges. A graph is bipartite if and only if it has no cycle with an odd number of edges. Hence, Step 5 determines $\overline{G}$ to have a loop or to be nonbipartite, and it concludes that A is not a network matrix. By Lemma (5.7.5), nonbalancedness of A implies that A is not totally unimodular, so the conclusion of Step 5 is correct.

Suppose that A is balanced. Assume that a cycle C of G contains an odd number of special edges. Select C to have a minimal number of nodes.

If C has a chord, then C plus that chord defines two cycles smaller than C, and at least one of these cycles must have an odd number of special edges. But then C is not minimal, a contradiction.

Thus, C has no chord and corresponds to a cycle submatrix of A whose entries sum to $2(\mathrm{mod}\ 4)$, a contradiction of the balancedness of A.

Hence, every cycle of G has an even number of special edges. Correspondingly, Step 5 determines $\overline{G}$ to have no loops and to be bipartite, and it declares A to be a network matrix.

It remains for us to show that A is totally unimodular. The nodes on one side of the bipartite graph $\overline{G}$ define a column subset of A. Suppose we scale the columns of A in that subset by -1, getting a matrix A''. It is easily checked that each row of A'' contains one $+1$ and one -1.

We claim that each square submatrix D of A'' has determinant 0, $+1$, or -1 and, hence, that both A'' and A are totally unimodular. The proof is by induction on the size of D. If D contains a row with at most one nonzero, then the determinant of D is 0, or we apply cofactor expansion and induction. In the remaining case, each row of D contains two nonzeros, that is, one $+1$ and one -1. But then the columns of D sum to 0, so the determinant of D is 0. □

The arguments validating Algorithm TEST NETWORK PROPERTY (5.7.9) imply the following result.

(5.7.10) Theorem. *Let A be a $\{0, \pm 1\}$ matrix with at most two nonzeros in each row or in each column. Then the following statements are equivalent.*

(i) *A is balanced.*
(ii) *A is totally unimodular.*
(iii) *A is a network matrix.*
(iv) *If each row (resp. column) of A has at most two nonzeros, then the columns (resp. rows) of A can be scaled by $\{\pm 1\}$ factors so that, in the scaled matrix A', each row (resp. column) with two nonzeros contains one $+1$ and one -1.*

Proof. The arguments for Algorithm TEST NETWORK PROPERTY (5.7.9) establish (i) $\Rightarrow$ (ii) $\Rightarrow$ (iii) $\Rightarrow$ (iv). We show (iv) $\Rightarrow$ (i). Clearly, the scaled matrix A' of (iv) cannot contain a nonbalanced cycle submatrix and thus is balanced. By Lemma (5.7.2), A is then balanced as well.   □

For later reference, we extract from Algorithm TEST NETWORK PROPERTY (5.7.9) the scaling steps that produce the matrix A' of Theorem (5.7.10)(iv).

(5.7.11) Algorithm SCALE NETWORK MATRIX. *Scales the columns (resp. rows) of a given* $\{0, \pm 1\}$ *network matrix* A *having at most two nonzeros in each row (resp. column) so that any row (resp. column) with two nonzeros is turned into a row (resp. column) with one* $+1$ *and one* -1.

Input: $\{0, \pm 1\}$ network matrix A, of size $m \times n$. The matrix has at most two nonzeros in each row or in each column.

Output: A matrix A' obtained from A by scaling with $\{\pm 1\}$ factors. The scaling involves the columns (resp. rows) of A if A has at most two nonzeros in each row (resp. column), and it results in an A' where each row (resp. column) with two nonzeros contains one $+1$ and one -1. In case both types of scaling are possible, preference is given to column scaling.

Complexity: $O(m + n)$.

Procedure:

1. If A has at least three nonzeros in some row: Replace A by its transpose.
2. Apply Algorithm TEST NETWORK PROPERTY (5.7.9) to A. The graph $\overline{G}$ of Step 5 of that algorithm necessarily has no loops and is bipartite. Scale the columns of A corresponding to the nodes of one side of $\overline{G}$ by -1 to obtain A'.
3. If in Step 1 the matrix A was replaced by its transpose: Replace A' by its transpose.
4. Output A', and stop.

Proof of Validity. Since A is assumed to be a network matrix, Algorithm TEST NETWORK PROPERTY (5.7.9) invoked in Step 2 must derive a bipartite graph $\overline{G}$. As shown in the proof of validity for that algorithm, the scaling of A based on the nodes of one side of $\overline{G}$ produces the desired A' or its transpose.   □

We turn to solution algorithms for the class of balanced matrices and its two subclasses. We first describe, without proof, several basic results of polyhedral combinatorics. Relevant references for that material are included in Section 5.9.

Some Results of Polyhedral Combinatorics

Define B to be an $m \times n$ rational matrix, and let b be an $m \times 1$ rational vector. Define P to be the set of $n \times 1$ rational vectors r satisfying the inequality system $B \cdot r \geq b$. Thus, $P = \{r \mid B \cdot r \geq b\}$. We say that P is the *polyhedron* defined by $B \cdot r \geq b$.

A vector $r \in P$ is an *extreme point* of P if for any two vectors r^1, $r^2 \in P$ satisfying $r = (r^1 + r^2)/2$, we necessarily have $r = r^1 = r^2$. A bounded and nonempty polyhedron always has at least one extreme point.

A bounded polyhedron is *integral* if each extreme point of the polyhedron is integral.

Suppose any solution for a given inequality system $B' \cdot r \geq b'$ is also a solution for the system $B \cdot r \geq b$ defining P. If P is bounded and if all extreme points of P satisfy $B' \cdot r \geq b'$, then the latter system also defines P.

For any $k \geq 1$, rational vectors $r^0, r^1, \ldots, r^k$ are *affinely independent* if the vectors $r^1 - r^0, r^2 - r^0, \ldots, r^k - r^0$ are linearly independent.

Let k be the largest integer such that P contains k affinely independent vectors. Then P is said to have *dimension k*.

Recall that the size of the vectors $r \in P$ is $n \times 1$. If the dimension of P is n, then P is said to be *full dimensional*. In that case, the inequality system $B \cdot r \geq b$ contains a subsystem $B' \cdot r \geq b'$ with the following features: $B' \cdot r \geq b'$ defines P, and any system $B'' \cdot r \geq b''$ that also defines P must contain $B' \cdot r \geq b'$ as a subsystem, up to a scaling of the inequalities of $B' \cdot r \geq b'$ with positive factors. Thus, $B' \cdot r \geq b'$ is up to such scaling the *unique minimal inequality system* defining P.

A *linear program*, abbreviated LP, is specified by rational arrays B, b, and c of size $m \times n$, $m \times 1$, and $n \times 1$, respectively. One must find an $n \times 1$ rational solution vector r^* that minimizes the linear function $c^t \cdot r$ over the polyhedron $P = \{r \mid B \cdot r \geq b\}$, or one must conclude that no such vector exists. The latter case arises in one of two ways. First, P may be empty, in which case the LP is *infeasible*. Second, the function $c^t \cdot r$ may take on arbitrarily small values on P, in which case the LP is *unbounded*. Note that unboundedness of the LP cannot occur when P is bounded.

We summarize the LP as follows, where s.t. stands for "subject to."

$$(5.7.12) \qquad \begin{aligned} \min \quad & c^t \cdot r \\ \text{s. t.} \quad & B \cdot r \geq b \end{aligned}$$

Assume that P is bounded and nonempty. Then P has an extreme point that may serve as minimizing r^*.

In a special LP case, the vector c is 0. Solving the LP is then equivalent to finding a vector in P or determining that P is empty.

The most popular method for solving LPs is the *Simplex Method*. The method is not polynomial, but has been shown to be very efficient for most

LPs arising from real-world problems. There also exist several polynomial algorithms for solving LPs—for example, the *Ellipsoid Method*. If a given LP has an extreme point solution, then such a solution can be found by any one of the cited methods.

A number of combinatorial optimization problems may be expressed as LPs where the underlying polyhedron is integral. In each such case, one may solve the problem in polynomial time by finding an optimal extreme point solution.

Integrality of the extreme points of the polyhedron P is obviously governed by the matrix B and the vector b. When B is a $\{0, \pm 1\}$ balanced matrix and certain assumptions are satisfied, then integrality of the extreme points is assured. The latter assumptions are not needed when B is totally unimodular. Below, we summarize these results about balanced and totally unimodular matrices. We omit the proofs. References are given in Section 5.9.

Extreme Point Results

For a given $m \times n$ $\{0, \pm 1\}$ matrix A, define $p(A)$ (resp. $q(A)$) to be the $m \times 1$ vector whose entry in position i is the number of $+1$s (resp. -1s) in row i of A. Thus, $p(A) + q(A)$ is the vector containing the number of nonzeros in each row of A.

(5.7.13) Theorem. *A $\{0, \pm 1\}$ matrix is balanced if and only if for each submatrix A of that matrix, the polyhedron*

$$(5.7.14) \qquad P(A) = \{r \mid A \cdot r \geq \underline{1} - q(A); \ 0 \leq r \leq \underline{1}\}$$

is integral.

Note the particular vector $\underline{1} - q(A)$ in the definition of $P(A)$ of (5.7.14). That vector may be replaced by an arbitrary integer vector b when one considers total unimodularity instead of balancedness as follows.

(5.7.15) Theorem. *A $\{0, \pm 1\}$ matrix is totally unimodular if and only if for each submatrix A of that matrix and for each integral vector b, the polyhedron*

$$(5.7.16) \qquad P(A, b) = \{r \mid A \cdot r \geq b; \ 0 \leq r \leq \underline{1}\}$$

is integral.

It turns out that one may solve a given SAT instance A over $\mathbb{B}$ very easily when the polyhedron $P(A)$ given by (5.7.14) is integral. Furthermore, any MINSAT instance (A, c) with integral $P(A)$ may be efficiently solved via some LP. We provide details next.

SAT, MINSAT, and Integral Polyhedra

We start with a simple SAT example. Consider the CNF system consisting of the single clause

(5.7.17) $\neg x_1 \lor x_2 \lor x_3 \lor \neg x_4$

The corresponding matrix A over $\mathbb{B}$ is

(5.7.18) $A = [-1, +1, +1, -1]$

with column index set $X = \{x_1, x_2, x_3, x_4\}$. Declare A to be over the rationals, and define $r = [r_1, r_2, r_3, r_4]^t$. For each r_i, we interpret $r_i = 1$ (resp. $r_i = 0$) to mean that we assign the value *True* (resp. *False*) to x_i.

Consider the inequality

(5.7.19) $(-1) \cdot (r_1 - 1) + (+1) \cdot r_2 + (+1) \cdot r_3 + (-1) \cdot (r_4 - 1) \geq 1$

The clause (5.7.17) evaluates to *True* if and only if x_1 is *False*, or x_2 is *True*, or x_3 is *True*, or x_4 is *False*. Correspondingly, the inequality (5.7.19) is satisfied by $\{0, 1\}$ r_i values if and only if $r_1 = 0$, or $r_2 = 1$, or $r_3 = 1$, or $r_4 = 0$.

We move the constant terms of the left-hand side of (5.7.19) to the right-hand side and get

(5.7.20) $(-1) \cdot r_1 + (+1) \cdot r_2 + (+1) \cdot r_3 + (-1) \cdot r_4 \geq 1 - 2$

Using A of (5.7.18) and the earlier defined vector $q(A)$, which here is equal to 2, we may express (5.7.20) in matrix notation as

(5.7.21) $A \cdot r \geq \underline{1} - q(A)$

Our example is readily extended to the general SAT case. Let A over $\mathbb{B}$ be the given SAT instance. Then solving the SAT problem for A is equivalent to finding a $\{0, 1\}$ solution vector r for the inequality of (5.7.21) or determining that no such vector exists. If the polyhedron $P(A)$ of (5.7.14) is integral, then the latter problem turns out to be very easy. The next theorem gives the reason.

(5.7.22) Theorem. *Let A with column index set $X = \{x_1, x_2, \ldots, x_n\}$ be a matrix over $\mathbb{B}$ with at least two nonzeros in each row. Suppose that the polyhedron $P(A)$, which according to (5.7.14) is*

(5.7.23) $P(A) = \{r \mid A \cdot r \geq \underline{1} - q(A); \ 0 \leq r \leq \underline{1}\}$

is integral. Then, for each column index x_j of A, there are two satisfying solutions for A, one of which assigns True to column x_j while the other one assigns False to that column.

Proof. Recall that $p(A)$ contains the number of $+1$s of each row of A. Since A has at least two nonzeros in each row, we have $A \cdot \underline{1} = p(A) + q(A) \geq 2 \cdot \underline{1}$, which implies that $A \cdot (\frac{1}{2} \cdot \underline{1}) = \frac{1}{2} \cdot [p(A) - q(A)] \geq \underline{1} - q(A)$. For the vector $r = \frac{1}{2} \cdot \underline{1}$, we thus have $A \cdot r = A \cdot (\frac{1}{2} \cdot \underline{1}) \geq \underline{1} - q(A)$. Accordingly, $r = \frac{1}{2} \cdot \underline{1}$ is in the polyhedron $P(A)$.

Consider the LP that asks for a vector $r \in P(A)$ that minimizes a given component r_j. Since $P(A)$ is bounded and nonempty, that LP has an optimal extreme point solution, say, r^*. Since A is balanced, any such extreme point is by Theorem (5.7.13) integral. Now the vector $r = \frac{1}{2} \cdot \underline{1}$ is in $P(A)$, so the integral r^* must have $r_j^* \leq \frac{1}{2}$, which implies $r_j^* = 0$. We conclude that r^* corresponds to a satisfying solution for A where column x_j receives the value *False*.

We repeat the above arguments, except that this time we use the LP that demands a vector $r \in P(A)$ with minimum $-r_j$. We conclude that there is an integral vector $r^* \in P$ with $r_j^* = 1$. Hence, A has a satisfying solution with the value *True* for column x_j.   □

Theorem (5.7.22) supports the following elementary algorithm for the SAT problem with integral $P(A)$. Given A, we first carry out the usual reductions. Hence, we may assume that each row of A has at least two nonzero entries. By Theorem (5.7.22), A is satisfiable, and we may arbitrarily fix the value for some column y to *True* or *False*, delete all rows now satisfied and column y, and solve the SAT problem for the reduced matrix.

The MINSAT case is a bit more complicated, but yields to the same approach. Let a matrix A over $\mathbb{B}$ and a rational nonnegative vector c define an instance (A, c). We still assume that $P(A)$ is integral. We solve the MINSAT instance (A, c) by locating an optimal extreme point solution for the LP

$$
\begin{array}{ll}
\min & c^t \cdot r \\
\text{s. t.} & A \cdot r \geq \underline{1} - q(A) \\
& -r \geq -\underline{1} \\
& r \geq 0
\end{array}
$$

(5.7.24)

with the earlier mentioned Ellipsoid Method, or with the Simplex Method, or with special methods applicable to particular matrices A. Since $P(A)$ is integral, that solution solves the MINSAT instance.

Solution Algorithms for Balanced, Totally Unimodular, and Network Matrices

The preceding discussion of SAT and MINSAT instances with integral $P(A)$ fully applies to matrices that are balanced, are totally unimodular, or have the network property. Indeed, by Lemma (5.7.5), totally unimodular and network matrices are balanced, and by Theorem (5.7.13), for any balanced matrix A the polyhedron $P(A)$ is integral. Thus, we may efficiently solve any SAT instance or MINSAT instance involving a balanced, totally unimodular, or network matrix. When the matrix of a MINSAT instance has the network property, then the LP (5.7.24) turns out to be equivalent to a *network flow problem* that may be solved by any one of several highly efficient network flow algorithms. References for these algorithms are included in Section 5.9.

The above observations validate the following algorithm.

(5.7.25) Algorithm SOLVE BALANCED SAT OR MINSAT.
Solves the SAT or MINSAT problem involving a given balanced matrix A over $\mathbb{B}$. *In the MINSAT case, the cost vector is a given rational nonnegative vector c.*

Input: Balanced matrix A over $\mathbb{B}$, of size $m \times n$. In the MINSAT case, a rational nonnegative vector c is also given.

Output: Either: A solution for the SAT instance A or the MINSAT instance (A, c), whichever applies. Or: "A is unsatisfiable."

Complexity: SAT case: $O(m + n + \text{count}(A))$. MINSAT case: Polynomial if a polynomial algorithm is used in Step 3.

Procedure:

1. If A has a zero row: Declare A to be unsatisfiable, and stop.
   If A has no rows: Assign *False* to each column of A. These values plus the earlier assigned *True/False* values, if any, constitute a solution for the SAT or MINSAT instance, whichever applies.
   If A has a row with just one nonzero, say, in column z: If that nonzero is $+1$ (resp. -1), assign *True* (resp. *False*) to column z, delete from A all rows now satisfied and column z, and repeat Step 1.

2. (A has at least two nonzeros in each row and by Theorem (5.7.22) is satisfiable.) In the SAT case, arbitrarily assign *True* or *False* to some column y of A, delete from A all rows now satisfied and column y, and go to Step 1.

3. (We have MINSAT case.) Solve the LP

$$
\begin{aligned}
\min \quad & c^t \cdot r \\
\text{s. t.} \quad & A \cdot r \geq \underline{1} - q(A) \\
& -r \geq -\underline{1} \\
& r \geq 0
\end{aligned}
$$

(5.7.26)

by a polynomial algorithm such as the Ellipsoid Method, or by the Simplex Method, or by any special method that exploits the particular structure of the LP. In particular, if A has exactly two nonzeros in each row, use the following polynomial method. First, apply Algorithm SCALE NETWORK MATRIX (5.7.11) to scale the columns of A so that in the scaled matrix each row with two nonzeros has one $+1$ and one -1. Second, for each column y for which a -1 scale factor was used, replace the element c_y of c by $-c_y$. Let A' be the scaled matrix, and c' be the scaled vector. Then the LP

$$
(5.7.27) \qquad
\begin{aligned}
\min \quad & (c')^t \cdot r \\
\text{s. t.} \quad & A' \cdot r \geq \underline{1} - q(A') \\
& -r \geq -\underline{1} \\
& r \geq 0
\end{aligned}
$$

may be solved by any one of several highly efficient network flow algorithms. For details, see the appropriate references given in Section 5.9.

The next theorem records the SAT and MINSAT centrality results proved above for balanced matrices, totally unimodular matrices, and matrices with the network property.

(5.7.28) Theorem.
(a) *The three classes consisting of the balanced matrices, the totally unimodular matrices, and the matrices with the network property are SAT central.*
(b) *The three classes of matrix/vector pairs where the matrices are balanced, are totally unimodular, or have the network property are MINSAT central.*

We discuss some applications of balanced matrices.

Applications of Balanced Matrices

Recall that the MIN2SAT problem of Section 5.4 is the MINSAT problem involving matrices with at most two nonzeros in each row. As argued in Section 5.4, MIN2SAT is $\mathcal{NP}$-hard. But when the matrix A of a MIN2SAT instance (A, c) is known to be balanced, then according to Algorithm SOLVE BALANCED SAT OR MINSAT (5.7.25) that instance can be transformed to a polynomially solvable network flow problem. We record this fact below.

(5.7.29) Theorem. *Let (A, c) be a MIN2SAT instance; that is, A has at most two nonzeros in each row, and c is a rational nonnegative vector. If A is balanced, then (A, c) is solved in polynomial time by Algorithm SOLVE BALANCED SAT OR MINSAT (5.7.25) as a network flow problem.*

A second application of balanced matrices arises from the following setting. Given is a complete bipartite graph G, say, with node subsets X

and Y for the two sides of G. Let $m = |X|$ and $n = |Y|$. We denote the edge connecting $x \in X$ with $y \in Y$ by (x, y). We want a subset of the edge set so that each node of X (resp. Y) has at most (resp. at least) one edge of that subset incident.

It is well known that this problem may be formulated as a network flow problem and thus may be efficiently solved. We present that formulation shortly. When this problem occurs as part of a SAT or MINSAT instance, the situation may be far from simple. We include details.

Let $p(x, y)$ defined on $X \times Y$ be the predicate that is *True* if the edge (x, y) is selected. The CNF system representing the constraints on the edge selection contains two types of clauses. First, for each $y \in Y$, there must be an $x \in X$ such that $p(x, y)$ has the value *True*. Thus, for each $y \in Y$, we have the clause

$$(5.7.30) \qquad \bigvee_{x \in X} p(x, y)$$

Second, for each $x \in X$, at most one $p(x, y)$ may have the value *True*. We express that condition as follows. For each $x \in X$ and for each pair of distinct $y, z \in Y$, we require

$$(5.7.31) \qquad \neg p(x, y) \vee \neg p(x, z)$$

Since $|X| = m$ and $|Y| = n$, there are n clauses of type (5.7.30) and $\frac{1}{2} \cdot m \cdot n \cdot (n-1)$ clauses of type (5.7.31). Define A to be the matrix over $\mathbb{B}$ representing these clauses. Thus, A has $m \cdot n$ columns and $n \cdot [1 + \frac{1}{2} \cdot m \cdot (n-1)]$ rows. Note that the number of rows is quadratic in n and thus grows rather rapidly with n.

If $m = n - 1$, then the desired edge subset cannot exist, and A is unsatisfiable. That case is informally known as the *pigeonhole problem*, where one is asked to fill n pigeonholes using at most $m = n - 1$ pigeons, an impossible task. Many satisfiability algorithms have a difficult time solving the pigeonhole problem even for reasonably small values of m. Accordingly, one may expect computational difficulties with these algorithms when that problem or variants occur embedded in logic formulations.

Let r be the $m \times n$ $\{0, 1\}$ vector with elements r_{xy}, for each $x \in X$ and $y \in Y$. Interpret $r_{xy} = 0$ (resp. $r_{xy} = 1$) to mean that the edge (r, y) is not (resp. is) selected. In terms of the vector r, the CNF system given by (5.7.30) and (5.7.31) can be compactly formulated by the following linear inequality system.

$$(5.7.32) \qquad \begin{aligned} \sum_{x \in X} r_{xy} &\geq 1, \quad \forall y \in Y \\ \sum_{y \in Y} r_{xy} &\leq 1, \quad \forall x \in X \end{aligned}$$

Thus, the SAT problem for (5.7.30) and (5.7.31) is equivalent to finding a $\{0,1\}$ vector r that satisfies (5.7.32) or concluding that no such vector exists.

Suppose we replace the condition that r must be a $\{0,1\}$ vector by the bounds $0 \le r \le 1$. When these bounds plus (5.7.32) are written in matrix notation, say, as a system $B \cdot r \ge b$, then B turns out to be totally unimodular. Indeed, one may rewrite the problem of solving $B \cdot r \ge b$ as a network flow problem that may be solved by the aforementioned very efficient algorithms. We conclude that the SAT problem, when seen this way, becomes a rather easy problem. The same conclusion applies to the MINSAT case of (5.7.30) and (5.7.31), so that problem may also be very efficiently solved with network flow algorithms.

The above observations are well known, and the reader may wonder why anyone would even consider the CNF formulation of (5.7.30) and (5.7.31). The reason is that (5.7.30) and (5.7.31) may occur as part of a larger logic formulation that defies translation to a network flow problem. But one might want to explore the following solution approach when faced with such a logic formulation. One breaks out the portion consisting of (5.7.30) and (5.7.31) using one of the decompositions of later chapters, solves that subproblem via (5.7.32), and finally combines that solution with one for the remainder of the problem. When properly implemented, that approach may result in an attractive solution algorithm.

The above approach requires that one identify the portion of a given CNF system corresponding to (5.7.30) and (5.7.31) and then define the equivalent (5.7.32). That step is simplified when the SAT instance is reformulated as an equivalent MINSAT instance where m new clauses play the role of the $\frac{1}{2} \cdot m \cdot n \cdot (n-1)$ clauses of (5.7.31). Details are as follows.

Recall that the clauses of (5.7.31) enforce that, for each $x \in X$, there is at most one $y \in Y$ such that $p(x,y)$ is *True*. We introduce an additional predicate $q(x)$ defined on X. We assign a cost of 1 to each $p(x,y)$ and to each $q(x)$. Then we replace the $\frac{1}{2} \cdot m \cdot n \cdot (n-1)$ clauses of (5.7.31) by the following m clauses, one for each $x \in X$.

$$(5.7.33) \qquad\qquad (\bigvee_{y \in Y} p(x,y)) \vee q(x)$$

By (5.7.33), any satisfying solution for the SAT instance is also a satisfying solution for the MINSAT instance. Indeed, due to the costs, any such solution of the SAT instance is optimal for the MINSAT instance, with total cost equal to m.

On the other hand, if the SAT instance is not satisfiable, then the MINSAT instance either is unsatisfiable or has an optimal solution with total cost greater than m.

The clauses of (5.7.30) and (5.7.33) and the costs are easily accommodated by a network flow formulation. We use the vector r defined earlier and introduce an $m \times 1$ vector s where, for each $x \in X$, the element s_x corresponds to $q(x)$. We assign the costs of $p(x, y)$ and $q(x)$ to r_{xy} and s_x, respectively, and represent (5.7.30) and (5.7.33) by the inequality system

(5.7.34)
$$\sum_{x \in X} r_{xy} \geq 1, \quad \forall\, y \in Y$$

$$\left(\sum_{y \in Y} r_{xy}\right) + s_x \geq 1, \quad \forall\, x \in X$$

That system, with costs as specified, defines a network flow problem that has a solution with total cost equal to m if and only if (5.7.30) and (5.7.31) are satisfiable. In the case of satisfiability, for each $x \in X$ and each $y \in Y$, $r_{xy} = 1$ (resp. $r_{xy} = 0$) implies that $p(x, y)$ is *True* (resp. *False*).

Resolution and Balanced Matrices

Resolution generally does not maintain balancedness. An example case involves the following balanced, indeed totally unimodular, matrix.

(5.7.35)

$$\begin{array}{ccc} y & a & b \\ \hline 1 & 1 & 1 \\ 1 & 1 & 0 \\ -1 & 0 & -1 \\ -1 & 0 & 0 \end{array}$$

Balanced matrix

When Algorithm RESOLUTION FOR MATRIX (5.4.4) is applied to eliminate column y, the following nonbalanced matrix results.

(5.7.36)

$$\begin{array}{cc} a & b \\ \hline 1 & 1 \\ 1 & -1 \\ 1 & 0 \end{array}$$

Nonbalanced matrix

The reader may object to the example matrix of (5.7.35), since it contains a row with one nonzero entry and a monotone column. But it is easy to embed that matrix into an example matrix that is balanced and does not have these shortcomings.

The above negative result notwithstanding, balancedness implies an important property of matrices A over $\mathbb{B}$ that is maintained by resolution. That property is the integrality of the polyhedron $P(A) = \{r \mid A \cdot r \geq \underline{1} - q(A); \ 0 \leq r \leq \underline{1}\}$. Before we state and prove that result, we introduce an algorithm for projecting polyhedra.

(5.7.37) Algorithm PROJECT POLYHEDRON. *Projects a polyhedron into lower dimensional space. Specifically, for a given rational matrix B, say, with column index set Y, and a given rational vector b, let P be the polyhedron $P = \{r \mid B \cdot r \geq b\}$. For a specified index $y \in Y$, the algorithm deduces a matrix B' with column index set $Y - \{y\}$ and a vector b' such that the polyhedron $P' = \{r' \mid B' \cdot r' \geq b'\}$ is a projection of P in the following sense. Each vector $r \in P$ becomes upon removal of the element r_y a vector $r' \in P'$, and each vector $r' \in P'$ can by addition of a suitable r_y be extended to a vector $r \in P$.*

Input: Rational matrix B with column index set Y and rational vector b defining $P = \{r \mid B \cdot r \geq b\}$. Column index y of B.

Output: Rational matrix B' with column index set $Y - \{y\}$ and rational vector b' defining $P' = \{r' \mid B' \cdot r' \geq b'\}$. The polyhedron P' is a projection of P; that is, each vector $r \in P$ becomes upon removal of the element r_y a vector $r' \in P'$, and each vector $r' \in P'$ can by addition of a suitable r_y be extended to a vector $r \in P$.

Complexity: Polynomial.

Procedure:

1. Scale the inequalities $B \cdot r \geq b$ with positive factors so that any term involving r_y becomes $+r_y$ or $-r_y$. Denote the scaled system again by $B \cdot r \geq b$.
   Initialize $B' \cdot r' \geq b'$ as the trivial inequality system where B' has no rows.
2. For each inequality of $B \cdot r \geq b$ that does not contain $+r_y$ or $-r_y$: Declare that inequality to be part of $B' \cdot r' \geq b'$.
3. For any two inequalities of $B \cdot r \geq b$ of the form $r_y + d^t \cdot r' \geq \alpha$ and $-r_y + e^t \cdot r' \geq \beta$: Declare the inequality

(5.7.38) $$(d^t + e^t) \cdot r' \geq \alpha + \beta$$

   to be part of $B' \cdot r' \geq b'$.
4. Output B' and b' defining $P' = \{r' \mid B' \cdot r' \geq b'\}$, and stop.

Proof of Validity. By the derivation of the inequality system $B' \cdot r' \geq b'$, the polyhedron P' contains all vectors r' derived from the vectors $r \in P$ by deletion of the element r_y.

Conversely, let $r' \in P'$ be given. The inequality $(d^t + e^t) \cdot r' \geq \alpha + \beta$ derived for $B' \cdot r' \geq b'$ in Step 3 implies that $e^t \cdot r' - \beta \geq -d^t \cdot r' + \alpha$. The latter inequality holds for the vectors d and e arising from any two inequalities of $B \cdot r \geq b$ with r_y and $-r_y$, respectively. Thus, there exists a scalar r_y such that, for all such d and e, $e^t \cdot r' - \beta \geq r_y \geq -d^t \cdot r' + \alpha$, which implies $r_y + d^t \cdot r' \geq \alpha$ and $-r_y + e^t \cdot r' \geq \beta$. Thus, the vector r composed of r' and r_y is in P. $\qquad\square$

Projection preserves boundedness and integrality of polyhedra as follows.

(5.7.39) Theorem. *Any polyhedron P' derived from a bounded integral polyhedron P by Algorithm PROJECT POLYHEDRON (5.7.37) is bounded and integral.*

Proof. Boundedness of P' directly follows from the boundedness of P. Suppose P' has a fractional extreme point r'. Assign to r_y the largest possible value so that r composed of r' and r_y is in P. By this choice of r_y and the fact that r' is an extreme point of P', the vector r is an extreme point of P. But r is fractional, which contradicts the integrality of P.   □

We are ready to state and prove the main result of this subsection.

(5.7.40) Theorem. *Let A be a matrix over $\mathbb{B}$. Suppose the polyhedron $P(A)$ given by*

$$(5.7.41) \qquad P(A) = \{r \mid A \cdot r \geq \underline{1} - q(A); \ 0 \leq r \leq \underline{1}\}$$

is integral. Then, for any matrix A' derived from A by Algorithm RESOLUTION FOR MATRIX (5.4.4), the polyhedron $P(A')$ is integral as well.

Proof. Let column y be eliminated by Algorithm RESOLUTION FOR MATRIX (5.4.4).

Suppose we use Algorithm PROJECT POLYHEDRON (5.7.37) to project out component r_y of the vectors r of the bounded and integral $P(A)$, say, getting a polyhedron $P' = \{r' \mid B' \cdot r' \geq b'\}$. By Theorem (5.7.39), boundedness and integrality of $P(A)$ implies that the polyhedron P' is integral. Hence, we are done if we can prove that $P(A') = P'$. We establish the latter equation by showing that the inequality system $B' \cdot r' \geq b'$ computed by Algorithm PROJECT POLYHEDRON (5.7.37) is essentially the same as the inequality system defining $P(A')$, which consists of $A' \cdot r' \geq \underline{1} - q(A')$ and $0 \leq r' \leq \underline{1}$.

We say "essentially the same," since we modify $B' \cdot r' \geq b'$ a bit for the comparison. The changes involve the removal of some redundant inequalities and a strengthening of some inequalities such that all extreme points of the integral P' still satisfy the strengthened inequalities. Thus, the system so derived from $B' \cdot r' \geq b'$ still defines P'. Details of the comparison are as follows.

Step 2 of Algorithm PROJECT POLYHEDRON (5.7.37) places all inequalities of $A \cdot r \geq \underline{1} - q(A)$ and $0 \leq r \leq \underline{1}$ not involving $+r_y$ or $-r_y$—in particular, the inequalities $0 \leq r' \leq \underline{1}$—into $B' \cdot r' \geq b'$. By Step 2 of Algorithm RESOLUTION FOR MATRIX (5.4.4), all such inequalities also occur in the definition of $P(A')$.

Step 3 of Algorithm PROJECT POLYHEDRON (5.7.37) considers pairs of inequalities, each of which is an inequality of $A \cdot r \geq \underline{1} - q(A)$ with

$+r_y$ or $-r_y$ or is one of the inequalities $r_y \geq 0$ and $-r_y \geq -1$. Three cases of such pairings are possible.

In the first case, an inequality of $A \cdot r \geq \underline{1} - q(A)$ containing $+r_y$, say, $r_y + d^t \cdot r' \geq 1 - \gamma$, is paired with one of $A \cdot r \geq \underline{1} - q(A)$ containing $-r_y$, say, $-r_y + e^t \cdot r' \geq 1 - \delta$. According to (5.7.38), the following inequality is then added to $B' \cdot r' \geq b'$.

$$(5.7.42) \qquad (d^t + e^t) \cdot r' \geq (1 - \gamma) + (1 - \delta) = 1 - (\gamma + \delta - 1)$$

Note that $\gamma + \delta - 1$ is the number of -1s in d and e and thus is nonnegative. We examine two particular subcases of (5.7.42).

If for some column index z, both d_z and e_z are nonzero and of opposite sign, then $-(\gamma + \delta - 1)$ is less than the sum of the negative entries of $d^t + e^t$, so (5.7.42) is satisfied by all $\{0, 1\}$ vectors r. Accordingly, the inequality is redundant and can be eliminated.

If $d^t + e^t$ contains a $+2$ (resp. -2), then we strengthen the inequality by reducing the $+2$ to $+1$ (resp. by increasing the -2 to -1 and decreasing $\gamma + \delta - 1$ by 1). It is easily checked that all extreme points of P' still satisfy the strengthened inequality.

Second, consider any pairing of an inequality of $A \cdot r \geq \underline{1} - q(A)$ with one of the inequalities $r_y \geq 0$ or $-r_y \geq -1$. In each one of the two possible cases, the resulting inequality holds for all $\{0, 1\}$ vectors r. Thus, the inequality is redundant and can be eliminated.

Third and last, the pairing of $r_y \geq 0$ with $-r_y \geq -1$ produces the trivial inequality $0 \geq -1$, which can be eliminated.

Suppose the inequalities produced by Step 3 of Algorithm PROJECT POLYHEDRON (5.7.37) have been revised as described above. The resulting inequalities are readily seen to be the inequalities established by Step 3 of Algorithm RESOLUTION FOR MATRIX (5.4.4) for $P(A')$. Thus, $P(A') = P'$. ☐

5.8 Comparison of Matrix Classes

The reader may wonder whether the matrix classes introduced in Sections 5.4–5.7 are really all that different. For example, one of the classes might essentially subsume all other classes or at least several of them. We address that concern in this section.

In the first part, we make precise the notion of one class of matrices subsuming another class. In the second part, we use that concept to compare the matrix classes of Sections 5.4–5.7. The conclusions are as follows. The class of hidden nearly negative matrices subsumes the classes of 2SAT and nearly negative matrices, and the class of balanced matrices subsumes the classes of totally unimodular and network matrices. Neither one of the two classes of hidden nearly negative matrices and balanced matrices subsumes the other one. So, effectively, there are just two matrix classes that are basically different: the class of hidden nearly negative matrices and the class of balanced matrices.

We begin the detailed discussion. Let S and S' be two CNF systems. We say that S *subsumes* S' if

(5.8.1)

    (i)   All variables of S' occur in S.

    (ii)   Each satisfying solution of S can, by deletion of *True/False* values for the variables of S not occurring in S', be reduced to a satisfying solution of S'.

    (iii) Each satisfying solution of S' can, by assignment of certain *True/False* values to the variables of S not occurring in S', be extended to a satisfying solution of S.

Conditions (ii) and (iii) of (5.8.1) imply that S and S' are either both satisfiable or both unsatisfiable.

A matrix A over $\mathbb{B}$ *subsumes* a matrix A' over $\mathbb{B}$ if the CNF system S producing A subsumes the CNF system S' producing A'. By (5.8.1), A subsumes A' if and only if

(5.8.2)

    (i)   All column indices of A' occur in A.

    (ii)   Each solution of $A \odot s = \underline{1}$ can, by deletion of the entries corresponding to the column indices of A not occurring in A', be reduced to a solution for $A' \odot s' = \underline{1}$.

    (iii) Each solution of $A' \odot s = \underline{1}$ can, by addition of the entries corresponding to the column indices of A not occurring in A', be extended to a solution for $A \odot s' = \underline{1}$.

Subsumption is related to resolution as follows.

(5.8.3) Lemma. *A matrix A over $\mathbb{B}$ subsumes another matrix A' over $\mathbb{B}$ if and only if repeated applications of Algorithm RESOLUTION FOR MATRIX (5.4.4) can reduce A to a matrix A'' having the same set of satisfying solutions as A'.*

Proof. To show the "only if" part, we apply Algorithm RESOLUTION FOR MATRIX (5.4.4) to A and eliminate all column indices not occurring in A'. Let A'' be the resulting matrix. If A subsumes A', then by Algorithm RESOLUTION FOR MATRIX (5.4.4) the matrix A'' must have the same set of satisfying solutions as A'.

Conversely, if the latter fact holds, then again by Algorithm RESO-
LUTION FOR MATRIX (5.4.4) the matrix A subsumes A'.   □

We apply the notion of subsumption to classes of matrices over $\mathbb{B}$ in
the following manner. Let C and C' be two such classes. Then C *subsumes*
C' if, for each matrix $A' \in C'$, there exists a matrix $A \in C$ that subsumes
A'.

We also define a notion of efficient subsumption. Let C and C' be
as before. Then C *polynomially subsumes* C' if there exists a polynomial
algorithm that, given a matrix $A' \in C'$ as input, produces a matrix $A \in C$
that subsumes A'.

We view "subsumes" and "polynomially subsumes" as binary relations.
We include a few lemmas about these relations.

(5.8.4) Lemma. *The two relations "subsumes" and "polynomially subsumes" are transitive. That is, if C subsumes (resp. polynomially subsumes) C' and if in turn C' subsumes (resp. polynomially subsumes) C'',
the C subsumes (resp. polynomially subsumes) C''.*

Proof. Let A'' be given. Since C' subsumes C'', there exists a matrix
$A' \in C'$ that by repeated application of Algorithm RESOLUTION FOR
MATRIX (5.4.4) can be reduced to a matrix with the same set of satisfying
solutions as A''. The analogous statement holds for some $A \in C$ and the
matrix A' just determined. But then A can by repeated application of
Algorithm RESOLUTION FOR MATRIX (5.4.4) be reduced to a matrix
with the same set of satisfying solutions as A''. Hence, A subsumes A'',
and C subsumes C''.

If C polynomially subsumes C' and if C' polynomially subsumes C'',
then the matrices A' and A can be determined in polynomial time. Hence,
C polynomially subsumes C''.   □

(5.8.5) Lemma. *The class of hidden nearly negative matrices does not
subsume the class of network matrices.*

Proof. Consider the network matrix A' given by

(5.8.6)
$$A' = \begin{array}{|ccc|} \hline 1 & 1 & 1 \\ -1 & -1 & -1 \\ \hline \end{array}$$

Network matrix

Clearly, a *True/False* vector is a satisfying solution for A' if and only if it
contains at least one *True* and one *False*. Suppose there exists a hidden
nearly negative matrix A that subsumes A'. Theorem (5.6.5) states that
Algorithm RESOLUTION FOR MATRIX (5.4.4) maintains the property
of being hidden nearly negative. Therefore, if A has more columns than
A', then we may reduce A with Algorithm RESOLUTION FOR MATRIX

(5.4.4) to a hidden nearly negative matrix that has as many columns as A' and that subsumes A'. Assume A already to be that reduced matrix. Thus, A has three columns and the same set of satisfying solutions as A'.

If A contains a row with at least one zero, then that row is not satisfied by a certain solution of A with at least one *True* and at least one *False*. But this contradicts the fact that A and A' have the same set of satisfying solutions.

Hence, each row of A contains three nonzeros. If A has a row with at least one $+1$ and at least one -1, then again that row is not satisfied by a certain solution with at least one *True* and at least one *False*, another contradiction.

Hence, each row of A has three $+1$s or three -1s. Indeed, for the solution sets of A and A' to match, both types of rows must occur in A. But then A cannot be scaled to become nearly negative and thus is not hidden nearly negative, a contradiction.   □

(5.8.7) Lemma. *The class of balanced matrices does not subsume the class of nearly negative 2SAT matrices.*

Proof. We use the following nearly negative 2SAT matrix A'.

(5.8.8)
$$A' = \begin{vmatrix} -1 & -1 & 0 \\ 0 & -1 & -1 \\ -1 & 0 & -1 \end{vmatrix}$$

Nearly negative 2SAT matrix

The satisfying solutions of A' are characterized by the requirement that they contain at most one *True* value. Let P be the integral polyhedron that contains precisely the $\{0,1\}$ vectors $r = [r_1, r_2, r_3]^t$ corresponding to the satisfying solutions of A'. It is easily checked that the inequalities

(5.8.9)
$$\begin{aligned} -\underline{1}^t \cdot r &\geq -1 \\ r &\geq 0 \end{aligned}$$

define P. Indeed, P is full-dimensional, and (5.8.9) is the unique minimal description $B \cdot r \geq b$ of P, up to scaling of the inequalities by positive factors.

We claim that the inequality $-\underline{1}^t \cdot r \geq -1$ of (5.8.9) cannot occur as part of an inequality system $A \cdot r \geq \underline{1} - q(A)$ with $\{0, \pm 1\}$ A. This is because r of (5.8.9) is a 3×1 vector, and $-\underline{1}^t \cdot r \geq -1$ as inequality of a system $A \cdot r \geq \underline{1} - q(A)$ would have the right-hand side value $1 - 3 = -2$ and not -1.

We conclude that there is no matrix A over $\mathbb{B}$ with three columns such that $P(A)$ is equal to P. Put differently, if a matrix A over $\mathbb{B}$ produces a polyhedron $P(A)$ with the same set of integer solutions as P, then $P(A)$ cannot be integral.

Suppose there exists a balanced matrix A that subsumes A' of (5.8.8). We reduce A with Algorithm RESOLUTION FOR MATRIX (5.4.4) to a matrix A'' with the same column index set and the same solution set as A'.

Theorem (5.7.13) states that balancedness of A implies integrality of $P(A)$. According to Theorem (5.7.40), integrality of $P(A)$ implies integrality for $P(A'')$. Finally, Lemma (5.8.3) says that A'' and A' must have the same set of satisfying solutions. But we have seen that the latter fact implies that $P(A'')$ cannot be integral, a contradiction.   □

The main result of this section follows.

(5.8.10) Theorem.
(a) *The class of hidden nearly negative matrices polynomially subsumes both the class of 2SAT matrices and the class of nearly negative matrices.*
(b) *The class of balanced matrices polynomially subsumes both the class of totally unimodular matrices and the class of network matrices.*
(c) *Neither one of the two classes of hidden nearly negative matrices and of balanced matrices subsumes the other one.*

Proof. We show parts (a) and (b). Evidently, the class of nearly negative matrices is contained in the class of hidden nearly negative matrices, and the classes of totally unimodular matrices and network matrices are contained in the class of balanced matrices. We complete the proof of parts (a) and (b) by showing that the class of hidden nearly negative matrices polynomially subsumes the class of 2SAT matrices. Let a 2SAT matrix A' be given. We use the polynomial Algorithm SOLVE 2SAT (5.4.1) to find a satisfying solution for A' or to conclude that A' is not satisfiable.

Assume the former case. Theorem (5.6.2) says that a satisfiable 2SAT matrix—in particular, the given A'—is hidden nearly negative. Thus, we choose $A = A'$ as the hidden nearly negative matrix that subsumes A'.

Suppose the latter case is at hand. We select as the hidden nearly negative matrix A the zero matrix with one row and with the same column index set as A'. The matrix A is unsatisfiable and thus subsumes A'.

We turn to part (c). According to Lemma (5.8.5), the class of hidden nearly negative matrices does not subsume the class of network matrices. On the other hand, the class of balanced matrices does subsume the class of network matrices. Since Lemma (5.8.4) establishes transitivity of "subsumes," we conclude that the class of hidden nearly negative matrices does not subsume the class of balanced matrices.

According to Lemma (5.8.7), the class of balanced matrices does not subsume the class of nearly negative 2SAT matrices. However, the class of hidden nearly negative matrices does subsume the class of nearly negative 2SAT matrices. Once more using the transitivity of "subsumes," we see that the class of balanced matrices does not subsume the class of hidden nearly negative matrices.   □

In the final section, we describe extensions and list references.

5.9 Extensions and References

The principle of dynamic programming mentioned in Section 5.2 is explained in any textbook on dynamic programming—for example, in Bellman (1957), Bellman and Dreyfus (1962), Dreyfus and Law (1977), or Bertsekas (1987).

The elementary reduction steps of Section 5.3 involving rows with at most one nonzero entry, monotone columns, etc. are used in virtually every algorithm for the SAT problem. It is interesting that these reductions may also be used in any analysis algorithm for SAT or MINSAT, as shown by Algorithms REDUCE SAT INSTANCE (5.3.4) and REDUCE MINSAT INSTANCE (5.3.5).

The linear time Algorithm SOLVE 2SAT (5.4.1) of Section 5.4 is due to Evan, Itai, and Shamir (1976). For other 2SAT solution algorithms and computational results, see Petreschi and Simeone (1980, 1991).

For the $\mathcal{NP}$-completeness proofs concerning the problems GRAPH K-COLORABILITY, INDEPENDENT SET, and VERTEX COVER, see Garey and Johnson (1979). The special case of INDEPENDENT SET solved in Section 5.4 via 2SAT is discussed in somewhat different form by Simeone (1985). Bagchi, Servatius, and Shi (1995) formulate the diagnosis of faulty processors in massively parallel computing systems as a 2SAT problem.

References about resolution are given in Chapter 1. We do not know who first stated Theorem (5.4.5), which says that resolution turns a given 2SAT instance into another 2SAT instance. At any rate, it is a well-known observation.

An efficient algorithm for solving nested 2SAT instances is given by Jaumard, Marchioro, Morgana, Petreschi, and Simeone (1990). The case of uniquely solvable 2SAT instances is treated in Hansen and Jaumard (1985).

Aspvall, Plass, and Tarjan (1979) provide a linear time algorithm for an extension of 2SAT where each clause is a CNF formula with quantification and at most two literals per clause. Hansen, Jaumard, and Plateau (1993) define a SAT-central class called *Extended Nested Satisfiability*. The linear time solution algorithm employs the 2SAT algorithms of Aspvall, Plass, and Tarjan (1979) and Hansen and Jaumard (1985) as subroutines. The recognition algorithm is also linear time. The class Extended Nested Satisfiability subsumes a class called *Nested Satisfiability* by Knuth (1990). The latter reference includes a linear time solution algorithm, but does not solve the recognition problem. In Chapter 11, it is shown that the class of Extended Nested Satisfiability is a particular case of augmented sum decomposition. Also related to the Nested Satisfiability class is a subclass of the MAXSAT problem discussed in Kratochvíl and Křivánek (1993).

Define MIN2SAT (resp. MAX2SAT) to be the subclass of MINSAT (resp. MAXSAT) where the matrix has 2SAT form. Approximation algorithms for MIN2SAT are discussed in Gusfield and Pitt (1992) and Hochbaum, Megiddo, Naor, and Tamir (1993). The latter reference includes an interesting MIN2SAT formulation of integer programs with bounded variables and with at most two variables occurring in each inequality. The assumption of bounds on the variables is important, since integer programs with at most two variables per inequality are shown in Lagarias (1985) to be $\mathcal{NP}$-complete. It is an open problem whether finding a feasible solution for such integer programs is difficult. Approximation algorithms for MAX2SAT are given by Johnson (1974), Lieberherr and Specker (1981), Poljak and Turzík (1982), Hansen and Jaumard (1990), Yannakakis (1992), Feige and Goemans (1995), Goemans and Williamson (1995), and Cheriyan, Cunningham, Tunçel, and Wang (1996).

The clauses represented by the nearly negative matrices of Section 5.5 were first investigated by Horn (1951) and are usually called *Horn clauses*. The SAT case of the linear Algorithm SOLVE NEARLY NEGATIVE SAT OR MINSAT (5.5.1) was first published by Itai and Makowsky (1982, 1987). Other linear time algorithms for that case are given by Dowling and Gallier (1984); see also Minoux (1988), Scutellà (1990), and Ghallab and Escalada-Imaz (1991). Dowling and Gallier (1984) also prove the minimality result of Theorem (5.5.2).

Gallo and Urbani (1989) and Gallo and Pretolani (1995) rely on nearly negative subproblems to prune the search tree while solving general SAT instances.

Minoux (1992), Berman, Franco, and Schlipf (1995), and Pretolani (1993b) give efficient algorithms for deciding unique satisfiability of nearly negative matrices. The third reference contains the fastest algorithm, which runs in linear time.

The problem of finding compact representations for satisfiability problems with nearly negative matrices is examined by Hammer and Kogan (1992, 1993, 1995, 1996), Boros and Čepek (1994), and Čepek (1995).

Additional results about nearly negative matrices are in Henschen and Wos (1974), Hooker (1989), Jeroslow and Wang (1989), Boros, Crama, and Hammer (1990), Heusch (1994), Boros and Čepek (1995), and Ekin, Hammer, and Peled (1997).

Further references for nearly negative matrices are included in Section 8.7.

We do not know who first observed the well-known Theorems (5.5.5) and (5.6.5), according to which resolution maintains near negativity and hidden near negativity, respectively.

Nearly negative matrices over the reals occur in areas quite removed from logic. We mention two example classes.

The first class consists of the *input–output matrices* pioneered by Leontief for economic analysis; see Dorfman, Samuelson, and Solow (1958), Leontief (1986), and Jeroslow, Martin, Rardin, and Wang (1992).

The second class consists of the Z *matrices* of linear complementarity theory. These matrices are real and square, and their off-diagonal elements are nonpositive. The matrices were first investigated by Chandrasekaran (1970), who gave a polynomial algorithm for the *linear complementarity problem* involving such matrices. For a comprehensive discussion of Z matrices, see Cottle, Pang, and Stone (1992).

We do not know who first proposed the use of column scaling as part of theorem-proving methods. An early reference is Meltzer (1965).

Lewis (1978) first gave a polynomial, indeed quadratic, recognition algorithm for the class of hidden nearly negative matrices of Section 5.6. The method uses the 2SAT formulation of that section. An improved linear algorithm, which also uses a 2SAT formulation, was subsequently given by Aspvall (1980). That reference includes Theorem (5.6.2), which links satisfiability of 2SAT instances with hidden near negativity.

Other recognition algorithms for hidden nearly negative matrices are due to Mannila and Mehlhorn (1985), Lindhorst and Shahrokhi (1989), and Hébrard (1994). A test for uniqueness of the column scaling factors that convert a hidden nearly negative matrix to a nearly negative matrix is presented by Hébrard (1995).

The characterization problem of hidden nearly negative matrices is treated in Chapter 6. Related references are Chandru, Coullard, and Montañez (1988) and Chandru, Coullard, Hammer, Montañez, and Sun (1990).

An interesting class of SAT semicentral matrices called *extended Horn* is given in Chandru and Hooker (1991). The class contains the class of hidden nearly negative matrices. The reference does not give a polynomial recognition algorithm for the class, but Swaminathan and Wagner (1995) describe a polynomial recognition algorithm for a certain subclass that properly includes the class of hidden nearly negative matrices. Thus, that subclass is SAT central.

One may extend Algorithm SOLVE NEARLY NEGATIVE SAT OR MINSAT (5.5.1) and Algorithm SOLVE HIDDEN NEARLY NEGATIVE SAT OR MINSAT (5.6.3) so that certain generalizations of SAT and MINSAT can be solved. Specifically, define SAT-b to be the following class of satisfiability problems. Each instance is specified by a CNF system S and an integral positive vector b. The elements of b are indexed by the clause index set X of S. One must find *True/False* values for the variables of S such that, for each $x \in X$, at least b_x literals of clause x evaluate to *True*. In matrix notation, an instance of SAT-b is given by a pair (A, b), where A is over $\mathbb{B}$ and has row index set X and where b is as before.

Analogously, we extend MINSAT to MINSAT-b, where each instance is given by a triple (A, b, c), where A and b are as before and where c is a rational nonnegative cost vector whose elements are indexed by the column index set Y of A.

When A is nearly negative, an elementary modification of Algorithm SOLVE NEARLY NEGATIVE SAT OR MINSAT (5.5.1) gives a solution algorithm for SAT-b and MINSAT-b. The changes of the steps of Algorithm SOLVE NEARLY NEGATIVE SAT OR MINSAT (5.5.1) are as follows.

1. If A has a row $x \in X$ with less than b_x nonzero entries: Declare the given SAT-b or MINSAT-b instance to be unsatisfiable, and stop.

   If A has no rows: Assign *False* to each column of A. These values plus the earlier assigned *True/False* values, if any, constitute a solution for the SAT-b or MINSAT-b instance, whichever applies.

   If A has a row x with exactly b_x nonzero entries, say, in columns $y \in \overline{Y} \subseteq Y$: Assign *True* (resp. *False*) to each column $y \in \overline{Y}$ with a $+1$ (resp. -1) in row x; for each $z \in X$, reduce b_z by the number of *True* values produced in row z by these assignments; delete all rows $z \in X$ for which the reduced b_z value is nonpositive; delete all columns $y \in \overline{Y}$; and repeat Step 1.

2. (A has in each row x more than b_x nonzeros and thus at least b_x -1s.) Assign *False* to each column of A. These values plus the earlier assigned *True/False* values, if any, constitute a solution for the SAT-b or MINSAT-b instance, whichever applies.

Validity of the revised algorithm is argued almost exactly as for Algorithm SOLVE NEARLY NEGATIVE SAT OR MINSAT (5.5.1). In particular, the analogue of Theorem (5.5.2) holds.

Similarly, one may derive from Algorithm SOLVE HIDDEN NEARLY NEGATIVE SAT OR MINSAT (5.6.3) a solution algorithm for SAT-b and MINSAT-b with hidden nearly negative A.

Theorems (5.4.5), (5.5.5), and (5.6.5) state that resolution maintains the 2SAT property, near negativity, and hidden near negativity, respectively. There is a fundamental difference, though, between the first result and the latter two. Since a 2SAT matrix with n columns has at most $O(n^2)$ distinct rows, the growth of the number of rows for 2SAT matrices under repeated application of resolution is polynomially bounded. Generally, this is not so for nearly negative or hidden nearly negative matrices. We give an example. For any $n \geq 1$, let B be the $(2n+2) \times n$ $\{0, \pm1\}$ matrix whose nonzero entries are, for $i = 1, 2$ and $j = 1, 2, \ldots, n$, $B_{i+2(j-1),j} = 1$ and $B_{i+2(j-1)+2,j} = -1$. Adjoin to B a negated identity matrix of appropriate size to get a nearly negative matrix $A = [B|(-I)]$. Apply resolution to A to eliminate the columns of B. It is easy to check that the resulting matrix has 2^{n+1} distinct rows. Thus, the growth of the number of rows cannot be bounded by a polynomial.

We turn to the balanced matrices of Section 5.7. Balancedness of $\{0, 1\}$ matrices was first defined and explored by Berge (1972, 1973). Early results about such matrices include Fulkerson, Hoffman, and Oppenheim (1974) and Truemper and Chandrasekaran (1978). A structure theory is given in Conforti, Cornuéjols, and Rao (1997).

The extension of balancedness to $\{0, \pm 1\}$ matrices is introduced in Truemper (1982); see also Truemper (1992). A structure theory of $\{0, \pm 1\}$ balanced matrices is developed in Conforti, Cornuéjols, Kapoor, and Vušković (1994a, 1994b). The second reference includes the polynomial recognition algorithm that in much abbreviated form is included as Algorithm TEST BALANCEDNESS (5.7.3). Additional results for $\{0, \pm 1\}$ balanced matrices are established in Conforti and Cornuéjols (1995). A comprehensive survey of balancedness results is given in Conforti, Cornuéjols, Kapoor, Vušković, and Rao (1994).

The pioneering decomposition of the so-called *regular matroids* by Seymour (1980) implicitly contains a polynomial recognition algorithm for totally unimodular matrices. An effective algorithm that uses the decomposition result of Seymour (1980) is given in Truemper (1990) and is the basis for Algorithm TEST TOTAL UNIMODULARITY (5.7.8).

Algorithm TEST NETWORK PROPERTY (5.7.9) is based on Heller and Tompkins (1956). An alternate method is given in Truemper (1976).

The polyhedral results and the algorithms for LPs cited in Section 5.7 may be found in Chvátal (1983), Schrijver (1986), Grötschel, Lovász, and Schrijver (1993), Nemhauser and Wolsey (1988), and Karloff (1991).

Theorem (5.7.13), which establishes the integrality of the polyhedron $P(A) = \{r \mid A \cdot r \geq \underline{1} - q(A); \ 0 \leq r \leq \underline{1}\}$, was first proved by Conforti and Cornuéjols (1992).

Theorem (5.7.15), which says that, for any totally unimodular matrix A and for any integral vector b, the polyhedron $P(A, b) = \{r \mid A \cdot r \geq b; \ 0 \leq r \leq \underline{1}\}$ is integral, is due to Hoffman and Kruskal (1956).

Both Theorems (5.7.13) and (5.7.15) are subsumed by a more general, yet easily proved, result of Conforti, Cornuéjols, and Truemper (1994). The result links balancedness and total unimodularity via the exclusion of classes of minimal non-totally unimodular matrices. The proof is based on Truemper and Chandrasekaran (1978), where the $\{0, 1\}$ matrix case is treated.

For a characterization of the integrality of the polyhedron $P(A) = \{r \mid A \cdot r \geq \underline{1} - q(A); \ 0 \leq r \leq \underline{1}\}$ for a certain class of matrices A over $\mathbb{B}$, see Hooker (1996). Results related to that reference are in Boros and Čepek (1995), Nobili and Sassano (1997), Conforti, Cornuéjols, and de Francesco (1997), and Guenin (1997). For a summarizing treatment, see Conforti, Cornuéjols, Kapoor, and Vušković (1996).

Theorem (5.7.22), which establishes the existence of certain *True/False* values for matrices A with integral polyhedron $P(A) = \{r \mid A \cdot r \geq \underline{1} - q(A); \ 0 \leq r \leq \underline{1}\}$, is due to Conforti and Cornuéjols (1992).

The basic reference for network flow matrices, problems, and algorithms is the classic book by Ford and Fulkerson (1962). Since that book appeared, a number of improved flow algorithms have been developed. For a complete treatment, see Ahuja, Magnanti, and Orlin (1993).

One may generalize the network matrices with at most two nonzeros in each column by dropping the requirement of total unimodularity. We call such matrices *matching matrices*, since the SAT and MINSAT instances involving them belong to the well-known class of *matching problems*. The pioneering work on matching problems by Edmonds (1965a, 1965b) produced efficient algorithms and important structural results. For a complete treatment, see Lawler (1976) and Lovász and Plummer (1986). Very efficient solution algorithms are given by Derigs and Metz (1991) and Applegate and Cook (1993).

Since SAT simple matrices do not have monotone columns and since a matching matrix without monotone columns is a network matrix, any SAT instance with a matching matrix can be reduced to one with a network matrix. Such a reduction generally is not possible for MINSAT instances involving matching matrices. Hence, such problems should be solved with the efficient algorithms of the cited references.

Solution approaches to the pigeonhole problem that are different from that of Section 5.7 are given by Cook (1976), Cook and Reckhow (1979), Buss (1987), Cook, Coullard, and Turán (1987), and Bibel (1990). Other results about the pigeonhole problem are given by Haken (1985), Buss and Turán (1988), Paris, Wilkie, and Woods (1988), Cook and Pitassi (1990), and Ajtai (1994).

Notions of SAT problem consequence and equivalence that are somewhat different from the concept of subsumption of Section 5.8 are provided by Kleine Büning (1990).

Algorithm PROJECT POLYHEDRON (5.7.37) is one step of the so-called Fourier–Motzkin *elimination method.* For a detailed discussion of that method, see Schrijver (1986). The connection between Fourier–Motzkin elimination and resolution used in the proof of Theorem (5.7.40) is well known. We do not know who established that result first.

Schlipf, Annexstein, Franco, and Swaminathan (1995) define a class of matrices A over $\mathbb{B}$ for which the SAT problem can be solved by a simple quadratic algorithm. The method can be made linear using parallel checking of certain cases. The class of matrices handled by the algorithm includes the extended Horn matrices of Chandru and Hooker (1991) and the balanced matrices. However, the class is not maintained under submatrix taking and thus is not SAT semicentral.

Additional central and semicentral classes of matrices and related references are included in Chapters 8–12. Chapter 14 summarizes and extends these results.

The next chapter contains characterizations of the class of hidden nearly negative matrices.

Chapter 6

Characterizations of Hidden Near Negativity

6.1 Overview

This chapter establishes some characterizations for the class of hidden nearly negative matrices. The results are not used in subsequent chapters, so the reader may skip this chapter without loss of continuity.

A class of matrices, say, C, is typically specified in one of two ways. First, C may be directly defined as the class of matrices having a certain property. Second, C may be indirectly defined as the class of matrices not having a certain other property. Regardless of the way C is defined, one may seek additional specifications that are mathematically interesting. Such alternate specifications usually are called *characterizations*. In particular, if one has a definition of the first (resp. second) kind, then it is interesting to determine characterizations of the second (resp. first) kind.

We pursue this idea for the classes of central matrices of Chapter 5. Theorem (5.8.10) says that those central classes are subsumed by the class of hidden nearly negative matrices and the class of balanced matrices. Hence, we restrict ourselves to the latter two classes.

Consider the class C of $\{0, \pm 1\}$ balanced matrices. Section 5.7 defines that class by the exclusion of cycle submatrices for which the sum of the entries is equal to 2(mod 4). Thus, we have a definition of C of the second kind.

Theorem (5.7.13) says that a $\{0, \pm 1\}$ matrix is balanced if and only if, for each submatrix A of that matrix, the polyhedron $P(A) = \{r \mid A \cdot r \geq \underline{1} - q(A); \ 0 \leq r \leq \underline{1}\}$ is integral. Evidently, Theorem (5.7.13) supplies a characterization of the first kind.

A more complicated characterization of C of the first kind is implicit in Algorithm TEST BALANCEDNESS (5.7.3). That algorithm utilizes certain decompositions. One may rewrite the underlying decomposition results into a characterization of C of the first kind.

We turn to the hidden nearly negative matrices, that is, to the $\{0, \pm 1\}$ matrices that can be column scaled to become nearly negative. Let C be the class of those matrices. Clearly, the definition of C is of the first kind.

We want a characterization of C of the second kind, that is, a characterization that demands a certain property to be absent. Specifically, we desire a characterization that establishes membership in C by the exclusion of certain matrix structures.

Since hidden near negativity is maintained under subregion taking and since we would like to obtain a compact characterization, we take it as our goal to identify the matrices that are not hidden nearly negative and that, subject to that condition, are minimal under subregion taking. We call these matrices the *minimal excluded subregions of hidden near negativity*. Locating these minimal matrices is surprisingly difficult. We carry out that task in the next two sections.

In Section 6.2, we review relevant definitions—in particular, the definition of the labeled, directed, bipartite graph DBG(A) arising from any $\{0, \pm 1\}$ matrix A and the definition of Boolean minor. We establish the minimal Boolean minors of the graphs DBG(A) that must be excluded if A is to be hidden nearly negative. These minors are the *minimal excluded Boolean minors of hidden near negativity*.

In Section 6.3, we derive from the minimal excluded Boolean minors the minimal excluded subregions of hidden near negativity. As a corollary, we derive a characterization of the satisfiable 2SAT matrices.

The reader may wonder why we do not establish the excluded subregion characterization of Section 6.3 directly, without the detour via minimal excluded Boolean minors. We have two reasons. First, the characterization of those minors is interesting in its own right. Second, the number of minimal excluded subregions is rather large. The detour via Boolean minors significantly reduces the effort to establish these minimal matrices.

In the final section, 6.4, we discuss related material and list references.

6.2 Minimal Excluded Boolean Minors

Suppose a given $\{0, \pm 1\}$ matrix A is not hidden nearly negative. In this section, we show that the labeled, directed, bipartite graph DBG(A) of A has one or more minors of a certain form that certify A to be not hidden nearly negative. The minors are minimal with respect to node deletion. Collectively, the minors constitute the minimal excluded Boolean minors of hidden near negativity. We first review relevant definitions.

Review of Definitions

According to Sections 5.5 and 5.6, a $\{0, \pm 1\}$ matrix A is nearly negative if each row contains at most one $+1$, and it is hidden nearly negative if A becomes nearly negative upon an appropriate column scaling by $\{\pm 1\}$ factors.

By Section 2.6, the labeled, directed, bipartite graph $\text{DBG}(A)$ is derived from a $\{0, \pm 1\}$ matrix A as follows. We start with the bipartite graph $\text{BG}(A)$. In the latter graph, each edge represents a $\{\pm 1\}$ entry of A. The graph $\text{BG}(A)$ does not differentiate between $+1$ and -1 entries of A. But we may encode that information for each nonzero entry A_{xy} of A by directing in $\text{BG}(A)$ the corresponding edge, which connects row node x with column node y. Specifically, if $A_{xy} = 1$ (resp. $A_{xy} = -1$), we direct that edge from row node x to column node y (resp. column node y to row node x). We convert the resulting directed, bipartite graph to a labeled, directed, bipartite graph by assigning the label 1 to each arc. The latter graph we declare to be $\text{DBG}(A)$, the "D" indicating "directed." Note that $\text{DBG}(A)$ has only 1s as arc labels, while a labeled, directed, bipartite graph in general may have 1s and 2s assigned to its arcs.

Section 2.5 defines the operation of Boolean minor taking for labeled, directed, bipartite graphs. Let H be such a graph. We scale a column node of H by reversing the direction of the arcs incident at that node. The labels of the arcs are not affected. Column scaling of H refers to possibly repeated scaling of column nodes of H. Suppose $G_1, G_2, \ldots, G_n$ are the strong components of H. Then we shrink H by first collapsing, for each G_k, $k = 1, 2, \ldots, n$, the row nodes of G_k to a new row node and collapsing the column nodes of G_k to a new column node. Of course, G_k may not have any row (resp. column) nodes. In that case, G_k has just one column (resp. row) node, and that node is not affected according to the rule for collapsing nodes. In the next step of the shrinking operation, we delete all arc labels and replace any instance of multiple arcs with the same endpoints and the same direction by just one arc each. Finally, in the reduced graph we assign to each arc the label 1 or 2, where the case of a 1 corresponds precisely to the following situation. Let the arc in question connect the row node r and the column node c of the reduced graph. Define R (resp. C) to be the set of row (resp. column) nodes of H that were collapsed to form r (resp. c). If in the reduced graph the arc in question goes from node r to node c (resp. from node c to node r), then that arc receives the label 1 if and only if in H every row node of R has exactly one arc outgoing to (resp. incoming from) the nodes of C and that arc has the label 1. The graph $\overline{H}$ resulting from these steps is the graph produced by shrinking from H.

We reduce H to a Boolean minor $\overline{H}$ using column scaling, shrinking, and deletion of column or row nodes. Any one of these operations may be omitted. But, modulo such omissions, we always consider these operations done in the specified order. The inverse operations are the addition of nodes, unshrinking, and column scaling, always done in that order. We demand adherence to the specified order, since a resequencing of reduction steps may produce different minors or may even lead to undefined situations.

Hidden Near Negativity of Graphs

We extend the notion of near negativity and hidden near negativity of matrices to labeled, directed, bipartite graphs. Let H be such a graph. Then H is *nearly negative* if the following condition is satisfied: Each row node has at most one leaving arc, and if such an arc is present, then that arc must have a 1 as label. The graph H is *hidden nearly negative* if it becomes nearly negative by appropriate column scaling. We emphasize that the above definitions apply not just to graphs of the form DBG(A), where each arc has a 1 as label, but to general labeled, directed, bipartite graphs, where both 1s and 2s may occur as arc labels.

The definitions of near negativity and hidden near negativity for labeled, directed, bipartite graphs are consistent with those for $\{0, \pm 1\}$ matrices, as shown in the following lemma.

(6.2.1) Lemma. *A $\{0, \pm 1\}$ matrix A is nearly negative (resp. hidden nearly negative) if and only if the graph DBG(A) has the same property.*

Proof. We prove the case of hidden near negativity. The matrix A has that property if and only if A can be column scaled such that each row of the scaled matrix contains at most one $+1$. The latter condition holds if and only if DBG(A) can be column scaled such that each row node of the scaled graph has at most one leaving arc. By the definition of DBG(A), any such arc must have a 1 as label. Thus, the scaling condition for DBG(A) is satisfied if and only if that graph is hidden nearly negative. □

We desire a simplified notation for the discussion to follow. So for the remainder of this chapter, we let H denote the labeled, directed, bipartite graph DBG(A) of a given $\{0, \pm 1\}$ matrix A. When the connection between the graph H and the matrix A is to be emphasized, we say that H *corresponds* to A. We also abbreviate "Boolean minor" to "minor."

Basic Results for Minors

We establish some elementary results for the minors of a given graph H.

(6.2.2) Lemma. *Let $\overline{H}$ be the minor of a graph H obtained by shrinking. Then every strong component of $\overline{H}$ has one or two nodes.*

Proof. Let $\overline{H}$ have a strong component with at least three nodes. Each node of that component corresponds to a node subset of H, and the union of these node subsets of H defines a strong component of H that is not among the strong components used in the shrinking process. But this contradicts the definition of the shrinking process. □

(6.2.3) Lemma. *Suppose a given graph H is column scaled and shrunk so that the resulting graph $\overline{H}$ has as few nodes as possible. Let a graph $\overline{H}'$ be derived from $\overline{H}$ by column scaling. Then H can be column scaled to a graph H' so that shrinking, when applied to H', produces $\overline{H}'$. Furthermore, the node sets of the strong components of $\overline{H}'$ are precisely the node sets of the strong components of $\overline{H}$.*

Proof. We may assume that $\overline{H}$ is obtained from H just by shrinking. Suppose we column scale $\overline{H}$, getting a graph $\overline{H}'$. We column scale H as follows. Suppose a column c of $\overline{H}$ was produced by collapsing a node subset C of H. We apply to each node of C in H the scale factor for node c of $\overline{H}$ that was used in the derivation of $\overline{H}'$. Let H' be the graph so produced from H. It is easily checked that any strong component of H corresponds to a strongly connected subgraph of H'.

Supppose H' contains a strong component that has no counterpart in H. Then shrinking reduces H' to a minor $\overline{\overline{H}}'$ that has fewer nodes than $\overline{H}$. Since $\overline{\overline{H}}'$ is also a minor of H, we have a contradiction of the minimality of $\overline{H}$. Thus, the node sets of the strong components of H are precisely the node sets of the strong components of H'. This fact implies that shrinking reduces H' to $\overline{H}'$ and that the node sets of the strong components of $\overline{H}'$ are precisely the node sets of the strong components of $\overline{H}$. □

Near negativity is maintained by shrinking and unshrinking as follows.

(6.2.4) Lemma. *Let $\overline{H}$ be a minor of a graph H obtained by shrinking. Then H is nearly negative if and only if this is so for $\overline{H}$.*

Proof. For proof of the "only if" part, suppose that $\overline{H}$ is not nearly negative. Thus, $\overline{H}$ has a row node r with at least two leaving arcs, each with the label 1, or with at least one leaving arc with the label 2. Let R be the subset of row nodes of H defining r. By the rules for labels of $\overline{H}$, the graph H must contain a row node in R having at least two leaving arcs with the label 1 or having at least one leaving arc with the label 2. Thus, H is not nearly negative. The converse part can be proved similarly. □

Theorem (6.2.6) below establishes that hidden near negativity holds for a graph H if and only if it holds for certain minors of H. We need the following lemma for the proof of that theorem.

(6.2.5) Lemma. *Assume that a graph H can be column scaled to become nearly negative. Suppose that a directed path of H begins at a column node* *and that the scaling factor for that node is selected to be a* -1. *Then near negativity can only be achieved if the scaling factor of each column node of the path is equal to* -1.

Proof. If the lemma is false, then the path has two successive arcs (i, k) and (k, j) where i is a column node with scaling factor -1, k is a row node, and j is a column node with scaling factor 1. The scaling converts the arc (i, k) to an arc (k, i), and leaves the arc (k, j) unchanged. Thus, the scaled graph has at least two arcs leaving row node k and cannot be nearly negative, a contradiction.   □

(6.2.6) Theorem. *Let $\overline{H}$ be a minor of H produced by shrinking. Then H can be column scaled to become nearly negative if and only if this is so for $\overline{H}$.*

Proof. For proof of the "if" part, suppose that $\overline{H}$ can be column scaled to become a nearly negative graph, say, $\overline{H}'$. We convert the scaling factors of $\overline{H}$ to ones for H as follows. Suppose a column node c of $\overline{H}$ is produced from H by collapsing of a column node subset C. Then, to each node of C in H, we assign as scaling factor the one for node c in $\overline{H}$. We claim that H', the graph derived from H by that scaling, is nearly negative, which establishes the "if" part. Suppose this is not so. Thus, a row node i of H' has at least two outgoing arcs or exactly one outgoing arc with the label 2. That node is part of a row node subset R that in the shrinking of H was collapsed to a row node r of $\overline{H}$. By direct checking of the few possible cases, we conclude that the row node r of $\overline{H}'$ has at least two outgoing arcs or exactly one outgoing arc with the label 2, a contradiction.

We turn to the "only if" part. Thus, we assume that H can be column scaled to become a nearly negative graph H'. We claim that all column nodes of a given strong component of H must be scaled with the same factor. Indeed, each column node of a strong component is joined to any other column node of that strong component by a directed path. By Lemma (6.2.5), all nodes of such a path must be scaled by the same factor if a nearly negative graph is to result. We thus may deduce scaling factors for the column nodes of $\overline{H}$ from the scaling factors for H as follows. If a column node subset C of H is collapsed to produce a column node c of $\overline{H}$, then we assign to node c the factor used for any one node of C in H. Direct checking of the few possible cases for the row nodes of $\overline{H}$ then establishes that these scaling factors convert $\overline{H}$ to a nearly negative graph.   □

(6.2.7) Corollary. *If H is hidden nearly negative, then this is so for each minor of H.*

Proof. A minor of H is produced by column scaling, shrinking, and deletion of nodes. Suppose H is hidden nearly negative. Clearly, column scaling and deletion of nodes maintain that property. By Theorem (6.2.6), shrinking does so as well.   □

Suppose that H is not hidden nearly negative. Then a given minor of H may or may not be hidden nearly negative. Hence, it makes sense to ask for a minor of H that is not hidden nearly negative and that, subject to that condition, has the least number of nodes. Such a minor is a *minimal excluded Boolean minor of hidden near negativity*, abbreviated *minimal minor*.

Auxiliary Graphs

For a compact display of the minimal minors, we introduce the following eight auxiliary graphs N_k' and N_k'', $1 \leq k \leq 4$. Here and later, we use the convention that any arc shown without a label implicitly has a 1 as label, and that row nodes are indicated by squares.

(6.2.8)

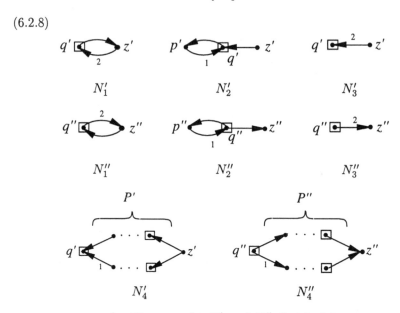

Auxiliary graphs N_k' and N_k'', $1 \leq k \leq 4$

Each arc labeled "≥ 1" may have a 1 or a 2 as label. Note that each N_k'' is a column scaled version of N_k'. The following conditions apply to N_4' and N_4''.

(6.2.9)

    (i)   Both N_4' and N_4'' have at least four nodes.

    (ii)   If the single arc of N_4' or N_4'' labeled "≥ 1" has a 2 as label, then the path P' of N_4' or P'' of N_4'' has just one arc.

Characterization of Minimal Minors

The desired minimal minors are the graphs L_0 and L_{kl}, $1 \leq k, l \leq 4$, of (6.2.10) below. Note that the two circles of the drawing for L_{kl} are labeled N'_k and N''_l. Those circles represent the graphs N'_k and N''_l of (6.2.8). In particular, the node z' (resp. z'') explicitly shown within the circle for N'_k (resp. N''_l) is the node z' of N'_k (resp. z'' of N''_l) shown in (6.2.8).

(6.2.10)

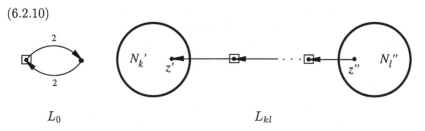

$$L_0 \qquad\qquad\qquad\qquad L_{kl}$$

Graphs L_0 and L_{kl}, $1 \leq k, l \leq 4$

(6.2.11) Theorem.
(a) *Let A be a $\{0, \pm1\}$ matrix, and let H be the graph corresponding to A. Then A is hidden nearly negative if and only if H does not have any minors of type L_0 or L_{kl}, $1 \leq k, l \leq 4$, of (6.2.10).*
(b) *The graphs L_0 and L_{kl}, $1 \leq k, l \leq 4$, are minimal in the sense that every proper subgraph obtained from these graphs by node deletions is hidden nearly negative.*

We accomplish the proof of Theorem (6.2.11) via a series of claims. We desire an abbreviated terminology for the graphs L_0 and L_{kl}, $1 \leq k, l \leq 4$. Thus, we refer to any one of these graphs as a graph of type L, or simply as an L *graph*.

Proof of Part (b)

We first handle the easy part (b) of Theorem (6.2.11).

Claim 1. *Each L graph cannot be column scaled to become nearly negative and is minimal with respect to that property.*

Proof. The result clearly holds for L_0. For L_{kl}, $1 \leq k, l \leq 4$, we argue as follows. First, N'_k is nearly negative, while N''_l is not. Second, near negativity of N'_k is lost if we scale the column node z' with a -1 factor, and it cannot be restored by scaling of other column nodes of N'_k. Third, near negativity is attained for N''_l only if we scale column node z'' plus possibly other column nodes of N''_l with a -1 factor. Fourth, suppose we scale the node z'' of L_{kl} with a -1 factor. If further column scaling is to result in a nearly negative graph, then by Lemma (6.2.5) we must also scale all other column nodes of the path from z'' to z'—in particular, the node z'. But we have already seen that such scaling of z' converts N'_k to a graph that cannot become nearly negative by additional column scaling. Thus, each L graph cannot be scaled to become nearly negative.

We turn to the minimality claim. Since L_0 has just two nodes, that graph is clearly minimal. By the above discussion, deletion of any node of the path of L_{kl} from z'' to z' results in a graph that can be column scaled to become nearly negative. Simple case checking reveals that the same conclusion is valid when we delete any one node of N'_k or N''_l.   □

Proof of Part (a)

We turn to part (a) of Theorem (6.2.11). For proof of the "only if" portion, we suppose that A is hidden nearly negative. Corollary (6.2.7) says that each minor of the corresponding graph H is hidden nearly negative. By part (b), each L graph is not hidden nearly negative and thus cannot be a minor of H.

For proof of the "if" portion of part (a), we assume that H does not have any L graph as minor, and we constructively show that H can be column scaled to become nearly negative.

We first column scale the graph H such that shrinking produces a minor $\overline{H}$ with minimum number of nodes. By Lemmas (6.2.2) and (6.2.3), the node sets of the strong components of $\overline{H}$ have at most two nodes and are precisely the node sets of the strong components of any graph $\overline{H}'$ derived by column scaling from $\overline{H}$. Indeed, by Lemma (6.2.3), we may freely column scale $\overline{H}$, since any graph $\overline{H}'$ so obtained may be produced from some scaled version H' of H and thus is a minor of H. As a matter of notational convenience, we always consider any such $\overline{H}'$ to be relabeled as $\overline{H}$, and H' to be relabeled as H, without explicit mention. In agreement with this scaling, we assume that the matrix A is column scaled to a matrix which we refer to as A again. Thus, at any point in time, the current H corresponds to the current A, and $\overline{H}$ may be obtained from H by shrinking only.

Later, we encounter in some scaled version of $\overline{H}$ a strong component with at least three nodes. We then say that we have found a *large strong component*. Of course, such a component cannot exist by Lemmas (6.2.2) and (6.2.3), and the situation at hand can be eliminated from consideration. In addition, we call a directed cycle with more than two arcs a *large directed cycle*. The existence of such a cycle in $\overline{H}$ also represents a contradiction.

Theorem (6.2.6) says that H can be column scaled to become nearly negative if and only if this is so for $\overline{H}$. Thus, we are done once we column scale $\overline{H}$ to a nearly negative graph.

If $\overline{H}$ can be reduced to an L graph by deletion of nodes, then H has that L graph as a minor. But this contradicts the assumed absence of L minors. Thus, deletion of nodes cannot reduce $\overline{H}$ to an L graph. The next claim strengthens that conclusion.

Claim 2. $\overline{H}$ cannot be reduced to an L graph by deletion of nodes and arcs; that is, $\overline{H}$ does not have any L graph as subgraph.

Proof. It suffices to show that presence of an L subgraph in $\overline{H}$ implies that an L graph may be obtained from $\overline{H}$ by column scaling and node deletions. We prove this as follows. We consider the addition of an arc to any one of the L graphs such that the arc does not have the same endpoints and the same direction as an already existing arc. We then show that, up to column scaling, the resulting graph either is again an L graph, or contains an L graph with fewer nodes, or cannot possibly be a subgraph of $\overline{H}$. We establish the third case by scaling column nodes so that a large strong component is created. A tedious but otherwise straightforward case analysis of the possible arc additions to each L graph confirms that each situation leads to one of the three conclusions. $\square$

In agreement with Claim 2, we suppose from now on that $\overline{H}$ and all column scaled versions of $\overline{H}$ do not have L subgraphs.

In subsequent claims and proofs, we utilize the notation of (6.2.8) to reference the nodes of the graphs N_1'–N_3' and N_1''–N_3''.

Claim 3.   *Without loss of generality,* $\overline{H}$ *does not contain* N_1''–N_3'' *subgraphs.*

Proof. Suppose there is such a subgraph, say, N. If a choice exists for N, we prefer a case of N_1'' or N_3'' over one of N_2''. Scale the node z'' of N with a -1 factor. We claim that this scaling does not introduce a new N_1'' or N_3'' (resp. N_1''–N_3'') subgraph if N is of type N_1'' or N_3'' (resp. is of type N_2''). Suppose there is such a new subgraph. Let M be the corresponding subgraph of $\overline{H}$.

Suppose N is of type N_1'' or N_3''. By assumption, the new subgraph introduced by the scaling is of type N_1'' or N_3'', so M must be an N_1' or N_3' subgraph, and the node z'' of N must be the node z' of M. Using M and N, one readily confirms that $\overline{H}$ contains a large strong component or an L subgraph, a contradiction.

In the remaining case, $\overline{H}$ has no N_1'' or N_3'' subgraphs, and N is of type N_2''. Straightforward checking confirms that M must be of type N_1', N_2', or N_3' and that node z'' of N must be the node z' of M. Once more, one then confirms that $\overline{H}$ contains a large strong component or an L subgraph, a contradiction.

Repeated column scaling thus can eliminate all instances of N_1''–N_3'' subgraphs from $\overline{H}$. $\square$

Due to Claim 3, we assume from now on that $\overline{H}$ has no $N_1''-N_3''$ subgraphs. Declare any node of $\overline{H}$ to be *forced* if it is the column node z' of a subgraph of type $N_1'-N_3'$. At this point, two situations must be considered, depending on whether $\overline{H}$ has at least one forced node. In Claims 4–7 we deal with the case where this is so. The second situation is treated in Claims 8–12.

Presence of Forced Nodes

Until stated otherwise, we assume that $\overline{H}$ has at least one forced node. To analyze $\overline{H}$, we carry out a *breadth first search* (BFS) as follows. We collect the forced nodes in a set called *layer* 1 and consider them *processed*. Inductively, suppose we have disjoint node sets called layers 1, 2, ..., $m-1$. Each layer contains just row nodes or just column nodes, and these nodes are considered processed. Let i be a node not processed so far. If there is an arc (i, j) connecting i with some node j of layer $m-1$, then we place node i into layer m and consider node i processed. We augment this rule as follows. Suppose layer $m-1$ is a layer of row nodes. Further, suppose that a not yet processed column node i has at least one arc incoming from the nodes of layer $m-1$, but has no arc outgoing to the latter nodes. Then we scale node i and thus make it eligible to become part of layer m.

Note that by the definition of the BFS, all odd-numbered (resp. even-numbered) layers consist of column (resp. row) nodes.

Claim 4.

(a) *From any node of any layer $m \geq 1$, there is a directed path to some node of layer 1. Any shortest such path contains one node each of layers $m, m-1, \ldots, 1$, in that order, and each arc of such a path has the label 1.*

(b) *For any odd $m \geq 3$, the layers $1, 3, \ldots, m$ of column nodes contain all neighbors of the row nodes in layers $2, 4, \ldots, m-1$.*

(c) *The column scaling of the BFS does not affect $N_1'-N_3'$ subgraphs and cannot create $N_1''-N_3''$ subgraphs.*

Proof. Parts (a) and (b) follow directly from the rules of the BFS, except for the claim about arc labels in (a). But any arc of $\overline{H}$ with the label 2, say, outgoing from a node i, constitutes an N_3' subgraph, and thus i is a forced column node and cannot be in any layer $m > 1$.

For part (c), we first note that the node z' of any $N_1'-N_3'$ subgraph is in layer 1 and thus cannot be column scaled by the BFS. Only one other column node must be considered for the possible loss of $N_1'-N_3'$ subgraphs, the node p' of N_2'. Suppose node p' is not in layer 1. Then the arc (p', q') of N_2' must have the label 1, since otherwise the nodes p' and q' define an N_1' subgraph, which implies p' to be in layer 1. But the label 1 of (p', q') implies that p' may be scaled by the BFS without affecting the N_2'

subgraph. Thus, the column scaling of the BFS does not affect N_1'-N_3' subgraphs. The above arguments also imply that the column scaling does not introduce any N_1''-N_3'' subgraphs.   □

Since by Claim 4(c) the column scaling of the BFS does not affect N_1'-N_3' subgraphs and does not introduce N_1''-N_3'' subgraphs, we may assume that the BFS does not involve any column scaling at all. When the BFS stops, the processed nodes induce a subgraph of $\overline{H}$, say, $\overline{H}_1$. If $\overline{H}_1 \neq \overline{H}$, then the rules of the BFS imply that the last layer processed in the BFS, say, layer m, must contain column nodes and that the arcs connecting these nodes with not yet processed nodes must be incoming into the latter nodes. Define $\overline{H}_2$ to be the graph induced by the not yet processed nodes. The following sketch depicts a typical situation.

(6.2.12)

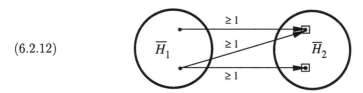

Graph $\overline{H}$ when BFS stops

Claim 5. $\overline{H}_1$ *is not nearly negative, or Theorem* (6.2.11)(a) *holds by induction.*

Proof. Suppose $\overline{H}_1$ is nearly negative. Let H_1 (resp. H_2) be the subgraphs of H corresponding to $\overline{H}_1$ (resp. $\overline{H}_2$). Then the drawing of (6.2.12) applies to H_1 and H_2 once we relabel $\overline{H}_1$ of the drawing to H_1 and relabel $\overline{H}_2$ to H_2. Lemma (6.2.4) says that a minor obtained by shrinking is nearly negative if and only if the original graph has that property. Hence, the near negativity of $\overline{H}_1$ implies that H_1 is nearly negative. Define X_1 (resp. Y_1) to be the set of row (resp. column) nodes of H_1, and define X_2 (resp. Y_2) to be the corresponding node sets of H_2. We claim that the sets X_1, X_2, Y_1, and Y_2 partition the current A as follows.

(6.2.13)

$$A = \begin{array}{c|c|c|} & Y_1 & Y_2 \\ \hline X_1 & A^1 & 0 \\ \hline X_2 & D & A^2 \\ \hline \end{array}$$

Matrix A

Indeed, since only column nodes of H_1 are linked to nodes of H_2, the submatrix indexed by X_1 and Y_2 is 0. Since H_1 is nearly negative, the submatrix A^1 indexed by X_1 and Y_1 is nearly negative. Finally, since all arcs connecting H_1 and H_2 are incoming into row nodes of H_2, the submatrix D indexed by X_2 and Y_1 does not contain any +1s. Hence, the matrix A can be column scaled to become nearly negative if and only if this is so for the submatrix A^2. Theorem (6.2.11)(a) clearly holds for small instances of A. Under an appropriate induction hypothesis, A^2 can be column scaled to become nearly negative, and thus Theorem (6.2.11)(a) holds by induction.   □

Due to Claim 5, we assume from now on that $\overline{H}_1$ is not nearly negative. Since $\overline{H}_1$ has no N_2'' subgraphs, it has a row node v with at least two outgoing arcs, say, (v, i) and (v, j). By (6.2.12), i and j are in $\overline{H}_1$. By Claim 4(a), there are directed paths from i and j to layer 1 nodes. Pick v, i, j, and the two paths so that the union of the two paths has a minimum number of nodes.

Suppose that node v occurs on one of the two paths, say, on the one from node i to a layer 1 node. Due to the BFS rules, this is possible only if node v is in some layer $m > 1$ and node i is in layer $m + 1$. Thus, the path has i and v as the first two nodes. Let w be the node on the path following v. Then the arcs (v, i), (i, v), and (v, w) constitute an N_2'' subgraph, a contradiction.

We enlarge the two paths on hand by adding arcs (v, i) and (v, j), obtaining two directed paths P_1 and P_2 that have node v as common endpoint and that have layer 1 nodes as second endpoints. Two subcases are possible, depending on whether the two paths have a common node other than v. Claim 6 below deals with the subcase where such a common node exists. Claim 7 handles the subcase when this is not so.

Claim 6. *Assume that the two paths have a common node other than v. Then, for some $1 \le k \le 3$, $\overline{H}$ has an L_{k4} subgraph.*

Proof. Among the common nodes different from v, let z'' be the one closest to v. Relabel v to q''. Let R be the subpath from z'' to the layer 1 node, say, z', of one of the two paths. We may have $z'' = z'$. Define P'' (resp. Q'') to be the subpath of P_1 from q'' to i to z'' (resp. of P_2 from q'' to j to z'').

If z'' is a row node, we column scale all nodes of P'' to convert $P'' \cup Q''$ to a large directed cycle, a contradiction. Thus, z'' is a column node. Then $P'' \cup Q''$ constitutes an N_4'' subgraph whose node z'' is connected by the directed path R to the forced node z'. By definition, the latter node is part of an N_1'–N_3' subgraph, say, M. According to the drawing of L_{kl} in (6.2.10), we thus have found, for some $1 \le k \le 3$, an L_{k4} subgraph, provided that $P'' \cup Q'' \cup R$ has only the node z' in common with M. The latter assumption is easily proved. As an example, we include details for the situation where $M = N_1'$. If the node q' of N_1' is in R, then the predecessor of q' in R is a forced node, a contradiction. If q' is in P'' and $z' \ne z''$, then the arc (z', q'), plus a part of P'', plus R, defines a large directed cycle, a contradiction. If q' is in P'' and $z' = z''$, then scaling of all column nodes of Q'' with a -1 factor produces a large directed cycle, again a contradiction.   □

Claim 7. *Assume that the two paths are node disjoint except for v. Then, for some $1 \leq k, l \leq 3$, $\overline{H}$ has an L_{kl} subgraph, up to column scaling.*

Proof. Let P_1 and P_2 be the two paths as before, and define z_1' and z_2', respectively, to be the layer 1 endpoints of these paths. By assumption, $z_1' \neq z_2'$.

Since z_1' (resp. z_2') is in layer 1, the graph $\overline{H}$ has an N_1'–N_3' subgraph M_1 (resp. M_2) with z_1' (resp. z_2') as forced node.

Suppose that M_1 (resp. M_2) has only the node z_1' (resp. z_2') in common with $P_1 \cup P_2$, and that M_1 and M_2 are node disjoint. Then we scale all column nodes of $M_1 \cup P_1$ with a -1 factor to get the desired L_{kl} subgraph. Thus, we are done once we prove the two assumptions just made.

First, if M_1 has a node other than z_1' in common with $P_1 \cup P_2$, then one readily shows that $\overline{H}$ has an N_2'' minor, or that a node of $P_1 \cup P_2$ other than z_1' or z_2' is forced, or that by column scaling one may obtain a large strong component.

Second, if the single row node of M_1 occurs also in M_2, then column scaling in $M_1 \cup M_2 \cup P_1 \cup P_2$ can create a large directed cycle. This leaves one case where M_1 and M_2 are both N_2' subgraphs that share the column node different from z_1' and z_2'. But then column scaling can create a large directed cycle.   □

We have completed the proof of Theorem (6.2.11)(a) for the case where $\overline{H}$ has at least one forced node.

Absence of Forced Nodes

From now on, we assume that $\overline{H}$ has no forced nodes.

Claim 8. *$\overline{H}$ has no arc with the label 2. For any strong component of $\overline{H}$ with two nodes, say, with row node q and column node p, the node q has only node p as neighbor.*

Proof. If $\overline{H}$ has an arc with the label 2, then $\overline{H}$ has an N_1' or N_3' subgraph. If the row node p of a strong component of $\overline{H}$ with two nodes has a neighbor other than p, then $\overline{H}$ has an N_2' or N_2'' subgraph. The N_2'' case is ruled out by Claim 3. We conclude that, if Claim 8 does not hold, then $\overline{H}$ has an N_1'–N_3' subgraph. But the absence of forced nodes rules out such a subgraph.   □

By Claim 8, the strong components of $\overline{H}$ on two nodes have no influence on whether or not $\overline{H}$ can be scaled to become nearly negative. Thus, for each such component, we delete its row node from $\overline{H}$. It is convenient for us to consider $\overline{H}$ itself to be that reduced graph.

Claim 9. $\overline{H}$ *cannot be column scaled so that the resulting graph contains a directed cycle.*

Proof. The initial graph $\overline{H}$ was derived from H by scaling and shrinking so that the resulting minor had a minimum number of nodes. Lemma (6.2.3) says that scaling of such a minor cannot introduce new strong components. The same conclusion must hold for the current graph $\overline{H}$. □

Claim 10. $\overline{H}$ *has an N_4' subgraph, or Theorem (6.2.11)(a) holds by induction.*

Proof. We apply the BFS to $\overline{H}$, except that this time we define layer 1 to contain just one arbitrarily selected column node s of $\overline{H}$. As before, the algorithm must stop with two subgraphs $\overline{H}_1$ and $\overline{H}_2$ as depicted in (6.2.12), and Claim 5 applies. That is, if $\overline{H}_1$ is nearly negative, then Theorem (6.2.11)(a) holds by induction. Hence, we assume that $\overline{H}_1$ is not nearly negative. Proceeding as before, we locate a row node v and two directed paths from v to the node s of layer 1. Let w be the node common to the two paths and closest to v. If w is a row node, then by column scaling we can obtain a directed cycle in $\overline{H}$, a contradiction of Claim 9. Thus, w is a column node. Evidently, the two subpaths from v to w constitute an N_4'' graph. We scale all column nodes of that subgraph to get the desired N_4' subgraph. □

From now on, we suppose that $\overline{H}$ itself has N_4' as subgraph. Once more we use the BFS. This time, layer 1 contains just the column node z' of N_4'. Also, the BFS ignores all nodes of N_4' other than z'. Equivalently, the BFS is done on the graph derived from $\overline{H}$ by deletion of all nodes of N_4' other than z'.

When the BFS stops, we have a subgraph $\overline{H}_1$ of $\overline{H}$ of processed nodes as well as a subgraph $\overline{H}_2$ containing the remaining nodes of $\overline{H}$ that are not in N_4'. The relationships between $\overline{H}_1$ and $\overline{H}_2$ are correctly depicted in (6.2.12). Note that Claim 4(a) and (b) apply to $\overline{H}_1$. Thus, from any node of any layer m of $\overline{H}_1$, there is a directed path to the node z' of layer 1. Any shortest such path contains one node each of layers m, $m-1$, ..., 1, in that order.

Claim 11. *Every arc of $\overline{H}$ having exactly one endpoint in $\overline{H}_1$ is of type (s,t) where s is a column node of $\overline{H}_1$ and where t is a row node of $\overline{H}_2$ or N_4'.*

Proof. By the BFS, the statement holds for arcs connecting $\overline{H}_1$ and $\overline{H}_2$. If an arc of $\overline{H}$ goes from a node $s \neq z'$ of N_4' to a node of $\overline{H}_1$, then that arc and directed paths in $\overline{H}_1$ and N_4' constitute a large directed cycle, a contradiction. Finally, if an arc of $\overline{H}$ goes from a row node s of $\overline{H}_1$ to a column node $t \neq z'$ of N_4', then by column scaling a large directed cycle can be produced, another contradiction. □

Claim 12. $\overline{H}$ *has an* L_{44} *subgraph, or Theorem* (6.2.11)(a) *holds by induction.*

Proof. If $\overline{H}_1$ is nearly negative, then we use Claim 11 and argue as in the proof of Claim 5 that Theorem (6.2.11)(a) holds by induction. So assume that $\overline{H}_1$ is not nearly negative. Thus, $\overline{H}_1$ has a row node v with two leaving arcs. We carry out the path construction used earlier in the proof of Claim 10, but without the final scaling step. Thus, we determine in $\overline{H}_1$ an N_4'' subgraph plus a directed path from the node z'' of the N_4'' subgraph to the node z'. The N_4'' subgraph, plus the directed path, plus the N_4' subgraph already on hand, comprises an L_{44} subgraph.   □

Finding a Minimal Minor

The above proof of Theorem (6.2.11) supports the following polynomial algorithm for finding one of the excluded minors.

(6.2.14) Algorithm EXCLUDED MINOR OF HIDDEN NEAR NEGATIVITY. *Derives one of the minimal excluded Boolean minors* L_0 *and* L_{kl}, $1 \le k, l \le 4$, *specified by* (6.2.10) *from the labeled, directed, bipartite graph* $H = \mathrm{DBG}(A)$ *of a matrix* A *that is not hidden nearly negative.*

Input: Matrix A over $\mathbb{B}$ that is not hidden nearly negative.

Output: A minor of the graph $H = \mathrm{DBG}(A)$ that is one of the minimal excluded minors L_0 and L_{kl}, $1 \le k, l \le 4$, of hidden near negativity.

Complexity: Polynomial.

Procedure:
1. Carry out shrinking for H, obtaining a minor $\overline{H}$.
2. Perform the steps implicit in the proofs of Claims 3–12 on $\overline{H}$. If during these steps it is determined that the graph H can by scaling and shrinking be reduced to a graph $\overline{H}'$ with fewer nodes than $\overline{H}$, then declare $\overline{H}'$ to be $\overline{H}$, and repeat the process with the new $\overline{H}$.

Proof of Validity. The steps implicit in the proofs of Claims 3–12 can clearly be carried out in polynomial time. If $\overline{H}$ is a minor with the least number of nodes obtained by scaling and shrinking from H, then these steps do produce the desired minor. Of course, we do not know that the initial $\overline{H}$ has the least number of nodes, but we proceed anyway. If during one of the steps it turns out that $\overline{H}$ does not have the least number of nodes, then we switch to a new $\overline{H}$ with fewer nodes and repeat all steps. The number of such repetitions is bounded by the number of nodes of H, so the entire scheme is polynomial.   □

In the next section, we rely on Theorem (6.2.11) to characterize hidden near negativity of $\{0, \pm 1\}$ matrices in terms of minimal excluded subregions.

6.3 Minimal Excluded Subregions

As before, let H be the labeled, directed, bipartite graph DBG(A) of a $\{0, \pm 1\}$ matrix A. In this section, we derive from the excluded minors of hidden near negativity for H given by Theorem (6.2.11) the minimal excluded subregions of hidden near negativity for A. As a corollary, we deduce a characterization of satisfiability for 2SAT matrices.

The characterization of the minimal excluded subregions is given by Theorem (6.3.4) below and involves $\{0, \pm 1\}$ matrices called V_1–V_9. We first describe the labeled, directed, bipartite graphs that correspond to these matrices. These graphs are the *minimal excluded subgraphs of hidden near negativity* and are called F_1–F_9. All arcs of the graphs have the label 1, as they must. The graphs contain a number of directed paths where all intermediate nodes have the degree 2. For notational convenience, we depict each such path by a line segment, with the direction of the path indicated by an arrowhead placed at the center of the line segment. Each such path contains at least one arc unless indicated otherwise.

Ladder

We need an auxiliary graph called a *ladder with end node pairs* (i'', i') *and* (j'', j'). Such a graph is of the form

(6.3.1)

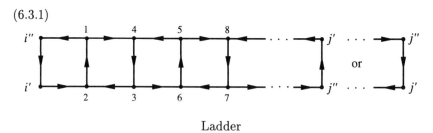

Ladder

The paths drawn vertically in (6.3.1) are the *rungs* of the ladder. Note the pattern of alternating directions of the rungs and of the horizontal paths as one moves from left to right along the ladder. That pattern uniquely determines which of the two right end cases must be present. The drawing of (6.3.1) implies that the explicitly shown nodes are column nodes, and that the row nodes, which are not shown, all have the degree 2. A *contracted ladder with end node pairs* (i'', i') *and* (j'', j') is obtained from the ladder of (6.3.1) by contracting any number of rungs—possibly none—to one node each that is declared to be a column node. Such contractions may involve the rung connecting i'' with i' (resp. j'' with j'), thus converting that rung to a single node labeled by both i'' and i' (resp. j'' and j'). Note that any contracted ladder is strongly connected.

Minimal Subgraphs

We are ready to present F_1–F_9.

(6.3.2)

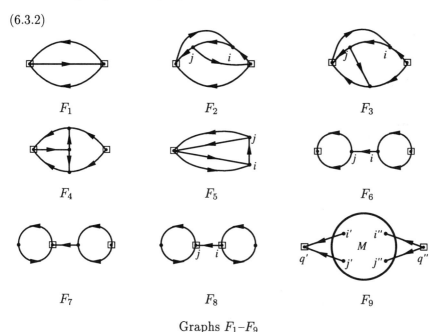

Graphs F_1–F_9

The following conditions are satisfied by the above graphs.

(6.3.3)

    (i)   Whenever nodes are labeled i and j, then $i = j$ is allowed; in that case, the displayed path i to j consists of just one node.

    (ii)   In F_9, the subgraph M is a contracted ladder with end node pairs (i'', i'), and (j'', j'); the case $i' = i''$ or $j' = j''$ is allowed.

It is easy to confirm that each graph of F_1–F_9 has an L_0 or L_{kl}, $1 \leq k, l \leq 4$, minor and thus is not hidden nearly negative. For example, the graph F_1 is strongly connected, and shrinking reduces that graph to the graph L_0.

Characterization of Minimal Subregions

We present the characterization of hidden near negativity in terms of minimal excluded subregions. Recall that the matrices producing the graphs F_1–F_9 are called V_1–V_9.

(6.3.4) Theorem.

(a) A $\{0, \pm 1\}$ matrix A is hidden nearly negative if and only if each column scaled version of A does not contain a subregion of type V_1–V_9.

(b) The matrices V_1–V_9 are minimal in the sense that every proper subregion extracted from these matrices is hidden nearly negative.

Characterization of 2SAT Satisfiability

Before we prove Theorem (6.3.4), we establish a corollary that characterizes 2SAT satisfiability.

(6.3.5) Corollary. Let A be the matrix of a 2SAT system with exactly two nonzero entries in each row. Then A is satisfiable if and only if no column scaled version of A has a submatrix of type V_6 or V_9. The excluded submatrices are minimal in the sense that every proper submatrix having exactly two nonzeros in each row—for example, any proper submatrix obtained by deletion of rows—is satisfiable.

Proof. Theorem (5.6.2) says that a 2SAT matrix with exactly two nonzero entries in each row is satisfiable if and only if A is hidden nearly negative. By Theorem (6.3.4), the latter conclusion is equivalent to the exclusion of subregions of type V_1–V_9. Except for F_6 and F_9, each one of the graphs F_1–F_9 has at least one row node of degree 3 or 4, so V_1–V_5, V_7, or V_8 cannot occur as subregions in A. On the other hand, each row node of F_6 and F_9 has the degree 2, so V_6 and V_9 are possible subregions of A. Let H be the graph corresponding to A. Since every row node of H has the degree 2, every connected subgraph of H without degree 1 nodes—in particular, F_6 and F_9—can be obtained from H just by node deletions. Accordingly, V_6 and V_9 can be derived from A by submatrix taking.

Minimality of V_6 and V_9 is argued as follows. By Theorem (6.3.4)(b), any proper submatrix of V_6 or V_9 is hidden nearly negative. If such a submatrix has exactly two nonzeros in each row, then by Theorem (5.6.2) it is satisfiable. □

We prove Theorem (6.3.4) in a series of claims. Let A be a $\{0, \pm 1\}$ matrix that is not hidden nearly negative and that is minimal in the sense that all proper subregions of A have that property. Define H to be the graph corresponding to A. By the proof of Theorem (6.2.11), column scaling and shrinking can reduce H to a minor $\overline{H}$ that has as few nodes as possible and that by subsequent node deletions becomes a graph $\overline{\overline{H}}$ of type L_0 or

L_{kl}, $1 \leq k, l \leq 4$. We need to show that H has, up to column scaling, one of F_1–F_9 as a subgraph. For convenient reference, we call any one graph of F_1–F_9 an *F graph*.

Partial and Complement Nodes

Lemma (6.2.2) says that each strong component of $\overline{H}$ has one or two nodes. Suppose a strong component of $\overline{H}$ with two nodes has just one node in $\overline{\overline{H}}$. Then we declare that node of $\overline{\overline{H}}$ to be a *partial node* and call the second node of that strong component a *complement node*. Note that the definition of partial and complement nodes is relative to $\overline{\overline{H}}$.

Claim 1. *Every node of $\overline{H}$ that is not in $\overline{\overline{H}}$ is the complement node of a unique partial node of $\overline{\overline{H}}$.*

Proof. Let t be a node of $\overline{H}$ that is not in $\overline{\overline{H}}$. If t is not a complement node, then a node set induced proper subgraph of H has $\overline{\overline{H}}$ as a minor, a contradiction of the minimality of A. Every strong component of $\overline{H}$ has at most two nodes, so the partial node of $\overline{\overline{H}}$ corresponding to t is unique.   □

Claim 2. *Let t be the complement node of a partial node s of $\overline{\overline{H}}$, and let R be the node set of a strong component of $\overline{\overline{H}} - \{s\}$. Then after any column scaling in $\overline{H}$, the subgraph of the scaled $\overline{H}$ induced by the node subset $(\{\text{nodes of } \overline{\overline{H}}\} - R) \cup \{t\}$ cannot contain any L_0 or L_{kl}, $1 \leq k, l \leq 4$, subgraph.*

Proof. If after some column scaling there is such an L_0 or L_{kl}, $1 \leq k, l \leq 4$, subgraph, then one such subgraph may be found by column scaling and just node deletions. By the derivation, the nodes of R cannot be complement nodes for that subgraph, which contradicts Claim 1.   □

Claim 3. *If $\overline{H}$ has an L_0 subgraph, then $\overline{\overline{H}}$ is that subgraph and $\overline{H} = \overline{\overline{H}}$.*

Proof. Neither of the two nodes of an L_0 subgraph can be partial relative to that subgraph. By Claim 1, we have $\overline{H} = \overline{\overline{H}}$.   □

Some Path Configurations

For several subsequent cases of H, we exhibit a certain configuration of paths. The next two claims establish that any such configuration implies an F subgraph in some column scaled version of H.

Claim 4. *Let H' be a subgraph of H consisting of two row nodes r and s, of two directed and internally node disjoint paths P_1 and P_2 from r to s, and of a third directed path P_3 from s to a node $v \neq s$ of P_1. Except for*

s and v, the path P_3 has no other node in common with $P_1 \cup P_2$. If node v is a row node, then H' is up to column scaling an F_1 graph.

Proof. Scaling of the column nodes in the subpath of P_1 from v to r proves the result. $\square$

Claim 5. Let H', P_1, P_2, and P_3 be as in Claim 4, but this time assume v to be a column node. Suppose H contains a path P_4 from a node $u \neq r$ of $P_1 \cup P_3$ to a node $w \neq s$ of P_2. Except for u and w, the path P_4 has no other node in common with $P_1 \cup P_2 \cup P_3$. Then, up to column scaling, $P_1 \cup P_2 \cup P_3 \cup P_4$ contains an F subgraph.

Proof. Enumeration of the few possible cases establishes the result. $\square$

Recall that $\overline{\overline{\overline{H}}}$ is equal to L_0 or to some L_{kl}, $1 \leq k, l \leq 4$. We analyze the two cases separately, beginning with $\overline{\overline{H}} = L_0$.

Case of $\overline{\overline{H}} = L_0$

Until stated otherwise, we assume that $\overline{\overline{H}} = L_0$. Claim 3 says that under that assumption we have $\overline{H} = \overline{\overline{H}}$. Hence, H is strongly connected and has a row node r (resp. s) with at least two outgoing (resp. incoming) arcs.

Claim 6. We may assume that necessarily $r = s$.

Proof. Assume the situation where $r \neq s$. Since H is strongly connected, there is a shortest directed path P_1 from r to s. Since r has at least two outgoing arcs and P_1 is shortest, there is an arc (r, i) where $i \notin P_1$. Similarly, there is an arc (j, s) where $j \notin P_1$. Since H is strongly connected, there exists a path $\overline{P}_2$ from i to j. Select r, s, P_1, i, j, and $\overline{P}_2$ so that $\overline{P}_2$ avoids the nodes of P_1 as much as possible.

Suppose $\overline{P}_2$ includes a node of P_1. Equivalently, we may assume that no two row nodes of H are connected by two internally node disjoint paths of the same direction. As we go from i to j along $\overline{P}_2$, let v be the first node of P_1 encountered, and let w be the last one. Define P_{21} (resp. P_{22}) to be the arc (r, i), (resp. (j, s)) plus the subpath of $\overline{P}_2$ from i to v (resp. w to j). Direct checking confirms that in each one of the possible cases of v and w, the graph $P_1 \cup P_{21} \cup P_{22}$ has, up to column scaling, an F_6, F_7, or F_8 subgraph, or r and s are joined by two internally node disjoint paths of the same direction, a contradiction.

We consider the second situation, where $\overline{P}_2$ is disjoint from P_1; that is, there exist two row nodes r and s that are joined by two internally node disjoint paths P_1 and P_2 of the same direction. Since H is strongly connected, there exists a path P_3 from s to r. If P_3 is internally node disjoint from P_1 and P_2, then $P_1 \cup P_2 \cup P_3$ is F_1. Otherwise, P_3 has proper subpaths P_{31} and P_{32} so that P_{31} contains s and an internal node v of P_1, say, and P_{32} contains a node $u \neq r, s$ of P_1 and a node $w \neq s$ of P_2. Except for s, u, v, and w, the subpaths P_{31} and P_{32} have no node in common with $P_1 \cup P_2$. By Claim 4, we may assume v to be a column node of P_1. Then, by Claim 5, $P_1 \cup P_2 \cup P_{31} \cup P_{32}$ has an F subgraph.   □

Claim 7. *H has F_5 or F_8 as subgraph.*

Proof. By Claim 6, some row node r has at least two outgoing and at least two incoming arcs, and every other row node has the degree 2. It is then easily checked that H has F_5, or F_8 with $i = j$, as subgraph.   □

We have completed the proof for the case of $\overline{\overline{H}} = L_0$ and turn to the situation where, for some $1 \leq k, l \leq 4$, $\overline{\overline{H}} = L_{kl}$.

Case of $\overline{\overline{H}} = L_{kl}$

From now on, we assume that $\overline{\overline{H}}$ is a graph of type L_{kl}, $1 \leq k, l \leq 4$. We suppose $\overline{\overline{H}}$ to be L_{kl} itself, and we use the notation of (6.2.8) and (6.2.10) to refer to nodes and subgraphs of that graph. Due to the symmetry under scaling, we may assume $k \leq l$. By Claim 3, at most one of the two arcs of any strong component of $\overline{\overline{H}}$ on two nodes may have a 2 as label.

Claim 8. *No row node of $\overline{\overline{H}}$ can be partial.*

Proof. Each instance of a partial row node is readily seen to contradict Claim 2, except possibly for the cases where k or $l = 3$ and where the row node of the subgraph N_3' or N_3'' is partial. By the symmetry, we only need to treat the N_3' case. Let Q' (resp. Z') be the node subsets of H corresponding to the node q' (resp. z') of N_3'. Since in $\overline{\overline{H}}$ the arc (z', q') of N_3' has the label 2, there is in H a node s of Q' having arcs incoming from at least two nodes of Z'. Except for s, delete all nodes of Q' from H. The resulting subgraph of H still has $\overline{\overline{H}}$ as a minor, a contradiction of the minimality of A.   □

Claim 9. *Suppose a strong component of $\overline{\overline{H}}$ has two nodes, one of which is a partial column node of $\overline{\overline{H}}$. Then both arcs of that component have a 1 as label.*

Proof. If one of the arcs has a 2 as label, then one readily produces a contradiction of Claim 2. Indeed, in the notation of Claim 2, for each case of L_{kl}, $1 \leq k, l \leq 4$, one can define a set R that contains the nodes of a strong component of N_k' or N_l''.   □

By Claims 8 and 9, each partial node of $\overline{\overline{H}}$ is a column node, and both arcs of the corresponding strong component of $\overline{\overline{H}}$ have a 1 as label.

Claim 10.   *Let s be a partial column node of $\overline{\overline{H}}$ of degree 2 or 3; if the degree is 3, then s is to have three neighbors. Then H has a proper subgraph H' that in turn has a minor with the same structure as $\overline{\overline{H}}$, except that the arcs formerly incident at s of $\overline{\overline{H}}$ have been subdivided into directed paths where each arc has the label 1.*

Proof. Suppose s has the degree 2 in $\overline{\overline{H}}$. Let i and j be the neighbors of s in $\overline{\overline{H}}$, and let t be the complement node of s. Suppose that (i, s) and (s, j) are the two arcs of $\overline{\overline{H}}$ incident at s. In H, let I, J, S, and T be the node subsets corresponding to i, j, s, and t, respectively, of $\overline{\overline{H}}$. Since $S \cup T$ defines a strong component of H, there is in H a directed path from some node of I to some node of J where all intermediate nodes are in $S \cup T$. Except for the nodes and arcs of that path, delete from H the nodes of $S \cup T$ and the arcs having at least one endpoint in $S \cup T$. The resulting proper subgraph of H has a minor like $\overline{\overline{H}}$ except that the arcs (i, s) and (s, j) have been subdivided into directed paths with label 1 arcs. Analogous arguments handle the cases where the two arcs incident at node s are (i, s) and (j, s) or where s is a degree 3 node of $\overline{\overline{H}}$. □

We emphasize that the replacement of arcs by paths of Claim 10 may replace an arc with the label 2 by a path where each arc has the label 1. We rely on that fact next.

Claim 11.   *If $\overline{\overline{H}} = L_{k4}$, then the arc labeled "$\geq 1$" in N_4'' of (6.2.8) has a 1 as label.*

Proof. If the arc labeled with "≥ 1" in N_4'' has a 2 as label, then the degree 2 column node of that arc must be partial. But then by Claim 10, H has a proper subgraph with an $\overline{\overline{H}}$ minor, a contradiction of the minimality of A. □

We analyze two subcases of $\overline{\overline{H}}$ depending on whether $z' \neq z''$ or $z' = z''$.

Subcase $z' \neq z''$

We begin with the case $z' \neq z''$.

Claim 12.   *If $z' \neq z''$, then, up to column scaling, H has an F subgraph.*

Proof. By Claim 10 and the minimality of A, at most the nodes z' and z'' of $\overline{\overline{H}}$ may be partial, and then only if $k = 3$ in case of a partial z', and only if $l = 3$ in case of a partial z''.

We show that the subgraph N_k' of $\overline{\overline{H}}$ gives rise to the left half of F_6 or F_7. Application of analogous arguments to N_l'' then proves the existence of F_6, F_7, or F_8. We examine the cases for k.

$k = 1$: In the directed path of $\overline{\overline{H}}$ from z'' to z', let r be the predecessor node of z'. By Claim 8, the row node r is not partial. Let Q' (resp. Z') be the node subset of H corresponding to q' (resp. z') of $\overline{\overline{H}}$. We claim that the subgraph H' of H induced by $\{r\} \cup Q' \cup Z'$ contains a directed path P from r to a node $\overline{z}'$ plus a cycle C, with the following structure: The node $\overline{z}'$ is the only node common to P and C; if $\overline{z}'$ is a column node, then C consists of two directed internally node disjoint paths from $\overline{z}'$ to a row node; if $\overline{z}'$ is a row node, then C can be column scaled to become directed. Put differently, $P \cup C$ is, up to column scaling of C, the left half of F_6 or F_7. To prove the existence of P and C, we only note that the label 2 arc (z', q') of N_1' implies that H' has a row node in Q' with two incoming arcs, say, with column nodes i and j as second endpoints. A simple analysis of the directed paths in H' from r to i and j then confirms the claim.

$k = 2$: It is easy to see that N_2' of $\overline{\overline{H}}$ leads to the left half of F_7 in H.

$k = 3$: Due to the label 2 on the arc (z', q') of N_3', the node z' must be partial. The arguments are then essentially those for $k = 1$, except that we define Q' to be the set of row nodes of H arising from q' and from the complement node of z'.

$k = 4$: By Claims 10 and 11, N_4' may be assumed to be the left-hand part of F_6. $\square$

A Result for Contracted Ladders

Before we turn to the case of $\overline{\overline{H}}$ with $z' = z''$, we establish the structure of a certain strongly connected graph M that we later need in connection with F_9.

Claim 13. *Let M be the labeled, directed, bipartite graph corresponding to a $\{0, \pm 1\}$ matrix. Suppose that M is strongly connected and that each row node of M has the degree 2. Further suppose that M has column nodes i', i'', j', and j'' that are distinct except that possibly $i' = i''$ or $j' = j''$. Finally, assume that M has no proper subgraph containing directed paths from each node of $\{i'', j''\}$ to each node of $\{i', j'\}$. Then, up to a relabeling of i'' to j'' and of j'' to i'', M is a contracted ladder with end node pairs (i'', i') and (j'', j').*

Proof. We say that a subgraph of M satisfies the *path condition* if that subgraph has directed paths from each node of $\{i'', j''\}$ to each node of $\{i', j'\}$. We say that M is *minimal* when no proper subgraph of M satisfies the path condition. Note that Claim 13 assumes M to be minimal.

Clearly, M has a directed path P_1 from a node of $\{i'', j''\}$ to one of $\{i', j'\}$ where each internal node is different from i', i'', j', and j''. By the relabeling condition of Claim 13, we may assume P_1 to go from j'' to j'.

Let P be any directed path of M from j'' to i'. We claim that P must visit j' and i'', in that order. Suppose P avoids i''. Choose two shortest directed paths, one from i'' to i' and the other one from i'' to j'. Then P_1, P, and the two paths constitute a subgraph $\overline{M}$ of M that satisfies the path condition and that is not strongly connected. Hence $\overline{M}$ is a proper subgraph of M, and M is not minimal, a contradiction. Similar arguments cover the situation where P avoids j', or where i'' is visited prior to j'.

Since P must visit j' and i'', in that order, we may consider P to be the union of P_1 and of two paths P_2 and P_3 going from j' to i'' and from i'' to i', respectively. Note that P_1 (resp. P_3) is just one node if $j' = j''$ (resp. $i' = i''$). Since M is strongly connected, there is a directed path P_4 from i' to j''. Indeed, by the minimality of M, we have $M = P_1 \cup P_2 \cup P_3 \cup P_4$. If P_4 has a node other than i' and j'' in common with $P_1 \cup P_3$, then we readily see that M is not minimal. Since all row nodes of M have the degree 2, the internal nodes common to P_2 and P_4 must be column nodes. Clearly, each such node has the degree 3 or 4. We label these internal common nodes as follows. We move from i'' to j' using the arcs of P_2; that is, we move against the direction of P_2. The first internal common node different from i'' we label with a 1 if its degree is 3, and with both 1 and 2 if its degree is 4. Suppose, inductively, that we most recently used the integer m, or the integers $m - 1$ and m, for labeling an internal common node. Then the next internal common node encountered receives the label $m + 1$ if its degree is 3, and the labels $m + 1$ and $m + 2$ if its degree is 4.

Assume that we have the case $i' \neq i''$ and $j' \neq j''$ and that all internal common nodes of P_2 and P_4 have the degree 3. It is then easily checked that, by the minimality of M, we must encounter the internal common nodes along P_4 in the order 2, 1, 4, 3, 6, 5,... This fact implies that $M = P_1 \cup P_2 \cup P_3 \cup P_4$ is the ladder of (6.3.1). If $i' = i''$ or $j' = j''$, or if some internal common nodes of P_2 and P_4 have the degree 4, then by similar arguments one confirms that M may be obtained from the ladder of (6.3.1) by the contraction of some rungs. Thus in all cases, M is a contracted ladder with end node pairs (i'', i') and (j'', j'). $\quad\square$

Subcase $z' = z''$

We are ready for the subcase of $\overline{\overline{H}}$ where $z' = z''$.

Claim 14. *If $z' = z''$, then, up to column scaling, H has an F subgraph.*

Proof. We examine the possible situations.

$k = l = 1$: This case is not possible, since then $\overline{\overline{H}}$ has a large component.

$k = 1$, $l = 2$: L_{12} cannot have partial nodes. Then one easily exhibits in H a scaled version of F_7 or of F_8.

$k = 1$, $l = 3$: L_{13} cannot have partial nodes. The arguments are very similar to those proving Claims 6 and 7 for the case $\overline{\overline{H}} = L_0$. They establish that, up to column scaling, an F graph must be present.

$k = 1$, $l = 4$: L_{14} cannot have partial nodes. Let Z' be the node subset of H corresponding to z' of L_{14}. The two paths in N_4'' of L_{14} from the node q'' to $z'' = z'$ correspond in H to two paths from q' to Z'. If there are two such paths in H that terminate at just one node of Z', then one readily deduces from H the graph F_6 or a scaled version of F_7. If two such paths terminate at two nodes of Z', then arguments almost identical to those for L_{13} provide the conclusion that, up to column scaling, an F graph is present in H.

$k = l = 2$: The arguments are essentially the same as for the corresponding case with $z' \neq z''$.

$k = 2$, $l = 3$: Due to the label 2 of the arc (q'', z''), the node $z' = z''$ must be partial. One then easily deduces from H the graph F_7 or a scaled version of F_8.

$k = 2$, $l = 4$: The arguments are essentially the same as for the corresponding case with $z' \neq z''$.

$k = l = 3$: The node $z' = z''$ must be partial. Thus, H clearly contains a graph of the form F_9 where $i' \neq i''$ and $j' \neq j''$ and where for the moment the subgraph M of F_9 satisfies the following conditions instead of (6.3.3)(ii): M is strongly connected, and each row node of M has the degree 2.

We claim that M satisfies the assumptions of Claim 13. We only need to show that M has no proper subgraph $\overline{M}$ satisfying the path condition. Suppppose such $\overline{M}$ exists. Delete all arcs from F_9 that are in M but not in $\overline{M}$, obtaining a graph $\overline{F}_9$. Suppose $\overline{F}_9$ can be column scaled to become nearly negative. Then at least one column neighbor of the row node q'' (see (6.3.2)) must be scaled by -1. Since $\overline{M}$ satisfies the path condition, there exist directed paths in $\overline{F}_9$ from that column node to the two column node neighbors of q'. By Lemma (6.2.5), the latter two column nodes must also be scaled by -1, which implies that the resulting graph is not nearly negative. Thus, $\overline{F}_9$ cannot be column scaled to become nearly negative, a contradiction of the minimality of A.

By Claim 13, M is thus a contracted ladder with end node pairs (i'', i') and (j'', j'), and F_9 satisfies (6.3.3)(ii) as desired.

$k = 3$, $l = 4$: The node $z' = z''$ must be partial. Let Z' be the node subset of H corresponding to z'. Then the two paths in N_4'' of L_{34} from q'' to $z' = z''$ correspond to two paths in H from q'' to Z'. If there are two paths in H terminating at one node of Z', we readily exhibit an F_6 subgraph. Otherwise, by the arguments for $k = l = 3$, we have F_9.

$k = l = 4$: If $z' = z''$ is not partial, we have F_6. Otherwise, we argue almost identically to the case $k = 3$, $l = 4$ to exhibit F_6 or F_9.

Claim 15. *Theorem (6.3.4) holds.*

Proof. Part (a) of Theorem (6.3.4) follows from Theorem (6.2.11)(a) and Claims 7, 12, and 14. For part (b), we must show that every proper subgraph of F_1–F_9 can be column scaled to become nearly negative. Routine checking completes the proof.   □

Finding a Minimal Subregion

The proof of Theorem (6.3.4) implies the following polynomial algorithm that derives one of the excluded subregions from any matrix A that is not hidden nearly negative.

(6.3.6) Algorithm EXCLUDED SUBREGION OF HIDDEN NEAR NEGATIVITY. *Derives one of the minimal excluded subregions V_1–V_9 from a $\{0, \pm 1\}$ matrix A that is not hidden nearly negative. The subregions V_1–V_9 are represented by the graphs F_1–F_9 of (6.3.2).*

Input: Matrix A over $\mathbb{B}$ that is not hidden nearly negative.

Output: A subregion of A that up to scaling is one of the minimal excluded subregions V_1–V_9 of hidden near negativity. The latter subregions are represented by the graphs F_1–F_9 of (6.3.2).

Complexity: Polynomial.

Procedure:
1. Execute Algorithm EXCLUDED MINOR OF HIDDEN NEAR NEGATIVITY (6.2.14) to $H = \text{DBG}(A)$ to locate a minimal excluded minor of hidden near negativity.
2. Apply the procedure that is implicit in the proof of Theorem (6.3.4), to the minor determined in Step 1 to obtain a minimal excluded subgraph from H. The subregion of A corresponding to that subgraph is up to scaling one of the desired minimal excluded subregions V_1–V_9.

Proof of Validity. Algorithm EXCLUDED MINOR OF HIDDEN NEAR NEGATIVITY (6.2.14) determines a minimal excluded minor in polynomial time. The proof of Theorem (6.3.4) clearly contains a polynomial procedure that deduces from that minor the desired subregion of A.   □

6.4 References

Chapter 5 contains basic material and references about hidden nearly negative matrices.

Chandru, Coullard, and Montañez (1988), and Chandru, Coullard, Hammer, Montañez, and Sun (1990) contain an excluded subregion characterization of hidden near negativity. In the notation of this chapter, that subregion is defined by two path substructures of the graph $H = \mathrm{DBG}(A)$. The subregion so determined may properly contain a subregion that also is not hidden nearly negative and thus may not be minimal.

Aspvall, Plass, and Tarjan (1979) characterize satisfiability of 2SAT systems using the following graph construction. Each variable x of a given 2SAT system produces two nodes labeled x and $\neg x$. Each clause of the 2SAT system is rewritten as two equivalent implications that in turn are represented by two directed arcs. Specifically, let $x \vee y$ be an arbitrary clause where x and y denote possibly negated variables. That clause is equivalent to each one of the implications $\neg x \Rightarrow y$ and $\neg y \Rightarrow x$, and the corresponding directed arcs are $(\neg x, y)$ and $(\neg y, x)$. Aspvall, Plass, and Tarjan (1979) show that a 2SAT system is unsatisfiable if and only if the directed graph constructed from the 2SAT system has a strongly connected component that contains, for some variable x, both the node x and the node $\neg x$.

The next chapter introduces a special class of matrices over $\mathbb{B}$ called Boolean closed.

Chapter 7

Boolean Closed Matrices

7.1 Overview

We introduce a class of matrices over $\mathbb{B}$ called Boolean closed. The matrices are used in Chapters 8 and 10. We begin with an informal discussion that motivates the subsequently given definition.

Suppose the satisfiability problem is to be solved for the following partitioned matrix A.

(7.1.1)

$$A = \begin{array}{c|c|c} & Y_1 & Y_2 \\ \hline X_1 & A^1 & 0 \\ \hline X_2 & D & A^2 \end{array}$$

Partitioned matrix A

Thus, we want to find a $\{\pm 1\}$ solution s vector for the inequality

(7.1.2) $$A \odot s \geq \underline{1}$$

or ascertain that such a vector s does not exist.

Let s^1 and s^2 be the subvectors of s indexed by the column index sets Y_1 and Y_2, respectively, of A. Using s^1 and s^2 instead of s, the inequality of (7.1.2) can be rewritten as

(7.1.3)
$$A^1 \odot s^1 \geq \underline{1}$$
$$(D \odot s^1) \oplus (A^2 \odot s^2) \geq \underline{1}$$

256

Suppose that s^{1*} and s^{2*} solve (7.1.3). Let d be the vector of subrange(D) defined by

(7.1.4) $$d = D \odot s^{1*}$$

Since s^{1*} and s^{2*} solve (7.1.3), they also solve

(7.1.5)
$$A^1 \odot s^1 \geq \underline{1}$$
$$D \odot s^1 \geq d$$
$$d \oplus (A^2 \odot s^2) \geq \underline{1}$$

According to (4.2.13), for any $\{0, 1\}$ vectors a, b, and c,

(7.1.6) $$a \oplus b \geq c \text{ if and only if } a \geq c \ominus b$$

Thus, the inequality $d \oplus (A^2 \odot s^2) \geq \underline{1}$ of (7.1.5) is equivalent to $A^2 \odot s^2 \geq \underline{1} \ominus d$, and (7.1.5) can be restated as

(7.1.7)
$$A^1 \odot s^1 \geq \underline{1}$$
$$D \odot s^1 \geq d$$
$$A^2 \odot s^2 \geq \underline{1} \ominus d$$

Conversely, suppose that for some $d \in$ subrange(D), (7.1.7) is solved by vectors s^{1**} and s^{2**}. By (4.2.17) and (4.2.20), for any $\{0, 1\}$ vectors a, b, c, and d,

(7.1.8) $$a \leq (a \ominus b) \oplus b$$

and

(7.1.9) $$a \geq b \text{ and } c \geq d \text{ imply } a \oplus c \geq b \oplus d$$

We insert s^{1**} and s^{2**} into (7.1.7) and apply (7.1.8) and (7.1.9) to obtain

(7.1.10)
$$A^1 \odot s^{1**} \geq \underline{1}$$
$$(D \odot s^{1**}) \oplus (A^2 \odot s^{2**}) \geq \underline{1}$$

Hence, s^{1**} and s^{2**} also solve the original inequality system (7.1.3).

The above observations support the following solution algorithm for (7.1.3). For each $d \in$ subrange(D), test whether

(7.1.11)
$$A^1 \odot s^1 \geq \underline{1}$$
$$D \odot s^1 \geq d$$

has a solution. Let R be the set of vectors $d \in \text{subrange}(D)$ for which this is so. If R is empty, then (7.1.3) is unsatisfiable. Otherwise, test whether

$$(7.1.12) \qquad\qquad A^2 \odot s^2 \geq \underline{1} \ominus d$$

has a solution for some $d \in R$. If (7.1.12) is unsatisfiable for all $d \in R$, then (7.1.3) is unsatisfiable. Otherwise, let (7.1.12) have a solution s^{2**} for some $d \in R$. Then s^{2**} and the solution s^{1**} of (7.1.11) with that d solve (7.1.3).

Computational effort for the above algorithm is as follows. For the determination of R, (7.1.11) must be solved for each $d \in \text{subrange}(D)$. Then (7.1.12) may have to be solved for each $d \in R$. Hence, (7.1.12) may have to be solved $r_0 = |\text{subrange}(D)|$ times.

Suppose that the matrix A^2 of (7.1.12) has a partition analogous to that of A in (7.1.1), and that we solve each instance of (7.1.12) using that partition of A^2 in the manner described above. Each instance of (7.1.12) is replaced by, say, r_1 subproblems. Hence, the solution of (7.1.12) for all $d \in R$ may require solving $r_0 \cdot r_1$ subproblems.

Continuing inductively, we see that repeated partitioning may result in an exponential growth of the number of subproblems that potentially must be solved. One is tempted to search for conditions on D and related modifications of the algorithm which avoid that calamity. For example, one might look for conditions under which, for any R, just one modified version of (7.1.12) must be solved. Let us pursue this notion. As argued earlier via (7.1.6), the inequality $A^2 \odot s^2 \geq \underline{1} \ominus d$ of (7.1.12) is equivalent to

$$(7.1.13) \qquad\qquad d \oplus (A^2 \odot s^2) \geq \underline{1}$$

Our goal is to decide whether, for a given $R \subseteq \text{subrange}(D)$, (7.1.13) has a solution for some $d \in R$. Suppose D has a column submatrix $\overline{D}$ such that $\text{subrange}(\overline{D}) = R$. Then (7.1.12) has a solution for some $d \in R$ if and only if the inequality

$$(7.1.14) \qquad\qquad (\overline{D} \odot \overline{s}) \oplus (A^2 \odot s^2) \geq \underline{1}$$

has a solution. We have successfully reduced the $r_0 = |R|$ cases of (7.1.12) to the problem (7.1.14), but at a price. That is, the approach works in general only if for any subset R of subrange(D), there is a column submatrix $\overline{D}$ such that subrange$(\overline{D}) = R$. The latter condition is very severe and is satisfied only by very simple matrices D. But maybe we can salvage the main idea by checking what is really needed.

First, suppose R contains two nested vectors d^1 and d^2, say, where $d^1 \leq d^2$. If (7.1.13) is satisfiable for $d = d^1$, then $d^1 \leq d^2$ implies that it is also satisfied for $d = d^2$. Similarly, if subrange($\overline{D}$) contains two nested vectors d^1 and d^2 with $d^1 \leq d^2$, then occurrence of d^1 in subrange($\overline{D}$) is irrelevant for deciding satisfiability of (7.1.14). Thus, we only need to assume that the maximal vectors of R are precisely the maximal vectors of subrange($\overline{D}$).

Second, instead of subrange($\overline{D}$), we may consider other sets T that may be generated from D and that include the subrange of column submatrices as a special case. For example, we may partition the column index set of D into disjoint subsets J_0, J_+, J_-, and $J_\pm$, define Q to be the set of vectors s satisfying

$$
(7.1.15) \qquad s_j = \begin{cases} 0 & \text{if } j \in J_0 \\ 1 & \text{if } j \in J_+ \\ -1 & \text{if } j \in J_- \\ \pm 1 & \text{if } j \in J_\pm \end{cases}
$$

and obtain a set

$$
(7.1.16) \qquad T = \{ d \mid d = D \odot s; \ s \in Q \}
$$

that might be substantially different from the subrange of any column submatrix $\overline{D}$ of D.

We combine the two considerations. Suppose for a given subset R of subrange(D), there exists a partition J_0, J_+, J_-, and $J_\pm$ of the column index set of D such that the maximal elements of R are precisely the maximal elements of T of (7.1.16). Then at least one case of the $|R|$ cases of (7.1.13) has a solution if and only if

$$
(7.1.17) \qquad \begin{array}{c} (D \odot s^1) \oplus (A^2 \odot s^2) \geq \underline{1} \\ s^1 \in Q \end{array}
$$

has a solution.

When a matrix D observes the above condition for all subsets R of subrange(D), we call it *column closed*. Similar arguments can be made for the case of a matrix derived from A of (7.1.1) by replacing the submatrix D by a zero matrix and by replacing the zero submatrix indexed by X_1 and Y_2 by a nonzero matrix E. The desired features of E constitute a property we call *row closedness*. The reader interested in the corresponding satisfiability algorithm should skip ahead to Chapter 10.

A matrix for which all submatrices are both column closed and row closed is called *Boolean closed*. This chapter contains a detailed investigation of the Boolean closed matrices.

Section 7.2 states in compact form the definitions of the three types of closedness.

Section 7.3 provides characterizations of the Boolean closed matrices by the exclusion of minimal submatrices and by a direct description.

Section 7.4 contains some properties of Boolean closed matrices. They concern the range, the subrange, and the inverse image of subrange vectors.

Section 7.5 consists of several algorithms, including a method for deciding whether a given matrix is Boolean closed.

Section 7.6 concludes the chapter with extensions.

7.2 Definitions

This section contains compact definitions of the three types of closedness introduced in the preceding section.

Column Closedness

A matrix A over $\mathbb{B}$ and with column index set Y is *column closed* if for any nonempty subset R of subrange(A) the following holds. There exists a partition of Y into sets J_0, J_+, J_-, and $J_\pm$ such that the set Q of vectors s satisfying

$$(7.2.1) \qquad s_j = \begin{cases} 0 & \text{if } j \in J_0 \\ 1 & \text{if } j \in J_+ \\ -1 & \text{if } j \in J_- \\ \pm 1 & \text{if } j \in J_\pm \end{cases}$$

produces a set

$$(7.2.2) \qquad T = \{b \mid b = A \odot s; \ s \in Q\}$$

whose maximal vectors are precisely the maximal vectors of R.

Row Closedness

A matrix A over $\mathbb{B}$ is *row closed* if, for any nonempty $R \subseteq$ subrange(A), there exists a $\{0,1\}$ vector b such that the set

$$(7.2.3) \qquad S_b = \{s \mid A \odot s \geq b; \ s_j \in \{\pm 1\}, \ \forall \, j\}$$

is equal to the set

$$(7.2.4) \qquad S_R = \cup_{f \in R}\{s \mid A \odot s \geq f; \ s_j \in \{\pm 1\}, \ \forall \, j\}$$

Boolean Closedness

A matrix A over $\mathbb{B}$ is *Boolean closed* if A and all submatrices of A are both column closed and row closed.

One may use the above definitions to directly show that the empty matrix as well as all trivial and zero matrices are Boolean closed. More interesting examples are given later.

Simplified Test for Row Closedness

The next lemma and corollary simplify testing for row closedness. The lemma tells how the vector b specified in the definition of row closedness may be chosen.

(7.2.5) Lemma. *A nonempty matrix A over $\mathbb{B}$ is row closed if and only if the following holds. For any nonempty subset R of subrange(A) and for b defined by*

$$(7.2.6) \qquad\qquad b_i = \min_{f \in R}\{f_i\}$$

the set S_b of (7.2.3) must be equal to the set S_R of (7.2.4).

Proof. The "if" part is trivial, since by assumption the vector b of (7.2.6) results in the equality $S_b = S_R$ demanded in the definition of row closedness.

For proof of the "only if" part, let R be any nonempty subset of subrange(A), and define b by (7.2.6). Then, for all $f \in R$, we have $b \le f$, and the definitions (7.2.3) and (7.2.4) for S_b and S_R imply that $S_b \supseteq S_R$.

Since A is row closed, there exists a $\{0,1\}$ vector c such that the set S_c defined analogously to S_b of (7.2.3) is equal to S_R. We claim that $b \ge c$. Assume the contrary. Then there exists an index i for which $c_i = 1$ and $b_i = 0$. Since $b_i = 0$, the definition of b by (7.2.6) implies that there exists a vector $f \in R$ with $f_i = 0$. Since R is a subset of subrange(A), there exists a $\{\pm 1\}$ vector s such that $A \odot s = f$. That vector s is in S_R, but not in S_c, so $S_c \ne S_R$, a contradiction. Hence, $b \ge c$, and $S_b \subseteq S_c$. We already know $S_b \supseteq S_R$ and $S_c = S_R$, so $S_b = S_c = S_R$. ☐

One may weaken the conditions of Lemma (7.2.5) and obtain a simpler test for row closedness.

(7.2.7) Corollary. *A nonempty matrix A over $\mathbb{B}$ is row closed if and only if the following holds. For any subset R of subrange(A) satisfying $|R| \ge 2$ and containing no nested vectors, for b defined by (7.2.6), and for any $\{\pm 1\}$ vector s satisfying $A \odot s \ge b$, the set R must contain a vector f such that $A \odot s \ge f$.*

Proof. We show that the conditions of the corollary are equivalent to the related ones of Lemma (7.2.5).

Suppose R contains two nested vectors f and g, say, $f \ge g$. If we delete f from R, then the vector b of (7.2.6) as well as the sets S_b and S_R of (7.2.3) and (7.2.4) remain unchanged. Hence, we may suppose that R does not contain nested vectors.

If $|R| = 1$, then b of (7.2.6) is equal to the single vector $f \in R$, and $S_b = S_R$ holds trivially. Hence, we may assume that $|R| \ge 2$.

By the definitions of S_b and S_R, the relation $S_b \supseteq S_R$ always holds. Thus, demanding $S_b \subseteq S_R$ is equivalent to requiring $S_b = S_R$.

Finally, a $\{\pm 1\}$ vector s satisfying $A \odot s \geq b$ also satisfies, for some $f \in R$, $A \odot s \geq f$, if and only if $S_b \subseteq S_R$.   □

7.3 Characterizations

We characterize the Boolean closed matrices by the exclusion of minimal submatrices and by a direct description.

Minimal Excluded Matrices

The matrices N^1–N^4 below turn out to be, up to column scaling, the minimal matrices whose exclusion produces Boolean closedness.

(7.3.1)

$$
N^1 = \begin{bmatrix} 1 & 0 \\ 0 & 1 \end{bmatrix} \quad
N^2 = \begin{bmatrix} 1 & 0 \\ 1 & 1 \\ -1 & 1 \end{bmatrix} \quad
N^3 = \begin{bmatrix} 1 & 1 \\ 1 & -1 \\ -1 & 1 \end{bmatrix} \quad
N^4 = \begin{bmatrix} 1 & 1 \\ -1 & 1 \\ -1 & -1 \end{bmatrix}
$$

Minimal excluded matrices N^1–N^4

We need some definitions for the direct description of the Boolean closed matrices.

Solid Staircase Matrix

According to Section 4.3, a $\{0,1\}$ matrix A is solid triangular if for all $i < j$, $A_{ij} = 0$, and for all $i \geq j$, $A_{ij} = 1$. When we add parallel or zero vectors any number of times to a solid triangular matrix, we get a solid staircase matrix.

We extend these definitions to $\{0, \pm 1\}$ matrices in the obvious way. Thus, a $\{0, \pm 1\}$ matrix is a *solid triangular* or *solid staircase matrix* if replacement of the -1s by $+1$s results in a $\{0,1\}$ matrix with the respective property. A typical $\{0, \pm 1\}$ solid staircase matrix has the following form.

(7.3.2)

Solid staircase matrix

Double Staircase Matrix

A $\{0, \pm 1\}$ matrix A is a *double staircase matrix* if it is of the following form.

(7.3.3)

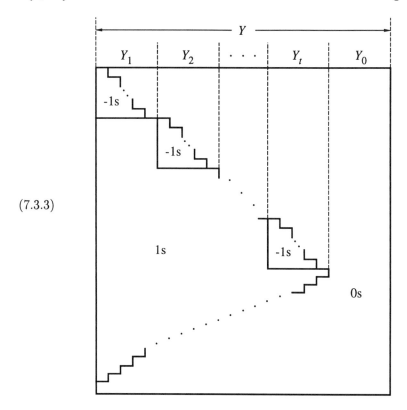

Double staircase matrix

Note that not all explicitly shown rows and columns need to be present. However, if the matrix is not a zero matrix, then we demand that one of the rows with maximum number of nonzero entries has no -1s. The latter requirement implies that the property "is a double staircase matrix" may not be maintained under submatrix taking. However, this is so for the property "is up to column scaling a double staircase matrix," as shown in the next lemma.

(7.3.4) Lemma. *If a $\{0, \pm 1\}$ matrix A is up to column scaling a double staircase matrix, then every submatrix of A has that property.*

Proof. We may assume that A itself is a double staircase matrix. Let $\overline{A}$ be a submatrix of A. If $\overline{A}$ is derived from A by deletion of a column, then at most a reordering of the rows of $\overline{A}$ proves $\overline{A}$ to be a double staircase matrix. If $\overline{A}$ is derived by deletion of a row, then the same conclusion applies unless in A a $\{0,1\}$ row with a maximum number of nonzero entries is deleted. The latter case may require column scaling plus a reordering of the rows to prove the conclusion.   □

Characterization of Boolean Closedness

The double staircase matrices turn out to be, up to column scaling, the Boolean closed matrices. That result plus the characterization of Boolean closedness by exclusion of N^1–N^4 of (7.3.1) constitutes the main result of this section. The precise statement is as follows.

(7.3.5) Theorem. *The following statements are equivalent for any matrix A over $\mathbb{B}$.*

  (i) *Matrix A is Boolean closed.*
 (ii) *Up to column scaling, A is a double staircase matrix.*
(iii) *Matrix A does not contain any column scaled version of any one of the matrices N^1–N^4 of (7.3.1) as submatrix.*

We prove Theorem (7.3.5) by showing (i)⇒(iii), (iii)⇒(ii), and (ii)⇒(i).

Part (i) implies (iii)

It suffices to show that a Boolean closed matrix cannot contain any one of the matrices N^1–N^4 of (7.3.1). Claim 1 below implies that result.

Claim 1. *The matrices N^1–N^4 of (7.3.1) are minimal matrices that are not Boolean closed.*

Proof. We use the notation employed in the definition of column and row closedness.

Let A be the 2×2 identity matrix N^1. Suppose A is Boolean closed and hence column closed. Select $R = \{[1\ 0]^t, [0\ 1]^t\}$. It is easy to verify that any sets J_0, J_+, J_-, and $J_\pm$ resulting in a set T of (7.2.2) with the same maximal vectors as R must satisfy $J_0 = J_+ = J_- = \emptyset$ and $J_\pm = Y$. The set Q defined by the latter sets via (7.2.1) is the set of all $\{0,1\}$ vectors with two entries, and the set T of (7.2.2) is equal to subrange(A). But then the vector $[1\ 1]^t$ is the unique maximal vector of T, yet that vector does not occur in R, a contradiction.

Similar arguments prove that N^2–N^4 are not column closed, using $R = \{[1\ 1\ 0]^t, [0\ 1\ 1]^t\}$ for N^2 and N^4 and using $R = \{[1\ 1\ 0]^t, [1\ 0\ 1]^t\}$ for N^3.

Finally, it is straightforward to show that all proper submatrices of N^1–N^4 are both column closed and row closed.   □

Part (iii) implies (ii)

Given is statement (iii), according to which A does not contain any column scaled version of N^1–N^4 as submatrix. We must demonstrate that A is up to column scaling a double staircase matrix. The next claim shows that A must be a solid staircase matrix.

Claim 2. *Matrix A is a solid staircase matrix.*

Proof. Lemma (4.3.21) states that a $\{0,1\}$ matrix is a solid staircase matrix if and only if it has no 2×2 identity submatrix. Hence, exclusion of all scaled versions of the 2×2 identity matrix N^1 assures A to be a $\{0,\pm 1\}$ solid staircase matrix. □

Due to Claim 2, column scaling, and trivial reductions, we may assume that A is a solid staircase matrix, has no zero rows or columns, and has only 1s in the last row. If A is a $\{0,1\}$ matrix, then it is a double staircase matrix, and we are done. Assume that A contains -1s.

Sort the columns of A while enforcing the solid staircase form such that each $\{0,1\}$ column is placed as far left as possible. Declare the resulting matrix to be A. The next claim establishes that the 1s of A are contiguous in a certain sense.

Claim 3. *For each row i and each column l, $A_{il} = 1$ implies that, for all $k < l$, $A_{ik} = 1$.*

Proof. If the claim does not hold, then there exist a row i and columns k and $l = k + 1$ such that $A_{il} = 1$ and $A_{ik} = 0$ or -1. The case of $A_{ik} = 0$ is not possible, since A is a solid staircase matrix. Thus, $A_{ik} = -1$.

If column l has a -1, say, in row q, then the 3×2 submatrix defined by the intersection of columns k and l with row i, row q, and the last row, is the matrix N^3 or N^4. Hence, column l of A is a $\{0,1\}$ vector. If columns k and l have the same support, that is, if they agree on the 0 entries, then column l must be to the left of column k due to the sorting assumption, a contradiction of the fact that $k < l$. Thus, there exists a row q for which $A_{ql} = 0$ and $A_{qk} = \pm 1$. Then, up to column scaling, the 3×2 submatrix of A defined by the intersection of columns k and l with row i, row q, and the last row, is the matrix N^2, a contradiction. □

Claim 3 implies that A can be partitioned according to (7.3.6) below, where the portions labeled B^1, $B^2, \ldots$ contain all -1s of A, possibly some 1s, but not 0s.

(7.3.6)

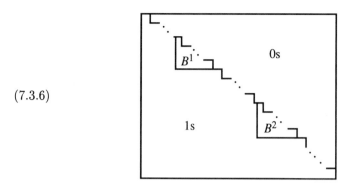

Structure of A

Choose B^1, B^2, ... such that the number of nonzeros contained in them is as small as possible.

Claim 4. *Each row of each B^p contains only 1s or only -1s.*

Proof. If the claim does not hold, then by Claim 3 there exist a row i and columns k and $l = k + 1$ in a B^p such that $B^p_{ik} = 1$ and $B^p_{il} = -1$. Arguing as in the proof of Claim 3, we must have, for all row indices q with $B^p_{ql} = \pm 1$, $B^p_{qk} = 1$. Thus, B^p must be

(7.3.7)

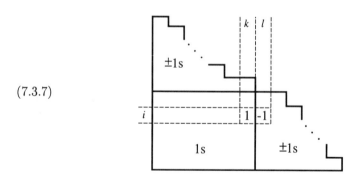

Structure of B^p

But then B^p can be partitioned into, say, $(B^p)'$ and $(B^p)''$ and a submatrix of 1s, which contradicts the minimality assumption of the number of nonzeros contained in B^1, B^2, ... □

Claim 5. *Matrix A is a double staircase matrix.*

Proof. By Claim 4, each row of each B^p in A of (7.3.6) contains only 1s or -1s. Thus, we may transfer the $\{0, 1\}$ rows of A to the bottom of the matrix to obtain the form of a double staircase matrix. □

Part (ii) implies (i)

Statement (ii) says that A is up to column scaling a double staircase matrix. We must prove that A is Boolean closed. Clearly, we may assume that A itself is a double staircase matrix and thus is given by (7.3.3).

We suppose that the jth column of A is indexed by j. Thus, for some $n \geq 1$, the column index set Y of A is equal to $\{1, 2, \ldots, n\}$, and the sets $Y_1, Y_2, \ldots, Y_t$, and Y_0 form a partition of that set. Due to this convention, it makes sense to refer to the smallest element of any one of the subsets Y_p of Y; that element is denoted by $y(p)$.

We introduce a color classification of the vectors of subrange(A) that simplifies later proofs.

Color Classification

For each vector b of subrange(A), we select a $\{\pm 1\}$ vector s such that $b = A \odot s$. If several choices exist for s, we arbitrarily select one from the possible candidates. The selected vector s is the *vector assigned to* b. The arbitrariness with which we have selected s shall not trouble us, since any one of the possible candidates would work for our purposes.

If the vector s assigned to b contains only -1s in the positions indexed by $Y - Y_0$, then b is declared to be *brown*.

Suppose s contains at least one 1 in the positions indexed by $Y - Y_0$. Let j be the smallest element of Y for which $s_j = 1$. For some $1 \leq p \leq t$, $j \in Y_p$. If j is equal to the smallest element $y(p)$ of Y_p, we declare b to be a *blue vector of Y_p*; otherwise, b is a *red vector of Y_p*.

Note that at most one vector of subrange(A) is declared to be brown.

Using the display of A in (7.3.3), one readily verifies that a blue vector b of Y_p can be expressed either as

$$(7.3.8) \qquad b = [\bigoplus_{j < y(p)} (A_{.j} \odot (-1))] \oplus [A_{.y(p)} \odot 1]$$

or, for some $z \in Y_p$ satisfying $z > y(p)$, as

$$(7.3.9) \quad b = [\bigoplus_{j < y(p)} (A_{.j} \odot (-1))] \oplus [\bigoplus_{y(p) \leq j < z} (A_{.j} \odot 1)] \oplus [A_{.z} \odot (-1)]$$

The second case applies if the vector s assigned to b contains at least one -1 with index in Y_p; the smallest such -1 of s is indexed by z.

Similarly, a red vector b of Y_p can be computed as follows. For the vector s assigned to b, let z be the smallest element of Y_p satisfying $s_z = 1$. Since b is red, $z > y(p)$. Then

$$(7.3.10) \qquad b = [\bigoplus_{j < z} (A_{.j} \odot (-1))] \oplus [A_{.z} \odot 1]$$

Finally, the brown vector is

(7.3.11) $$b = [\bigoplus_{j < y(0)} (A_{.j} \odot (-1))]$$

The color classification supports the following characterization of nested vectors of subrange(A).

Claim 6.
(a) For all $1 \leq p \leq t$: Any two blue vectors of Y_p are nested.
(b) For all $1 \leq p < q \leq t$: Any red vector of Y_p and any blue vector of Y_q are nested.
(c) Any two red vectors are nested.
(c) Any red vector and the brown vector are nested.

Proof. The result may be established by direct checking using the display of A in (7.3.3). □

Claim 6 implies the following result.

Claim 7. Let R be a subset of subrange(A) containing no nested vectors and satisfying $|R| \geq 2$. Then, for some $r \geq 1$ and for some indices $1 \leq p(1) < p(2) < \cdots < p(r) \leq t$, R contains exactly one blue vector each of $Y_{p(1)}, Y_{p(2)}, \ldots, Y_{p(r)}$, plus possibly, for some $q \geq p(r)$, one red vector of Y_q, plus possibly the brown vector, provided that a red vector is not present.

Proof. Claim 6 narrows down the choices for R to the stated cases. □

Boolean Closedness

We are ready to prove that A of (7.3.3) is Boolean closed. We first establish column closedness.

Claim 8. Matrix A of (7.3.3) is column closed.

Proof. Let R be a nonempty subset of subrange(A) that does not contain nested vectors. We must determine a partition of the column index set Y into J_0, J_+, J_-, and $J_\pm$ so that the set Q defined by these sets via (7.2.1) produces an instance T of (7.2.2) whose maximal vectors are precisely the vectors of R.

The following algorithm constructs the desired partition. It relies on the characterization of R by Claim 7.

1. If R contains just one vector b: Take any $\{\pm 1\}$ solution s for $A \odot s = b$, and define $J_+ = \{j \mid s_j = 1\}$, $J_- = \{j \mid s_j = -1\}$, $J_0 = J_\pm = \emptyset$, and stop.
2. Assign to each vector b of R a $\{\pm 1\}$ vector s such that $A \odot s = b$. Using the assigned vectors s, classify each vector of R either as a blue

vector of some Y_p, or as a red vector of some Y_q, or as brown. Let r be the number of blue vectors. (At this point, R has been determined to have, for some $1 \leq p(1) < p(2) < \cdots < p(r) \leq t$, one blue vector of each of the sets $Y_{p(1)}, Y_{p(2)}, \ldots, Y_{p(r)}$; possibly, for some $q \geq p(r)$, a red vector of Y_q; and, possibly, a brown vector.)

3. If R has no red vector, define $q = p(r)$.
4. Initialize $J_+ = \emptyset$, $J_- = \{y(i) \mid 1 \leq i \leq q;\ i \neq p(1), p(2), \ldots, p(r)\}$, and $J_\pm = \{y(i) \mid i = p(1), p(2), \ldots, p(r)\}$.
5. If $q \neq p(r)$: Add $y(q)$ to J_-.
6. If R has only blue vectors: Remove $y(p(r))$ from $J_\pm$, and add it to J_+.
7. If there is a red vector: Let s be the vector assigned to that red vector. Add the smallest index z of Y_q for which $s_z = 1$ to J_+.
8. Do for $i = p(1), p(2), \ldots, p(r)$: Let s be the vector assigned to the blue vector of Y_i. Declare $j(i)$ to be the smallest index of Y_i with $s_{j(i)} = -1$. If $j(i)$ is defined, add it to J_-.
9. If z was defined in Step 7, and $j(r)$ was defined in Step 8, and $j(r) = z$: Remove z from J_+, remove $j(r)$ from J_-, and add $j(r)$ to $J_\pm$.
10. If the brown vector is in R: For all $i > p(r)$, add $y(i)$ to J_-.
11. Define J_0 to contain the indices of Y not present in any one of the sets J_+, J_-, and $J_\pm$.

Simple but tedious checking proves that R is the set of maximal vectors of the set T of (7.2.2). We cannot include details here, but should mention a display of the vectors of R that we have found to be very helpful for the proof and that the reader may want to employ as well. We draw the blue, red, and brown vectors of R, in color, into the matrix A of (7.3.3). Specifically, a blue vector of Y_p overlays the column $y(p)$ of A, the red vector overlays the column z of A, and the brown vector is drawn as one additional column to the right of A.

Let $b \in T$; that is, for some $s \in Q$, $b = A \odot s$. It turns out that the smallest index j for which $s_j = 1$ completely determines b, with one exception. In the exceptional case, z was defined in Step 7, $j(r)$ was defined in Step 8, and $j(r) = z$. In that situation, b is determined by the sign of s_z.

One then confirms by case analysis that each vector $f \in R$ occurs in T and that, for each $b \in T$, there exists an $f \in R$ satisfying $b \leq f$. Thus, the vectors of R are precisely the maximal vectors of T. $\square$

Claim 9. *Matrix A of (7.3.3) is row closed.*

Proof. According to Corollary (7.2.7), A is row closed if and only if for any subset R of subrange(A) satisfying $|R| \geq 2$ and containing no nested vectors, for b defined by

$$(7.3.12) \qquad\qquad b_i = \min_{f \in R}\{f_i\}$$

and for any $\{\pm 1\}$ vector s satisfying $A \odot s \geq b$, the set R must contain a vector f such that $A \odot s \geq f$.

So let R be given, and define b by (7.3.12). Similarly to Claim 8, the proof can be accomplished by a tedious case analysis that we omit here. The combined display of the vectors of R as described in the proof of Claim 8 should be very helpful.

The proof proceeds roughly as follows. Suppose a $\{\pm 1\}$ vector s satisfies $A \odot s \geq b$. Let $g = A \odot s$. Add g, in the color induced by s, to the display of A, which already contains the blue, red, and brown vectors of R. Since $A \odot s \geq b$, we must have $g_i = 1$ in every row i in which all vectors $f \in R$ have a 1. By an analysis of the possible cases, one then confirms that, for some $f \in R$, $g \geq f$ and thus $A \odot s \geq f$. That conclusion proves A to be row closed. □

The final claim establishes Boolean closedness of A.

Claim 10. *Matrix A of (7.3.3) is Boolean closed.*

Proof. We must show that every submatrix of A is both column closed and row closed.

Lemma (7.3.4) states that the property of being column scalable to a double staircase matrix is maintained under submatrix taking. Claims 8 and 9 say that any double staircase matrix is both column closed and row closed. These facts imply that A is Boolean closed. □

The next section establishes properties of Boolean closed matrices.

7.4 Properties

We describe properties of Boolean closed matrices that arise from a certain partial order of vectors, and we investigate the inverse image of the subrange as well as the cardinalities of the range and subrange. We also identify the $\{0, \pm 1\}$ matrices with low GF(3)-rank that are Boolean closed, and we show that certain subregions of Boolean closed matrices are closed.

Doubly Nested Vectors

We begin with the partial order. Define a binary relation called "doubly nests" on the set of all $\{0, \pm 1\}$ vectors as follows. Let a and b be two $\{0, \pm 1\}$ vectors. Then a *doubly nests* b if a and b have the same length, and if at least one of (7.4.1)(i) and (ii) below holds.

(7.4.1)

(i)   There exists a $\delta = 1$ or -1 such that, for all i, $b_i \neq 0$ implies $a_i = \delta$.

(ii) There exists a $\delta = 1$ or -1 such that, for all i, $b_i \neq 0$ implies $a_i = \delta \cdot b_i$.

(7.4.2) **Lemma.** *Suppose $\{0, \pm 1\}$ vectors a and b of the same length are scaled by $\{\pm 1\}$ factors, resulting in a' and b', respectively. Then a doubly nests b if and only if a' doubly nests b'.*

Proof. One only needs to verify that (7.4.1)(i) and (ii) are invariant under scaling of a and b by $\{\pm 1\}$ factors. □

(7.4.3) **Theorem.** *The binary relation "doubly nests" is a partial order for any set of $\{0, \pm 1\}$ vectors that does not contain parallel vectors.*

Proof. We must show that "doubly nests" is reflexive, antisymmetric, and transitive on the given set of vectors.

Reflexive: Let a be any vector. Condition (7.4.1)(ii) trivially holds if $b = a$, so a doubly nests a.

Antisymmetric: For $a \neq b$, suppose that a doubly nests b and that b doubly nests a. It is easy to deduce from (7.4.1) that a and b must have the same support and indeed must be parallel. But this contradicts the assumption that the given set of vectors does not contain parallel vectors.

Transitive: For distinct a, b, and c, suppose that a doubly nests b and b doubly nests c. A straightforward analysis of the cases arising from (7.4.1) proves that a doubly nests c. □

The partial order "doubly nests" supports yet another characterization of Boolean closedness.

(7.4.4) **Theorem.** *Let A be a matrix over $\mathbb{B}$ without parallel columns, and define C to be the set consisting of the column vectors of A. Then A is Boolean closed if and only if the partial order "doubly nests" when restricted to C is a total order for C.*

Proof. Let A be Boolean closed. Scale the columns of A so that the double staircase matrix of (7.3.3) results. For any two consecutive column vectors a and b of the latter matrix, clearly (7.4.1)(i) or (ii) holds, so a doubly nests b. Lemma (7.4.2) says that scaling does not affect the ordering under "doubly nests." Hence, the set C is totally ordered.

Conversely, suppose "doubly nests" induces a total order for C. Theorem (7.3.5) states that A is Boolean closed if and only if A does not contain any column scaled version of N^1–N^4 of (7.3.1). Hence, we are done once we show that no column scaled version of N^1–N^4 is present. Suppose this is so, say, for columns a and b of A. Direct checking of (7.4.1) for the matrices N^1–N^4 confirms that neither a doubly nests b nor b doubly nests a, which contradicts the assumption that C is totally ordered. □

(7.4.5) **Corollary.** *Suppose the columns of a Boolean closed matrix A are arranged in increasing order using "doubly nests," with the smallest column in the leftmost position. Then one can permute the rows and scale the columns of A such that the double staircase matrix of (7.3.3) results.*

Proof. The ordering given by "doubly nests" is unique up to exchanges of parallel columns or zero columns. The result then follows from the proof of Theorem (7.4.4).   □

Representative Solutions

As the next topic, we investigate the inverse image of the subrange of Boolean closed matrices. We begin with a definition.

Let A be a matrix over $\mathbb{B}$ with $n \geq 1$ columns. For some $k \geq 1$, let s^1, s^2, ..., s^k be $\{0, \pm 1\}$ vectors that have n entries each and satisfy the following two conditions.

First, for any $b \in$ subrange(A) and for any $\{\pm 1\}$ solution vector s of $A \odot s = b$, there must exist at least one vector s^i such that each nonzero element s^i_j of s^i satisfies $s^i_j = s_j$ and such that $A \odot s^i = b$. Any such s^i is said to *represent* s. Second, for any s^i, the vector $b = A \odot s^i$ must be in subrange(A).

Since the vectors s^1, s^2, ..., s^k collectively represent all vectors of the inverse image of subrange(A), we call s^1, s^2, ..., s^k *representative solution vectors* for subrange(A).

For a general matrix A over $\mathbb{B}$, a minimum set of representative vectors for subrange(A) may have rather large cardinality even if subrange(A) is a small set. For example, let A be the $\{0, 1\}$ matrix with six rows and $6n \geq 6$ columns where each column is one of the six possible unit vectors and where each such unit vector occurs exactly n times. Then $|$subrange(A)$| = 64$, while any set of representative solution vectors for subrange(A) has cardinality larger than n^6.

In contrast, the cardinality of a minimum set of representative vectors for the subrange of any Boolean closed matrix cannot differ much from the cardinality of the subrange. Details are given by the next theorem and corollary. Recall from Section 2.6 that an array is monotone if the entries are all nonnegative or all nonpositive. Evidently, the nonmonotone columns of a double staircase matrix are precisely the columns containing at least one -1.

(7.4.6) Theorem. *Let A be a Boolean closed matrix with n_1 nonzero columns of which n_2 are not monotone. Then there exists a set of representative solution vectors for subrange(A) with cardinality at most $n_1 + \min\{n_2, 1\}$.*

Proof. For the moment, assume that A is the double staircase matrix of (7.3.3) and has n columns. Define s^1 to be the vector that has -1s in the positions indexed by $Y - Y_0$ and that has 0s in all other positions. The remaining s^i are constructed as follows.

For $1 \leq p \leq t$, create the $\{0, \pm 1\}$ vectors with n entries that have -1s in the positions $j < y(p)$ and that have all remaining nonzero entries indexed by some subset of Y_p. The latter nonzero entries are defined according to any one of the following three choices. Note that all vectors satisfying any one of the choices must be constructed.

In the first choice, a 1 is assigned to the smallest position $y(p)$ of Y_p and to all subsequent positions $j \in Y_p$ for which column j of A is not monotone.

The second choice is possible only if A has at least two nonmonotone columns with indices in Y_p. In that choice, a nonempty sequence of 1s starting at position $y(p)$ is assigned and is followed by one -1; the sequence must be so selected that the -1 occurs in a position $j \in Y_p$ for which column j of A is not monotone.

The third choice requires that $|Y_p| \geq 2$. In that choice, a nonempty sequence of -1s starting at position $y(p)$ is assigned and is followed by one 1; the sequence must be so selected that the 1 occurs in a position $j \in Y_p$.

Direct counting shows that the above process creates a total of $k \leq n_1 + \min\{n_2, 1\}$ vectors $s^1, s^2, \ldots, s^k$. To show that these vectors are representative solution vectors for subrange(A), let s be an $\{\pm 1\}$ vector with n entries, and $b = A \odot s$. View s as the vector assigned to b. Thus, s induces the color brown, red, or blue for b.

If b is brown, then s has -1s in the positions indexed by $Y - Y_0$, and s^1 represents s.

If b is a red vector, say, of Y_p, let z be the smallest index of Y_p satisfying $s_z = 1$. Since b is red, $z > y(p)$. The third choice above produces, among others, a vector s^i such that $s^i_z = 1$. For that s^i, each nonzero element s^i_j satisfies $s^i_j = s_j$. Using (7.3.3), it is easy to verify that $b = A \odot s^i$.

The case of a blue b is handled analogously to the situation of red. This time, the vector s^i is produced by the first or second choice above.

We still must show that, for each s^i, the vector $b = A \odot s^i$ is in subrange(A). To prove this, we arbitrarily replace the 0s of s^i by ± 1s, getting a vector s. It is easy to check that $A \odot s^i = A \odot s$, so $b = A \odot s^i$ is in subrange(A).

We have completed the proof for the case of a double staircase matrix. In the general case, the given Boolean closed matrix A can be column scaled to become a staircase matrix. We carry out such scaling, derive representative solution vectors for the subrange of the scaled matrix as described above, and finally apply the same scaling factors to the elements of the vectors. The result is a set of representative solution vectors for subrange(A). $\qquad\Box$

Cardinalities of Range and Subrange

In yet another change of topic, we examine the cardinalities of the range and subrange of Boolean closed matrices.

For arbitrary $n \geq 1$, let A be the $n \times n$ matrix that has -1s on the diagonal, 0s above the diagonal, and 1s below the diagonal. According to the matrix of (7.3.3), A is a column scaled version of a double staircase matrix and thus is Boolean closed. It is easy to see that, for any $n \times 1$ $\{0,1\}$ vector b, the equation $A \odot s = b$ has a $\{0,\pm1\}$ solution s. Hence, range(A) consists of all such b, and $|\text{range}(A)| = 2^n$.

In contrast, the following corollary of Theorem (7.4.6) shows that the subrange of any Boolean closed matrix is always a small set.

(7.4.7) Corollary. *Let A be a Boolean closed matrix with n_1 nonzero columns of which n_2 are monotone. Then $|\text{subrange}(A)| \leq n_1 + \min\{n_2, 1\}$.*

Proof. Theorem (7.4.6) says that the Boolean closed matrix A has a set of $k \leq n_1 + \min\{n_2, 1\}$ representative solution vectors for subrange(A). By the definition of these vectors, there exists for each $b \in$ subrange(A) a vector s^i such that $b = A \odot s^i$. Hence, $|\text{subrange}(A)| \leq k \leq n_1 + \min\{n_2, 1\}$.   □

Closed GF(3) Matrices

Section 4.4 contains results linking the matrices over $\mathbb{B}$ to the matrices over GF(3). In the spirit of that investigation, we include a characterization of the low rank matrices over GF(3) that are Boolean closed.

(7.4.8) Theorem. *Let A be a $\{0, \pm1\}$ matrix, considered to be over $\mathbb{B}$ or GF(3) as appropriate.*

(a) *If GF(3)-rank(A) ≤ 1, then A is Boolean closed.*
(b) *If GF(3)-rank(A) $= 2$, then A is Boolean closed if and only if column scaling followed by deletion of duplicate columns and rows can reduce A to a matrix that has GF(3)-rank equal to 2 and that is a submatrix of one of the matrices F^1–F^3 below.*

(7.4.9)

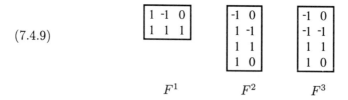

$$F^1 \qquad\qquad F^2 \qquad\qquad F^3$$

Boolean closed matrices F^1–F^3

Proof. If A has GF(3)-rank(A) ≤ 1, then A is a zero matrix, or all nonzero columns of A are identical up to column scaling. Thus, A can be column scaled to become a double staircase matrix and is Boolean closed.

To prove part (b), we first observe that the matrices F^1–F^3 of (7.4.9) are double staircase matrices. Let A be a matrix that by column scaling and deletion of duplicate rows and columns can be reduced to a submatrix of one of F^1–F^3. Then A itself can be column scaled to become a double staircase matrix and is Boolean closed.

Conversely, suppose that A is a Boolean closed matrix and that GF(3)-rank$(A) = 2$. Then A contains a 2×2 GF(3)-nonsingular submatrix with two, three, or four nonzeros. We analyze the three cases.

The first case implies that A contains a column scaled version of N^1 of (7.3.1), which is not possible.

Straightforward enumeration shows that each matrix of the remaining two cases can at most be extended to a column scaled version of F^1–F^3 of (7.4.9) if one rules out duplicate rows, duplicate columns, and submatrices that up to column scaling are equal to N^2–N^4 of (7.3.1). $\qquad\Box$

Closed Subregions

By the very definition of Boolean closedness, that property is maintained under submatrix taking. The next result shows that certain subregion taking maintains that property as well.

(7.4.10) Theorem. *Let A be a Boolean closed matrix with column index set Y. Define α_j, $j \in Y$, to be $\{\pm 1\}$ scaling factors that convert A to a double staircase matrix. Suppose we replace all entries A_{ij} satisfying $A_{ij} = -\alpha_j$, $j \in Y$, by 0s. Then the subregion of A so obtained is Boolean closed.*

Proof. Let B be the double staircase matrix derived from A by scaling with the α_j factors. Thus, any entry B_{ij} of B satisfies $B_{ij} = \alpha_j \cdot A_{ij}$, and that entry is equal to -1 if and only if $A_{ij} = -\alpha_j$.

Suppose in B we replace all -1s by 0s. According to (7.3.3), this change must result in a subregion B' of B that is a solid staircase matrix. Hence, B' is Boolean closed.

Replace in A all entries A_{ij} satisfying $A_{ij} = -\alpha_j$ by 0s, getting a subregion A'. Since the -1s of B correspond to the entries of A satisfying $A_{ij} = -\alpha_j$, the Boolean closed B' is a column scaled version of A'. Hence, A' is Boolean closed. $\qquad\Box$

The next section presents several algorithms that solve problems connected with Boolean closed matrices.

7.5 Algorithms

The results and proofs of the preceding sections imply efficient algorithms that test for Boolean closedness, that partition a matrix into Boolean closed column submatrices, and that compute certain vectors and sets connected with Boolean closed matrices. We provide these algorithms here.

Test Boolean Closedness

We begin with the test for Boolean closedness. Recall from Section 2.6 that count(A) is the number of nonzero entries of A.

(7.5.1) Algorithm TEST BOOLEAN CLOSEDNESS. *Tests a matrix A over $\mathbb{B}$ for Boolean closedness.*

Input: Matrix A over $\mathbb{B}$, of size $m \times n$.

Output: Either: "A is Boolean closed," together with column scaling factors and permutations of rows and columns that convert A to an instance of the double staircase matrix of (7.3.3). Or: "A is not Boolean closed."

Complexity: $O(m + n + \text{count}(A))$.

Procedure:

1. Select a row of A with the largest number of nonzero entries. Scale each column of A having a -1 in that row by -1.

2. Arrange the rows of A in increasing order of the number of nonzero row entries. Thus, the first row has a minimum number of nonzeros. Arrange the columns of the resulting matrix in decreasing order of the number of nonzero entries and such that each $\{0, 1\}$ column is placed to the left of any nonmonotone column with the same support. Thus, the first column has a maximum number of nonzeros.

3. Transfer the $\{0, 1\}$ rows of A to the bottom of the matrix in inverted order. Thus, a $\{0, 1\}$ row with a minimum number of 1s becomes the last row.

4. Either confirm that A is of the form (7.3.3), or declare that the input matrix is not Boolean closed. In the former case, declare the input matrix to be Boolean closed, and output the column scaling factors of Step 1 and the row and column permutations of Step 2 that convert the input matrix to an instance of (7.3.3).

Proof of Validity. The steps are a direct implementation of the proofs of Claims 2–5, which are part of the proof of Theorem (7.3.5). The steps can be accomplished with the claimed complexity when a bucket sort is used for the sorting of rows and columns. □

Partition into Closed Matrices

In Chapter 8, we desire a partition of an arbitrary matrix over $\mathbb{B}$ into Boolean closed column submatrices. Since each column of A constitutes a Boolean closed column submatrix, the task is trivial unless additional conditions are imposed. For example, one might demand that the number of such column submatrices be minimum. That case is solved by the next algorithm. Recall from Theorem (7.4.3) that the relation "doubly nests" is a partial order for any set of $\{0, \pm 1\}$ vectors that does not contain parallel vectors.

(7.5.2) Algorithm **BOOLEAN CLOSED PARTITION.** *Partitions a matrix A over $\mathbb{B}$ into a minimum number of Boolean closed column submatrices.*

Input: Matrix A over $\mathbb{B}$, with column index set Y.

Output: A partition of A into Boolean closed column submatrices A^1, $A^2, \ldots, A^k$ where k is minimum.

Complexity: Polynomial.

Procedure:
1. Do for each set of parallel columns of A: Delete the columns of that set except one from A.
2. Construct a directed graph G as follows. Each column y of A defines a node y of G. To determine the directed arcs of G, do for all distinct y and z of Y: If column y of A doubly nests column z, then introduce a directed arc from node y to node z. (Since "doubly nests" is a partial order on the columns of A, the graph G is acyclic.)
3. Use Algorithm PATH COVER (2.5.17) to find a minimum number of directed paths of G, say, $P_1, P_2, \ldots, P_k$, that cover the nodes of G.
4. Let $A^1, A^2, \ldots, A^k$ be the column submatrices of A whose columns are indexed by the nodes of $P_1, P_2, \ldots, P_k$, respectively. Delete columns from $A^1, A^2, \ldots, A^k$ until each column of A occurs in exactly one of $A^1, A^2, \ldots, A^k$.
5. Do for each column j deleted from the input matrix in Step 1: If A^i contains a column that is parallel to column j, then add column j to A^i. The resulting $A^1, A^2, \ldots, A^k$ constitute the desired output.

Proof of Validity. Suppose that the input matrix A does not have parallel columns. Theorem (7.4.4) implies that any column submatrix of A is Boolean closed if and only if the binary relation "doubly nests" is a total order of the columns of that submatrix. Thus, each Boolean closed submatrix of A corresponds to a directed path of the graph G constructed in Step 2.

Algorithm PATH COVER (2.5.17) finds a minimum number of directed paths covering the nodes of G. The column submatrices corresponding to these paths constitute a minimum number of column submatrices that include all columns of A. Hence, the matrices $A^1, A^2, \ldots, A^k$ on hand at the end of Step 4 constitute the desired partition of A.

If A has parallel columns, then Step 5 appropriately adjusts the A^1, $A^2, \ldots, A^k$.

Algorithm PATH COVER (2.5.17) used in Step 3 is polynomial. The other steps clearly can be done with polynomial effort. Hence, the entire algorithm is polynomial. $\qquad\square$

Find Representative Solutions

The next algorithm assembles representative solution vectors for the subrange of Boolean closed matrices.

(7.5.3) Algorithm REPRESENTATIVE SOLUTIONS. *Finds representative solution vectors for the subrange of a Boolean closed matrix A.*

Input: Boolean closed matrix A of size $m \times n$, with n_1 nonzero columns of which n_2 are nonmonotone. (Below, the notation is that of Section 7.3. In particular, $Y = \{1, 2, \ldots, n\}$ is assumed to be the column index set of A, and, for $1 \le p \le t$, $y(p)$ is defined to be the smallest element of the subset Y_p of Y.)

Output: For some $k \le n_1 + \min\{n_2, 1\}$, representative solution vectors s^1, $s^2, \ldots, s^k$ for subrange(A).

Complexity: $O(m + n^2 + \text{count}(A))$.

Procedure:

1. Using Algorithm TEST BOOLEAN CLOSEDNESS (7.5.1), column scale A so that A becomes the double staircase matrix of (7.3.3).
2. Define s^1 to be the vector that has -1s in the positions indexed by $Y - Y_0$ and has 0s in all other positions.
3. Do for $1 \le p \le t$: Create the $\{0, \pm 1\}$ vectors with n entries that have -1s in the positions $j < y(p)$ and that have all remaining nonzero entries indexed by some subset of Y_p. The latter nonzero entries are defined according to any one of the following three choices. Note that all vectors satisfying any one of the choices must be constructed.
   First choice: Assign a 1 to the smallest position $y(p)$ of Y_p and to all subsequent positions $j \in Y_p$ for which column j of A is not monotone.
   Second choice (requires that A has at least two nonmonotone columns with indices in Y_p): Assign a nonempty sequence of 1s starting at position $y(p)$, then append one -1; the sequence must be so selected that the -1 occurs in a position $j \in Y_p$ for which column j of A is not monotone.
   Third choice (requires $|Y_p| \ge 2$): Assign a nonempty sequence of -1s starting at position $y(p)$, then append one 1; the sequence must be so selected that the 1 occurs in a position $j \in Y_p$.
4. Scale the elements of the vectors created in Steps 2 and 3 by the column scaling factors determined in Step 1. Declare the resulting vectors to be $s^1, s^2, \ldots, s^k$. These vectors constitute the desired output.

Proof of Validity. The steps are taken almost verbatim from the proof of Theorem (7.4.6), which establishes the existence of the desired representative solution vectors for subrange(A).

The claimed complexity is achieved by Algorithm TEST BOOLEAN CLOSEDNESS (7.5.1) in Step 1 and by a suitable implementation of Steps 2–4. □

Find Subrange

Algorithm REPRESENTATIVE SOLUTIONS (7.5.3) is utilized in the following scheme for computing the subrange of Boolean closed matrices.

(7.5.4) Algorithm SUBRANGE OF BOOLEAN CLOSED MATRIX. *Finds* subrange(A) *for a Boolean closed matrix* A.

Input: Boolean closed matrix A of size $m \times n$, with n_1 nonzero columns of which n_2 are nonmonotone.

Output: subrange(A).

Complexity: $O(m \cdot n + n^2)$.

Procedure:
1. Using Algorithm REPRESENTATIVE SOLUTIONS (7.5.3), find, for some $k \leq n_1 + \min\{n_2, 1\}$, representative solution vectors $s^1, s^2, \ldots, s^k$ for subrange(A).
2. For $1, 2, \ldots, k$, compute $b^i = A \odot s^i$.
3. Eliminate duplicate vectors from $b^1, b^2, \ldots, b^k$. The remaining vectors constitute subrange(A).

Proof of Validity. By the definition of the representative solution vectors for subrange(A), the vectors produced at the end of Step 3 are indeed the vectors of subrange(A).

The claimed complexity is achieved by Algorithm REPRESENTATIVE SOLUTIONS (7.5.3) in Step 1 and by a suitable implementation of Steps 2 and 3. □

Find J-Sets

The final algorithm carries out the construction of the sets J_0, J_+, J_-, and $J_\pm$ of Claim 8, which is part of the proof of Theorem (7.3.5).

(7.5.5) Algorithm J-SETS. *Finds a partition* J_0, J_+, J_-, *and* $J_\pm$ *for the column index set* Y *of a double staircase matrix* A *over* $\mathbb{B}$ *and for a given subset* R *of* subrange(A), *such that the following holds. Let* Q *be the set of* $\{0, \pm 1\}$ *vectors* s *observing*

(7.5.6)
$$ s_j = \begin{cases} 0 & \text{if } j \in J_0 \\ 1 & \text{if } j \in J_+ \\ -1 & \text{if } j \in J_- \\ \pm 1 & \text{if } j \in J_\pm \end{cases} $$

Then the maximal vectors of the set T given by

(7.5.7) $T = \{b \mid b = A \odot s;\ s \in Q\}$

are precisely the maximal vectors of R.

Input: Double staircase matrix A over $\mathbb{B}$, of size $m \times n$. A subset R of subrange(A). For each $b \in R$, a $\{\pm 1\}$ vector s such that $b = A \odot s$. (Below, the notation is that of Section 7.3. In particular, s is called the vector assigned to b; $Y = \{1, 2, \dots, n\}$ is assumed to be the column index set of A; for $1 \le p \le t$, $y(p)$ is defined to be the smallest element of the subset Y_p of Y.)

Output: Partition J_0, J_+, J_-, and $J_\pm$ of Y.

Complexity: $O(m \cdot n + n^2)$.

Procedure:

1. Delete all nonmaximal vectors from R. If the reduced R contains just one vector b, use the vector s assigned to b to define the output sets $J_+ = \{j \mid s_j = 1\}$, $J_- = \{j \mid s_j = -1\}$, $J_0 = J_\pm = \emptyset$, and stop.

2. Using the assigned vectors s, classify each vector of R as follows. If the vector s assigned to b contains only -1s in the positions indexed by $Y - Y_0$, then b is declared to be brown.

   Suppose s contains at least one 1 in the positions indexed by $Y - Y_0$. Let j be the smallest element of Y for which $s_j = 1$. For some $1 \le p \le t$, $j \in Y_p$. If j is equal to the smallest element $y(p)$ of Y_p, declare b to be a blue vector of Y_p; otherwise, b is a red vector of Y_p.

   Let r be the number of blue vectors. (At this point, R has been determined to have, for some $1 \le p(1) < p(2) < \cdots < p(r) \le t$, one blue vector of each of the sets $Y_{p(1)}, Y_{p(2)}, \dots, Y_{p(r)}$; possibly, for some $q \ge p(r)$, a red vector of Y_q; and, possibly, a brown vector.)

3. If R has no red vector, define $q = p(r)$.

4. Initialize $J_+ = \emptyset$, $J_- = \{y(i) \mid 1 \le i \le q;\ i \ne p(1), p(2), \dots, p(r)\}$, and $J_\pm = \{y(i) \mid i = p(1), p(2), \dots, p(r)\}$.

5. If $q \ne p(r)$: Add $y(q)$ to J_-.

6. If R has only blue vectors: Remove $y(p(r))$ from $J_\pm$, and add it to J_+.

7. If there is a red vector: Let s be the vector assigned to that red vector. Add the smallest index z of Y_q for which $s_z = 1$ to J_+.

8. Do for $i = p(1), p(2), \dots, p(r)$: Let s be the vector assigned to the blue vector of Y_i. Declare $j(i)$ to be the smallest index of Y_i with $s_{j(i)} = -1$. If $j(i)$ is defined, add it to J_-.

9. If z was defined in Step 7, and $j(r)$ was defined in Step 8, and $j(r) = z$: Remove z from J_+, remove $j(r)$ from J_-, and add $j(r)$ to $J_\pm$.

10. If the brown vector is in R: For all $i > p(r)$, add $y(i)$ to J_-.

11. Define J_0 to contain the indices of Y not present in any one of the sets J_+, J_-, and $J_\pm$. The sets J_0, J_+, J_-, and $J_\pm$ constitute the desired partition of Y.

Proof of Validity. The steps are taken from the proof of Claim 8, which is part of the proof of Theorem (7.3.5).

The complexity can be attained by a suitable implementation of the steps. $\square$

7.6 Extensions

The efficiency of Algorithms (7.5.3)–(7.5.5) can be significantly improved by a special encoding of double staircase matrices and their subrange sets. Instead of an explicit representation of each nonzero entry of a double staircase matrix, one stores for each row or column the first and last position of each sequence of consecutive 1s or -1s contained in that row or column. It is easily checked using (7.3.3) that each subrange vector of a double staircase matrix contains at most two sequences of consecutive 1s. Hence, the subrange vectors can also be compactly encoded. The related modifications of Algorithms (7.5.3)–(7.5.5) are straightforward, so we leave them to the reader.

In the next chapter, we decompose matrices over $\mathbb{B}$ into closed Boolean matrices and matrices of the classes introduced in Chapter 5.

Chapter 8

Closed Subregion Decomposition

8.1 Overview

Suppose we are to solve the SAT problem for a matrix A that does not belong to any well-solved class we already know, but that nevertheless is similar to a matrix A' of such a class. Because of the similarity of the matrices, one would be tempted to modify the satisfiability algorithm for A' to obtain one for A. In this chapter, we formalize and extend this intuitive notion. Before we go into details, we review some definitions and material of earlier chapters.

SAT and MINSAT Centrality

According to (5.2.1), a class C of matrices over $\mathbb{B}$ is SAT central if the following conditions are satisfied.

(8.1.1)

    (i)   If $A \in C$, then any submatrix of A is also in C.

    (ii)   There is a polynomial algorithm for solving the SAT instances given by the matrices of C.

    (iii)   There is a polynomial algorithm for recognizing the matrices of C.

The class C is SAT semicentral if it observes (8.1.1)(i) and (ii).

According to (5.2.2), a class C of matrix/vector pairs (A, c) is MINSAT central if (8.1.2) below holds.

(8.1.2)

(i)   If $(A, c) \in C$, then any submatrix pair of (A, c) is also in C.

(ii) There is a polynomial algorithm for solving the MINSAT instances given by the matrix/vector pairs of C.

(iii) There is a polynomial algorithm for recognizing the matrix/vector pairs of C.

The class C is MINSAT semicentral if it observes (8.1.2)(i) and (ii).

Chapter 5 contains several SAT central and MINSAT central classes. With the exception of two cases, which concern the classes of balanced or totally unimodular matrices, both the solution algorithms and the recognition algorithms for these classes are very fast.

Subregion Cover

According to Section 2.6, a subregion is obtained from a given matrix by first taking a submatrix and then replacing in that submatrix some nonzero entries by zeros.

Section 4.4 declares a subregion cover of a matrix A to be a finite collection of subregions of A, say, $A^1, A^2, \ldots, A^k$, having the same size as A and observing the following condition. For each nonzero entry A_{ij} of A, there is at least one matrix A^l containing that entry. Any such matrix A^l is said to cover the entry A_{ij}.

Closed Subregion Decomposition

Section 7.2 defines the property of Boolean closedness of matrices. We shall not repeat the rather technical definition here. We use the concept in the following definition. Let $A^0, A^1, \ldots, A^q$ be a subregion cover of a given matrix A. If $A^1, A^2, \ldots, A^q$ are Boolean closed, then $A^0, A^1, \ldots, A^q$ is a *closed subregion decomposition* of A. Note that no conditions are imposed on A^0.

We are ready to sketch the main idea of this chapter. Suppose we must solve the SAT problem for a matrix A that is similar to a matrix of some SAT semicentral class C. The word "similar" admits numerous interpretations. For present purposes, it is to mean that A has a closed subregion decomposition $A^0, A^1, \ldots, A^q$ where A^0 belongs to the SAT semicentral class C and where the number of Boolean closed subregions, q, is small. We show in this chapter that for any such decomposition one can combine the satisfiability algorithm of C for A^0 with algorithms of Chapter 7 for the Boolean closed matrices A^1, A^2,..., A^q to obtain a satisfiability algorithm for A. The complexity of the latter algorithm depends on the complexity of the algorithm for A^0, the number of columns of A that are entirely contained in A^0, and the number of Boolean closed matrices, q. If q is considered bounded by a constant, then the algorithm is polynomial.

The same ideas apply to the MINSAT problem. Let a matrix/vector pair (A, c) be the given MINSAT instance. Analogously to the previous situation, assume that a closed subregion decomposition A^0, A^1,..., A^q of A is at hand where (A^0, c) belongs to a MINSAT semicentral class C. The solution algorithm of C for (A^0, c) can then be combined with algorithms of Chapter 7 for A^1, A^2,..., A^q to obtain a solution algorithm for (A, c). The latter algorithm is polynomial if q is bounded by a constant.

Finding a closed subregion decomposition that leads to a fast solution algorithm may be quite difficult. Indeed, even if the given semicentral class is one of the central classes of Chapter 5, the task of finding a best decomposition is difficult. Yet, closed subregion decompositions where A^0 is in one of the central classes of Chapter 5 are very useful from a practical viewpoint, so we feel compelled to devise polynomial and practically effective algorithms that search for attractive but not necessarily optimal decompositions.

We describe three such algorithms in detail and sketch others. Each of the schemes is based on a heuristic method for solving certain integer optimization problems called integer programs. The latter method has uses that go beyond the decomposition task at hand. For example, we rely on it in Chapter 13 to compute approximate solutions for MINSAT instances.

The presentation proceeds in the following manner. In Section 8.2, we develop the solution algorithm for the SAT and MINSAT problems involving matrices with a given closed subregion decomposition.

Section 8.3 contains the polynomial heuristic method for solving integer programs.

In Sections 8.4–8.6, we use that heuristic method to construct polynomial algorithms for finding closed subregion decompositions where A^0 is a 2SAT matrix, or is hidden nearly negative relative to a given submatrix, or is a network matrix. Recall that such matrices define certain SAT or MINSAT central classes of Chapter 5.

The final section, 8.7, contains extensions and references.

In the discussion to follow, we repeatedly use equations and inequalities of Lemmas (4.2.4), (4.2.8), and (4.2.14) without explicitly referencing them. Of particular use are the following two results. For any $\{0, 1\}$ vectors a, b, c, and d, $a \oplus b \geq c$ holds if and only if $a \geq c \ominus b$, and $a \geq b$ plus $c \geq d$ implies $a \oplus c \geq b \oplus d$.

8.2 Algorithm for SAT and MINSAT

The goal of this section is a solution algorithm for the SAT and MINSAT instances where the matrix A over $\mathbb{B}$ has a closed subregion decomposition. For convenient reference, we repeat the definition of that decomposition.

Closed Subregion Decomposition

Let A be a matrix over $\mathbb{B}$. Then a collection of matrices A^0, A^1, ..., A^q is a *closed subregion decomposition* of A if the following conditions are satisfied.

(8.2.1)   (i) $A^0, A^1, \ldots, A^q$ constitute a subregion cover of A.
            (ii) $A^1, A^2, \ldots, A^q$ are Boolean closed.

Representative Solutions

We review some results of Chapter 7. According to Section 7.4, any $\{0, \pm 1\}$ vectors s^1, s^2, ..., s^k are representative solution vectors for the subrange of a matrix A over $\mathbb{B}$ if the following two conditions are satisfied.

First, for any $b \in \mathrm{subrange}(A)$ and for any $\{\pm 1\}$ solution vector s of $A \odot s = b$, there must exist at least one vector s^i such that each nonzero element s^i_j of s^i satisfies $s^i_j = s_j$, and such that $A \odot s^i = b$. Any such s^i is said to represent s. Second, for any s^i, the vector $b = A \odot s^i$ must be in $\mathrm{subrange}(A)$.

Theorem (7.4.6) says that a Boolean closed matrix A, with n_1 nonzero columns of which n_2 are not monotone, has a set of representative solution vectors s^1, s^2, ..., s^k for $\mathrm{subrange}(A)$ where $k \leq n_1 + \min\{n_2, 1\}$. Algorithm REPRESENTATIVE SOLUTIONS (7.5.3) efficiently constructs such vectors.

Mutually Consistent Vectors

Let a collection of vectors of the same size be given. Then the vectors are *mutually consistent* if for any two vectors a and b of the collection and for any index j, $a_j \neq 0$ and $b_j \neq 0$ imply $a_j = b_j$. The first condition listed above in the definition of representative solution vectors can then be rephrased as follows. For any $b \in \mathrm{subrange}(A)$ and for any $\{\pm 1\}$ solution vector s of $A \odot s = b$, there exists at least one vector s^i that is mutually consistent with s and that satisfies $A \odot s^i = b$.

Basic Inequality

The next theorem establishes an important inequality. It is the basis for the solution algorithm to come.

(8.2.2) Theorem. *Suppose a matrix A over $\mathbb{B}$ has a closed subregion decomposition into matrices A^0, $A^1, \ldots, A^q$. For $1 \le p \le q$, let $s^{p,1}$, $s^{p,2}, \ldots, s^{p,k(p)}$ be representative solution vectors for subrange(A^p). Let s be a $\{\pm 1\}$ vector satisfying $A \odot s \ge \underline{1}$. Then, for $1 \le p \le q$, one may select one vector, say, $s^{p,i(p)}$, from the set of representative solution vectors for subrange(A^p) such that the selected vectors $s^{1,i(1)}$, $s^{2,i(2)}, \ldots, s^{q,i(q)}$ are mutually consistent, and such that s and the vector*

$$(8.2.3) \qquad\qquad b = \bigoplus_{p=1}^{q} (A^p \odot s^{p,i(p)})$$

satisfy

$$(8.2.4) \qquad\qquad A^0 \odot s \ge \underline{1} \ominus b$$

Proof. Since A^0, $A^1, \ldots, A^q$ are a subregion cover of A, each nonzero of A occurs in at least one of the matrices A^0, $A^1, \ldots, A^q$. Hence, $A \odot s \ge \underline{1}$ implies

$$(8.2.5) \qquad\qquad [A^0 \odot s] \oplus [\bigoplus_{p=1}^{q} (A^p \odot s)] \ge \underline{1}$$

By the definition of representative solution vectors, for $1 \le p \le q$, there exists a representative solution vector $s^{p,i(p)}$ for subrange(A^p) that is mutually consistent with s and that satisfies $A^p \odot s^{p,i(p)} = A^p \odot s$. We use the latter equation and the definition of b by (8.2.3) to deduce from (8.2.5) the inequality $(A^0 \odot s) \oplus b \ge \underline{1}$. That inequality is equivalent to $A^0 \odot s \ge \underline{1} \ominus b$ of (8.2.4). $\qquad\qquad\square$

Solution Algorithm

We are ready to present the solution algorithm. The scheme essentially searches for the mutually consistent vectors $s^{1,i(1)}$, $s^{2,i(2)}, \ldots, s^{q,i(q)}$ of Theorem (8.2.2) by enumerating the possible combinations.

(8.2.6) Algorithm SOLVE CLOSED SUBREGION DECOMPO-SITION SAT OR MINSAT. *Solves the SAT or MINSAT problem involving a given matrix A over $\mathbb{B}$ that has a closed subregion decomposition. In the MINSAT case, the cost vector is a given rational nonnegative vector c.*

Input: Matrix A over $\mathbb{B}$, of size $m \times n$. In the MINSAT case, rational nonnegative vector c.

Matrices A^0, $A^1, \ldots, A^q$ over $\mathbb{B}$ that constitute a closed subregion decomposition of A. For $1 \leq p \leq q$, A^p has n_{p1} nonzero columns of which n_{p2} are nonmonotone.

An algorithm for solving the SAT or MINSAT instances involving any submatrix $\overline{A}^0$ of A^0 and involving in the MINSAT case the corresponding subvector $\overline{c}$ of c. The algorithm is assumed to require at most β (resp. γ) effort in the SAT (resp. MINSAT) case.

Output: Either: A solution for the SAT instance A or the MINSAT instance (A, c), whichever applies. Or: "A is unsatisfiable."

Complexity: $O(\alpha(\beta + \varphi))$ in the SAT case and $O(\alpha(\gamma + \varphi))$ in the MINSAT case, where $\alpha = \prod_{p=1}^{q}(n_{p1} + \min\{n_{p2}, 1\})$ and $\varphi = m + n + \sum_{p=1}^{q} \text{count}(A^p)$. The effort is polynomial if β or γ, whichever applies, is polynomially bounded and if q is bounded by a constant.

Procedure:

1. Declare L to be an empty list. MINSAT case only: Initialize an integer z as $z = \infty$.

2. Use Algorithm REPRESENTATIVE SOLUTIONS (7.5.3) to determine, for $1 \leq p \leq q$, representative solution vectors $s^{p,1}$, $s^{p,2}, \ldots,$ $s^{p,k(p)}$ for subrange(A^p).

3. Do for each q-tuple $l = (i(1), i(2), \ldots, i(q))$ satisfying $1 \leq i(p) \leq k(p)$ for $1 \leq p \leq q$:
   Check if $s^{1,i(1)}$, $s^{2,i(2)}, \ldots, s^{q,i(q)}$ are mutually consistent. If this is so, add l to L.

4. Do for each q-tuple $l = (i(1), i(2), \ldots, i(q))$ of L: Compute

$$b^l = \bigoplus_{p=1}^{q}(A^p \odot s^{p,i(p)})$$

(8.2.7)

$$J^l_+ = \{j \mid \exists p \text{ such that } s^{p,i(p)}_j = 1\}$$

$$J^l_- = \{j \mid \exists p \text{ such that } s^{p,i(p)}_j = -1\}$$

Go to Step 5 (resp. Step 6) if a SAT (resp. MINSAT) problem is to be solved.

5. (SAT case) Do for each q-tuple $l = (i(1), i(2), \ldots, i(q))$ of L: With the assumed SAT algorithm, either find a $\{\pm1\}$ vector s that satisfies

(8.2.8)

$$A^0 \odot s \geq \underline{1} \ominus b^l$$

$$s_j = 1, \quad j \in J^l_+$$

$$s_j = -1, \quad j \in J^l_-$$

or conclude that no such solution exists. As soon as the former case is encountered, output s as the solution vector for the SAT problem of A, and stop. If for all $l \in L$, (8.2.8) is found to have no satisfying solution, declare A to be unsatisfiable, and stop.

6. (MINSAT case) Do for each q-tuple $l = (i(1), i(2), \ldots, i(q))$ of L: With the assumed MINSAT algorithm, either find a $\{\pm 1\}$ vector s that satisfies (8.2.8) and that, subject to that condition, minimizes total cost, or conclude that (8.2.8) has no solution. If the former case applies and if total cost is less than z, then redefine z to be equal to the total cost, and declare s to be the current solution candidate.

7. (MINSAT case) If $z = \infty$, declare A to be unsatisfiable, and stop. Otherwise, output the current solution candidate as the optimal solution. The total cost of that solution is z.

Proof of Validity. We first consider the SAT case. We must show that a given matrix A has a satisfying solution if and only if the algorithm produces such a solution. For proof of the "only if" part, let s be a $\{\pm 1\}$ solution vector. Theorem (8.2.2) says that there exist representative solution vectors $s^{1,i(1)}$, $s^{2,i(2)}, \ldots, s^{q,i(q)}$ that are mutually consistent and that for $b = \bigoplus_{p=1}^{q}(A^p \odot s^{p,i(p)})$ satisfy $A^0 \odot s \geq \underline{1} \ominus b$. Hence, Step 3 places the q-tuple $l = (i(1), i(2), \ldots, i(q))$ into L. For this l, Step 4 computes a vector b^l that is equal to b, plus sets J_+^l and J_-^l. By the derivation of b^l, J_+^l, and J_-^l, the given vector s satisfies the problem (8.2.8) in Step 5. Thus, Step 5 must output a vector that, for some q-tuple l of L, solves (8.2.8).

For proof of the "if" part, suppose that the algorithm produces in Step 5, for some $l = (i(1), i(2), \ldots, i(q))$ of L, a vector s satisfying (8.2.8). The definition of b^l in (8.2.7), plus the inequality $A^0 \odot s \geq \underline{1} \ominus b^l$ of (8.2.8), implies that

$$(8.2.9) \qquad (A^0 \odot s) \oplus [\bigoplus_{p=1}^{q}(A^p \odot s^{p,i(p)})] = (A^0 \odot s) \oplus b^l \geq \underline{1}$$

By the construction of L in Step 3, the vectors $s^{1,i(1)}$, $s^{2,i(2)}, \ldots, s^{q,i(q)}$ are mutually consistent. The conditions imposed in (8.2.8) on the elements of s by J_+^l, and J_-^l guarantee that s and any one of the vectors $s^{p,i(p)}$ are mutually consistent. Hence, for $1 \leq p \leq q$, $A^p \odot s^{p,i(p)} \leq A^p \odot s$. The latter inequality may be combined with (8.2.9) to

$$(8.2.10) \qquad (A^0 \odot s) \oplus [\bigoplus_{p=1}^{q}(A^p \odot s)] \geq \underline{1}$$

Since $A^0, A^1, \ldots, A^q$ constitute a subregion cover of A, (8.2.10) implies $A \odot s \geq \underline{1}$. We conclude that s is a satisfiable solution for the matrix A.

We have shown that the algorithm solves SAT instances correctly. For the MINSAT case, the additional cost considerations are handled appropriately by Steps 1, 6, and 7.

We establish the claimed complexity. For $1 \leq p \leq q$, Algorithm REP-RESENTATIVE SOLUTIONS (7.5.3) used in Step 2 needs to focus only on nonzero columns of A^p and thus finds at most $n_{p1} + \min\{n_{p2}, 1\}$ representative solution vectors for A^p with $O(m + (n_{p1})^2 + \text{count}(A^p))$ effort. Thus, Step 3 produces a set L having at most $\alpha = \prod_{p=1}^{q}(n_{p1} + \min\{n_{p2}, 1\})$ q-tuples. The latter step as well as Step 4 can clearly be done with $O(\varphi \cdot \alpha)$ effort, where $\varphi = m + n + \sum_{p=1}^{q} \text{count}(A^p)$.

The SAT or MINSAT problem given by (8.2.8) and the cost vector c, if applicable, is nothing but a SAT or MINSAT problem involving a submatrix of A^0 or (A^0, c). Accordingly, total effort for Step 5 (resp. Step 6) is bounded by $|L| \cdot \beta \leq \alpha \cdot \beta$ (resp. $|L| \cdot \gamma \leq \alpha \cdot \gamma$). The overall bound $O(\alpha(\beta + \varphi))$ (resp. $O(\alpha(\gamma + \varphi))$) stated for the SAT (resp. MINSAT) case dominates the bounds derived above for the various steps and thus is valid. That overall bound is clearly polynomial if β or γ, whichever applies, is polynomially bounded and if q is bounded by a constant.   □

Semicentral Classes

Chapter 5 includes a number of SAT or MINSAT central classes. We use the notion of closed subregion decomposition to extend these classes to significantly larger SAT or MINSAT semicentral classes. We first treat the general case.

(8.2.11) Theorem. *Let C be a class of matrices A (resp. matrix/vector pairs (A, c)) each of which belongs to a given SAT (resp. MINSAT) semicentral class C' or has, for some q bounded by a constant, a closed subregion decomposition into $A^0, A^1, \ldots, A^q$ where A^0 (resp. (A^0, c)) is in C'. Then C is SAT or MINSAT semicentral, whichever applies.*

Proof. We first consider the SAT case. We must confirm (8.1.1)(i) and (ii). That is, C must be maintained under submatrix taking, and there must be a polynomial algorithm that solves the SAT problem for all matrices of C.

Let A be a matrix of C. In the nontrivial case, A has a closed subregion decomposition into $A^0, A^1, \ldots, A^q$ where A^0 is in C', where $A^1, A^2, \ldots, A^q$ are Boolean closed, and where q is bounded by a constant. By the very definition of Boolean closedness, that property is maintained under submatrix taking.

Let $\overline{A}$ be a submatrix of A. Evidently, the corresponding submatrices $\overline{A}^0, \overline{A}^1, \ldots, \overline{A}^q$ of $A^0, A^1, \ldots, A^q$ constitute a subregion cover of $\overline{A}$ where $\overline{A}^0$ is in C', and where $\overline{A}^1, \overline{A}^2, \ldots, \overline{A}^q$ are Boolean closed. Hence, $\overline{A}$ is in C, and (8.1.1) holds.

The polynomial SAT algorithm for C' given by the SAT semicentrality of C' and Algorithm SOLVE CLOSED SUBREGION DECOMPOSITION SAT OR MINSAT (8.2.6) solve the SAT problem for A in polynomial time. Hence, (8.1.1)(ii) holds, and C has been proved to be SAT semicentral.

The MINSAT case is handled by almost identical arguments using (8.1.2)(i) and (ii). $\qquad\Box$

(8.2.12) Corollary.
(a) *Let C be the following class of matrices A over $\mathbb{B}$. Each $A \in C$ is a 2SAT matrix, or is hidden nearly negative, or is balanced, or has a closed subregion decomposition into A^0, $A^1, \ldots, A^q$ where A^0 has one of the first three properties and where q is bounded by a constant. Then C is SAT semicentral.*
(b) *Let C be the following class of matrix/vector pairs (A, c) where A is a matrix over $\mathbb{B}$ and where c is a rational nonnegative vector. Each $(A, c) \in C$ is hidden nearly negative relative to the column submatrix of A corresponding to the zero entries of c, or is balanced, or has a closed subregion decomposition into A^0, $A^1, \ldots, A^q$ where A^0 has one of the first two properties and where q is bounded by a constant. Then C is MINSAT semicentral.*

Proof. The arguments for parts (a) and (b) are virtually identical, so we just establish (a). Theorems (5.4.2), (5.6.4), and (5.7.28) say that the classes of 2SAT matrices, hidden nearly negative matrices, and balanced matrices are SAT central. Lemma (5.3.2) states that a union of a finite number of SAT central classes is SAT central. These results plus Theorem (8.2.11) prove that the class C of part (a) is SAT semicentral. $\qquad\Box$

Finding Decompositions

So far, we have concentrated on solution algorithms for the SAT or MIN-SAT instances where the matrix A has a suitable closed subregion decomposition. We now turn to the problem of finding such decompositions.

Ideally, we would like to identify decompositions that result in minimum computational effort when Algorithm SOLVE CLOSED SUBREGION DECOMPOSITION SAT OR MINSAT (8.2.6) solves the SAT or MINSAT problem using that decomposition. The search for such decompositions seems very difficult. Instead, we might want to look for decompositions that minimize the complexity bound for Algorithm SOLVE CLOSED SUBREGION DECOMPOSITION SAT OR MINSAT (8.2.6). That bound is $O(\alpha(\beta + \varphi))$ in the SAT case and is $O(\alpha(\gamma + \varphi))$ in the MINSAT case, where $\alpha = \prod_{p=1}^{q}(n_{p1} + \min\{n_{p2}, 1\})$ and $\varphi = m + n + \sum_{p=1}^{q} \text{count}(A^p)$. Finding best decompositions according to these bounds still seems to be quite difficult. Hence, we simplify the situation even further and look for decompositions A^0, $A^1, \ldots, A^q$ where A^0 belongs to a given SAT central class C or (A^0, c) belongs to a given MINSAT central class C, where the number of nonzero columns of A that are entirely contained in A^0 is maximum, and where, subject to these conditions, q is minimum. Optimization according to the latter criteria may be viewed as approximate minimization of the cited complexity bounds.

Optimal Decomposition

We call a decomposition that is best according to the above criteria *optimal relative to C*. If C is characterized by a certain property P, we also say that the decomposition is *optimal for P*. For example, if C is the class of 2SAT matrices, we say that the *decomposition is optimal for* 2SAT.

Our pragmatic approach has simplified the search for decompositions somewhat. But we see later that finding a decomposition that is optimal according to our definition may nevertheless be $\mathcal{NP}$-hard. On the other hand, real-world applications demand that we find decompositions for large matrices rapidly and reliably. Hence, we opt for polynomial heuristic methods for finding decompositions. We call such decompositions *good*, since they may not be optimal.

In Sections 8.4–8.6, we develop such heuristic methods for three important classes of closed decompositions. They differ by the conditions imposed on the matrix A^0 of the decompositions. In Section 8.4, A^0 is required to be a 2SAT matrix. In Section 8.5, A^0 must be hidden nearly negative relative to a specified column submatrix of A^0. In Section 8.5, A^0 is demanded to be a network matrix.

The heuristic methods of Sections 8.4–8.6 rely on a general heuristic scheme for a class of integer optimization problems called integer programs. The next section provides details about that scheme.

8.3 Heuristic for Integer Programs

We consider integer optimization problems that are given by a rational matrix B and rational vectors b and c. One is to find a $\{0,1\}$ vector r that satisfies the inequality $B \cdot r \geq b$, and that, subject to that condition, minimizes the objective function $c^t \cdot r$.

Integer Program

Let "s. t." stand for "subject to." The problem is therefore

$$
\begin{array}{lll}
& \min & c^t \cdot r \\
(8.3.1) & \text{s. t.} & B \cdot r \geq b \\
& & r \text{ is a } \{0,1\} \text{ vector}
\end{array}
$$

Each instance of such a problem is an *integer program*, abbreviated IP.

Later, we encounter IPs where the objective function calls for maximization and where some constraints are given by equations. Such cases are brought into the form (8.3.1) by multiplying the objective function by -1 and by replacing each equation by two inequalities. For example, max $r_1 + r_2$ is replaced by min $-r_1 - r_2$, and an equation $r_1 + r_2 = \beta$ is replaced by $r_1 + r_2 \geq \beta$ and $-r_1 - r_2 \geq -\beta$.

Linear Program

When we weaken the requirement of (8.3.1) that r be a $\{0, 1\}$ vector to the inequality $0 \leq r \leq 1$, we get the following *linear program* (LP).

$$
\begin{aligned}
&\text{min} \quad c^t \cdot r \\
(8.3.2) \quad &\text{s. t.} \quad B \cdot r \geq b \\
&\qquad\quad 0 \leq r \leq 1
\end{aligned}
$$

Linear Programming Results

We review and adapt some material about LPs given in Section 5.7.

A feasible solution r of the LP (8.3.2) is an *extreme point solution* if for any feasible solutions r^1 and r^2 satisfying $r = (r^1 + r^2)/2$, we necessarily have $r = r^1 = r^2$.

A basic linear programming result says that if the LP (8.3.2) has a feasible solution, then it must have an optimal solution that is an extreme point solution.

LPs are well solved. Most popular is the *Simplex Method*. It is not polynomial, but has proven to efficiently solve most LPs arising from real-world problems. There also exist polynomial solution methods—for example, the *Ellipsoid Method*. If (8.3.2) has a feasible solution, then any method producing an optimal solution is readily transformed into one producing an optimal extreme point solution. The Simplex Method does this naturally, since it examines extreme point solutions only. At any rate, we assume that any method selected below for the solution of LPs produces optimal extreme point solutions.

Assumptions about Integer Program

Numerous solution strategies for solving IPs of the form (8.3.1) have been published. Appropriate references are included in Section 8.7. Many of the schemes rely on the LP (8.3.2) or on modified versions of that LP. The heuristic method described below does so as well. It uses a set J and a subroutine Q that we define first.

The set J may be any subset of the column index set of the matrix B for which the following holds. If in the LP (8.3.2) the variables r_j with index j in J are fixed to arbitrary $\{0,1\}$ values and if the modified LP has a feasible solution, then all extreme point solutions of that LP must be integral.

The subroutine Q is needed for the following task. Suppose arbitrary $\{0,1\}$ values have been assigned to the variables r_j with index j in an arbitrary subset $\overline{J}$ of J. Then subroutine Q is to decide whether one can assign $\{0,1\}$ values to the remaining variables r_j such that the resulting $\{0,1\}$ vector r is a feasible solution for the IP (8.3.1).

Heuristic Solution Method

The heuristic method proceeds as follows. We determine with subroutine Q whether (8.3.1) has a feasible solution. If this is not so, we declare (8.3.1) to have no solution and stop. Assume that this case does not apply.

We solve the LP (8.3.2). If the optimal extreme point solution of the LP is integral, we output it as an optimal solution of the IP (8.3.1) and stop. Assume that the LP solution is not integral.

We select a nonempty subset $\overline{J}$ of J that indexes variables with fractional solution values. The selection is such that fractional values close to $\frac{1}{2}$ are preferred to values close to 0 or 1. For each possible way of fixing the variables r_j with index j in $\overline{J}$ to 0 or 1, we determine with subroutine Q whether the fixed values can be extended to a feasible solution of the IP (8.3.1). For each case where Q supplies an affirmative answer, we solve the LP (8.3.2) with the additional constraint that the r_j with index j in $\overline{J}$ must have the given $\{0,1\}$ values. Using the solutions for the various LPs, we select one variable r_{j^*} with index $j^* \in \overline{J}$ and permanently fix that variable to either 0 or 1. The fixing is done so that the IP (8.3.1) still has a feasible solution. The permanent fixing of r_{j^*} allows us to effectively eliminate the variable r_{j^*} from the IP (8.3.1) and from the LP (8.3.2).

We apply the above steps recursively to the reduced IP (8.3.1) and the reduced LP (8.3.2), with two minor changes. First, the reduced IP (8.3.1) has by its derivation a feasible solution, so we need not check for this with subroutine Q. Second, if the optimal extreme point solution of the reduced LP (8.3.2) is integral, then that solution plus the $\{0,1\}$ values already assigned to variables constitutes a good but not necessarily optimal solution for the original IP. The heuristic method outputs that solution and then stops.

By the definition of the set J, the recursive process stops at the latest when all variables r_j with index j in J have been permanently fixed.

We can estimate the quality of the solution found for the IP (8.3.1) as follows. When the original LP (8.3.2) has been solved, we save its optimal objective function value. Denote that value by β. Clearly, β is a lower bound on the optimal objective function value of the IP (8.3.1). When the recursive process stops, we compare β with the objective function value of the solution obtained for the IP (8.3.1). If the difference is small (resp. 0), then that solution is close to optimal (resp. is indeed optimal).

Details of the method are presented next.

(8.3.3) Heuristic SOLVE IP. *Finds a good but not necessarily optimal solution for the* IP

$$
\begin{aligned}
\text{min} \quad & c^t \cdot r \\
(8.3.4) \qquad \text{s. t.} \quad & B \cdot r \geq b \\
& r \text{ is a } \{0,1\} \text{ vector}
\end{aligned}
$$

or declares the IP *to have no feasible solution.*

Input: Rational matrix B with column index set Y, rational vectors b and c, and a positive integer k.

A subset J of Y such that fixing of the variables r_j with index j in J to any $\{0,1\}$ values reduces the LP

$$
\begin{aligned}
\text{min} \quad & c^t \cdot r \\
(8.3.5) \qquad \text{s. t.} \quad & B \cdot r \geq b \\
& 0 \leq r \leq 1
\end{aligned}
$$

to an LP that either has no feasible solution or has only integral extreme point solutions.

A subroutine Q that carries out the following task. Suppose arbitrary $\{0,1\}$ values have been assigned to the variables r_j with index j in some subset $\bar{J}$ of J. Then subroutine Q is to decide whether one can assign $\{0,1\}$ values to the remaining variables r_j such that the resulting $\{0,1\}$ vector r is a feasible solution for the IP (8.3.4). Subroutine Q is assumed to require at most σ effort.

A subroutine R that finds an optimal extreme point solution for any one of the following modified versions of the LP (8.3.5). Each version is obtained from the LP (8.3.5) by fixing the variables r_j with index j in some subset $\bar{J}$ of J to some $\{0,1\}$ values such that the modified LP still has a feasible solution. Subroutine R is assumed to require at most λ effort.

Output: Either: A good but not necessarily optimal solution for the IP (8.3.4), plus a rational number β that is a lower bound on the optimal objective function value of the IP (8.3.4). If the difference between β and the objective function value of the solution is small (resp. 0), then that solution is close to optimal (resp. is indeed optimal). Or: "The IP (8.3.4) has no feasible solution."

Complexity: $O(2^k \cdot (|J| + 1) \cdot (\sigma + \lambda))$. The effort is polynomial if σ and λ are polynomially bounded and if k is bounded by a constant.

Procedure:

1. (Check feasibility) Use subroutine Q to decide whether the IP (8.3.4) has a feasible solution. If this is not so, output that conclusion, and stop.

2. (Solve LP) Use subroutine R to solve the LP (8.3.5). Let $\tilde{r}$ be the solution vector. If this is the first time this step is executed, define $\beta = c^t \cdot \tilde{r}$.

3. (Termination test) If $\tilde{r}$ is integral, output $\tilde{r}$ plus the r_{j*} values assigned in earlier passes through Step 6, if any, as solution vector for the original IP, and stop.

4. (Select variables) Place the indices $j \in J$ for which $\gamma_j = \min\{1-\tilde{r}_j, \tilde{r}_j\}$ is positive into a set $\overline{J}$. If $|\overline{J}| > k$, sort the indices j of $\overline{J}$ in decreasing order of the values γ_j, then delete from $\overline{J}$ all indices except for the first k indices of the sorted list. (We have $0 \leq \gamma_j \leq \frac{1}{2}$; the minimum (resp. maximum) is attained if and only if $\tilde{r}_j = 0$ or 1 (resp. $\tilde{r}_j = \frac{1}{2}$). Thus, $\overline{J}$ contains up to k indices j for which $\tilde{r}_j$ is as close to $\frac{1}{2}$ as possible.)

5. (Temporarily fix variables) Do for each possible way of defining a $\{0,1\}$ value α_j for each $j \in \overline{J}$:

   For each $j \in \overline{J}$, fix r_j to α_j; test with subroutine Q if the IP (8.3.4) with these fixed values has a feasible solution; if this is so, use subroutine R to solve the LP (8.3.5) while enforcing, for all $j \in \overline{J}$, the value α_j for r_j.

   Number the LPs that have been solved in the do loop, say, as $1, 2, \ldots,$ n. For $1 \leq i \leq n$, declare β_i to be the objective function value of the ith LP. For each $j \in \overline{J}$ and for each $1 \leq i \leq n$, let α_j^i be the value to which the variable r_j was fixed in the ith LP.

6. (Permanently fix one variable) Select an i^*, $1 \leq i^* \leq n$, such that $\beta_{i^*} = \min\{\beta_i \mid 1 \leq i \leq n\}$. For each $j \in \overline{J}$, define $\delta_j = \min\{\beta_i \mid$ i such that $\alpha_j^i \neq \alpha_j^{i^*}\}$. Select a j^* from $\overline{J}$ such that $\delta_{j^*} = \max\{\delta_j | j \in \overline{J}\}$. Permanently fix r_{j*} to the value $\alpha_{j*}^{i^*}$. (Assume that β_{i^*} is approximately equal to the optimal objective function value of the IP (8.3.4). Then δ_j may be viewed to be a reasonable estimate of the increase in objective function value of the IP (8.3.4) when the variable r_j is fixed to the $\{0,1\}$ value different from $\alpha_j^{i^*}$. So if δ_j is large, then quite likely the variable r_j has the value $\alpha_j^{i^*}$ in any optimal solution of the IP (8.3.4). The argument is most convincing for the index j^*, since it maximizes δ_j. Accordingly, we fix r_{j*} to $\alpha_{j*}^{i^*}$.)

7. (Reduce problem) Reduce the IP (8.3.4) and the LP (8.3.1) using $r_{j*} = \alpha_{j*}^{i^*}$, and go to Step 2.

Proof of Validity. We assume the nontrivial case where the subroutine Q determines in Step 1 that the IP (8.3.4) has a feasible solution. Inductively, assume that the IP on hand as Step 2 is entered has a feasible solution.

Step 2 produces an optimal solution $\tilde{r}$ for the LP (8.3.5). If this is the first pass through Step 2, then the optimal objective function value of the LP, which is recorded under β, is clearly a lower bound for the objective function value of the IP (8.3.4). If $\tilde{r}$ is integral, then $\tilde{r}$ plus the r_{j^*} values permanently assigned in earlier passes through Step 6 is a good solution for the original IP (8.3.4), and we stop in Step 3. Assume that $\tilde{r}$ contains noninteger entries.

We claim that at least one of the $\tilde{r}_j$ with j in J is fractional. If this is not so, then by the assumption on J any optimal extreme point solution of the LP (8.3.5)—in particular, $\tilde{r}$—must be integral, a contradiction.

Step 4 selects up to k indices for the subset $\overline{J}$ of J. The indices correspond to fractional values of the solution that are as close to $\frac{1}{2}$ as possible.

Step 5 enumerates all possible ways of fixing the variables r_j with index j in $\overline{J}$ to 0 or 1. Recall the inductive assumption that the IP (8.3.4) is feasible. Hence, for at least one such fixing of the r_j variables, the subroutine Q must determine that the corresponding IP has a solution. The LP solutions are computed by the subroutine R.

Step 6 permanently fixes one variable r_{j^*} with index j^* in $\overline{J}$ to the value $\alpha_{j^*}^{i^*}$. The selection of the index j^* is based on heuristic arguments that guess the IP (8.3.4) to have an optimal solution where r_{j^*} has the assigned value.

By the determination of j^* and of the value $\alpha_{j^*}^{i^*}$ for r_{j^*}, the IP (8.3.4) does have a feasible solution where r_{j^*} takes on the assigned value. Hence, we may reduce the IP (8.3.4) and the LP (8.3.5) according to that fixing of r_{j^*} and are assured that the inductive assumption is satisfied as we return to Step 2.

In the complexity formula $O(2^k \cdot (|J| + 1) \cdot (\sigma + \lambda))$, the factor 2^k is a bound on the number of applications of the subroutines Q and R in a given pass through Step 5. By the assumption on J, the LP solution $\tilde{r}$ of Step 2 must be integral when all variables with index in J have been fixed to $\{0, 1\}$ values. Each pass through Steps 2–7 results in the permanent fixing of exactly one variable, so the number of passes through these steps is bounded by $|J| + 1$. Thus, total effort is $O(2^k \cdot (|J| + 1) \cdot (\sigma + \lambda))$ as claimed.   □

For computational efficiency, one would want to sequence the computations of Step 5 so that consecutive cases differ as little as possible. That way, the optimal solution for one LP is an attractive starting solution for the next LP. We show that such sequencing is easily determined.

Suppose we collect in a vector the $\{0, 1\}$ values α_j to which the variables r_j are fixed in a given iteration of the do loop of Step 5. That vector has $n = |\overline{J}|$ entries. Then each iteration of the do loop corresponds to a $\{0, 1\}$ vector with n entries, with all possible cases occurring. Furthermore, sequencing the computations of Step 5 so that consecutive cases differ as little as possible corresponds to sorting the $\{0,1\}$ vectors with n entries so that consecutive vectors of the resulting sorted list differ by exactly one entry. Such sorting may be done as follows.

Let $n \geq 2$. Inductively, suppose a sorted list L is on hand that consists of all $\{0,1\}$ vectors with $n-1$ entries, with consecutive vectors differing by exactly one entry. We compute a list for the $\{0,1\}$ vectors with n entries in two steps.

First, take each vector of L, and add a 1 as the nth entry. Collect the resulting vectors with n entries in a list L'.

Second, process the vectors of L once more, but this time in reverse order, and add to each vector a 0 as the nth entry. Adjoin the resulting vectors, in the order of processing, at the bottom of L'. It is easily checked that the resulting L' is the desired sorted list of the $\{0,1\}$ vectors with n entries.

With Heuristic SOLVE IP (8.3.3) at hand, we are ready to tackle the problem of finding, for given A over $\mathbb{B}$, an attractive closed subregion decomposition A^0, A^1, ..., A^q where A^0 belongs to one of the following three matrix classes: the class of 2SAT matrices, the class of matrices that are hidden nearly negative relative to a specified column submatrix, or the class of network matrices. The three cases are treated in Sections 8.4–8.6. Additional cases and extensions are covered in Section 8.7.

8.4 Decomposition for 2SAT

Recall from Sections 5.4 and 8.2 that a 2SAT matrix has at most two nonzero entries in each row, and that a closed subregion decomposition for 2SAT, say, A^0, A^1, ..., A^q, is optimal if the following requirements are met. The matrix A^0 must be a 2SAT matrix that maximizes the number of columns that are entirely contained in A^0; subject to that condition, the number of Boolean closed subregions, q, must be minimum.

We desire, for a given matrix A over $\mathbb{B}$, a good if not optimal closed subregion decomposition for 2SAT.

Complexity

We first show that the problem of finding an optimal decomposition is $\mathcal{NP}$-hard, using the following result of Bartholdi (1982). Define a *permutation matrix* to be a matrix that up to the permutation of rows—or, equivalently, columns—is an identity matrix.

(8.4.1) Theorem. *Let P be a property of $\{0,1\}$ matrices that is maintained under submatrix taking. Assume that P holds for all permutation matrices and that it fails for an infinite number of $\{0,1\}$ matrices. Then the problem of finding a maximum submatrix or a maximum column submatrix that has P is $\mathcal{NP}$-hard.*

Clearly, the 2SAT property is maintained under submatrix taking, it holds for all permutation matrices, and it fails for an infinite number of $\{0,1\}$ matrices. Theorem (8.4.1) implies that under these conditions the problem of finding a maximum 2SAT column submatrix in a given $\{0,1\}$ matrix is $\mathcal{NP}$-hard. The latter problem is solved by any optimal closed subregion decomposition for 2SAT, so finding such decompositions is $\mathcal{NP}$-hard as well.

Decomposition Algorithm

We use a rather simple heuristic that in two steps finds good closed subregion decompositions for 2SAT. In the first step, we search heuristically for a 2SAT subregion A^0 that contains a large number of nonzero columns of A. In the second step, we divide up the remaining nonzero columns of A among a minimum number of Boolean closed subregions, say, A^1, A^2, ..., A^q.

We formulate the first step as an IP. Let Y be the column index set of the given matrix A. Define D to be the support matrix of A. Thus, D is the $\{0,1\}$ matrix of the same size as A whose 1s correspond to the nonzeros of A. If a column y of A is placed into the subregion A^0, we say that the column is *assigned to A^0*.

For each $y \in Y$, define a $\{0,1\}$ variable r_y as follows.

$$(8.4.2) \qquad r_y = \begin{cases} 1 & \text{if column } y \text{ is assigned to } A^0 \\ 0 & \text{otherwise} \end{cases}$$

Collect the variables r_y in a vector r. The condition that A^0 is a 2SAT matrix is equivalent to the requirement that

$$(8.4.3) \qquad D \cdot r \leq 2 \cdot \underline{1}$$

Hence, we find a subregion A^0 that contains as many nonzero columns of A as possible, by determining a $\{0,1\}$ vector that solves

$$(8.4.4) \qquad \max \underline{1}^t \cdot r$$

subject to (8.4.3). As we have seen, finding an optimal A^0, and hence solving the IP of (8.4.3) and (8.4.4), may be difficult. So we use Heuristic SOLVE IP (8.3.3) to find a good but not necessarily optimal A^0.

In the second step, we collect in a matrix E all nonzero columns of A that do not occur in A^0, and we apply Algorithm BOOLEAN CLOSED PARTITION (7.5.2) to partition E into a minimum number of Boolean closed submatrices, say, E^1, $E^2, \ldots, E^q$. For $p = 1, 2, \ldots, q$, we define A^p to be the Boolean closed subregion of A whose nonzero columns are precisely the nonzero columns of E^p. Then A^0, $A^1, \ldots, A^q$ is a good closed subregion decomposition of A for 2SAT.

We summarize the above steps.

(8.4.5) Heuristic DECOMPOSITION FOR 2SAT. *Finds a good but not necessarily optimal closed subregion decomposition A^0, $A^1, \ldots, A^q$ of a matrix A over $\mathbb{B}$ where A^0 is a 2SAT matrix.*

Input: Matrix A over $\mathbb{B}$, with column index set Y.

Output: A good but not necessarily optimal closed subregion decomposition A^0, $A^1, \ldots, A^q$ of A where A^0 is a 2SAT submatrix.

Complexity: Polynomial if in Step 1 a polynomial version of Heuristic SOLVE IP (8.3.3) is used.

Procedure:

1. (Find A^0) Let D be the support matrix of A. Use Heuristic SOLVE IP (8.3.3) to obtain a good but not necessarily optimal solution for the IP given by (8.4.3) and (8.4.4). The input for that heuristic method consists of the IP, any $k \geq 1$, $J = Y$, the subroutine Q specified next, and a subroutine R for solving LPs. Subroutine Q decides whether $\{0, 1\}$ values assigned to some specified variables r_y can be extended to a feasible solution for the inequality $D \cdot r \leq 2 \cdot \underline{1}$ of (8.4.3). Since $D \geq 0$, that task is simple. Indeed, subroutine Q declares that an extension to a feasible solution is possible if and only if the given values for the specified variables plus zeros for the remaining variables constitute a feasible solution for (8.4.3).

   Let r^* be the solution produced by Heuristic SOLVE IP (8.3.3). Obtain A^0 from A by replacing each column y for which $r^*_y = 0$ by a zero column.

2. (Find A^1, $A^2, \ldots, A^q$) Collect in a matrix E all nonzero columns of A that do not occur in A^0. Use Algorithm BOOLEAN CLOSED PARTITION (7.5.2) to partition E into a minimum number of Boolean closed column submatrices, say, E^1, $E^2, \ldots, E^q$. For $p = 1, 2, \ldots, q$, define A^p to be the Boolean closed subregion of A whose nonzero columns are precisely the nonzero columns of E^p. Output A^0, $A^1, \ldots, A^q$ as the desired decomposition, and stop.

Proof of Validity. As argued earlier, Steps 1 and 2 produce a closed subregion decomposition for 2SAT. Since Algorithm BOOLEAN CLOSED

PARTITION (7.5.2) is polynomial, the entire procedure is polynomial if a polynomial version of Heuristic SOLVE IP (8.3.3) is used in Step 1.   □

We turn to the situation where A^0 is to be hidden nearly negative relative to a given column submatrix.

8.5 Decomposition for Hidden Near Negativity

By Section 5.5, a matrix A over $\mathbb{B}$ is nearly negative if each row of A contains at most one 1. Let D be any column submatrix of A. Section 5.6 defines A to be hidden nearly negative relative to D if scaling of the columns of D with $\{\pm 1\}$ factors can convert A to a nearly negative matrix.

Let A with column submatrix D be given. We desire a good, if not optimal, closed subregion decomposition of A into A^0, $A^1, \ldots, A^q$ where A^0 is hidden nearly negative relative to the submatrix of A^0 indexed by the column indices of D.

Complexity

We first show that finding an optimal decomposition is $\mathcal{NP}$-hard, no matter which columns of A occur in D. We reduce the problem EXACT COVER BY 3-SETS, which is known to be $\mathcal{NP}$-complete, to the problem at hand. An instance of EXACT COVER BY 3-SETS is given by a set X, with $|X| = 3k$, for some $k \geq 1$, and a collection Y of 3-element subsets of X. The question to be answered is: Does Y contain an exact cover of X? That is, does Y contain disjoint subsets of X whose union is equal to X? Evidently, an equivalent question is: Does Y contain k disjoint subsets of X?

We encode a given instance by a matrix B with row index set X and column index set Y. The entry of B in row $x \in X$ and column $y \in Y$ is 1 if y is a subset of X containing x and is 0 otherwise. In terms of B, the question has become: Does B have a column submatrix with k columns where each row of the submatrix contains exactly one 1 or, equivalently, at most one 1?

Derive a matrix A from B by replacing each 0 entry of B by -1. The question can now be rephrased as: Does A have a nearly negative column submatrix with k columns?

Assume the nontrivial case where $k \geq 3$. Let D be any column submatrix of A. We claim that the question may be worded as: Does A have a column submatrix with k columns that is hidden nearly negative relative to D? Indeed, if a column of any selected hidden nearly negative submatrix requires scaling by -1 to achieve near negativity of the submatrix, then the scaled column contains $3(k-1)$ 1s. But then the submatrix contains at most $2 < k$ columns. Thus, selecting a column submatrix of A with k columns that is hidden nearly negative relative to D is equivalent to choosing a nearly negative column submatrix of A with k columns.

Finally, suppose an optimal solution A^0, A^1, ..., A^q of the decomposition problem is at hand. The above arguments imply that, regardless of the form of D, the matrix A^0 is a nearly negative subregion of A that maximizes the number of columns of A entirely contained in A^0. That number is k if and only if Y contains an exact cover of X.

Decomposition Algorithm

Analogously to the 2SAT case of Section 8.4, we use a simple heuristic that in two steps finds good closed subregion decompositions for hidden near negativity. In the first step, we search heuristically for a hidden nearly negative subregion A^0 that contains a large number of nonzero columns of A. In the second step, we divide up the remaining columns of A among a minimum number of Boolean closed subregions, say, A^1, A^2, ..., A^q.

We formulate the first step as an IP. Let $A = [C \mid D]$ be the given matrix, with row index set X. Suppose that the columns of A (resp. C, D) are indexed by a set Y (resp. Y_C, Y_D). As in Section 8.4, we say that a column y of A is assigned to A^0 if that column is placed into A^0.

The variables of the IP are

(8.5.1)
$$r_y = \begin{cases} 1 & \text{if column } y \text{ is assigned to } A^0 \text{ and scaled by } 1 \\ 0 & \text{otherwise} \end{cases}$$
$$s_y = \begin{cases} 1 & \text{if column } y \text{ is assigned to } A^0 \text{ and scaled by } -1 \\ 0 & \text{otherwise} \end{cases}$$

Collect the variables r_y and s_y in vectors r and s, respectively.

Since $r_y = 1$ and $s_y = 1$ correspond to mutually exclusive cases and since the columns of C may not be scaled by -1, we enforce

(8.5.2)
$$r + s \leq 1$$
$$s_y = 0, \quad \forall \, y \in Y_C$$

The scaling must convert A^0 to a nearly negative matrix. Thus, the scaled matrix may have at most one 1 in each row. The latter requirement is expressed by

(8.5.3)
$$\sum_{y \ni A_{xy}=1} r_y + \sum_{y \ni A_{xy}=-1} s_y \leq 1, \quad \forall \, x \in X$$

Since the goal is assignment of a maximum number of columns of A to A^0, the objective function is

$$(8.5.4) \qquad\qquad \max \underline{1}^t \cdot (r + s)$$

We have seen that finding an optimal A^0 may be difficult, so solving the IP of (8.5.2)–(8.5.4) may be difficult as well. Accordingly, we use Heuristic SOLVE IP (8.3.3) to find a good but not necessarily optimal A^0.

The second step is the same as for the 2SAT case of Section 8.4. That is, we collect in a matrix E all nonzero columns of A that do not occur in A^0, and we apply Algorithm BOOLEAN CLOSED PARTITION (7.5.2) to partition E into a minimum number of Boolean closed submatrices, say, $E^1, E^2, \ldots, E^q$. For $p = 1, 2, \ldots, q$, we define A^p to be the Boolean closed subregion of A whose nonzero columns are precisely the nonzero columns of E^p. Then $A^0, A^1, \ldots, A^q$ is a good closed subregion decomposition of A for hidden near negativity.

We summarize the above steps.

(8.5.5) Heuristic DECOMPOSITION FOR HIDDEN NEAR NEGATIVITY. *Finds a good but not necessarily optimal closed subregion decomposition $A^0, A^1, \ldots, A^q$ of a matrix A over $\mathbb{B}$ where A^0 is hidden nearly negative relative to a specified submatrix.*

Input: Matrix $A = [C \mid D]$ over $\mathbb{B}$, with row index set X and column index set Y. Denote the column index sets of C and D by Y_C and Y_D, respectively.

Output: A good but not necessarily optimal closed subregion decomposition $A^0, A^1, \ldots, A^q$ of A where A^0 is hidden nearly negative relative to the column submatrix indexed by Y_D.

Complexity: Polynomial if in Step 1 a polynomial version of Heuristic SOLVE IP (8.3.3) is used.

Procedure:

1. (Find A^0) Use Heuristic SOLVE IP (8.3.3) to obtain a good but not necessarily optimal solution for the IP (8.5.2)–(8.5.4). The input for that heuristic method consists of the IP, any $k \geq 1$, a set J containing the indices of all variables, the subroutine Q specified next, and a subroutine R for solving LPs. Subroutine Q decides whether $\{0, 1\}$ values assigned to some specified variables r_y and s_y can be extended to a feasible solution of the IP (8.5.2)–(8.5.4). That task is simple. Indeed, subroutine Q declares that an extension to a feasible solution is possible if and only if the given values for the specified variables plus zeros for the remaining variables constitute a feasible solution for the IP (8.5.2)–(8.5.4).

Let r^* and s^* be the solution produced by Heuristic SOLVE IP (8.3.3). Obtain A^0 from A by replacing each column y for which $r_y^* = s_y^* = 0$ by a zero column.

2. (Find A^1, A^2,..., A^q) Collect in a matrix E all nonzero columns of A that do not occur in A^0. Use Algorithm BOOLEAN CLOSED PARTITION (7.5.2) to partition E into a minimum number of Boolean closed column submatrices, say, E^1, E^2,..., E^q. For $p = 1, 2,..., q$, define A^p to be the Boolean closed subregion of A whose nonzero columns are precisely the nonzero columns of E^p. Output A^0, A^1,..., A^q as the desired decomposition, and stop.

Proof of Validity. As argued earlier, Steps 1 and 2 produce a closed subregion decomposition for hidden near negativity. Algorithm BOOLEAN CLOSED PARTITION (7.5.2) is polynomial, so the entire procedure is polynomial if a polynomial version of Heuristic SOLVE IP (8.3.3) is used in Step 1. □

The next section concerns the case where the matrix A^0 of the decomposition is to be a network matrix.

8.6 Decomposition for Network Property

By Section 5.7, a matrix A over $\mathbb{B}$ is balanced if it does not contain certain cycle submatrices, and it is totally unimodular if every square submatrix, when viewed over the rationals, has its determinant in $\{0, \pm 1\}$. A matrix A over $\mathbb{B}$ is a network matrix, or has the network property, if A is totally unimodular and has at most two nonzero entries in each column or in each row.

We desire, for a given matrix A over $\mathbb{B}$, a good, if not optimal, closed subregion decomposition A^0, A^1,..., A^q where A^0 is a network matrix.

The reader may wonder why we restrict ourselves to the network property, and why we do not treat the more general case of balancedness or total unimodularity. The reason is that we do not know of a compact representation of the latter properties by linear inequalities with $\{0,1\}$ variables.

Since any network matrix with at most two nonzeros in each row is a 2SAT matrix and since closed subregion decompositions for 2SAT have been covered in Section 8.4, it may seem that we need not consider such network matrices here. But according to Sections 5.4 and 5.7, the MINSAT problem for 2SAT matrices is generally difficult, while the MINSAT problem for network matrices is easy. Hence, for MINSAT instances one may want to search for a decomposition A^0, A^1,..., A^q where A^0 is a network matrix with at most two nonzeros in each row.

Complexity

We first show that finding an optimal closed subregion decomposition for the network property is $\mathcal{NP}$-hard. Evidently, the network property is maintained under submatrix taking, it holds for all permutation matrices, and it fails for an infinite number of $\{0, 1\}$ matrices. Theorem (8.4.1) implies that under these conditions the problem of finding in a given $\{0, 1\}$ matrix a maximum column submatrix with the network property is $\mathcal{NP}$-hard. The latter problem is solved by any optimal closed subregion decomposition with the network property, so finding such decompositions is $\mathcal{NP}$-hard as well.

Decomposition Algorithm

Since the network property demands at most two nonzeros in each column or in each row, we find a good but not necessarily optimal decomposition with the network property by considering two cases. In the first (resp. second) case, we demand that each column (resp. row) has at most two nonzeros. We compare the two decompositions so determined and select the more attractive one.

The IPs of the two cases rely on Theorem (5.7.10)(iii) and (iv), according to which a matrix A is a network matrix if and only if the following holds. If each row (resp. column) of A has at most two nonzeros, then the columns (resp. rows) of A can be scaled by $\{\pm 1\}$ factors so that, in the scaled matrix A', each row (resp. column) with two nonzeros contains one $+1$ and one -1.

First Integer Program

Let the given matrix A have row index set X and column index set Y. The first IP treats the case where A^0 is to be a network matrix with at most two nonzeros in each column. The variables representing the assignment of the columns y of A to A^0 are

$$(8.6.1) \qquad r_y = \begin{cases} 1 & \text{if column } y \text{ is assigned to } A^0 \\ 0 & \text{otherwise} \end{cases}$$

Collect the variables r_y in a vector r. We need variables that represent $\{\pm 1\}$ scaling factors for the rows $x \in X$. Note that we actually never scale A or A^0 and that the variables specified next are solely used to enforce total unimodularity of A^0. For this reason, we call the scaling factors *conceptual*. The variables are, for $x \in X$,

$$(8.6.2) \qquad g_x = \begin{cases} 1 & \text{if row } x \text{ is scaled by } -1 \\ 0 & \text{if row } x \text{ is scaled by } 1 \end{cases}$$

Define Y_1 to be the set of $y \in Y$ for which column y of A contains at most one nonzero entry. Define Y_0 (resp. Y_2) to be the subset of the elements $y \in Y$ satisfying the following conditions. Column y of A must have exactly two nonzero entries, and in the case of Y_0 (resp. Y_2) these entries must sum to 0 (resp. ± 2). Let $Y_3 = Y - (Y_0 \cup Y_1 \cup Y_2)$.

Each column $y \in Y_1$ has at most one nonzero and thus can be assigned to A^0. Hence,

$$(8.6.3) \qquad\qquad r_y = 1, \quad \forall\, y \in Y_1$$

On the other hand, each column $y \in Y_3$ has at least three nonzeros and thus cannot be assigned to A^0. Hence,

$$(8.6.4) \qquad\qquad r_y = 0, \quad \forall\, y \in Y_3$$

For the remaining columns $y \in (Y_0 \cup Y_2)$, we enforce the scaling condition by the following inequalities, where we assume that the two nonzero entries in column y reside in rows $x(y)$ and $z(y)$. We justify the equations momentarily.

$$(8.6.5)\qquad
\begin{aligned}
g_{x(y)} - g_{z(y)} + r_y &\le 1, \quad \forall\, y \in Y_0 \\
-g_{x(y)} + g_{z(y)} + r_y &\le 1, \quad \forall\, y \in Y_0 \\
-g_{x(y)} - g_{z(y)} + r_y &\le 0, \quad \forall\, y \in Y_2 \\
g_{x(y)} + g_{z(y)} + r_y &\le 2, \quad \forall\, y \in Y_2
\end{aligned}$$

Suppose, for some $y \in Y$, $g_{x(y)} = g_{z(y)} = 0$ or $g_{x(y)} = g_{z(y)} = 1$. Then both rows $x(y)$ and $z(y)$ are conceptually scaled by 1 or -1. Furthermore, after the scaling, each column $y \in Y_0$ (resp. $y \in Y_2$) contains two nonzeros with opposite signs (resp. the same sign). Accordingly, column y can be assigned to A^0 if and only if $y \in Y_0$. The inequalities of (8.6.5) are readily seen to enforce the latter condition. Similarly, if one of $g_{x(y)}$ and $g_{z(y)}$ is equal to 0 while the other one is equal to 1, then (8.6.5) enforces that column y can be assigned to A^0 if and only if $y \in Y_2$.

The objective function for the IP is

$$(8.6.6) \qquad\qquad \max \, \underline{1}^t \cdot r$$

Second Integer Program

We turn to the second IP, where A^0 is to be a network matrix with at most two nonzeros in each row. Selection and conceptual scaling of the columns y of A is handled by the variables

$$(8.6.7)\qquad
\begin{aligned}
r_y &= \begin{cases} 1 & \text{if column } y \text{ is assigned to } A^0\text{, with scaling factor } 1 \\ 0 & \text{otherwise} \end{cases} \\[2mm]
s_y &= \begin{cases} 1 & \text{if column } y \text{ is assigned to } A^0\text{, with scaling factor } -1 \\ 0 & \text{otherwise} \end{cases}
\end{aligned}$$

Denote the vector containing the r_y (resp. s_y) variables by r (resp. s). Since $r_y = 1$ and $s_y = 1$ are mutually exclusive cases, we require

$$(8.6.8) \qquad\qquad r + s \leq \underline{1}$$

Let D be the support matrix of A. Since A^0 must have at most two nonzero entries in each row, we enforce

$$(8.6.9) \qquad\qquad D \cdot (r + s) \leq 2 \cdot \underline{1}$$

The conceptual column scaling must transform a row of A^0 with two nonzeros into a row with one 1 and one -1. It is easy to confirm that this condition is expressed by

$$(8.6.10) \qquad\qquad -\underline{1} \leq A \cdot (r - s) \leq \underline{1}$$

The objective function for the IP is

$$(8.6.11) \qquad\qquad \max \underline{1}^t \cdot (r + s)$$

The following heuristic algorithm uses the two IPs to determine a good but not necessarily optimal closed subregion decomposition of a matrix A into $A^0, A^1, \ldots, A^q$ where A^0 is a network matrix.

(8.6.12) Heuristic DECOMPOSITION FOR NETWORK PROPERTY. *Finds a good but not necessarily optimal closed subregion decomposition $A^0, A^1, \ldots, A^q$ of a matrix A over $\mathbb{B}$ where A^0 is a network matrix.*

Input: Matrix A over $\mathbb{B}$, with row index set X and column index set Y.

Output: A good but not necessarily optimal closed subregion decomposition $A^0, A^1, \ldots, A^q$ of A where A^0 is a network matrix.

Complexity: Polynomial if in Step 1 a polynomial version of Heuristic SOLVE IP (8.3.3) is used.

Procedure:

1. (Find A^0 using first IP) Use Heuristic SOLVE IP (8.3.3) to obtain a good but not necessarily optimal solution for the IP (8.6.3)–(8.6.6). The input for that heuristic method consists of the IP, any $k \geq 1$, a set J containing the indices of the variables of type g_x, and a subroutine R for solving LPs. The subroutine Q is not needed.
   Let r^* be the solution produced by Heuristic SOLVE IP (8.3.3). Obtain A^0 from A by replacing each column y for which $r_y^* = 0$ by a zero column.

2. (Find $A^1, A^2, \ldots, A^q$) Collect in a matrix E all nonzero columns of A that do not occur in A^0. Use Algorithm BOOLEAN CLOSED

PARTITION (7.5.2) to partition E into a minimum number of Boolean closed column submatrices, say, E^1, E^2, ..., E^q. For $p = 1, 2, \ldots, q$, define A^p to be the Boolean closed subregion of A whose nonzero columns are precisely the nonzero columns of E^p. Suppose for $1 \leq p \leq q$, E^p has n_{p1} columns in total and has n_{p2} nonmonotone columns. Compute $\alpha = \prod_{p=1}^{q}(n_{p1} + \min\{n_{p2}, 1\})$.

3. (Find A^0, A^1, ..., A^q using second IP) Repeat Steps 1 and 2, except that this time the IP of (8.6.8)–(8.6.11) is solved in Step 1. The set J for the Heuristic SOLVE IP (8.3.3) is the index set of all variables, and the subroutine Q consists of the following test. Given $\{0, 1\}$ values for some specified variables, subroutine Q checks whether these values plus zeros assigned to the remaining variables constitute a solution for (8.6.8)–(8.6.10). If this is so, the fixed values can be extended to a feasible solution of the IP. Otherwise, such an extension is not possible.

4. (Select decomposition) Compare the closed subregion decomposition of A produced by the two passes through Steps 1 and 2. If α of the second pass is greater than or equal to (resp. is less than) α of the first pass, output A^0, A^1, ..., A^q obtained in the first (resp. second) pass as the desired closed subregion decomposition of A.

Proof of Validity. It is easily verified that the omission of subroutine Q in the first pass, as well as the use of the simple subroutine Q of the second pass, is appropriate. Thus, Heuristic SOLVE IP (8.3.3) produces solutions for the two IPs. The remaining arguments match those for Heuristic DECOMPOSITION FOR 2SAT (8.4.5) or Heuristic DECOMPOSITION FOR HIDDEN NEAR NEGATIVITY (8.5.5), except for the straightforward comparison of the results of the two passes using the two values for α. $\qquad\qquad\Box$

8.7 Extensions and References

Algorithms for solving the IP (8.3.1) or the LP (8.3.2) are given in Hu (1969), Garfinkel and Nemhauser (1972), Chvátal (1983), Schrijver (1986), Grötschel, Lovász, and Schrijver (1993), Nemhauser and Wolsey (1988), and Karloff (1991).

Folklore has it that IPs such as (8.3.1) can be reasonably well solved by rounding the LP solution of (8.3.2). That is, one iteratively solves the LP and rounds down (resp. up) fractional values that are close to 0 (resp. 1). In contrast, Heuristic SOLVE IP (8.3.3) fixes variables whose fractional LP solution values are as far as away from 0 and 1 as possible. Of course, the fixing is done only when its effect has been established by a reasonable enumerative effort. In computational comparisons of the two methods as well as hybrid methods, we have found that the approach proposed here is usually superior to the traditional rounding. We should mention, though, that our test comparisons were confined to IPs arising from decomposition problems and MINSAT instances. Accordingly, the claim of superior performance applies only to IPs arising from such problems. The test problems had up to several thousand variables and inequalities. In each case, the Simplex Method was selected as subroutine R. The crucial parameter k ranged from 1 to 5. Various aspects of integer rounding for the solution of integer optimization problems are discussed in Nemhauser and Wolsey (1988). A number of results are known for rounding of specially structured IPs; for example, see Nemhauser and Trotter (1975), Bartholdi, Orlin, and Ratliff (1980), Agarwal, Sharma, and Mittal (1982), Chandrasekaran (1984), Hochbaum, Megiddo, Naor, and Tamir (1993), Lakshminarayanan and Chandrasekaran (1994), and Chandrasekaran, Kabadi, and Lakshminarayanan (1996). Chandru and Hooker (1991) use the rounding result of Chandrasekaran (1984) to solve the SAT problem for *extended Horn* matrices. Additional information about the latter matrices is included in Section 5.9.

Theorem (8.4.1) is proved by Bartholdi (1982) using work of Lewis and Yannakakis (1980) and of Yannakakis (1981). A $\{0, \pm 1\}$ matrix version of Yannakakis (1981) is given in Crama, Ekin, and Hammer (1997). Chandru and Hooker (1992) prove the result of Section 8.5 that the problem of finding a maximum hidden nearly negative column submatrix is $\mathcal{NP}$-hard. For the $\mathcal{NP}$-completeness proof of EXACT COVER BY 3-SETS, see Garey and Johnson (1979).

The decomposition algorithms of Sections 8.4–8.6 can be refined by allowing additional choices of taking subregions. We summarize the main ideas of such refinements using the case of hidden near negativity. Thus, we consider the situation where a decomposition of a given matrix A over $\mathbb{B}$ into A^0, $A^1, \ldots, A^q$ is desired where A^0 is hidden nearly negative relative to a specified submatrix.

We want a decomposition that minimizes the solution effort by Algorithm SOLVE CLOSED SUBREGION DECOMPOSITION SAT OR MINSAT (8.2.6). Let us take the upper bound on the performance of that algorithm as an indicator of worst-case performance. That bound is $O(\alpha(\beta + \varphi))$ in the SAT case and $O(\alpha(\gamma + \varphi))$ in the MINSAT case, where $\alpha = \prod_{p=1}^{q} (n_{p1} + \min\{n_{p2}, 1\})$ and $\varphi = m + n + \sum_{p=1}^{q} \text{count}(A^p)$. For the case of hidden near negativity, the factor α essentially determines whether the upper bound is small or not.

Heuristic DECOMPOSITION FOR HIDDEN NEAR NEGATIVITY (8.5.5) searches for an A^0 corresponding to a maximum hidden nearly negative column submatrix of A, in the hope that the resulting α is small. Instead, one could search for an A^0 corresponding to a submatrix that is derived from A by both column and row deletions. Let us pursue the latter idea.

For the moment, we assume a worst-case α where each deleted column and each deleted row produces one closed subregion. Thus, for each deleted column, the corresponding factor in α is equal to 2, while for each deleted row x, the factor is the number of nonzero entries in that row plus 1. We denote the latter factor by n_x.

Minimization of α is equivalent to minimization of the logarithm of α. Declare $\log_2(2)$, which is equal to 1, to be the cost of deleting any column. Define $\log_2(n_x)$ to be the cost of deleting any row x. A submatrix A' of A then minimizes the logarithm of α if and only if the total cost of the corresponding deletion of columns and rows is minimum.

We are ready to formulate the problem of selecting A', and thus A^0, as an IP. In agreement with the terminology of Sections 8.4–8.6, we say that the columns and rows defining the submatrix A' are *assigned to* A^0. Note that a nonzero of A, say, in row x and column y, is present in A^0 if and only if both row x and column y have been assigned to A^0.

Let A have row index set X and column index set Y. For a simplified exposition, assume that each row of A is nonzero.

The variables of the IP consist of the variables r_y and s_y of (8.5.1); that is, for $y \in Y$,

(8.7.1)
$$r_y = \begin{cases} 1 & \text{if column } y \text{ is assigned to } A^0 \text{ and scaled by } 1 \\ 0 & \text{otherwise} \end{cases}$$
$$s_y = \begin{cases} 1 & \text{if column } y \text{ is assigned to } A^0 \text{ and scaled by } -1 \\ 0 & \text{otherwise} \end{cases}$$

plus, for each $x \in X$, a variable

(8.7.2)
$$g_x = \begin{cases} 1 & \text{if row } x \text{ is assigned to } A^0 \\ 0 & \text{otherwise} \end{cases}$$

In agreement with (8.5.2), we enforce

(8.7.3)
$$r + s \leq 1$$
$$s_y = 0, \quad \forall \, y \in Y_C$$

The scaling must convert A^0 to a nearly negative matrix. That requirement is expressed by

(8.7.4)
$$\sum_{y \ni A_{xy}=1} r_y + \sum_{y \ni A_{xy}=-1} s_y + n_x \cdot (g_x - 1) \leq 1, \quad \forall \, x \in X$$

The term $n_x \cdot (g_x - 1)$ in (8.7.4) assures that the inequality is effective if and only if $g_x = 1$.

The goal is an assignment of the columns and rows to A^0 that minimizes the total cost of the deleted columns and rows. That total cost is equal to $\sum_{y \in Y}(1 - r_y - s_y) + \sum_{x \in X}(\log_2(n_x) \cdot (1 - g_x))$. Minimization of that function is equivalent to

$$(8.7.5) \qquad\qquad \max \underline{1}^t \cdot (r + s) + \sum_{x \in X} \log_2(n_x) \cdot g_x$$

The solution of the IP by Heuristic SOLVE IP (8.3.3) is straightforward, so we omit details. Once a good solution is at hand, say, r^*, s^*, and g^*, we deduce A' from A by deleting all columns y with $r_y^* = s_y^* = 0$ and all rows x with $g_x^* = 0$. The nonzeros of A' are then assigned to the subregion A^0.

Let E be obtained from A by the replacement of all nonzeros that have been assigned to A^0, by zeros. As in Heuristic DECOMPOSITION FOR HIDDEN NEAR NEGATIVITY (8.5.5), we use Algorithm BOOLEAN CLOSED PARTITION (7.5.2) to partition E into a minimum number of Boolean closed column submatrices, say, $E^1, E^2, \ldots, E^q$. For $p = 1, 2, \ldots,$ q, we define A^p to be the Boolean closed subregion of A whose nonzero columns are precisely the nonzero columns of E^p. The matrices $A^0, A^1, \ldots,$ A^q so obtained give the desired decomposition.

The above method may result in an unnecessarily large α, since a judicious transfer of some nonzeros from E to A^0 might significantly reduce α. Hence, one should optimize the choice of nonzeros that may be so transferred. This can be done approximately by the solution of yet another IP that aims at minimizing α while maintaining A^0 to be hidden nearly negative relative to the specified column submatrix. It would take up too much space if we were to discuss details of that IP. Suffice it to say the following. Algorithm BOOLEAN CLOSED PARTITION (7.5.2) essentially solves a max flow problem when it invokes Algorithm PATH COVER (2.5.17). One first formulates that flow problem as an IP. Then one enlarges that IP by considering the effect of possible transfers of nonzeros from E to A^0 to obtain the desired formulation.

One could refine the above approach even further by formulating the selection of all nonzeros of A for A^0 as well as the selection of the Boolean closed subregions $A^1, A^2, \ldots, A^q$ in one IP. Because of space limitations we omit a detailed discussion. We should mention, though, that the latter approach produces large IPs even for modestly sized matrices A and thus may be of limited utility.

Analogously to the above discussion, one may enhance the heuristic methods of Section 8.4 and 8.6. Given the above discussion, the reader should have no difficulty constructing the improved algorithms.

Recall from Section 5.9 that a matching matrix is a $\{0, \pm 1\}$ matrix with at most two nonzeros in each column. Matching matrices generalize network matrices with at most two nonzeros in each column, since matching matrices do not demand total unimodularity. Section 5.9 shows that this extension of network matrices is of interest only for the MINSAT problem and not for the SAT problem. That section also references efficient algorithms for solving such MINSAT instances. The above ideas and methods are readily adapted so that one may determine closed subregion decompositions $A^0, A^1, \ldots, A^q$ where A^0 is a matching matrix. Given the treatment of the network case in Section 8.6 and the above discussion of extensions, we omit details.

A special case of closed subregion decomposition is treated by Yamasaki and Doshita (1983) and is improved upon by Arvind and Biswas (1987) and by Chandru, Coullard, Hammer, Montañez, and Sun (1990). The matrix A is decomposed into a nearly negative matrix A^0 and a $\{0,1\}$ subregion A^1 that is a solid staircase matrix as depicted in (7.3.2). According to Theorem (7.3.5), solid staircase matrices are Boolean closed, so A^0 and A^1 are a special case of closed subregion decomposition of A. The cited references contain efficient algorithms for finding such a decomposition if it exists and also include efficient solution algorithms for the SAT case. The third reference also contains a compact characterization in terms of excluded submatrices.

Gallo and Scutellà (1988) create a sequence of matrix classes based on the above decomposition by Yamasaki and Doshita (1983). The classes are indexed by the nonnegative integers. The class with index 0 consists of the nearly negative matrices. The class with index 1 consists of the matrices having the decomposition due to Yamasaki and Doshita (1983). For $q \geq 2$, the class with index q consists of matrices that, in our terminology, have a closed subregion decomposition into $A^0, A^1, \ldots, A^q$ where A^0 is nearly negative and where $A^1, A^2, \ldots, A^q$ are solid staircase matrices. Gallo and Scutellà (1988) supply, for any $q \geq 0$, recognition and solution algorithms for the matrices in the class with index q. The algorithms are polynomial if q is bounded by a constant. For related results, see Kleine Büning (1993).

Suppose one enlarges the matrix classes of Gallo and Scutellà (1988) so that the specification involves hidden nearly negative matrices instead of nearly negative matrices. Eiter, Kilpeläinen, and Mannila (1995) show that recognizing membership in any revised class with index $q \geq 1$ is $\mathcal{NP}$-complete.

Dalal and Etherington (1992) as well as Pretolani (1993a, 1996) extend the work of Gallo and Scutellà (1988) by enlarging the class numbered 0 and generalizing the definition of the other classes.

One may view the results of the latter references as decompositions that sometimes are more elaborate than those of this chapter. On the other hand, the decompositions of these references only apply to SAT, and they may fail to uncover decompositions that are detected by the methods described in this chapter and that may make solution of some SAT cases very easy.

Gallo and Urbani (1989) rewrite any SAT instance into a formulation that is equivalent to a closed subregion decomposition A^0, A^1, ..., A^q where A^0 is nearly negative and where each matrix of A^1, A^2, ..., A^q has exactly one nonzero row.

The next four chapters treat several sums of matrices over $\mathbb{B}$. First, we cover monotone sums.

Chapter 9

Monotone Sum

9.1 Overview

This chapter and Chapters 10–12 treat several sums of matrices over $\mathbb{B}$. We rely on these sums to decompose SAT or MINSAT instances into smaller instances. We summarize the main idea below. For a detailed discussion, the reader should review Sections 4.7 and 4.8.

Let B be a matrix over $\mathbb{B}$, and let b be a $\{0,1\}$ vector. According to Section 4.7, the matrix B is b-satisfiable if there exists a $\{\pm 1\}$ vector s such that $B \odot s \geq b$.

With one exception that we ignore for the moment, each sum decomposition of B produces two component matrices, say, B^1 and B^2. The components are so selected that, for any b, B is b-satisfiable if and only if for $i = 1$, 2, there exists a $\{0,1\}$ vector b^i such that a certain column submatrix $\overline{B}^i$ of B^i is b^i-satisfiable. Assume that we are given B and b, as well as B^1 and B^2. We want to decide whether B is b-satisfiable. If we knew b^1 and b^2, then we could reduce the b-satisfiability problem for B to the b^1-satisfiability problem for $\overline{B}^1$ and the b^2-satisfiability problem for $\overline{B}^2$. Unfortunately, b^1 or b^2 are not always easily determined. But it turns out that we can always carry out the following alternate process.

First, we determine certain vectors b^1 and solve for these vectors the b^1-satisfiability problem for $\overline{B}^1$.

Second, given the results of those computations, we construct certain vectors b^2 and solve for these vectors the b^2-satisfiability problem for $\overline{B}^2$. At that point, we can decide whether B is b-satisfiable.

Finally, if B is found to be b-satisfiable, we combine the solution of one of the b^2-satisfiability problems for $\overline{B}^2$ in a backtracking step with the solution of one of the b^1-satisfiability problems for $\overline{B}^1$ to a solution for the b-satisfiability problem for B.

Section 4.7 classifies each sum according to worst-case upper bounds on the number of b^1- and b^2-satisfiability problems for $\overline{B}^1$ and $\overline{B}^2$ that may have to be solved by the SAT algorithm we have developed for that sum. If that upper bound is 1 for both $\overline{B}^1$ and $\overline{B}^2$, the sum is said to be of type I. If the upper bound is at least 2 for $\overline{B}^1$ and is 1 for $\overline{B}^2$, then the sum is of type II. In the remaining case, where both upper bounds are at least 2, the sum is of type III.

It is important for our purposes that any sum be maintained under submatrix taking. That is, if B is a certain sum, say, of B^1 and B^2, then any submatrix of B must be a sum of the same type, and the component matrices must be submatrices of B^1 and B^2. These requirements significantly restrict the choice of sums, but also support the construction of large SAT and MINSAT central or semicentral matrix classes.

This chapter concerns sums called monotone. The presentation proceeds as follows. In Section 9.2, we define the monotone sum and the related decomposition and composition and prove several properties.

In Section 9.3, we describe a polynomial algorithm for finding monotone decompositions that are best in a certain sense.

In Section 9.4, we give a solution algorithm for the SAT or MINSAT problem involving a monotone sum B with components B^1 and B^2. The algorithm transforms a given problem into one SAT or MINSAT problem involving B^1 and into a second SAT or MINSAT problem involving B^2. Accordingly, the monotone sum is of type I.

The final section, 9.5, contains extensions and references.

9.2 Definitions and Properties

In this section, we define the monotone sum and the related monotone decomposition and composition, and we prove basic properties. In particular, we introduce a partial order for the possible monotone decompositions of a given matrix B over $\mathbb{B}$, and we show that there is an essentially unique monotone decomposition that is largest under that partial order.

Let B be a matrix over $\mathbb{B}$, with row index set X and column index Y. By Section 5.5, B is nearly negative if each row of B contains at most one 1. Let C be a column submatrix of B. By Section 5.6, B is hidden nearly negative relative to C if scaling of the columns of C with $\{\pm 1\}$ factors can convert B to a nearly negative matrix.

Monotone Decomposition and Separation

A *monotone decomposition* of B is achieved by the following three steps.

First, in B the columns of the submatrix C are scaled with $\{\pm1\}$ factors.

Second, the row index set X and column index set Y of the scaled matrix are partitioned into X_1, X_2 and Y_1, Y_2, respectively, so that the four submatrices defined by the partitions have the following properties. The submatrix A^1 indexed by X_1 and Y_1 is nearly negative. The submatrix D defined by X_2 and Y_1 is nonpositive. The submatrix indexed by X_1 and Y_2 is zero. Finally, no constraint is imposed on the submatrix A^2 defined by X_2 and Y_2; for this reason, we call A^2 *unconstrained* and demand that the scaling factors for the columns of B containing A^2 are all equal to 1. We display the partitioned matrix below.

(9.2.1)

	Y_1	Y_2
X_1	A^1	0
X_2	D	A^2

Scaled and partitioned matrix

Note that any one of the submatrices of (9.2.1) is allowed to be trivial or empty. Define $(X_1 \cup Y_1, X_2 \cup Y_2)$ to be a *monotone separation* of B relative to C.

Third, we decompose B by declaring the *component* B^1 (resp. B^2) to be equal to the submatrix A^1 (resp. $[D|A^2]$) of the matrix of (9.2.1). Thus,

(9.2.2)

$$B^1 = \begin{array}{c|c} & Y_1 \\ \hline X_1 & A^1 \end{array} \qquad B^2 = \begin{array}{c|c|c} & Y_1 & Y_2 \\ \hline X_2 & D & A^2 \end{array}$$

Components B^1 and B^2 of monotone decomposition

We say that B has been decomposed by a *monotone decomposition relative to C*.

Monotone Composition

Conversely, suppose B^1 and B^2 of (9.2.2) are given, where A^1 is nearly negative and D is nonpositive. We may combine these matrices to the matrix of (9.2.1), scale in the latter matrix some columns by $\{\pm1\}$ factors, and obtain B again. We say that B has been created by a *monotone composition* of B^1 and B^2.

Monotone Sum

We say that B is a *monotone sum* of B^1 and B^2 if the latter matrices are the components of a monotone decomposition of B or, equivalently, if B is created from B^1 and B^2 by a monotone composition. We denote that situation by $B = B^1 \boxplus_m B^2$.

A monotone sum is *proper* if in the corresponding matrix (9.2.1) both submatrices A^1 and A^2 are nontrivial and nonempty, that is, if both submatrices contain at least one entry each.

Note that the definitions given above extend those of Section 4.7. In particular, the latter definitions do not consider any scaling of B.

Maximum Monotone Decomposition and Separation

Let $(X_1 \cup Y_1, X_2 \cup Y_2)$ and $(X_1' \cup Y_1', X_2' \cup Y_2')$ be two monotone separations of B relative to C. We say that the first separation is greater than or equal to the second one if $X_1 \cup Y_1 \supseteq X_1' \cup Y_1'$. The binary relation so defined is reflexive, antisymmetric, and transitive and thus establishes a partial order on the set of monotone separations relative to C. Evidently, $(\emptyset, X \cup Y)$ is the unique minimal separation under that partial order. Less obvious is the following result.

(9.2.3) Theorem. *Any matrix B over $\mathbb{B}$ has a unique maximal monotone separation relative to a given column submatrix C of B.*

Proof. Suppose B has two distinct maximal monotone separations relative to C, say, $(X_1 \cup Y_1, X_2 \cup Y_2)$ and $(X_1' \cup Y_1', X_2' \cup Y_2')$.

Each one of the two separations implies a certain column scaling and partitioning of B. Suppose that we scale the columns of B indexed by Y_1 as in the first decomposition and that we scale the columns indexed by $Y_1' - Y_1$ as in the second decomposition. For the matrix so scaled, it is easily verified that $X_1 \cup X_1'$ and $Y_1 \cup Y_1'$ index a nearly negative submatrix, that $X - (X_1 \cup X_1')$ and $Y_1 \cup Y_1'$ index a nonpositive submatrix, and that $X_1 \cup X_1'$ and $Y - (Y_1 \cup Y_1')$ index a zero submatrix. Thus, $(X_1 \cup X_1' \cup Y_1 \cup Y_1', (X \cup Y) - (X_1 \cup X_1' \cup Y_1 \cup Y_1'))$ is a monotone separation relative to C that is larger than $(X_1 \cup Y_1, X_2 \cup Y_2)$, a contradiction. $\qquad\square$

Theorem (9.2.3) justifies that we call any maximal monotone separation of a matrix B relative to C the *maximum monotone separation relative to C*. The monotone decomposition or sum corresponding to a maximum monotone separation need not be unique due to possible differences in the column scaling. But, up to column scaling, that monotone decomposition or sum is unique and is called a *maximum monotone decomposition* or *sum*.

One may rephrase Theorem (9.2.3) in terms of monotone decompositions as follows.

(9.2.4) Corollary. *Suppose a given matrix B has a monotone decomposition relative to some column submatrix C. Denote by A^1 (resp. A^2) the nearly negative (resp. unconstrained) submatrix of the decomposition. Let $A^{1'}$ (resp. $A^{2'}$) be the nearly negative (resp. unconstrained) submatrix of any maximum monotone decomposition of B relative to C. Then, $A^{2'}$ is a submatrix of A^2, and, up to column scaling, A^1 is a submatrix of $A^{1'}$.*

Proof. By Theorem (9.2.3), A^1 is up to column scaling a submatrix of $A^{1'}$. Hence, $A^{2'}$ is a submatrix of A^2.   □

Let P be a matrix property that is inherited under submatrix taking. Corollary (9.2.4) supports the following result.

(9.2.5) Corollary. *Suppose a given matrix B has a monotone decomposition relative to some column submatrix C of B. If the unconstrained submatrix of that decomposition has property P, then this is so for every maximum monotone decomposition of B relative to C.*

Proof. Let a monotone decomposition of B be given where the unconstrained matrix, say, A^2, has property P. Let $A^{2'}$ be the unconstrained submatrix of a maximum monotone decomposition of B relative to C. By Corollary (9.2.4), $A^{2'}$ is a submatrix of A^2, and it has P, since A^2 has P.   □

In the next section, we establish a polynomial algorithm for finding maximum monotone decompositions.

9.3 Decomposition Algorithm

This section contains a polynomial algorithm that, for a given matrix B, finds a maximum monotone decomposition relative to a given column submatrix C of B. We outline the algorithm.

If B has zero rows, we first remove all such rows. The indices of these rows will eventually be in the set $X_1 \subseteq X$ of the maximum monotone decomposition of B.

From the matrix B on hand, we arbitrarily select a column y of B. To simplify the discussion, we temporarily assume that column y is in C. Hence, both 1 and -1 are permitted as scaling factors for that column. We also assume that column y contains both 1s and -1s.

First, we define the scaling factor for column y to be 1 and check whether y can possibly occur in the column index set Y_1 of some monotone decomposition of B. If the answer is yes, we reduce the decomposition to a smaller case. If the answer is no, we change the scaling factor for column y to -1 and again check whether y can possibly occur in the column index set

Y_1 of some monotone decomposition of B. If the answer is yes, we reduce the decomposition problem. If the answer is no, we know that in any monotone decomposition the index y must occur in the column index set Y_2. Hence, we select another column and repeat the above process. Once all columns have been processed, a simple backtracking scheme constructs the maximum monotone decomposition.

Straightforward implementation of the algorithm results in quadratic run time. One may speed up performance as follows. When it has been determined that an index $y \in Y$ must be in the set Y_2 of any monotone decomposition, then the computations producing that conclusion imply that certain other indices z must also be in Y_2, or can only be in Y_1 if column z is scaled in a certain way. Using that additional information, one streamlines the algorithm so that each column is examined at most twice. Hence, the improved algorithm runs in linear time.

We present details of the algorithm. We delete all zero rows from B and select an arbitrary column $y \in Y$ of the resulting matrix.

We tentatively fix the scaling factor for column y to 1 and declare A^1 to be the submatrix of B that precisely contains the 1s of column y. With A^1 so determined, we define D, E, and A^2 according to (9.3.1) below.

(9.3.1)

$$B = \begin{array}{c|c|c} & Y_1 & Y_2 \\ \hline X_1 & A^1 & E \\ \hline X_2 & D & A^2 \end{array}$$

Scaled and partitioned matrix

Note that each row of A^1 contains exactly one 1 and that D is nonpositive. Below, we assume inductively that this condition is satisfied.

If E is zero, then we remove A^1, D, and E from B and recursively apply the algorithm to the reduced B. Hence, we suppose that E is nonzero.

Select any nonzero column z of E. For the time being, let us assume that column z of E can be scaled to become nonpositive. We scale column z of E so that the scaled column is nonpositive, and we scale the corresponding column z of A^2 with the same factor. We repartition B in two steps. First, we shift the scaled column z of E to A^1 and correspondingly shift the scaled column z of A^2 to D. Since D was nonpositive prior to the shifting, the new D has at most one 1 in each row. In the second step, we shift all rows of D containing one 1 from D to A^1, and we shift the corresponding rows of A^2 from A^2 to E. The new A^1 has exactly one 1 in each row, and the new D is nonpositive. Thus, the inductive assumption is satisfied, and we may recursively continue with the new A^1, D, E, and A^2. We check if E is nonzero, etc.

So far, we have assumed that column z of E can be scaled to become nonpositive. Now suppose that this is not so, either because that column contains 1s and -1s or because that column is nonnegative and cannot be scaled by -1. The next lemma establishes that certain monotone decompositions can be ruled out on the basis of the current partition of the current matrix B. For any $v \in Y$, define v to be *usable* if there exists a monotone decomposition of the current matrix B where, in the notation of (9.2.1), v is scaled by 1 and is in Y_1. We extend this terminology using the notation $\bar{v}$, by saying that $\bar{v}$ is *usable* if there exists a monotone decomposition of the current matrix B where column v is scaled by -1 and is in Y_1.

The scaling restriction imposed on the columns v outside the column submatrix C of the current matrix B can now be rephrased. For each such column v, we have $\bar{v}$ unusable.

(9.3.2) Lemma. *Suppose the above algorithm reaches the situation where a nonzero column z of E cannot be scaled to become nonpositive. Let (9.3.1) display the matrix B on hand at that time. Then the following statements hold.*

(a) *The submatrix A^1 has exactly one 1 in each row.*
(b) *The submatrix D is nonpositive.*
(c) *y is unusable.*
(d) *Assume that $\bar{y}$ is unusable. Then, for each $v \in (Y_1 - \{y\})$, $\bar{v}$ is unusable, and both z and $\bar{z}$ are unusable.*

Proof. Parts (a) and (b) have already been established by the above discussion.

For the proof of part (c), consider y to be usable. One may view the scaling and shifting of vectors of the above steps to be a direct consequence of that assumption. But that scaling and shifting produces a contradictory situation for column z, so y must be unusable.

We prove part (d) for the elements $v \in (Y_1 - \{y\})$ by induction. The base case, where $Y_1 - \{y\}$ is empty, is trivial. For the inductive step, we once more apply the algorithm to the matrix B at hand. That is, we select column y with scaling factor 1 and process the columns in the order in which they were originally processed. Thus, the same steps are carried out, except that all scaling factors are equal to 1.

Assume inductively that the claim holds for the columns processed up to a certain point and that column v is processed next. The matrix on hand is of the form (9.3.1). By parts (a) and (b), A^1 has exactly one 1 in each row, and D is nonpositive. Since all scaling factors are equal to 1, column v of E is nonzero and nonpositive.

Suppose $\bar{v}$ is usable. Thus, there is a monotone decomposition where column v is scaled by -1 and where the scaled column v is in the nearly negative column submatrix of the decomposition. The scaling converts the nonzero and nonpositive column v of E to a nonnegative vector, say, with a 1 in row x. Since A^1 has a 1 in each row, it has a 1 in row x, say, in column w. The assumed monotone decomposition requires that column w be scaled by -1 and that the scaled column be in the nearly negative column submatrix of the decomposition. Thus $\overline{w}$ is usable. If $w = y$, then $\overline{w}$ is not usable by the assumption of part (d), a contradiction. If $w \neq y$, then $\overline{w}$ is not usable by induction, another contradiction.

It remains to be shown for part (d) that both z and $\overline{z}$ are unusable. We know that column z of E on hand cannot be scaled to become nonpositive. There are two possible causes.

First, column z of E may contain both 1s and -1s. We then argue as in the inductive proof for $v \in (Y_1 - \{y\})$ above that both z and $\overline{z}$ are unusable.

Second, column z of E may be nonnegative, and the scaling restriction on the original matrix rules out scaling by -1. Arguing as before, z is unusable. The scaling restriction causes $\overline{z}$ to be unusable as well.   □

We continue the description of the algorithm. Lemma (9.3.2)(c) says that y is unusable. If $\overline{y}$ is possibly usable, we scale column y by -1, apply the above steps, and either reduce the matrix or determine $\overline{y}$ to be unusable.

Thus, at most two applications of the above steps yield a reduced matrix or establish both y and $\overline{y}$ to be unusable.

Recursive application of the above scheme to all columns of the matrix results in any number of reductions, plus eventually the conclusion that for any column v of the remaining matrix, both v and $\overline{v}$ are unusable. At that time, we declare the reduced matrix to be a column scaled version of the unconstrained matrix of a maximum monotone decomposition. We undo any scaling for that unconstrained matrix and adjoin the deleted columns and rows to get the matrix (9.2.1) of the decomposition. From the latter matrix, we define the components B^1 and B^2 of the decomposition via (9.2.2).

The complexity of the algorithm as described is quadratic. We speed up the scheme by the following changes. We determine whether y or $\overline{y}$ is usable by carrying out the above process for column y and column y scaled by -1 in parallel. If either case results in a reduction, we stop the parallel processing, carry out the reduction, and begin with another column of the reduced matrix. If neither case leads to a reduction, we record both y and $\overline{y}$ as unusable. Assume the latter situation. Lemma (9.3.2)(d) implies that for each one of the two cases certain v and $\overline{v}$ are unusable. Indeed, for each column processed in the two cases, we have by Lemma (9.3.2)(d) at least one such conclusion. We record that insight and later skip processing steps for any column v for which v or $\overline{v}$ is already known to be unusable.

When these ideas are properly implemented, a linear algorithm results. That algorithm records the unusable v or $\overline{v}$ in a set N. While processing column y, each v or $\bar{v}$ that is unusable according to Lemma (9.3.2)(d) is stored in a set L.

In the case of a reduction, the set L is discarded. In the remaining situation, where both y and $\bar{y}$ are proved to be unusable, the elements of L as well as y and $\bar{y}$ are added to N.

The rules for handling N and L in the algorithm include that an element v or $\bar{v}$ in N or L must be changed whenever column v is scaled by -1. That way, the sets N and L remain correct relative to the current matrix.

Finally, the interpretation of N is not affected by any reductions. Indeed, if v (resp. $\bar{v}$) is usable after a reduction, then v (resp. $\bar{v}$) must have been usable prior to that reduction.

We summarize the algorithm below.

(9.3.3) Algorithm MONOTONE DECOMPOSITION. *Finds a maximum monotone decomposition for a given matrix B over $\mathbb{B}$.*

Input: Matrix B of size $m \times n$, with row index set X and column index set Y. A column submatrix C of B. The columns of C may be scaled by $\{\pm 1\}$ factors, while columns outside C may only be scaled by 1.

Output: A maximum monotone decomposition of B relative to C.

Complexity: $O(m + n + \text{count}(B))$ if the two scaling cases for column y, which are handled by Steps 5–10 below, are evaluated in parallel.

Procedure:

1. (Initialization) Initialize a set N as the set of elements $\bar{y}$ for which column y is not in C.

2. (Remove zero rows.) Remove all zero rows from B.

3. (Termination) If, for each column y of the current B, both y and $\bar{y}$ are in N: Undo any prior scaling for the current B, and declare the resulting matrix to be A^2, say, with row index set X_2 and column index set Y_2. Adjoin to A^2 the rows and columns deleted in earlier passes through Step 2 or 6. Partition the resulting matrix into A^1, D, A^2, and a zero submatrix in agreement with (9.2.1). The partitioned matrix represents a maximum monotone decomposition of the original matrix B relative to C. The component matrices B^1 and B^2 are given by (9.2.2). Output that decomposition, and stop.

4. (Select candidate column y.) Select any column y of the current B such that N does not contain both y and $\bar{y}$. Initialize a set L as the empty set. Define $\alpha = 1$ (α denotes the current scaling factor for y.) If N contains y, go to Step 11.

5. (Search for a decomposition where $y \in Y_1$.) Declare A^1 to be the subvector of column y that precisely contains the 1s of column y. With A^1 determined, define D, E, and A^2 according to (9.3.1).

6. (D is nonpositive. $E = 0$ implies that a decomposition is at hand.) If $E = 0$: Remove A^1, D, and E from B, and go to Step 2.

7. (E is nonzero.) Select any nonzero column z of E.

8. (Shift column z from E if possible.) If $z \notin N$, and if column z of E is nonpositive: Shift column z of E from E to A^1, and correspondingly shift column z of A^2 from A^2 to D. Shift all rows of D with exactly one 1 from D to A^1, and shift the corresponding rows of A^2 from A^2 to E. Add $\bar{z}$ to L. Go to Step 6.

9. (Scale and shift column z from E if possible.) If $\bar{z} \notin N$, and if column z of E is nonnegative: Scale column z of E and column z of A^2 by -1. Shift the scaled column z of E from E to A^1, and correspondingly shift the scaled column z of A^2 from A^2 to D. Shift all rows of D with exactly one 1 from D to A^1, and shift the corresponding rows of A^2 from A^2 to E. If $z \in N$, replace it by $\bar{z}$. If $\bar{z} \in L$, replace it by z. Add $\bar{z}$ to L. Go to Step 6.

10. (Decomposition attempt is unsuccessful.) Add both z and $\bar{z}$ to L.

11. (Try alternate scaling factor.) If $\bar{y} \notin N$ and $\alpha = 1$: Scale column y by -1. Set $\alpha = -1$. If $y \in N$, replace it by $\bar{y}$. Go to Step 5.

12. (Decomposition attempt is unsuccessful for both scaling cases of column y.) Add both y and $\bar{y}$ to N. Add all elements of L to N. Go to Step 3.

We turn to the solution algorithm for the SAT or MINSAT instances where the given matrix has a monotone decomposition.

9.4 Solution Algorithm

Let B be a monotone sum $B = B^1 \boxplus_m B^2$. Let b be a $\{0, 1\}$ vector of appropriate size. From now on, we denote the SAT instance consisting of the b-satisfiability problem for B by (B, b).

In this section, we develop an algorithm that solves the SAT instance (B, b) by transforming it into a certain SAT instance (B^1, b^1) and into a second SAT instance (B^2, b^2).

The algorithm also handles MINSAT instances where a b-satisfying solution for B is to be found that minimizes costs given by a rational nonnegative vector c. The MINSAT instance is $\min \sum c_y$ such that $B \odot s \geq b$, where the summation is over the $y \in Y$ for which $s_y = 1$. From now on, we denote that MINSAT instance by (B, b, c).

In our solution approach for (B, b, c), the vector c restricts the scaling of the monotone decomposition. Specifically, the scaling of columns of B in the decomposition must be confined to the columns y of B for which $c_y = 0$.

The algorithm is based on Theorem (5.5.2), which we repeat below. Recall from Section 5.5 that a satisfying solution for a given matrix A over $\mathbb{B}$ is minimum with respect to *True* if the following condition holds. If the solution has the value *True* for a column of A, then every satisfying solution for A must have *True* for that column.

(9.4.1) Theorem. *Let A be a nearly negative matrix that is satisfiable. Then A has a satisfying solution that is minimum with respect to True, and this solution is found by Algorithm SOLVE NEARLY NEGATIVE SAT OR MINSAT (5.5.1).*

SAT Case

We first develop the algorithm for the SAT instance (B, b). Let X be the row index set of B, and let Y be the column index set. Suppose B has a monotone decomposition as given by (9.2.1) and (9.2.2). To simplify the discussion, we assume that the decomposition does not involve any column scaling. Thus, B itself has the form

(9.4.2)

$$
B = \begin{array}{c|c|c}
 & Y_1 & Y_2 \\
\hline
X_1 & A^1 & 0 \\
\hline
X_2 & D & A^2 \\
\end{array}
$$

Partitioned matrix B

and has as submatrices the component matrices

(9.4.3)

$$
B^1 = \begin{array}{c|c}
 & Y_1 \\
\hline
X_1 & A^1 \\
\end{array}
\qquad
B^2 = \begin{array}{c|c|c}
 & Y_1 & Y_2 \\
\hline
X_2 & D & A^2 \\
\end{array}
$$

Components B^1 and B^2 of monotone decomposition

Recall that the submatrix A^1 is nearly negative and that the submatrix D is nonpositive.

The SAT instance (B, b) demands solution of $B \odot s \geq b$. Partition b into b^1 and b^2 corresponding to the index sets X_1 and X_2, respectively, of (9.4.2), and partition s into s^1 and s^2 corresponding to the index sets Y_1 and Y_2, respectively. By (9.4.2) and (9.4.3), the inequality $B \odot s \geq b$ may be rewritten as

(9.4.4)

$$
B^1 \odot s^1 = A^1 \odot s^1 \geq b^1
$$
$$
B^2 \odot [s^1/s^2] = (D \odot s^1) \oplus (A^2 \odot s^2) \geq b^2
$$

Since $B^1 = A^1$ is nearly negative, we may use Algorithm SOLVE NEARLY NEGATIVE SAT OR MINSAT (5.5.1) either to find a solution s^{1*} for $B^1 \odot s^1 \geq b^1$ or to conclude that (B^1, b^1) is unsatisfiable. By (9.4.4), the latter case implies that B is unsatisfiable, and we stop with that conclusion. In the former case, Theorem (9.4.1) says that the satisfying solution so found for B^1 is minimum with respect to *True*. Equivalently, one may say that any element y of s^{1*} is equal to 1 only if every solution of $B^1 \odot s^1 \geq b^1$ has the value 1 for that element. The latter result and the nonpositivity of D imply that, for any vector s^1 satisfying $B^1 \odot s^1 \geq b^1$,

(9.4.5)
$$D \odot s^{1*} \geq D \odot s^1$$

By (4.2.19), the inequality of (9.4.5) and the second inequality of (9.4.4) imply, for any vector s^1 satisfying $B^1 \odot s^1 \geq b^1$,

(9.4.6) $(D \odot s^{1*}) \oplus (A^2 \odot s^2) \geq (D \odot s^1) \oplus (A^2 \odot s^2) \geq b^2$

Suppose we have an algorithm that, for any $\{0, 1\}$ vector a^2 of appropriate size, solves $A^2 \odot s^2 \geq a^2$. We use that algorithm to solve

(9.4.7) $A^2 \odot s^2 \geq b^2 \ominus (D \odot s^{1*})$

which, by (4.2.13), is equivalent to $(D \odot s^{1*}) \oplus (A^2 \odot s^2) \geq b^2$ of (9.4.6). If a solution, say, s^{2*}, is found, then $s^* = [s^{1*}/s^{2*}]$ solves (9.4.4). If no solution exists, then (9.4.6) has no solution for any s^1 satisfying $B^1 \odot s^1 \geq b^1$, and thus $B \odot s \geq b$ has no solution.

We may summarize the above process as follows. Solve the SAT instance (A^1, b^1) with Algorithm SOLVE NEARLY NEGATIVE SAT OR MINSAT (5.5.1). If no solution exists, then (B, b) has no solution. Otherwise, let s^{1*} be the solution. Next, solve the SAT instance $(A^2, b^2 \ominus (D \odot s^{1*}))$. If no solution exists, (B, b) has no solution. Otherwise, $s^* = [s^{1*}/s^{2*}]$ is a solution for (B, b).

MINSAT case

We turn to the MINSAT case. Let (B, b, c) be the given instance, where b is as before and where c is a rational nonnegative vector. Define C to be the column submatrix of B indexed by the $y \in Y$ for which $c_y = 0$. Suppose B has a monotone decomposition relative to C. Thus, the columns of the submatrix C of B may be scaled so that B becomes the matrix of (9.4.2). As before, let us assume that B itself is the latter matrix. Partition c into c^1 and c^2 corresponding to the index sets Y_1 and Y_2, respectively, of (9.4.2). Arguments that are virtually identical to those for the SAT case justify the following solution algorithm.

First, solve the MINSAT instance (A^1, b^1, c^1) with Algorithm SOLVE NEARLY NEGATIVE SAT OR MINSAT (5.5.1). If no solution exists, (B, b, c) is unsatisfiable. Otherwise, let s^{1*} be the solution. Next, solve the MINSAT instance $(A^2, b^2 \ominus (D \odot s^{1*}), c^2)$. If no solution exists, (B, b) has no solution. Otherwise, $s^* = [s^{1*}/s^{2*}]$ solves (B, b, c).

Solution Algorithm

The above discussion validates the following algorithm.

(9.4.8) Algorithm SOLVE MONOTONE SUM SAT OR MIN-SAT. *Solves* SAT *instance* (B, b) *or* MINSAT *instance* (B, b, c) *where* B *is a monotone sum,* b *is a* $\{0, 1\}$ *vector, and* c *is a rational nonnegative vector.*

Input: Matrix B over $\mathbb{B}$ of size $m \times n$, with row index set X and column index set Y. A $\{0, 1\}$ vector b with m entries. In the MINSAT case, a rational nonnegative vector c with n entries.

A monotone decomposition relative to the following column submatrix C of B. In the SAT case, $C = B$. In the MINSAT case, C is the column submatrix of B indexed by the $y \in Y$ for which $c_y = 0$. The decomposition is displayed by (9.2.1) and (9.2.2). Consider b to be partitioned into b^1 and b^2 according to the index sets X_1 and X_2, respectively, and c to be partitioned into c^1 and c^2 according to the index sets Y_1 and Y_2, respectively, of the decomposition.

An algorithm that in the SAT (resp. MINSAT) case solves, for any $\{0, 1\}$ vector a^2 of appropriate size, the SAT instance (A^2, a^2) (resp. MINSAT instance (A^2, a^2, c^2)) in at most β (resp. γ) time.

Output: Either: A solution s^* for (B, b) or (B, b, c), whichever applies. Or: "The given instance has no solution."

Complexity: $O(m + n + \text{count}(B) + \beta)$ in the SAT case, and $O(m + n + \text{count}(B) + \gamma)$ in the MINSAT case.

Procedure:

1. In the SAT (resp. MINSAT) case, use Algorithm SOLVE NEARLY NEGATIVE SAT OR MINSAT (5.5.1) to solve the SAT instance (A^1, b^1) (resp. MINSAT instance (A^1, b^1, c^1)). If the instance is unsatisfiable, then declare the original problem to be unsatisfiable, and stop. Otherwise, let s^{1*} be the solution.

2. Using the algorithm given in the input, solve in the SAT (resp. MINSAT) case the instance $(A^2, b^2 \ominus (D \odot s^{1*}))$ (resp. $(A^2, b^2 \ominus (D \odot s^{1*}), c^2)$). If the instance is unsatisfiable, declare the original problem to be unsatisfiable, and stop. Otherwise, let s^{2*} be the solution.

3. Let $s^* = [s^{1*}/s^{2*}]$. Scale s^* by the scaling factors that were used in the monotone decomposition of B. Output the scaled vector as the solution vector.

Proof of Validity. The prior arguments establish validity for Steps 1 and 2. The scaling of s^* in Step 3 accounts for the scaling of B in the monotone decomposition. □

Algorithm SOLVE MONOTONE SUM SAT OR MINSAT (9.4.8) evidently solves one SAT or MINSAT instance for each one of the components

B^1 and B^2. Hence, the monotone sum is of type I. We record this fact for future reference.

(9.4.9) Theorem. *The monotone sum is of type I.*

SAT and MINSAT Centrality

We establish a certain centrality result for monotone sums. We review relevant definitions of Section 5.2.

Let C be a class of matrices over $\mathbb{B}$. Then C is SAT central if the following conditions are satisfied.

(9.4.10)
	(i)	If $A \in C$, then any submatrix of A is also in C.
	(ii)	There is a polynomial algorithm for solving the SAT instances given by the matrices of C.
	(iii)	There is a polynomial algorithm for recognizing the matrices of C.

The class C is SAT semicentral if it observes (9.4.10)(i) and (ii).

Let C be a set of matrix/vector pairs (A, c), where A is a matrix over $\mathbb{B}$ and c is a rational nonnegative vector. We declare C to be MINSAT central if the following conditions are satisfied.

(9.4.11)
	(i)	If $(A, c) \in C$, then any submatrix pair of (A, c) is also in C.
	(ii)	There is a polynomial algorithm for solving the MINSAT instances given by the matrix/vector pairs of C.
	(iii)	There is a polynomial algorithm for recognizing the matrix/vector pairs of C.

The class C is MINSAT semicentral if it observes (9.4.11)(i) and (ii).

(9.4.12) Theorem.
(a) *Let C_0 be a class of SAT central (resp. semicentral) matrices. Then the class C of monotone sums $B = B^1 \boxplus_m B^2$ where the submatrix A^2 of B^2 is in C_0 is SAT central (resp. semicentral).*
(b) *Let C_0 be a class of MINSAT central (resp. semicentral) matrix/vector pairs. Then the class C of matrix/vector pairs (B, c) for which B is a monotone sum $B = B^1 \boxplus_m B^2$ and for which the submatrix A^2 of B^2 and the related subvector c^2 of c form a matrix/vector pair (A^2, c^2) of C_0, is MINSAT central (resp. semicentral).*

Proof. We establish part (a). First, assume that the class C_0 is SAT central. We prove (9.4.10), and thus SAT centrality, for C using the notation of (9.2.1) and (9.2.2).

Let $B = B^1 \boxplus_m B^2$ be in C, and let $\overline{B}$ be any submatrix of B. Define $\overline{A}^2$, $\overline{B}^1$, and $\overline{B}^2$ to be the submatrices of A^2, B^1, and B^2, respectively, that occur in $\overline{B}$. Since the conditions imposed by the monotone sum on the submatrices A^1 and D are maintained under submatrix taking and since A^2 and $\overline{A}^2$ are in the SAT central C_0, we have $\overline{B} = \overline{B}^1 \boxplus_m \overline{B}^2$ in C. Thus, (9.4.10)(i) holds.

We solve the SAT problem for any $B = B^1 \boxplus_m B^2$ in C in polynomial time using the polynomial Algorithm SOLVE MONOTONE SUM SAT OR MINSAT (9.4.8), with the polynomial SAT algorithm for the SAT central C_0 as subroutine. Thus, (9.4.10)(ii) holds.

We turn to (9.4.10)(iii). We decide membership in C in polynomial time using the polynomial Algorithm MONOTONE DECOMPOSITION (9.3.3) and the polynomial recognition algorithm for the SAT central C_0 as follows. Let a matrix B be given. We apply Algorithm MONOTONE DECOMPOSITION (9.3.3) to find a maximum monotone decomposition of B, and test if the submatrix $A^{2'}$ of that decomposition corresponding to A^2 of (9.2.1) is in C_0. If this is so, we have proved B to be in C.

Assume that $A^{2'}$ is not in C_0. We claim that B is not in C. Indeed, if B is in C, then it must have a monotone decomposition of the form (9.2.1) where A^2 is in C_0. Corollary (9.2.4) says that $A^{2'}$ is a submatrix of A^2, so by the SAT centrality of C_0, $A^{2'}$ is in C_0, a contradiction.

We have proved that C is SAT central. A portion of the above arguments proves SAT semicentrality of C if C_0 is SAT semicentral. Judicious modification of the arguments also proves part (b).   □

The final section covers extensions and references.

9.5 Extensions and References

Boros, Crama, and Hammer (1990), and Boros, Hammer, and Sun (1994) treat the SAT case where, in our terminology, the given matrix B is a monotone sum $B = B^1 \boxplus_m B^2$ for which the submatrix A^2 of B^2 is a 2SAT matrix. That is, A^2 has at most two nonzero entries in each row. The references define the CNF system represented by the matrix B to be q-Horn. In the discussion below, we apply that term to the matrix B as well.

The cited references give a linear algorithm for solving the SAT problem involving q-Horn CNF systems. In our terminology, the algorithm consists of Algorithm SOLVE MONOTONE SUM SAT OR MINSAT (9.4.8) and Algorithm SOLVE 2SAT (5.4.1).

Boros, Hammer, and Sun (1994) describe a linear algorithm for recognizing q-Horn matrices. The more general Algorithm MONOTONE DE-

COMPOSITION (9.3.3) detects maximum monotone decompositions in linear time, with no restriction imposed on the submatrix A^2. On the other hand, Corollary (9.2.5) implies that any maximum monotone decomposition determined with Algorithm MONOTONE DECOMPOSITION (9.3.3) has a 2SAT submatrix A^2 if and only if the given matrix is q-Horn. Thus, the latter algorithm may be used to test for the q-Horn property in linear time.

Boros, Crama, Hammer, and Saks (1994) base a complexity index for SAT on the idea of q-Horn matrices. Pretolani (1993a) gives a linear algorithm for deciding unique satisfiability for q-Horn matrices. Boros and Čepek (1995) link q-Horn matrices and so-called perfect $\{0, \pm 1\}$ matrices.

Boros and Hammer (1992) use q-Horn matrices to effect a decomposition of matrices that is more general than the monotone decomposition. To simplify the discussion, let us assume that the given matrix has at least two nonzero entries in each row. In our terminology, the decomposition of Boros and Hammer (1992) involves column scaling and partitioning of the given matrix, and results in

(9.5.1)

	Y_1	Y_2
X_1	A^1	E
X_2	D	A^2

Scaled and partitioned matrix

where A^1 is nearly negative and has in each row at least two nonzero entries and where D is nonpositive. The matrices E and A^2 are arbitrary. The conditions imposed on A^1 imply that A^1 has at least one -1 in each row. The latter fact and the requirement that D is nonpositive imply that one may set all variables indexed by Y_1 to *False* and thus obtain a reduced matrix that is satisfiable if and only if the original matrix is satisfiable.

In contrast to monotone decompositions, the decomposition of (9.5.1) generally is not maintained under submatrix taking. Thus, results analogous to the SAT or MINSAT centrality or semicentrality statements of Theorem (9.4.12), which are crucial for our subsequent use of the monotone decomposition, do not hold.

But suppose we want to obtain the decomposition (9.5.1) to reduce SAT or MINSAT instances. Boros and Hammer (1992) describe an algorithm that finds such a decomposition in linear time if certain choices are made appropriately. It is easy to modify Algorithm MONOTONE DECOMPOSITION (9.3.3) so that it finds such decompositions in linear time, too. We sketch the changes. In Step 6, the condition $E = 0$ is replaced by the requirement that A^1 has in each row at least two nonzero entries. In

Step 7, a nonzero column z of E is selected according to some heuristic rule with the following aim. When column z is scaled and shifted in Step 8 or 9, then the new A^1 must be nearly negative and, subject to that condition, should have at least two nonzeros in as many rows as possible.

Suppose we want to make the choices in the algorithm of Boros and Hammer (1992) or in the modified version of Algorithm MONOTONE DECOMPOSITION (9.3.3) so that a decomposition of the form (9.5.1) is found that is best according to some reasonable measure. For example, given a decomposition of the form (9.5.1), let $\overline{X} \subseteq X_1 \cup X_2$ index the nonzero rows of the column submatrix $[A^1/D]$. One might declare a decomposition of the form (9.5.1) to be best if some monotone increasing function of $|\overline{X}|$ and $|Y_1|$—for example, $|Y_1|$ or $|\overline{X}| + |Y_1|$—is maximum.

We show that finding a decomposition that is best according to any one of these measures is $\mathcal{NP}$-hard. Indeed, for $\{\pm 1\}$ matrices, the problem becomes one of finding a maximum hidden nearly negative column submatrix, which in Section 8.5 is shown to be $\mathcal{NP}$-hard. Accordingly, making appropriate choices in the algorithm of Boros and Hammer (1992) or making appropriate heuristic selections in the modified version of Algorithm MONOTONE DECOMPOSITION (9.3.3), so that a decomposition is found that is best according to any one of the above measures, is $\mathcal{NP}$-hard as well.

We should mention that choices can be specified in the algorithm of Boros and Hammer (1992) so that it produces a maximum monotone decomposition. However, such choices may result in quadratic run time.

The monotone decomposition may be defined and computed for any class of matrices whose entries are classified as either positive, negative, or zero. Therefore, the decomposition may be useful in other areas involving nearly negative matrices. Section 5.9 mentions two such areas, economic input–output analysis and linear complementarity theory.

The next chapter investigates a sum called closed sum.

Chapter 10

Closed Sum

10.1 Overview

In this chapter, we utilize the Boolean closed matrices of Chapter 7 to effect a composition and decomposition called closed sum. We have already used Boolean closed matrices. In Chapter 8, we employed them to obtain the closed subregion decomposition. The latter decomposition is quite different from the closed sum discussed here.

Sections 4.7 and 4.8 include a general discussion of sums of matrices and their uses. That material appears in condensed form in Section 9.1. Given that presentation, we only mention that a sum decomposition of a given matrix B, say, with components B^1 and B^2, permits us to solve any b-satisfiability problem for B by solving some b^1-satisfiability problems for a certain column submatrix $\overline{B}^1$ of B^1 and some b^2-satisfiability problems for a certain column submatrix $\overline{B}^2$ of B^2. Some sums—for example, the monotone sum of Chapter 9—may be used to solve MINSAT instances analogously to the SAT case. But this is not so for the closed sum discussed here.

Section 4.7 classifies each sum according to worst-case upper bounds on the number of b^1- and b^2-satisfiability problems for $\overline{B}^1$ and $\overline{B}^2$ that may have to be solved by the SAT algorithm we have developed for that sum. If that upper bound is 1 for both $\overline{B}^1$ and $\overline{B}^2$, the sum is said to be of type I. If the upper bound is at least 2 for $\overline{B}^1$ and is 1 for $\overline{B}^2$, then the sum is of type II. In the remaining case, where both upper bounds are at least 2, the sum is of type III. It turns out that the closed sum is of type II.

330

The chapter is organized as follows. In Section 10.2, we review Boolean closed matrices and define the closed sum and related concepts.

In Section 10.3, we present algorithms for identifying certain closed sums.

Section 10.4 concerns the solution algorithm for SAT instances involving closed sums.

The final section, 10.5, contains extensions.

10.2 Review and Definitions

We review relevant material of Chapter 7 and introduce the closed sum and related concepts. We begin with the definition of Boolean closed matrices given in Section 7.2.

Column Closedness

A matrix A over $\mathbb{B}$ and with column index set Y is column closed if for any nonempty subset R of subrange(A) the following holds. There exists a partition of Y into sets J_0, J_+, J_-, and $J_\pm$ such that the set Q of vectors s satisfying

$$(10.2.1) \qquad s_j = \begin{cases} 0 & \text{if } j \in J_0 \\ 1 & \text{if } j \in J_+ \\ -1 & \text{if } j \in J_- \\ \pm 1 & \text{if } j \in J_\pm \end{cases}$$

produces a set

$$(10.2.2) \qquad T = \{b \mid b = A \odot s; \ s \in Q\}$$

whose maximal vectors are precisely the maximal vectors of R.

Row Closedness

A matrix A over $\mathbb{B}$ is row closed if, for any nonempty $R \subseteq$ subrange(A), there exists a $\{0, 1\}$ vector b such that the set

$$(10.2.3) \qquad S_b = \{s \mid A \odot s \geq b; \ s_j \in \{\pm 1\}, \ \forall \, j\}$$

is equal to the set

$$(10.2.4) \qquad S_R = \cup_{f \in R}\{s \mid A \odot s \geq f; \ s_j \in \{\pm 1\}, \ \forall \, j\}$$

Boolean Closedness

A matrix A over $\mathbb{B}$ is Boolean closed if A and all submatrices of A are both column closed and row closed. For such a matrix A, let R be an arbitrary subset of subrange(A). Lemma (7.2.5) implies that the vector b for which the set S_b of (10.2.3) is equal to the set S_R of (10.2.4) is given by

$$(10.2.5) \qquad\qquad b_i = \min_{f \in R}\{f_i\}$$

Furthermore, Algorithm J-SETS (7.5.5) finds a partition J_0, J_+, J_-, and $J_\pm$ of the column index of A such that the maximal vectors of the set T defined by (10.2.1) and (10.2.2) are precisely the maximal vectors of R.

We are ready to define the closed sum and related concepts. The definitions given below extend those of Section 4.7.

Closed Decomposition and Separation

Let B be a matrix over $\mathbb{B}$ of the form

(10.2.6)

$$B = \begin{array}{c|c|c|} & Y_1 & Y_2 \\ \hline X_1 & A^1 & 0 \\ \hline X_2 & D & A^2 \\ \hline \end{array}$$

Matrix B with closed separation

where D is Boolean closed. Note that any one of the submatrices of B is allowed to be trivial or empty. Define $(X_1 \cup Y_1, X_2 \cup Y_2)$ to be a *closed separation* of B.

We decompose B into *components* B^1 and B^2 in one of two ways. In the first case, B^1 (resp. B^2) is the column (resp. row) submatrix of B indexed by Y_1 (resp. X_2). Thus,

(10.2.7)

$$B^1 = \begin{array}{c|c|} & Y_1 \\ \hline X_1 & A^1 \\ \hline X_2 & D \\ \hline \end{array} \qquad\qquad B^2 = \begin{array}{c|c|c|} & Y_1 & Y_2 \\ \hline X_2 & D & A^2 \\ \hline \end{array}$$

Components B^1 and B^2 of closed sum B, first case

In the second case, the roles of B^1 and B^2 of (10.2.7) are reversed. That is, B^1 (resp. B^2) is the row (resp. column) submatrix of B indexed by X_2 (resp. Y_1). Thus,

(10.2.8)

$$
B^1 = \begin{array}{c|c|c|}
 & Y_1 & Y_2 \\
\hline
X_2 & D & A^2 \\
\hline
\end{array}
\qquad\qquad
B^2 = \begin{array}{c|c|}
 & Y_1 \\
\hline
X_1 & A^1 \\
\hline
X_2 & D \\
\hline
\end{array}
$$

Components B^1 and B^2 of closed sum B, second case

In both cases, we say that B has been decomposed by a *closed decomposition*.

The reader may wonder why we consider the two decomposition cases (10.2.7) and (10.2.8), instead of just one of them, say, (10.2.7). The reason is that the decompositions may be employed with different efficiency to solve the SAT problem for B. Details are included in Section 10.4 below.

We emphasize that the closed decomposition is quite different from the closed subregion decomposition of Chapter 8. In that decomposition, a given matrix A is decomposed into a subregion cover $A^0, A^1, \ldots, A^q$ where $A^1, A^2, \ldots, A^q$ are Boolean closed.

Closed Composition

Suppose that the matrices B^1 and B^2 of (10.2.7) or (10.2.8) are given, and that the submatrix D present in both matrices is Boolean closed. In either case, we may combine these matrices according to (10.2.6) and obtain B again. We say that B has been created by a *closed composition* of B^1 and B^2.

Closed Sum

We say that B is a *closed sum* of B^1 and B^2 if the latter matrices are the components of a closed decomposition of B or, equivalently, if B is created from B^1 and B^2 by a closed composition. We denote that situation by $B = B^1 \boxplus_c B^2$.

A closed sum is *proper* if the corresponding matrix (10.2.6) satisfies the following condition. If the submatrix D is a zero (resp. nonzero) matrix, then both submatrices A^1 and A^2 must be nonempty (resp. nontrivial and nonempty). Equivalently, in the case of a zero matrix D, both sets $X_1 \cup Y_1$ and $X_2 \cup Y_2$ must be nonempty, while in the case of a nonzero D the four sets X_1, X_2, Y_1, and Y_2 must be nonempty.

Classification Using GF(3)

We classify proper closed sums using the field GF(3). The definitions are based on the concepts of matroid separation and matroid sum discussed in

Sections 3.4 and 3.6. Suppose B is a proper closed sum given by (10.2.6). If for some positive integer k

$$(10.2.9) \qquad \text{GF}(3)\text{-rank}(D) = k - 1$$

then B has a *closed k-separation*. The closed decomposition, composition, and sum that correspond to a closed k-separation are called *closed k-decomposition*, *closed k-composition*, and *closed k-sum*, respectively. We stress that these definitions are based on a closed sum that is proper.

If $k = 1$, then $\text{GF}(3)\text{-rank}(D) = 0$, and thus $D = 0$. For that case, the closed 1-sum is the 1-sum of Section 4.7.

Submatrix Taking

Since we intend to employ closed sums for the solution of satisfiability problems, we would want them to be maintained under submatrix taking. Indeed, we have the following stronger conclusion.

(10.2.10) Theorem. *Let B be a closed k-sum with components B^1 and B^2. Then any submatrix $\overline{B}$ of B is contained in B^1 or B^2 or is, for some $\overline{k} \leq k$, a closed $\overline{k}$-sum whose components $\overline{B}^1$ and $\overline{B}^2$ are submatrices of B^1 and B^2, respectively.*

Proof. Let B be given by (10.2.6). We take $\overline{B}$ to have the structure of (10.2.6), but with bars added to all symbols.

If $\overline{B}$ is contained in one of the submatrices $[A^1/D]$ or $[D|A^2]$ of B, then, by (10.2.7) and (10.2.8), $\overline{B}$ is contained in B^1 or B^2.

Assume neither case applies. Thus, both index sets $\overline{X}_1$ and $\overline{Y}_2$ of $\overline{B}$ are nonempty.

If both index sets $\overline{X}_2$ and $\overline{Y}_1$ are nonempty as well, then $\overline{B}$ clearly is a closed $\overline{k}$-sum for some $\overline{k} \leq k$. If at least one of the latter sets is empty, then the submatrix $\overline{D}$ of D is trivial or empty, and $\overline{B}$ is a closed 1-sum.   □

In the next section, we establish polynomial algorithms for finding certain closed k-separations.

10.3 Decomposition Algorithms

In this section and Section 10.5, we use algorithms and results of Section 3.5 to find closed k-separations for any given k. The resulting methods are polynomially bounded provided that k is bounded by a constant. However, the schemes are practically useful only if $k = 1$, 2, or 3. We cover details for the latter cases here and treat the case of general k in Section 10.5.

We rely on Theorem (7.4.8), which we repeat below.

(10.3.1) Theorem. *Let A be a $\{0, \pm 1\}$ matrix, considered to be over $\mathbb{B}$ or $GF(3)$ as appropriate.*

(a) *If $GF(3)$-rank$(A) \leq 1$, then A is Boolean closed.*
(b) *If $GF(3)$-rank$(A) = 2$, then A is Boolean closed if and only if column scaling followed by deletion of duplicate columns and rows can reduce A to a matrix which has $GF(3)$-rank equal to 2 and which is a submatrix of one of the matrices F^1–F^3 below.*

(10.3.2)

$$
F^1 = \begin{array}{|ccc|} 1 & -1 & 0 \\ 1 & 1 & 1 \end{array}
\qquad
F^2 = \begin{array}{|cc|} -1 & 0 \\ 1 & -1 \\ 1 & 1 \\ 1 & 0 \end{array}
\qquad
F^3 = \begin{array}{|cc|} -1 & 0 \\ -1 & -1 \\ 1 & 1 \\ 1 & 0 \end{array}
$$

$$F^1 \qquad\qquad F^2 \qquad\qquad F^3$$

Boolean closed matrices F^1–F^3

Closed 1-Separation

Since the closed 1-sum is the 1-sum of Section 3.5, the case $k = 1$, where $GF(3)$-rank$(D) = 0$ and thus $D = 0$, is detected by Algorithm 1-SEPARATION (3.5.1).

Closed 2-Separation

Finding a closed 2-separation for a given matrix B requires partitioning of B as

(10.3.3)

$$
B = \begin{array}{c|c|c} & Y_1 & Y_2 \\ \hline X_1 & A^1 & 0 \\ \hline X_2 & D & A^2 \end{array}
$$

Matrix B with closed separation

such that D has $GF(3)$-rank$(D) = 1$ and is Boolean closed and such that the submatrices A^1 and A^2 contain at least one entry each. According to Theorem (10.3.1)(a), $GF(3)$-rank$(D) = 1$ implies that D is Boolean closed. We conclude that the desired closed 2-separation is at hand if $GF(3)$-rank$(D) = 1$ and if A^1 and A^2 observe the stated condition.

The algorithm introduced below for locating closed 2-separations is similar to Algorithm $GF(3)$-2-SEPARATION (3.5.26). We decided on a separate treatment for two reasons. First, the latter algorithm requires the given matrix B to be connected and simple. That condition is not imposed here. Second, the discussion motivating the algorithm of this section sets the stage for the subsequent, more complicated case involving closed 3-separations.

The main tool for finding the desired separation is Algorithm IN-DUCED $\mathcal{F}$-SEPARATION (3.5.14). We review that algorithm and related material of Section 3.5. Let B be a matrix over a field $\mathcal{F}$, with row index set X and column index set Y. Define $\overline{B}$ to be a submatrix of B of the form

(10.3.4)

$$\overline{B} = \begin{array}{c|c|c} & \overline{Y}_1 & \overline{Y}_2 \\ \hline \overline{X}_1 & \overline{A}^1 & \overline{D}^2 \\ \hline \overline{X}_2 & \overline{D}^1 & \overline{A}^2 \end{array}$$

Submatrix $\overline{B}$ of B

Let $k \geq 1$ be given. Assume that for some $l \geq k$,

(10.3.5) $$|\overline{X}_1 \cup \overline{Y}_1|, |\overline{X}_2 \cup \overline{Y}_2| \geq l$$

and that

(10.3.6) $$\mathcal{F}\text{-rank}(\overline{D}^1) + \mathcal{F}\text{-rank}(\overline{D}^2) = k - 1$$

Then $(\overline{X}_1 \cup \overline{Y}_1, \overline{X}_2 \cup \overline{Y}_2)$ is an exact k-separation of $\overline{B}$. That separation of $\overline{B}$ induces one for B if X and Y can be partitioned into X_1, X_2 and Y_1, Y_2, respectively, such that, for $i = 1, 2$, $X_i \supseteq \overline{X}_i$ and $Y_i \supseteq \overline{Y}_i$, and such that $(X_1 \cup Y_1, X_2 \cup Y_2)$ is an exact k-separation of B.

Algorithm INDUCED $\mathcal{F}$-SEPARATION (3.5.14) decides whether a given exact k-separation of $\overline{B}$ induces one for B. Lemma (3.5.15) contains two observations about the output of that algorithm as follows.

(10.3.7) Lemma.
(a) *Any k-separation produced by Algorithm INDUCED $\mathcal{F}$-SEPARATION (3.5.14) has $X_1 \cup Y_1$ minimal and $X_2 \cup Y_2$ maximal, in the sense that any other k-separation $(X_1' \cup Y_1', X_2' \cup Y_2')$ of B induced by the exact k-separation $(\overline{X}_1 \cup \overline{Y}_1, \overline{X}_2 \cup \overline{Y}_2)$ of $\overline{B}$ observes $X_1 \subseteq X_1'$, $X_2 \supseteq X_2'$, $Y_1 \subseteq Y_1'$, and $Y_2 \supseteq Y_2'$.*
(b) *Let $(\overline{X}_1 \cup \overline{Y}_1, \overline{X}_2 \cup \overline{Y}_2)$ be an exact k-separation of $\overline{B}$, except that $|\overline{X}_2 \cup \overline{Y}_2|$ may be equal to $k - 1$. If B has a k-separation $(X_1' \cup Y_1', X_2' \cup Y_2')$ where, for $i = 1, 2$, $\overline{X}_i \subseteq X_i'$ and $\overline{Y}_i \subseteq Y_i'$, then one such k-separation of B is found by Algorithm INDUCED $\mathcal{F}$-SEPARATION (3.5.14).*

The algorithm for finding closed 2-separations views a given matrix B over $\mathbb{B}$ to be over $GF(3)$. It enumerates all $\overline{B}$ submatrices of the form (10.3.4) where $GF(3)$-$rank(\overline{D}^1) = 1$ and $|\overline{X}_1| = |\overline{X}_2| = |\overline{Y}_1| = |\overline{Y}_2| = 1$. For each such $\overline{B}$, it checks with Algorithm INDUCED $\mathcal{F}$-SEPARATION (3.5.14) if $\overline{B}$ induces a 2-separation of B. If no $\overline{B}$ induces a 2-separation of B, then, in agreement with Lemma (10.3.7)(b), the algorithm declares that B does not have a closed 2-separation. Otherwise, the algorithm stops as soon as a 2-separation of B is found.

Using Lemma (10.3.7)(b), one may speed up the algorithm by replacing the condition $|\overline{Y}_2| = 1$ imposed on $\overline{B}$ by $|\overline{Y}_2| = 0$. We have not done so to simplify the subsequent discussion of a variant of the algorithm.

(10.3.8) Algorithm CLOSED 2-SEPARATION. *Finds a closed 2-separation for a matrix B over $\mathbb{B}$ or declares that such a separation does not exist.*

Input: Matrix B over $\mathbb{B}$, with row index set X and column index set Y.

Output: Either: A closed 2-separation $(X_1 \cup Y_1, X_2 \cup Y_2)$ of B. Or: "B does not have a closed 2-separation."

Complexity: Polynomial.

Procedure:
1. If $|X| \leq 1$ or $|Y| \leq 1$ or if B is a zero matrix, then declare that B does not have a closed 2-separation, and stop.
2. (Enumerate all choices of $\overline{B}$.)
   Do for all possible disjoint sets $|\overline{X}_1| \subseteq X$, $|\overline{X}_2| \subseteq X$, $|\overline{Y}_1| \subseteq Y$, and $|\overline{Y}_2| \subseteq Y$ where each one of the sets has exactly one element, and where the 1×1 submatrix of B indexed by $\overline{X}_2$ and $\overline{Y}_1$ (resp. $\overline{X}_1$ and $\overline{Y}_2$) is nonzero (resp. zero):
   Let $\overline{B}$ be the submatrix of B defined by $\overline{X}_1$, $\overline{X}_2$, $\overline{Y}_1$, and $\overline{Y}_2$. Do Algorithm INDUCED $\mathcal{F}$-SEPARATION (3.5.14) with B and $\overline{B}$ as input. If a 2-separation is found, output it as a closed 2-separation of B, and stop.
3. Declare that B does not have a closed 2-separation, and stop.

Closed 3-Separation

The algorithm for finding closed 3-separations is a more complicated version of Algorithm CLOSED 2-SEPARATION (10.3.8). As before, we view B to be over $GF(3)$ whenever this is appropriate.

Suppose a closed 3-separation is at hand, say, given by (10.3.3) with $GF(3)$-$rank(D) = 2$. Theorem (10.3.1)(b) implies that column scaling followed by deletion of duplicate rows and columns can reduce the submatrix D of B to a matrix that has $GF(3)$-rank equal to 2 and that is a submatrix of one of the matrices F^1–F^3 of (10.3.2), say, F^i. That submatrix of F^i contains a 2×2 GF(3)-nonsingular submatrix, say, $\overline{D}^1$ indexed by some sets $\overline{X}_2 \subseteq X_2$ and $\overline{Y}_1 \subseteq Y_1$. Since the submatrix A^1 (resp. A^2) of B contains at least one entry, the index set X_1 (resp. Y_2) is nonempty, and we may select a subset $\overline{X}_1 \subseteq X_1$ (resp. $\overline{Y}_2 \subseteq Y_2$) containing just one element. Define $\overline{B}$ to be the submatrix of B with row index set $\overline{X}_1 \cup \overline{X}_2$ and column index set $\overline{Y}_1 \cup \overline{Y}_2$. Partition $\overline{B}$ as in (10.3.4). Thus, $\overline{B}$ consists of the submatrices $\overline{A}^1$, $\overline{A}^2$, $\overline{D}^1$, and $\overline{D}^2$.

Suppose we know F^i and $\overline{B}$, but do not know the closed 3-separation of B. Given that limited knowledge, we locate a closed 3-separation of B by the following process.

First, we use Theorem (10.3.1)(b) to decide that certain rows and columns of B cannot possibly intersect the yet to be found closed submatrix D of the closed 3-separation. Accordingly, we enlarge $\overline{A}^1$, $\overline{A}^2$, and $\overline{D}^2$ and redefine $\overline{B}$.

Second, we use Algorithm INDUCED $\mathcal{F}$-SEPARATION (3.5.14) to find a 3-separation induced by $\overline{B}$. That separation is the desired closed 3-separation of B.

Validity of the method follows directly from Theorem (10.3.1)(b) and Lemma (10.3.7)(b).

Of course, the algorithm for finding closed 3-separations does not have the prior knowledge of F^i or $\overline{B}$ assumed above. Instead, the algorithm enumerates all possible cases and applies the above process to each instance.

Analogously to Algorithm CLOSED 2-SEPARATION (10.3.8), the algorithm can be speeded up by replacing the condition $|\overline{Y}_2| = 1$ imposed on $\overline{B}$ by $|\overline{Y}_2| = 0$. We have not done so to simplify the discussion of a variant of the algorithm.

Here is the algorithm.

(10.3.9) Algorithm CLOSED 3-SEPARATION. *Finds a closed 3-separation for a matrix B over $\mathbb{B}$ or declares that such a separation does not exist.*

Input: Matrix B over $\mathbb{B}$, with row index set X and column index set Y.

Output: Either: A closed 3-separation $(X_1 \cup Y_1, X_2 \cup Y_2)$ of B. Or: "B does not have a closed 3-separation."

Complexity: Polynomial.

Procedure:

1. If $|X| \leq 2$ or $|Y| \leq 2$, or if GF(3)-rank$(B) \leq 1$, declare that B does not have a closed 3-separation, and stop.

2. (Enumerate all choices of F^i and $\overline{B}$.) Do Steps 3–8 below for each possible choice of F^i, $\overline{X}_1$, $\overline{X}_2$, $\overline{Y}_1$, and $\overline{Y}_2$ satisfying the following conditions: F^i must be one of the matrices F^1–F^3 of (10.3.2); $\overline{X}_1 \subseteq X$ and $\overline{Y}_2 \subseteq Y$ must index a 1×1 zero submatrix of B; $\overline{X}_2 \subseteq (X - \overline{X}_1)$

and $\overline{Y}_1 \subseteq (Y - \overline{Y}_2)$ must index a 2×2 GF(3)-nonsingular submatrix of B that up to column scaling is a submatrix of F^i. For each such choice, define $\overline{B}$ to be the submatrix of B with row index set $\overline{X}_1 \cup \overline{X}_2$ and column index set $\overline{Y}_1 \cup \overline{Y}_2$. Partition $\overline{B}$ as in (10.3.4), so that $\overline{B}$ consists of the submatrices $\overline{A}^1$, $\overline{A}^2$, $\overline{D}^1$, and $\overline{D}^2$.

If none of the choices produce a closed 3-separation of B in Steps 3–8, declare that B does not have such a separation, and stop.

3. (Partition B.) Scale the two columns of B containing $\overline{D}^1$ such that the new $\overline{D}^1$ is a submatrix of F^i. Redefine $\overline{A}^1$ and $\overline{B}$ accordingly. Partition B as

(10.3.10)

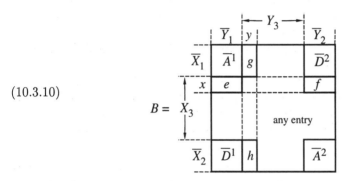

Partition of B induced by $\overline{A}^1$, $\overline{A}^2$, $\overline{D}^1$, and $\overline{D}^2$ of $\overline{B}$

4. (Shift rows based on Theorem (10.3.1)(b).) Shift each row $x \in X_3$ from X_3 to $\overline{X}_1$ for which the subvector e is nonzero, is not a duplicate of a row of $\overline{D}^1$ if $F^i = F^1$, and is not a duplicate of a row of F^i if $F^i = F^2$ or F^3.

5. (Shift columns based on Theorem (10.3.1)(b).) Shift each column $y \in Y_3$ from Y_3 to $\overline{Y}_2$ for which the subvector h is nonzero, is not a column of F^1 up to column scaling if $F^i = F^1$, and is not a column of $\overline{D}^1$ up to column scaling if $F^i = F^2$ or F^3.

6. (Redefine $\overline{B}$.) Enlarge $\overline{A}^1$, $\overline{A}^2$, and $\overline{D}^2$ in agreement with the increases of $\overline{X}_1$ and $\overline{Y}_2$ in Steps 4 and 5. Redefine $\overline{B}$ so that it is composed of the revised $\overline{A}^1$, $\overline{A}^2$, $\overline{D}^1$, and $\overline{D}^2$.

7. (Check sufficient condition for termination.) If $\overline{D}^2$ is nonzero, go to Step 2, and select the next choice.

8. (Check for induced separation.) Applying Algorithm INDUCED $\mathcal{F}$-SEPARATION (3.5.14), either find a 3-separation induced by $\overline{B}$ or conclude that no such induced separation exists. In the former case, output the separation as a closed 3-separation of B, and stop. In the latter case, begin the next choice of Step 2.

Proof of Validity. Given the above discussion, we only need to prove the complexity claim. The number of possible choices in Step 2 is polynomial.

Algorithm INDUCED $\mathcal{F}$-SEPARATION (3.5.14) is polynomial as well, so the entire algorithm is polynomial as claimed.   □

Minimality of Closed Separations

Let $i = 1$ or 2, and define k to be a positive integer. Declare a closed k-separation of B as in (10.3.3) to have A^i *minimal* (resp. *maximal*) relative to a specified set of closed k-separations of B if, for any one of the specified closed k-separations of B, the submatrix that corresponds to A^i is not properly contained in (resp. does not properly contain) A^i. When A^i is minimal (resp. maximal) relative to all closed k-separations of B, we simply say that A^i is *minimal* (resp. *maximal*).

Clearly, minimality (resp. maximality) of A^1 corresponds to maximality (resp. minimality) of A^2.

The algorithms for finding closed 1-, 2-, and 3-separations are easily modified so that they find all separations with minimal or maximal A^1, and hence with maximal or minimal A^2. For the case $k = 1$, that task is elementary. For $k = 2$ or 3, Theorems (10.3.11) and (10.3.12) below provide details.

(10.3.11) Theorem. (Case of minimal A^1 and maximal A^2) *Change Algorithm CLOSED 2-SEPARATION (10.3.8) (resp. Algorithm CLOSED 3-SEPARATION (10.3.9)) such that the modified scheme produces, for $k = 2$ (resp. $k = 3$), all closed k-separations of B that are induced by the possible choices of $\overline{B}$ (resp. of F^i and $\overline{B}$) in Step 2 of the algorithm. Subsequently, the scheme outputs the closed k-separations that have A^1 minimal relative to the closed k-separations so found. The scheme so constructed obtains, for the applicable $k = 2$ or 3, all closed k-separations with A^1 minimal and A^2 maximal. If properly implemented, the scheme is polynomial in the size of B.*

Proof. To prove the case $k = 2$, assume that a closed 2-separation of the form (10.3.3) is at hand where A^1 is minimal relative to all possible closed 2-separations of B. Among the candidate submatrices that may be chosen in Step 2 of Algorithm CLOSED 2-SEPARATION (10.3.8), there is at least one, say, $\overline{B}$, whose submatrices $\overline{A}^1$, $\overline{A}^2$, and $\overline{D}^1$ are contained in A^1, A^2, and D, respectively, of B. Lemma (10.3.7)(a) says that Algorithm INDUCED $\mathcal{F}$-SEPARATION (3.5.14) produces a 2-separation where, given the choice of $\overline{B}$, the submatrix containing $\overline{A}^1$ is as small as possible. Thus, $\overline{A}^1$ is contained in A^1 and indeed must be equal to A^1 due to the assumed minimality of A^1. We conclude that the modified scheme produces all 2-separations of B having A^1 minimal and A^2 maximal. Clearly, the scheme has an implementation that is polynomial in the size of B.

The proof for the case $k = 3$ and Algorithm CLOSED 3-SEPARATION (10.3.9) is the same as for $k = 2$, except that one selects not only

$\overline{B}$, but also an appropriate F^i for the assumed closed 3-separation with minimal A^1.   □

(10.3.12) Theorem. (Case of maximal A^1 and minimal A^2) *Suppose Algorithm* CLOSED 2-SEPARATION (10.3.8) *(resp. Algorithm* CLOSED 3-SEPARATION (10.3.9)) *is changed as in Theorem* (10.3.11) *and is further modified so that Step 2 (resp. Step 8) of the algorithm applies Algorithm* INDUCED $\mathcal{F}$-SEPARATION (3.5.14) *to a rearranged B and $\overline{B}$ where $\overline{A}^1$ and $\overline{A}^2$ have traded places and where $\overline{D}^1$ and $\overline{D}^2$ have traded places as well. Then the revised scheme is polynomial in the size of B, and it finds, for the applicable $k = 2$ or 3, all closed k-separations with A^1 maximal and A^2 minimal.*

Proof. The result follows from the proof of Theorem (10.3.11) and the fact that, according to Lemma (10.3.7)(a), Algorithm INDUCED $\mathcal{F}$-SEPARATION (3.5.14), when given the modified input, searches for a separation with maximal A^1 and minimal A^2.   □

The next section presents an algorithm that solves the SAT instances B with a closed separation.

10.4 Solution Algorithm

Let B be a matrix over $\mathbb{B}$ with a closed decomposition given by

(10.4.1)

$$
B = \begin{array}{c|c|c}
 & Y_1 & Y_2 \\
\hline
X_1 & A^1 & 0 \\
\hline
X_2 & D & A^2 \\
\end{array}
$$

Matrix B with closed separation

For a given $\{0, 1\}$ vector b, we want to solve the b-satisfiability problem for B, which involves the solution of $B \odot s \geq b$. We denote that SAT instance by (B, b). In this section, we provide a solution method that utilizes one of the two ways (10.2.7) and (10.2.8) of decomposing B.

    Partition b into b^1 and b^2 corresponding to the index sets X_1 and X_2, respectively, of (10.4.1), and partition s into s^1 and s^2 corresponding to the index sets Y_1 and Y_2, respectively. With these vectors, the inequality $B \odot s \geq b$ may be rewritten as

(10.4.2)
$$
A^1 \odot s^1 \geq b^1
$$
$$
(D \odot s^1) \oplus (A^2 \odot s^2) \geq b^2
$$

Let $d \in$ subrange(D). Consider the following inequality system.

$$A^1 \odot s^1 \geq b^1$$
(10.4.3)
$$D \odot s^1 \geq d$$
$$A^2 \odot s^2 \geq b^2 \ominus d$$

The next lemma establishes a certain equivalence between the two inequality systems (10.4.2) and (10.4.3).

(10.4.4) Lemma. *The inequalities of* (10.4.2) *have a solution if and only if, for some* $d \in$ subrange(D), *this is so for the inequalities of* (10.4.3).

Proof. If s^{1*} and s^{2*} solve (10.4.2), then $d = D \odot s^{1*}$ is in subrange(D), and s^{1*} and s^{2*} solve (10.4.3) for this choice of d.

Conversely, let (10.4.3) have a solution s^{1*} and s^{2*} for some $d \in$ subrange(D). Adding the second and third inequality of (10.4.3) with $s^1 = s^{1*}$ and $s^2 = s^{2*}$, we obtain $(D \odot s^{1*}) \oplus (A^2 \odot s^{2*}) \geq d \oplus (b^2 \ominus d) \geq b^2$. Thus, s^{1*} and s^{2*} solve (10.4.2). □

First Decomposition Case

Consider B decomposed into B^1 and B^2 according to (10.2.7). Thus,

(10.4.5)

$$B^1 = \begin{array}{c|c|} & Y_1 \\ \hline X_1 & A^1 \\ \hline X_2 & D \\ \hline \end{array} \qquad\qquad B^2 = \begin{array}{c|c|c|} & Y_1 & Y_2 \\ \hline X_2 & D & A^2 \\ \hline \end{array}$$

Components B^1 and B^2 of closed sum B, first case

We employ the following solution strategy. First, we find all vectors $d \in$ subrange(D) for which $B^1 \odot s^1 \geq [b^1/d]$ or, equivalently,

(10.4.6)
$$A^1 \odot s^1 \geq b^1$$
$$D \odot s^1 \geq d$$

has a solution, say, $s^1(d)$. Let $R \subseteq$ subrange(D) be the set of such vectors d. If R is empty, then (10.4.3) has no solution, and we stop. Assume that R is nonempty.

Second, we determine whether, for some $d \in R$, the inequality $A^2 \odot s^2 \geq b^2 \ominus d$ or, equivalently,

(10.4.7)
$$d \oplus (A^2 \odot s^2) \geq b^2$$

has a solution. If this is so, we have solved (10.4.3). Otherwise, (10.4.3) has no solution.

It clearly is sufficient that we search for a solution of (10.4.7) using just the maximal vectors of R. We use this fact and the Boolean closedness of D to simplify that search. With Algorithm J-SETS (7.5.5), we produce a partition J_0, J_+, J_-, and $J_\pm$ of Y_1 such that the set Q of vectors s^1 given by

$$(10.4.8) \qquad s_j^1 = \begin{cases} 0 & \text{if } j \in J_0 \\ 1 & \text{if } j \in J_+ \\ -1 & \text{if } j \in J_- \\ \pm 1 & \text{if } j \in J_\pm \end{cases}$$

defines a set

$$(10.4.9) \qquad T = \{d \mid d = D \odot s^1;\ s^1 \in Q\}$$

whose maximal vectors are precisely the maximal vectors of R. Thus, searching for a solution of (10.4.7) for some maximal vector of R is equivalent to solving

$$(10.4.10) \qquad (D \odot s^1) \oplus (A^2 \odot s^2) \geq b^2,\quad s^1 \in Q$$

The restriction $s^1 \in Q$ is equivalent to deleting the columns $j \in J_0$ and fixing, for $j \in J_+$ (resp. $j \in J_-$), s_j^1 to 1 (resp. -1). Hence, (10.4.10) represents just one SAT instance. If that instance has no solution, then (10.4.3) has no solution either. Otherwise, let $[\tilde{s}^1/s^{2*}]$ be a solution. By the above derivation of R and Q, (10.4.6) with $d = D \odot \tilde{s}^1$ is solved by $s^1(d)$. Then $s^{1*} = s^1(d)$ and s^{2*} solve (10.4.3) for that d.

Second Decomposition Case

A similar process works for the second decomposition case. Here,

$$(10.4.11)$$

$$B^1 = \begin{array}{c} \\ X_2 \end{array} \begin{array}{|c|c|} \hline Y_1 & Y_2 \\ \hline D & A^2 \\ \hline \end{array} \qquad\qquad B^2 = \begin{array}{c} \\ X_1 \\ \\ X_2 \end{array} \begin{array}{|c|} \hline Y_1 \\ \hline A^1 \\ \hline D \\ \hline \end{array}$$

Components B^1 and B^2 of closed sum B, second case

First, we find all vectors $d \in \text{subrange}(D)$ for which

$$(10.4.12) \qquad A^2 \odot s^2 \geq b^2 \ominus d$$

has a solution, say, $s^2(d)$. Let $R \subseteq \text{subrange}(D)$ be the set of vectors d for which this is so. As a matter of notational clarity, we use f below to denote any vector of R. If R is empty, then (10.4.3) has no solution, and we stop. Assume otherwise.

Second, we determine whether, for some $f \in R$,

(10.4.13)
$$A^1 \odot s^1 \geq b^1$$
$$D \odot s^1 \geq f$$

has a solution. If this is so, we have solved (10.4.3). Otherwise, (10.4.3) has no solution.

A vector s^1 satisfies the second inequality of (10.4.13) if and only if s^1 is in the set

(10.4.14) $$S_R = \cup_{f \in R} \{s^1 \mid D \odot s^1 \geq f; \ s^1 = \{\pm 1\} \text{ vector}\}$$

Select a vector d by

(10.4.15) $$d_i = \min_{f \in R} \{f_i\}$$

Since D is Boolean closed, Lemma (7.2.5) implies that S_R is equal to the set S_d given by

(10.4.16) $$S_d = \{s^1 \mid D \odot s^1 \geq d; \ s_j^1 \in \{\pm 1\}, \ \forall \ j\}$$

Hence, (10.4.13) has a solution for some $f \in R$ if and only if, for d defined by (10.4.15),

(10.4.17)
$$A^1 \odot s^1 \geq b^1$$
$$D \odot s^1 \geq d$$

has a solution.

If (10.4.17) has no solution, then (10.4.3) has no solution either. Otherwise, let s^{1*} solve (10.4.17). By the above discussion, (10.4.12) with $d = D \odot s^{1*}$ is solved by $s^2(d)$. Then s^{1*} and $s^{2*} = s^2(d)$ solve (10.4.3) for that d.

Solution Algorithm

We summarize the above solution processes for the two decomposition cases.

(10.4.18) Algorithm SOLVE CLOSED SUM SAT. *Solves SAT instance* (B, b) *where* B *is a closed sum and where* b *is a* $\{0, 1\}$ *vector.*

Input: Matrix B over $\mathbb{B}$ of size $m \times n$, with row index set X and column index set Y. A $\{0, 1\}$ vector b with m entries.

A closed separation of B as displayed by (10.4.1). Consider b to be partitioned into b^1 and b^2 according to the index sets X_1 and X_2, respectively, of the separation. For $i = 1, 2$, define $m_i = |X_i|$ and $n_i = |Y_i|$.

An algorithm that solves the SAT instance of any submatrix of $[A^1/D]$ in at most β_1 time.

If decomposition case (10.4.5) is selected: A second algorithm that solves the SAT instance of any submatrix of $[D|A^2]$ in at most β_2' time.

If decomposition case (10.4.11) is selected: A second algorithm that solves the SAT instance of any submatrix of A^2 in at most β_2'' time.

Output: Either: A solution s^* for (B, b). Or: "The given instance has no solution."

Complexity: If decomposition case (10.4.5) is selected: $O(m_2 \cdot n_1 + n_1^2 + n_1 \cdot \beta_1 + \beta_2')$. If decomposition case (10.4.11) is selected: $O(m_2 \cdot n_1 + n_1^2 + n_1 \cdot \beta_2'' + \beta_1)$.

Procedure:

1. Obtain subrange(D) with Algorithm SUBRANGE OF BOOLEAN CLOSED MATRIX (7.5.4). If the decomposition case (10.4.11) is selected, go to Step 5.

2. (Decomposition case (10.4.5)) For each $d \in$ subrange(D), solve the SAT instance $([A^1/D], [b^1/d])$ with the appropriate given algorithm. Let $R \subseteq$ subrange(D) be the set of vectors d for which a solution, say, $s^1(d)$, is found. If R is empty, declare that (B, b) has no solution, and stop.

3. Use Algorithm J-SETS (7.5.5) to determine the sets J_0, J_+, J_-, and $J_\pm$. From the latter sets, determine the set Q via (10.4.8).

4. Solve with the appropriate given SAT algorithm the SAT instance $([D|A^2], b^2)$ under the restriction that the component s^1 corresponding to the submatrix D must be in Q. If that SAT instance does not have a solution, output that (B, b) has no solution, and stop. Otherwise, let $[\tilde{s}^1/s^{2*}]$ be a solution for $([D|A^2], b^2)$. Define $d = D \odot \tilde{s}^1$, and $s^{1*} = s^1(d)$. Output s^* composed of s^{1*} and s^{2*} as a solution for (B, b), and stop.

5. (Decomposition case (10.4.11)) For each $d \in$ subrange(D), solve the SAT instance $(A^2, b^2 \ominus d)$ with the appropriate given SAT algorithm. Let $R \in$ subrange(D) be the set of vectors d for which a solution, say, $s^2(d)$, is found. If R is empty, declare that (B, b) has no solution, and stop.

6. Define d by $d_i = \min_{f \in R}\{f_i\}$.

7. Solve with the appropriate given SAT algorithm the SAT instance

$([A^1/D], [b^1/d])$. If that SAT instance does not have a solution, output that (B, b) has no solution, and stop. Otherwise, let s^{1*} be a solution for $([A^1/D], [b^1/d])$. Compute $d = D \odot s^{1*}$, and define $s^{2*} = s^2(d)$. Output s^* composed of s^{1*} and s^{2*} as a solution for (B, b), and stop.

Proof of Validity. Given the prior discussion, we only need to establish the claimed complexity. First, assume the decomposition case (10.4.5).

In Step 1, Algorithm SUBRANGE OF BOOLEAN CLOSED MATRIX (7.5.4) determines subrange(D) in $O(m_2 \cdot n_1 + n_1^2)$ time. Corollary (7.4.7) implies that the cardinality of subrange(D) is $O(n_1)$. Thus, the $|\text{subrange}(D)|$ SAT instances of Step 2 are solved in $O(n_1 \cdot \beta_1)$ time. In Step 3, Algorithm J-SETS (7.5.5) finds the J-sets for Q in $O(m_2 \cdot n_1 + n_1^2)$ time. The single SAT instance of Step 4 requires at most β_2' time.

Combining the above time bounds, we get for the decomposition case (10.4.5) an overall bound of $O(m_2 \cdot n_1 + n_1^2 + n_1 \cdot \beta_1 + \beta_2')$.

The complexity for the decomposition case (10.4.11) is handled by very similar arguments, so we omit details. ☐

For the decomposition case (10.4.5) or (10.4.11), Algorithm SOLVE CLOSED SUM SAT (10.4.18) generally solves several SAT instances involving B^1 and one SAT instance involving B^2. Thus, the closed sum is of type II. We record this fact below.

(10.4.19) Theorem. *The closed sum is of type II.*

SAT Centrality

We prove a SAT centrality result for closed sums. Recall from Section 5.2 that a class C of matrices over $\mathbb{B}$ is SAT central if the following conditions are satisfied.

(10.4.20)

(i)   If $A \in C$, then any submatrix of A is also in C.
(ii)   There is a polynomial algorithm for solving the SAT instances given by the matrices of C.
(iii) There is a polynomial algorithm for recognizing the matrices of C.

Let C_0 be a given class of SAT central matrices. Construct a class C of matrices over $\mathbb{B}$ by the following recursive process. Initialize $C = C_0$. In the recursive step, select two matrices $[A^1/D]$ and $[D|A^2]$ such that one of the two matrices is in C_0, while the other one is in C. In addition, the submatrix D occurring in the two matrices must be Boolean closed and, for some $k \leq 3$, must have GF(3)-rank$(D) = k - 1$. Add the closed sum of the two matrices, as given by (10.4.1), to C. If the closed sum added to C is larger than $[A^1/D]$ as well as $[D|A^2]$, then it is readily checked that the closed sum is a closed k-sum with $k \leq 3$.

We associate a *level* with each matrix of C. The matrices of C that are in C_0 have level 0. For any other matrix B of C, the level is equal to the least number of recursive construction steps needed to create B.

We mean the above construction process when we say that C is *created* from C_0 by repeated closed k-sum steps with $k \leq 3$.

The next theorem provides a SAT centrality result for the class C created from a given SAT central class C_0 by repeated closed k-sum steps.

(10.4.21) Theorem. *Let C_0 be a SAT central class of matrices. Define C to be the class of matrices created from C_0 by repeated closed k-sum steps with $k \leq 3$. Then C is SAT central.*

Proof. We must prove (10.4.20)(i)–(iii).

We begin with (10.4.20)(i). We must show that C is maintained under submatrix taking. Let $B \in C$ have level l. Define $\overline{B}$ to be an arbitrary submatrix of B.

The proof is by induction on the level l of B. If $l = 0$, then $B \in C_0$, and the SAT centrality of C_0 implies that $\overline{B} \in C_0 \subseteq C$.

For the inductive step, let $l \geq 1$. By the recursive construction of C and the definition of level, B is, for some $k \leq 3$, a closed k-sum where one component of the sum is in C_0, while the other one is a matrix of C at level $l - 1$.

If $\overline{B}$ is a submatrix of one of the two components, then $\overline{B}$ is in C by induction. Otherwise, Theorem (10.2.10) confirms that $\overline{B}$ is, for some $\overline{k} \leq k$, a closed $\overline{k}$-sum where the components are submatrices of the components of the k-sum. Thus, one component of the $\overline{k}$-sum is in C_0, while the other is by induction in C. We conclude that $\overline{B}$ is in C.

We skip ahead to the proof of (10.4.20)(iii), which demands a polynomial recognition algorithm for membership in C. Since C_0 is SAT central, we are given a polynomial recognition algorithm for testing membership in C_0.

Let a matrix B be given. We test for membership of B in C_0 with the given algorithm. If B is determined to be in C_0, then $B \in C$, and we are done.

Otherwise, we apply Algorithm 1-SEPARATION (3.5.1) and the polynomial schemes of Theorems (10.3.11) and (10.3.12) to find, for each $k \leq 3$, first all closed k-separations with minimal A^1 and then all closed k-separations with minimal A^2. For each k-separation with minimal A^1 (resp. A^2), we test the component $[A^1/D]$ (resp. $[D|A^2]$) for membership in C_0; if $[A^1/D]$ (resp. $[D|A^2]$) is in C_0, then we reduce B to the remaining component $[D|A^2]$ (resp. $[A^1/D]$) and proceed recursively with the reduced B.

Suppose the above method reduces the initial B to a matrix that is not in C_0 and that cannot be reduced by the above process. We claim that B is not in C. Since C is closed under submatrix taking, we may assume, for the purposes of the proof, that B itself is the irreducible matrix.

Suppose that $B \in C$. Since $B \notin C_0$, it is, for some $k \leq 3$, a closed k-sum where one of the two components is in C_0. For example, let that component be $[A^1/D]$. Since C_0 is maintained under submatrix taking, we may pick the closed k-sum so that A^1 is as small as possible. But then the above process would have detected that closed k-sum and would have reduced B. The case where the component $[D|A^2]$ is in C_0 is argued using a case with minimal A^2.

We prove (10.4.20)(ii) with the following polynomial solution algorithm. Let $B \in C$ be given. If $B \in C_0$, then the SAT centrality of C_0 supplies a polynomial algorithm for the SAT problem for B. If $B \notin C_0$, then B is, for some $k \leq 3$, a closed k-sum where one of the components is in C_0. We solve the SAT instance for B with Algorithm SOLVE CLOSED SUM SAT (10.4.18) by solving several SAT instances involving the component in C_0 and solving one SAT instance involving the second component. Since any closed k-sum is proper, that second component is smaller than B, and we can invoke recursion. The entire solution algorithm is polynomial, since it invokes polynomial algorithms a polynomial number of times.   ☐

10.5 Extensions

Section 10.3 notes that Algorithm CLOSED 2-SEPARATION (10.3.8) and Algorithm CLOSED 3-SEPARATION (10.3.9) may be speeded up by replacing the condition $|\overline{Y}_2| = 1$ imposed on the candidate matrices $\overline{B}$ by $|\overline{Y}_2| = 0$. Validity of the change follows from Lemma (10.3.7)(b).

There is another algorithm for finding closed 2-separations that is faster than Algorithm CLOSED 2-SEPARATION (10.3.8) even when the above-mentioned improvement is made in the latter scheme. Recall from Section 2.6 that a matrix is simple if no row or column has less than two nonzeros and if there are no parallel rows or columns. Let us assume that the given matrix B has no 1-separation. The alternate algorithm for finding closed 2-separations consists of reduction of B to a simple matrix and subsequent application of Algorithm GF(3)-2-SEPARATION (3.5.26). We omit details, but point out that the reduction of B to a simple matrix may already produce a closed 2-separation.

The reader may wonder why Section 10.3 does not give the above alternate algorithm instead of Algorithm CLOSED 2-SEPARATION (10.3.8). The alternate algorithm involves several special cases plus reduction and expansion steps, and it seemingly does not lend itself to a short description. In addition, Algorithm CLOSED 2-SEPARATION (10.3.8) is a nice stepping stone toward the more complicated Algorithm CLOSED 3-SEPARATION (10.3.9), and it is convenient for the proof of the SAT centrality result of Section 10.4.

The algorithms of Section 10.3 for finding closed k-separations for $k \leq 3$, as well as the SAT centrality result of Section 10.4 for matrix classes created by repeated closed k-sums with $k \leq 3$, may be extended to the case where k is bounded by some constant. We sketch the arguments.

The finiteness of the field GF(3) supports the following claim.

(10.5.1) Lemma. *For any given $k \geq 1$, there exist a finite number of Boolean closed matrices over GF(3) with GF(3)-rank equal to $k - 1$, say, $F^{k,1}$, $F^{k,2}, \ldots, F^{k,n(k)}$, such that any Boolean closed matrix over GF(3) with GF(3)-rank equal to $k - 1$ is, up to column scaling and removal of duplicate rows and columns, a submatrix of at least one of the matrices $F^{k,i}$.*

One may use the matrices $F^{k,1}$, $F^{k,2}, \ldots, F^{k,n(k)}$ instead of the matrices F^1, F^2, and F^3 of (10.3.2) in an appropriately adapted version of Algorithm CLOSED 3-SEPARATION (10.3.9) to identify closed k-separations. The algorithm is polynomial if k is bounded by a constant, but is not practically useful. We omit details of the algorithm, but record the claim of polynomiality.

(10.5.2) Theorem. *There is an algorithm that, for any $k \geq 1$ bounded by a constant, either finds a closed k-separation for a given input matrix B or declares that B does not have such a separation. The algorithm is polynomial in the size of B if k is bounded by a constant.*

One may adapt Theorems (10.3.11) and (10.3.12) to the case at hand, getting the following result. We omit the proof, since it is along the line of the proofs of Theorems (10.3.11) and (10.3.12).

(10.5.3) Theorem. *There is an algorithm that finds all closed k-separations of a given input matrix B with A^1 minimal and A^2 maximal, as well as all closed k-separations with A^1 maximal and A^2 minimal. The algorithm is polynomial in the size of B if k is bounded by a constant.*

Theorem (10.5.3) and the discussion of Section 10.4 validate the following extension of Theorem (10.4.21).

(10.5.4) Theorem. *Let C_0 be a SAT central class of matrices. Define C to be the class of matrices created from C_0 by repeated closed k-sum steps where k is bounded by a constant. Then C is SAT central.*

In the next chapter, we meet a sum called augmented sum that is more complex than the sums described so far.

Chapter 11

Augmented Sum

11.1 Overview

The component matrices of the sums of Chapters 9 and 10 are submatrices of the given matrix B. In this chapter, we use a more elaborate way of constructing the component matrices B^1 and B^2 from a partitioned B. That is, we replace certain submatrices of B by other matrices. In the case of B^2, the replacement matrices generally do not occur in B. For this reason, we call B an augmented sum of B^1 and B^2.

Section 4.7 classifies each sum according to worst-case upper bounds on the number of b^1- and b^2-satisfiability problems for certain column submatrices $\overline{B}^1$ and $\overline{B}^2$ of B^1 and B^2 that may have to be solved by the SAT algorithm we have developed for that sum. If that upper bound is 1 for both $\overline{B}^1$ and $\overline{B}^2$, the sum is said to be of type I. If the upper bound is at least 2 for $\overline{B}^1$ and is 1 for $\overline{B}^2$, then the sum is of type II. In the remaining case, where both upper bounds are at least 2, the sum is of type III. It turns out that the augmented sum is of type II.

We emphasize that the augmented sum does not apply to the MINSAT problem.

The presentation proceeds as follows. Section 11.2 contains precise definitions of the augmented sum and related concepts.

Section 11.3 presents an algorithm for finding augmented sums.

Section 11.4 provides a solution algorithm for the SAT instances involving augmented sums. There it is shown that the augmented sum is of type II.

The final section, 11.5, supplies extensions and references.

11.2 Definitions

We define the augmented sum and related concepts.

Augmented Separation

Let B be a matrix over $\mathbb{B}$ of the form

(11.2.1)

$$
B = \begin{array}{c|c|c}
 & Y_1 & Y_2 \\
\hline
X_1 & A^1 & E \\
\hline
X_2 & D & A^2
\end{array}
$$

Matrix B with augmented separation

where the submatrices A^1 and A^2 are nonempty. Then $(X_1 \cup Y_1, X_2 \cup Y_2)$ constitutes an *augmented separation* of B.

Augmented Decomposition

Collect the nonzero rows of the submatrix D of B, say, indexed by $X_{21} \subseteq X_2$, in a matrix D^1, and collect the nonzero columns of E, say, indexed by $Y_{21} \subseteq Y_2$, in a matrix E^1. Let $X_{22} = X_2 - X_{21}$ and $Y_{22} = Y_2 - Y_{21}$. Hence,

(11.2.2)

$$
D = \begin{array}{c|c}
 & Y_1 \\
\hline
X_2 \begin{array}{c} X_{21} \\ X_{22} \end{array} & \begin{array}{c} D^1 \\ 0 \end{array}
\end{array}
\qquad
E = \begin{array}{c|cc}
 & \multicolumn{2}{c}{Y_2} \\
 & Y_{21} & Y_{22} \\
\hline
X_1 & E^1 & 0
\end{array}
$$

Partition of D and E

Define k to be equal to the number of rows of D^1, plus the number of columns of E^1, plus 1. Thus,

(11.2.3)
$$
k = |X_{21}| + |Y_{21}| + 1
$$

Let $\tilde{D}^1$ be the $|X_{21}| \times |X_{21}|$ identity matrix with rows indexed by X_{21} and with columns indexed by an arbitrarily selected set $\tilde{Y}_1$. Note that this choice of $\tilde{D}^1$ guarantees that, regardless of the form of D^1, we have

(11.2.4)
$$
\text{range}(\tilde{D}^1) \supseteq \text{range}(D^1)
$$

That feature is needed later.

We adjoin an appropriately sized zero matrix to $\tilde{D}^1$ to get the following matrix $\tilde{D}$.

(11.2.5)
$$\tilde{D} = \begin{array}{c} \\ X_2 \end{array} \begin{array}{c} \tilde{Y}_1 \\ \hline \begin{array}{c|c} X_{21} & \tilde{D}^1 \\ \hline X_{22} & 0 \end{array} \end{array}$$

Matrix $\tilde{D}$

Define F to be the $\{\pm 1\}$ matrix whose rows consist of all possible $\{\pm 1\}$ vectors with $|\tilde{Y}_1| + |Y_{21}|$ entries. Clearly, F has $2^{|\tilde{Y}_1| + |Y_{21}|}$ rows. We partition F into column submatrices $\tilde{F}$ and $\tilde{E}^1$, then add index sets $\tilde{X}_1$, $\tilde{Y}_1$, and Y_{21}, getting

(11.2.6)
$$F = \begin{array}{c} \\ \tilde{X}_1 \end{array} \begin{array}{cc} \tilde{Y}_1 & Y_{21} \\ \hline \tilde{F} & \tilde{E}^1 \end{array}$$

Partitioned matrix F

We adjoin a zero matrix to the submatrix $\tilde{E}^1$ of F to obtain the following matrix $\tilde{E}$.

(11.2.7)
$$\tilde{E} = \begin{array}{c} \\ \\ \tilde{X}_1 \end{array} \begin{array}{c} Y_2 \\ \hline \begin{array}{c|c} Y_{21} & Y_{22} \\ \hline \tilde{E}^1 & 0 \end{array} \end{array}$$

Matrix $\tilde{E}$

Derive a matrix B^1 from B of (11.2.1) by replacing the submatrices D and E by D^1 and E^1, respectively, defined above and by replacing A^2 by a zero matrix of suitable size.

Obtain a matrix B^2 from B by replacing the submatrices A^1, D, and E by the matrices $\tilde{F}$, $\tilde{D}$, and $\tilde{E}$ given by (11.2.6), (11.2.5), and (11.2.7), respectively. Accordingly,

(11.2.8)
$$B^1 = \begin{array}{c} \\ X_1 \\ \hline X_{21} \end{array} \begin{array}{cc} Y_1 & Y_{21} \\ \hline A^1 & E^1 \\ \hline D^1 & 0 \end{array} \qquad B^2 = \begin{array}{c} \\ \tilde{X}_1 \\ \hline X_2 \end{array} \begin{array}{cc} \tilde{Y}_1 & Y_2 \\ \hline \tilde{F} & \tilde{E} \\ \hline \tilde{D} & A^2 \end{array}$$

Matrices B^1 and B^2

We pause to motivate the above definitions. Suppose for a given vector b we want to decide b-satisfiability of B. We prove in Section 11.4 that this may be accomplished as follows. We solve a number of satisfiability problems involving B^1. Based on the outcomes of the latter satisfiability problems, we construct and solve one satisfiability problem for B^2. The solution of the latter problem, plus the solution for one of the satisfiability problems involving B^1, turns out to be a solution for the original satisfiability problem for B.

For the subsequent discussion, it is convenient that we display B, B^1, and B^2 of (11.2.1) and (11.2.8) so that the relationships between these matrices and the matrices D^1, E^1, $\tilde{D}^1$, $\tilde{E}^1$, $\tilde{F}$ are simultaneously exhibited. The matrix B becomes

(11.2.9)

$$
B = \begin{array}{c|c|cc}
 & Y_1 & \multicolumn{2}{c}{Y_2} \\
 & & Y_{21} & Y_{22} \\
\hline
X_1 & A^1 & E^1 & 0 \\
\hline
X_2\,\begin{matrix}X_{21}\\X_{22}\end{matrix} & \begin{matrix}D^1\\0\end{matrix} & \multicolumn{2}{c}{A^2} \\
\end{array}
$$

Matrix B

The matrix B^1 of (11.2.8) is unchanged, while B^2 of (11.2.8) is subdivided. We have

(11.2.10)

$$
B^1 = \begin{array}{c|c|c}
 & Y_1 & Y_{21} \\
\hline
X_1 & A^1 & E^1 \\
\hline
X_{21} & D^1 & 0 \\
\end{array}
\qquad
B^2 = \begin{array}{c|c|cc}
 & \tilde{Y}_1 & \multicolumn{2}{c}{Y_2} \\
 & & Y_{21} & Y_{22} \\
\hline
\tilde{X}_1 & \tilde{F} & \tilde{E}^1 & 0 \\
\hline
X_2\,\begin{matrix}X_{21}\\X_{22}\end{matrix} & \begin{matrix}\tilde{D}^1\\0\end{matrix} & \multicolumn{2}{c}{A^2} \\
\end{array}
$$

Matrices B^1 and B^2

The matrices B^1 and B^2 are the *components* of an *augmented decomposition* of B.

Augmented Composition

We may create B of (11.2.9) from B^1 and B^2 of (11.2.10). The matrix B is then obtained by an *augmented composition* of B^1 and B^2.

Augmented Sum

A matrix B is an *augmented sum* of B^1 and B^2 if the latter matrices are the components of an augmented decomposition of B or, equivalently, if B is created from B^1 and B^2 by an augmented composition. We denote that situation by $B = B^1 \boxplus_a B^2$.

We define *proper* augmented sums using $k = |X_{21}| + |Y_{21}| + 1$ of (11.2.3). If $k = 1$, we impose no additional conditions beyond those obeyed by augmented sums. That is, the submatrices A^1 and A^2 of B must be nonempty. If $k \geq 2$, we enforce conditions that assure that both components B^1 and B^2 have fewer rows and fewer columns than B. This is so for B^1 if both X_{22} and Y_{22} are nonempty. For B^2, the situation is a bit more complicated. The desired conditions are

$$(11.2.11) \qquad \begin{aligned} |X_1| > |\tilde{X}_1| \\ |Y_1| > |\tilde{Y}_1| \end{aligned}$$

We want sufficient conditions that imply the inequalities of (11.2.11) and that depend only on the index sets of B and related parameters. We determine such conditions as follows.

Since $\tilde{D}^1$ is an identity matrix and thus square, we have

$$(11.2.12) \qquad |X_{21}| = |\tilde{Y}_1|$$

Recall that F has $|\tilde{X}_1| = 2^{|\tilde{Y}_1| + |Y_{21}|}$ rows. Using $|X_{21}| = |\tilde{Y}^1|$ of (11.2.12) and $k = |X_{21}| + |Y_{21}| + 1$ of (11.2.3), we conclude

$$(11.2.13) \qquad 2^{k-1} = 2^{|X_{21}| + |Y_{21}|} = 2^{|\tilde{Y}_1| + |Y_{21}|} = |\tilde{X}_1|$$

Suppose we enforce

$$(11.2.14) \qquad \begin{aligned} |X_1| > 2^{k-1} \\ |Y_1| > |X_{21}| \\ |X_{22}| > 0 \\ |Y_{22}| > 0 \end{aligned}$$

Then X_{22} and Y_{22} are nonempty, and by (11.2.12) and (11.2.13) the inequalities of (11.2.11) hold.

To summarize, an augmented sum with $k = 1$ is always *proper*, while an augmented sum with $k \geq 2$ is *proper* if the inequalities of (11.2.14) are satisfied.

Classification Using k

We define a proper augmented sum with k given by (11.2.3) to be an *augmented k-sum*. The corresponding separation, decomposition, and composition are an *augmented k-separation*, *augmented k-decomposition*, and *augmented k-composition*, respectively. Evidently, the augmented 1-sum is the 1-sum of Section 4.7.

One may restate the conditions for an augmented k-separation of B with $k \geq 2$ in terms of the bipartite graph $\mathrm{BG}(B)$ as follows.

(11.2.15) Lemma. *Let B be a matrix over $\mathbb{B}$, with row index set X and column index set Y. For any $k \geq 2$, B has an augmented k-separation if and only if conditions (a)–(c) below are satisfied.*

(a) *The index sets X and Y of B can be partitioned into X_1, X_{21}, X_{22} and Y_1, Y_{21}, Y_{22}, respectively, such that*

$$
\begin{aligned}
|X_1| &> 2^{k-1} \\
|Y_1| &> |X_{21}| \\
|X_{22}| &> 0 \\
|Y_{22}| &> 0 \\
|X_{21}| + |Y_{21}| &= k - 1
\end{aligned}
$$

(11.2.16)

(b) *In the bipartite graph $\mathrm{BG}(B)$, each node of $X_{21} \cup Y_{21}$ is connected by at least one arc to a node of $X_1 \cup Y_1$.*

(c) *Deletion of the nodes of $X_{21} \cup Y_{21}$ from $\mathrm{BG}(B)$ disconnects the nodes of $X_1 \cup Y_1$ from those of $X_{22} \cup Y_{22}$.*

Proof. For the proof of the "only if" part, assume that B has an augmented k-separation. The corresponding partition given by (11.2.9) supplies index sets X_1, X_{21}, X_{22}, Y_1, Y_{21}, Y_{22}. By the definition of augmented k-separation, these index sets observe (11.2.16), so (a) holds. Also, the submatrix D^1 (resp. E^1) of B does not have zero rows (resp. columns), which implies (b). Finally, deletion of the rows indexed by X_{21} and of the columns indexed by Y_{21} converts B into a block diagonal matrix where each block resides within A^1 or within the reduced A^2. Hence, (c) holds. The proof of the "if" part is just as easy, and we omit details. □

Submatrix Taking

For the purpose of solving satisfiability problems, one would desire that augmented sums are in some sense maintained under submatrix taking. This is indeed so.

(11.2.17) Theorem. *Let B be an augmented sum with components B^1 and B^2. Then any submatrix $\overline{B}$ of B is contained in the submatrix A^1 of B^1 or in the submatrix A^2 of B^2, or is an augmented sum whose components $\overline{B}^1$ and $\overline{B}^2$ are submatrices of B^1 and B^2, respectively. In the third case, the index sets X_{21} and Y_{21} of B^1 and B^2 as shown in (11.2.8) and the corresponding index sets $\overline{X}_{21}$ and $\overline{Y}_{21}$ of $\overline{B}^1$ and $\overline{B}^2$ satisfy $|X_{21} \cup Y_{21}| \geq |\overline{X}_{21} \cup \overline{Y}_{21}|$.*

Proof. We use the notation of (11.2.9) and (11.2.10). Suppose that $\overline{B}$ intersects both submatrices A^1 and A^2 of B in nonempty submatrices, say, $\overline{A}^1$ and $\overline{A}^2$, respectively. We must show that $\overline{B}$ is an augmented sum whose components are submatrices of B^1 and B^2. Note that $\overline{A}^1$ or $\overline{A}^2$ may be trivial.

Define $\overline{D}^1$ (resp. $\overline{E}^1$) to be the intersection of $\overline{B}$ with D^1 (resp. E^1), minus all zero rows (resp. columns) of that intersection.

Let $\overline{X}_{21} \subseteq X_{21}$ be the row index set of $\overline{D}^1$, and let $\overline{Y}_{21} \subseteq Y_{21}$ be the column index set of $\overline{E}^1$. Thus, $|X_{21} \cup Y_{21}| \geq |\overline{X}_{21} \cup \overline{Y}_{21}|$.

Let $\tilde{D}^1$ be the $|\overline{X}_{21}| \times |\overline{X}_{21}|$ identity submatrix of $\tilde{D}^1$ with row index set $\overline{X}_{21} \subseteq X_{21}$. Let $\tilde{Y}_1 \subseteq \tilde{Y}_1$ be the column index set of $\tilde{D}^1$.

From the column submatrix of F indexed by $\tilde{Y}_1 \cup \overline{Y}_{21}$, delete duplicate rows to obtain a matrix $\overline{F}$. Partition the latter matrix into column submatrices $\tilde{F}$ and $\tilde{E}^1$, where $\tilde{E}^1$ is indexed by $\overline{Y}_{21}$.

Analogously to (11.2.10), compose $\overline{B}^1$ from $\overline{A}^1$, $\overline{D}^1$, and $\overline{E}^1$, and compose $\overline{B}^2$ from $\overline{A}^2$, $\tilde{D}^1$, $\tilde{E}^1$, and $\tilde{F}$.

By the derivation, $\overline{B}^1$ and $\overline{B}^2$ are submatrices of B^1 and B^2, respectively. It is easily checked that $\overline{B}$ is an augmented sum with components $\overline{B}^1$ and $\overline{B}^2$. □

Note that $\overline{B}$ of Theorem (11.2.17) need not be a proper augmented sum even if this is so for B.

In the next section, we describe an algorithm for finding augmented k-sums.

11.3 Decomposition Algorithm

Since the augmented 1-sum is the 1-sum of Section 4.7, Algorithm 1-SEPARATION (3.5.1) may be used to determine whether a given matrix has an augmented 1-separation.

Searching for augmented k-separations with $k \geq 2$ is more complicated. In this section, we provide a decomposition algorithm for that task. The algorithm is polynomial if k is bounded by a constant.

For an augmented k-sum to be useful, the effort for solving the satisfiability problems involving B^1 and B^2 should be less than that for the

satisfiability problem involving B. Verifying that condition while searching for augmented k-sums or, equivalently, for the related augmented k-
separations, seems to be a difficult task. So instead, one may want to
impose the restriction that k be small. Indeed, since the dense and difficult-
to-handle submatrix $F = [\tilde{F}|\tilde{E}^1]$ of B^2 has $k - 1$ columns and 2^{k-1} rows,
one should restrict the search in practical applications to augmented k-
separations where $k \leq 4$. For that case, we include enhancements of the
decomposition algorithm that result in a fast and effective method.

We outline the decomposition algorithm. The main tool is the bipartite graph $\mathrm{BG}(B)$. Lemma (11.2.15) says that B has an augmented
k-separation for given $k \geq 2$ if and only if a partition of the vertex set
of $\mathrm{BG}(B)$ into X_1, X_{21}, X_{22}, Y_1, Y_{21}, and Y_{22} exists such that conditions
(a)–(c) of Lemma (11.2.15) are satisfied.

The algorithm enumerates all possible sets $X_{21} \subseteq X$ and $Y_{21} \subseteq Y$
satisfying $|X_{21} \cup Y_{21}| = k-1$. For a given case of X_{21} and Y_{21} satisfying that
condition, the algorithm analyzes the graph $\mathrm{BG}(B)$ and either determines
sets X_1, X_{22}, Y_1, and Y_{22} such that these sets plus the already selected
X_{21} and Y_{21} fulfill Lemma (11.2.15)(a)–(c) or concludes that such sets X_1,
X_{22}, Y_1, and Y_{22} do not exist. In the first case, the algorithm has found
an augmented k-separation and stops. In the latter case, it proceeds to the
next choice of X_{21} and Y_{21}.

If none of the possible choices of X_{21} and Y_{21} results in an augmented
k-separation of B, the algorithm correctly concludes that B does not have
such a separation.

Details of the algorithm are as follows.

(11.3.1) Algorithm AUGMENTED k-SEPARATION. *Finds an
augmented k-separation with $k \geq 2$ for a matrix B over* $\mathbb{B}$ *or declares
that such a separation does not exist.*

Input: Matrix B over $\mathbb{B}$, with row index set X and column index set Y.
An integer $k \geq 2$. It is known that, for any $l \leq k - 1$, B does not have
an augmented l-separation. In particular, B does not have an augmented
1-separation and thus is connected.

Output: Either: An augmented k-separation $(X_1 \cup Y_1, X_2 \cup Y_2)$ of B, plus
the component matrices B^1 and B^2 of the corresponding augmented k-
decomposition of B. Or: "B does not have an augmented k-separation."

Complexity: Polynomial if k is bounded by a constant.

Procedure:
1. (Enumerate all choices of X_{21} and Y_{21}.) Do Steps 2 and 3 below for
   all subsets $X_{21} \subseteq X$ and $Y_{21} \subseteq Y$ satisfying $|X_{21} \cup Y_{21}| = k - 1$.
2. (Reduce $\mathrm{BG}(B)$.) Delete the nodes of $X_{21} \cup Y_{21}$ from $\mathrm{BG}(B)$, getting
   a graph G.
3. (Find a partition of G.) Search for a partition of G into disjoint graphs

H_1 and H_2, say, with node set $X_1 \cup Y_1$ (resp. $X_{22} \cup Y_{22}$) where $X_1 \subseteq X$, $Y_1 \subseteq Y$ (resp. $X_{22} \subseteq X$, $Y_{22} \subseteq Y$), such that $|X_1| > 2^{k-1}$, $|Y_1| > |X_{21}|$, $|X_{22}| > 0$, and $|Y_{22}| > 0$. If such a partition is found, go to Step 5.

4. (B has no augmented k-separation.) Declare that B has no augmented k-separation, and stop.

5. (Have augmented k-separation.) Partition B as in (11.2.9), using the sets X_1, X_{21}, X_{22}, Y_1, Y_{21}, and Y_{22} on hand. Define $\tilde{D}^1$ to be the $|X_{21}| \times |X_{21}|$ identity matrix with rows indexed by X_{21} and with columns indexed by an arbitrarily selected set $\tilde{Y}_1$. Define the submatrices $\tilde{F}$ and $\tilde{E}^1$ of B^2 via $F = [\tilde{F}|\tilde{E}^1]$, where F has column index set $\tilde{Y}_1 \cup Y_{21}$ and where F consists of all possible $\{\pm 1\}$ row vectors with $|\tilde{Y}_1 \cup Y_{21}|$ entries. Output $(X_1 \cup Y_1, X_2 \cup Y_2)$ as an augmented k-separation of B, together with component matrices B^1 and B^2 defined by the submatrices of B and the matrices $\tilde{D}^1$, $\tilde{F}$, and $\tilde{E}^1$ just computed, and stop.

Proof of Validity. We prove validity in four steps.

First, we show that any separation produced by the algorithm is indeed an augmented k-separation by verifying that conditions (a)–(c) of Lemma (11.2.15) hold. The selection of X_{21} and Y_{21} in Step 1 and the conditions imposed in Step 3 on H_1 and H_2 assure that Lemma (11.2.15)(a) and (c) are satisfied. Lemma (11.2.15)(b) demands that in BG(B) each node of $X_{21} \cup Y_{21}$ is connected by at least one arc to a node of $X_1 \cup Y_1$. Suppose this is not so. We move all nodes of $X_{21} \cup Y_{21}$ that have no arc going to $X_1 \cup Y_1$, from $X_{21} \cup Y_{21}$ to $X_{22} \cup Y_{22}$, getting, say, $X'_{21} \cup Y'_{21}$ and $X'_{22} \cup Y'_{22}$. The sets X_1, X'_{21}, X'_{22}, Y_1, Y'_{21}, and Y'_{22} define, for some $l \leq k - 1$, an augmented l-separation, which contradicts the assumption that the input matrix B does not have such a separation.

Second, we show that termination in Step 4 implies that B does not have an augmented k-separation. Assume otherwise. Let X_1, X_{21}, X_{22}, Y_1, Y_{21}, and Y_{22} define an augmented k-separation. The sets X_{21} and Y_{21} constitute one of the cases of Step 1, and the sets $X_1 \cup Y_1$ and $X_{22} \cup Y_{22}$ define one of the possible partitions of G in Step 3. Hence, Step 3 must determine an augmented k-separation, a contradiction.

Third, the construction of $\tilde{D}^1$, $\tilde{F}$, and $\tilde{E}^1$ in Step 5 directly implies that these matrices plus the submatrices of B define the desired component matrices B^1 and B^2.

Fourth, if k is bounded by a constant, then clearly all steps can be implemented in polynomial time. □

We sketch enhancements of the algorithm for the practically important cases $2 \leq k \leq 4$. The improvements reduce the number of cases of $X_{21} \cup Y_{21}$ that need to be evaluated in Steps 2 and 3, and they produce augmented k-separations that are best in a certain sense.

Reduction of the Number of Cases

If the graph G determined in Step 2 is to produce a separation, then according to Step 3 G must be disconnected. Hence, Step 1 only needs to consider cases of $X_{21} \cup Y_{21}$ resulting in a disconnected graph G. For $2 \leq k \leq 4$, the candidate sets $X_{21} \cup Y_{21}$ satisfying that condition can be readily determined as follows.

$k = 2, 3$: Step 1 must consider node sets $X_{21} \cup Y_{21}$ of BG(B) of cardinality equal to 1 or 2 whose removal produces a disconnected graph. The candidate node sets $X_{21} \cup Y_{21}$ may be found efficiently by *depth first search*; see Tarjan (1972) and Hopcroft and Tarjan (1973).

$k = 4$: The candidate node sets of Step 1, which must have cardinality equal to 3, may be efficiently found as follows. Iteratively remove one node from BG(B), getting, say, G', then find all node pairs whose removal disconnects G' by depth first search as for the case $k = 3$.

Finding a Best Augmented Separation

We still assume that $2 \leq k \leq 4$. We define best augmented k-separations via certain SAT central classes of matrices.

Let C be the class of matrices where each matrix is block diagonal and where each block is in one of the SAT central classes of Chapter 5 with very fast recognition algorithms. That is, each block is a 2SAT matrix, a hidden nearly negative matrix, a network matrix, or an extension of one of these matrices as described in Section 5.3.

Take each matrix A of C, and create from A all matrices that consist of A plus up to $k - 1$ additional rows and columns. Let C' be the class of matrices so produced. One may solve SAT instances of C' efficiently—for example, employing subregion decomposition for the adjoined rows and columns and using the SAT algorithms of Chapter 5 for the remaining block diagonal submatrix. Theorem (8.2.11) says that the SAT centrality of C implies that C' is SAT semicentral. Clearly, membership in C' can be tested in polynomial time, so C' is actually a SAT central class of matrices.

Define an augmented k-decomposition to be *best* if the corresponding component matrix B^1 has the submatrix A^1 in C and if, subject to that condition, the length of A^1 is maximum. Note that A^1 in C implies that B^1 is in the SAT central class C'. Thus, satisfiability of any submatrix of B^1 can be efficiently decided.

We describe how a best decomposition can be found. We may assume that the given matrix B is connected, since otherwise we apply the method described below to each block of B.

Assume that a candidate set $X_{21} \cup Y_{21}$ has been selected in Step 1 and that Step 2 has reduced BG(B) to G. Step 3 attempts to compose H_1 and H_2 from the connected components of G while observing certain conditions.

Note that the submatrices corresponding to the graph components in H_1 make up A^1 of B^1.

We add to Step 3 the following requirement. The submatrix A^1 of B corresponding to H_1 is to be in C, and, subject to that condition, H_1 is to have a maximum number of nodes. Clearly, H_1 and H_2 selected by these rules correspond to an augmented k-decomposition that is best relative to the choice of $X_{21} \cup Y_{21}$.

To find the overall best decomposition, we carry out Step 2 and the revised Step 3 for all candidate sets $X_{21} \cup Y_{21}$, using the earlier described method for an efficient selection of these sets.

If none of the candidate sets $X_{21} \cup Y_{21}$ produces an augmented k-separation, then we declare that the input matrix B does not have an augmented k-decomposition where the submatrix A^1 of the component B^1 is in C. Otherwise, from the augmented k-separations found, we choose one for which the length of A^1 is maximum, thus getting a best augmented k-decomposition.

The revised Algorithm AUGMENTED k-SEPARATION (11.3.1) is polynomial—in fact, very efficient—for any $2 \le k \le 4$.

The next section presents an algorithm for solving the satisfiability problem for augmented sums.

11.4 Solution Algorithm

Let B be an augmented sum with components B^1 and B^2, as given by (11.2.9) and (11.2.10). In this section, we describe an algorithm that solves any b-satisfiability problem of B by first solving several satisfiability problems involving B^1 and then solving one satisfiability problem involving B^2. The proof of validity rests on a reduction theorem given next that links any satisfiability problem involving B to one involving B^2, using certain sets and vectors that are defined via B^1.

For convenient reference, we display again B, B^1, and B^2 of (11.2.9) and (11.2.10).

(11.4.1)

$$
B =
\begin{array}{c|c|c|c}
 & Y_1 & Y_{21} & Y_{22} \\
\hline
X_1 & A^1 & E^1 & 0 \\
\hline
X_2 \begin{array}{c} X_{21} \\ X_{22} \end{array} & \begin{array}{c} D^1 \\ 0 \end{array} & \multicolumn{2}{c}{A^2} \\
\end{array}
$$

Matrix B

$$(11.4.2) \qquad B^1 = \begin{array}{c|cc} & Y_1 & Y_{21} \\ \hline X_1 & A^1 & E^1 \\ \hline X_{21} & D^1 & 0 \end{array} \qquad\qquad B^2 = \begin{array}{c|ccc} & \tilde{Y}_1 & Y_{21} & Y_{22} \\ \hline \tilde{X}_1 & \tilde{F} & \tilde{E}^1 & 0 \\ \hline X_2 \begin{array}{c} X_{21} \\ X_{22} \end{array} & \begin{array}{c} \tilde{D}^1 \\ 0 \end{array} & & A^2 \end{array}$$

Matrices B^1 and B^2

Reduction Theorem

We need a few definitions to state the result. The definitions utilize the index sets and submatrices of B of (11.4.1) and of B^1 and B^2 of (11.4.2).

Define $b = [b^1/b^2]$ to be any $\{0,1\}$ vector whose subvectors b^1 and b^2 are indexed by X_1 and X_2, respectively.

Take R to be the set of $\{0,1\}$ vectors $[d^1/e^1]$ with $d^1 \in \text{subrange}(D^1)$ and $e^1 \in \text{subrange}(E^1)$ for which some $\{\pm1\}$ vector s^1 satisfies

$$(11.4.3) \qquad \begin{aligned} A^1 \odot s^1 &\geq b^1 \ominus e^1 \\ D^1 \odot s^1 &\geq d^1 \end{aligned}$$

Let S be the set of $\{\pm1\}$ vectors $[\tilde{s}^1/s^{21}]$ for which a vector $[d^1/e^1] \in R$ exists such that

$$(11.4.4) \qquad \begin{aligned} \tilde{D}^1 \odot \tilde{s}^1 &= d^1 \\ E^1 \odot s^{21} &= e^1 \end{aligned}$$

Since $\tilde{D}^1$ is an identity matrix, the determination of $\tilde{s}^1$ is trivial.

Recall from (11.2.6) that $F = [\tilde{F}|\tilde{E}^1]$. Define f to be the following $\{0,1\}$ vector. The elements f_i of f are indexed by $\tilde{X}_1$ and are determined by the row vector $F_{i.}$ of F and by S via

$$(11.4.5) \qquad f_i = \begin{cases} 0 & \text{if } -(F_{i.})^t \in S \\ 1 & \text{otherwise} \end{cases}$$

The theorem below links the b-satisfiability problem of B with a certain satisfiability problem involving B^2.

(11.4.6) Theorem. *Let B, B^1, and B^2 be the matrices of (11.4.1) and (11.4.2), and let $b = [b^1/b^2]$ be a $\{0,1\}$ vector whose subvectors b^1 and b^2 are indexed by X_1 and X_2, respectively. Define a $\{0,1\}$ vector f via R and S using (11.4.3)–(11.4.5). Then, for any $\{\pm1\}$ vector $s^2 = [s^{21}/s^{22}]$, with s^{21} and s^{22} indexed by Y_{21} and Y_{22}, respectively, the following statements are equivalent.*

(i) *There exists a* $\{\pm1\}$ *vector* s^1 *such that*

(11.4.7)
$$(A^1 \odot s^1) \oplus (E^1 \odot s^{21}) \geq b^1$$
$$([D^1/0] \odot s^1) \oplus (A^2 \odot s^2) \geq b^2$$

(ii) *There exists a* $\{\pm1\}$ *vector* $\tilde{s}^1$ *such that*

(11.4.8)
$$(\tilde{F} \odot \tilde{s}^1) \oplus (\tilde{E}^1 \odot s^{21}) \geq f$$
$$([\tilde{D}^1/0] \odot \tilde{s}^1) \oplus (A^2 \odot s^2) \geq b^2$$

Proof. Assume that (i) holds. Let $s = [s^1/s^2] = [s^1/s^{21}/s^{22}]$. Since s is a $\{\pm1\}$ vector, $d^1 = D^1 \odot s^1$ is in subrange(D^1), and $e^1 = E^1 \odot s^{21}$ is in subrange(E^1). By (11.4.7), $A^1 \odot s^1 \geq b^1 \ominus e^1$, and trivially $D^1 \odot s^1 \geq d^1$. Using (11.4.3), we conclude that the vector $[d^1/e^1]$ is in R. Let $\tilde{s}^1$ be the unique $\{\pm1\}$ vector satisfying $\tilde{D}^1 \odot \tilde{s}^1 = d^1$.

To prove (ii), we show that $\tilde{s}^1$ and $s^2 = [s^{21}/s^{22}]$ satisfy (11.4.8). Since $d^1 = D^1 \odot s^1 = \tilde{D}^1 \odot \tilde{s}^1$, the second inequality of (11.4.7) implies the second inequality of (11.4.8). We confirm the first inequality of (11.4.8) as follows. The facts $[d^1/e^1] \in R$, $\tilde{D}^1 \odot \tilde{s}^1 = d^1$, and $E^1 \odot s^{21} = e^1$ imply by (11.4.4) that the vector $[\tilde{s}^1/s^{21}]$ is in S. By (11.4.5), the element f_i corresponding to the row vector $F_{i.}$ of F for which $-(F_{i.})^t = [\tilde{s}^1/s^{21}]$ is equal to 0. Since F contains all possible $\{\pm1\}$ row vectors, we have, for any $j \neq i$, $-(F_{j.})^t \odot [\tilde{s}^1/s^{21}] = 1$. Hence, $F \odot [\tilde{s}^1/s^{21}] \geq f$ or, equivalently, $(\tilde{F} \odot \tilde{s}^1) \oplus (\tilde{E}^1 \odot s^{21}) \geq f$, so the first inequality of (11.4.8) is satisfied.

We have shown that (i) implies (ii). For the proof of the converse, assume that (ii) holds. We claim that $[\tilde{s}^1/s^{21}]$ is in S. Suppose that this is not so. By (11.4.5), the row vector $F_{i.}$ of F for which $-(F_{i.})^t = [\tilde{s}^1/s^{21}]$ then defines f_i to be equal to 1 and satisfies $-(F_{i.})^t \odot [\tilde{s}^1/s^{21}] = 0$. Hence, the first inequality of (11.4.8) is not satisfied, a contradiction.

Since $[\tilde{s}^1/s^{21}] \in S$, there exists by (11.4.4) a vector $[d^1/e^1] \in R$ for which $\tilde{D}^1 \odot \tilde{s}^1 = d^1$ and $E^1 \odot s^{21} = e^1$. The definition of R via (11.4.3) implies that there exists a $\{\pm1\}$ vector s^1 for which the two inequalities of (11.4.3) hold. By (4.2.13), the first of the two inequalities of (11.4.3) and $E^1 \odot s^{21} = e^1$ imply the first inequality of (11.4.7).

Using $\tilde{D}^1 \odot \tilde{s}^1 = d^1$ and (4.2.13), we rewrite the second inequality of (11.4.8) as $A^2 \odot s^2 \geq b^2 \ominus [d^1/0]$. Using (4.2.20), we add the latter inequality to the inequality $[D^1/0] \odot s^1 \geq [d^1/0]$ implied by the second inequality of (11.4.3). Simplification using (4.2.17) produces $([D^1/0] \odot s^1) \oplus (A^2 \odot s^2) \geq (b^2 \ominus [d^1/0]) \oplus [d^1/0] \geq b^2$, which proves the second inequality of (11.4.7). Hence (i) holds. □

(11.4.9) Corollary. *Let B, b, R, and S be as in Theorem (11.4.6). If the SAT instance (B, b) is satisfiable, then both R and S are nonempty.*

Proof. Assume that (B, b) is satisfiable. The first part of the proof of Theorem (11.4.6) shows that R is nonempty. That conclusion and the definition of S by (11.4.4) imply that S is nonempty as well. $\qquad\square$

In the first part of the proof of Theorem (11.4.6), we needed, for any $[d^1/e^1] \in R$, a $\{\pm 1\}$ vector $\tilde{s}^1$ satisfying $\tilde{D}^1 \odot \tilde{s}^1 = d^1$. In the definition of the augmented sum in Section 11.2, we declared $\tilde{D}^1$ to be an identity matrix to guarantee the existence of that solution vector. Suppose that instead we let $\tilde{D}^1$ be a $\{0, 1\}$ matrix satisfying the condition range($\tilde{D}^1$) $\supseteq$ range(D^1) of (11.2.4). Since the range of a matrix always contains the subrange, with equality holding for $\{0, 1\}$ matrices, we also have subrange($\tilde{D}^1$) $\supseteq$ subrange(D^1). The latter relation guarantees existence of the solution $\tilde{s}^1$ for $\tilde{D}^1 \odot \tilde{s}^1 = d^1$, as desired. Hence, one could generalize the results of this chapter by replacing the definition that $\tilde{D}^1$ is an identity matrix by the requirement that $\tilde{D}^1$ is a $\{0, 1\}$ matrix satisfying range($\tilde{D}^1$) $\supseteq$ range(D^1). We have not done so to simplify the presentation.

We are ready for the solution algorithm.

Solution Algorithm

The scheme accepts a given satisfiability problem for B of the form (11.4.7) as input; computes the sets R, S, and the vector f of (11.4.3)–(11.4.5); solves one satisfiability problem involving B^2 of the form (11.4.8); and finally deduces from that information a solution for the input problem.

(11.4.10) Algorithm SOLVE AUGMENTED SUM SAT. *Solves SAT instance (B, b) where B is an augmented sum and where b is a $\{0, 1\}$ vector.*

Input: Matrix B over $\mathbb{B}$, with row index set X and column index set Y. A $\{0, 1\}$ vector b with $|X|$ entries.

An augmented decomposition of B with components B^1 and B^2, as displayed by (11.4.1) and (11.4.2). Consider b to be partitioned into b^1 and b^2 according to the index sets X_1 and X_2, respectively, of the decomposition. Define $k = |X_{21}| + |Y_{21}| + 1$.

An algorithm that solves the SAT instance of any submatrix of $[A^1/D^1]$ of B^1 in at most β_1 time. A second algorithm that solves the SAT instance of any submatrix of B^2 in at most β_2 time.

Output: Either: A solution s^* for (B, b). Or: "The given instance has no solution."

Complexity: $O(2^k \cdot k \cdot \beta_1 + \beta_2)$.

Procedure:

1. For the submatrices D^1 and E^1 of B, use Algorithm RANGE (4.3.11) to determine subrange(D^1) and subrange(E^1). Initialize $R = S = \emptyset$.

2. Do for all vectors $d^1 \in$ subrange(D^1) and $e^1 \in$ subrange(E^1):
   Solve with the appropriate given SAT algorithm the SAT instance
   $([A^1/D^1], [(b^1 \ominus e^1)/d^1])$. If the instance has a solution, then add
   $[d^1/e^1]$ to R, and store the solution as $s^1(d^1, e^1)$.

3. If $R = \emptyset$, declare that (B, b) has no solution, and stop.

4. Do for each $[d^1/e^1] \in R$:
   Add to S all $\{\pm 1\}$ vectors $[\tilde{s}^1/s^{21}]$ for which $\tilde{D}^1 \odot \tilde{s}^1 = d^1$ and $E^1 \odot s^{21} = e^1$.

5. Determine the entries of a $\{\pm 1\}$ vector f indexed by $\tilde{X}_1$ as follows.
   For $i \in \tilde{X}_1$, set $f_i = 0$ if the row vector $F_{i.}$ of F satisfies $-(F_{i.})^t \in S$,
   and set $f_i = 1$ otherwise.

6. Solve with the appropriate given SAT algorithm the SAT instance
   $(B^2, [f/b^2])$. If no solution exists, declare that (B, b) has no solution,
   and stop. Otherwise, partition the solution vector as $[\tilde{s}^{1*}/s^{2*}]$, where
   $\tilde{s}^{1*}$ is indexed by $\tilde{Y}_1$ and where s^{2*} is indexed by Y_2. Let s^{21*} be the
   subvector of s^{2*} indexed by Y_{21}.

7. Compute $d^1 = \tilde{D}^1 \odot \tilde{s}^{1*}$ and $e^1 = E^1 \odot s^{21*}$. Define $s^{1*} = s^1(d^1, e^1)$.
   Output $s^* = [s^{1*}/s^{2*}]$ as a satisfying solution for (B, b), and stop.

Proof of Validity. Steps 1, 2, 4, and 5 compute R, S, and f in agreement
with (11.4.3)–(11.4.5).

Corollary (11.4.9) validates the claim of Step 3 that $R = \emptyset$ implies
(B, b) to be unsatisfiable.

Step 6 either finds a solution $[\tilde{s}^{1*}/s^{2*}]$ for the inequality system (11.4.8)
of Theorem (11.4.6)(ii) or concludes that none exists. If that inequality
system has no solution, then by the equivalence of Theorem (11.4.6)(i) and
(ii), (B, b) has no solution. Suppose a solution is found in Step 6.

Step 7 relies on the second part of the proof of Theorem (11.4.6) to
deduce from the solution of Step 6 a solution for (B, b).

The complexity claim is argued as follows. By definition, $k = |X_{21}| + |Y_{21}| + 1$. Let A be an $m \times n$ matrix over $\mathbb{B}$. Since the BG-rank of A can-
not exceed $\min\{m, n\}$, Theorem (4.4.19) implies that Algorithm RANGE
(4.3.11) determines the subrange of A with $O(2^{\min\{m,n\}} \cdot m \cdot n)$ effort. We apply that conclusion to the computation of subrange(D^1) and subrange(E^1).
We observe that D^1 has at most k rows, that E^1 has at most k columns,
and that the number of columns of D^1 and the number of rows of E^1 must
be $O(\beta_1)$. Hence, Algorithm RANGE (4.3.11) calculates subrange(D^1) and
subrange(E^1) with $O(2^k \cdot k \cdot \beta_1)$ effort.

By Corollary (4.3.36), the cardinality of the subrange of an $m \times n$
matrix A is bounded from above by $2^{\min\{m,n\}}$, so $|\text{subrange}(D^1)| \le 2^{|X_{21}|}$,
$|\text{subrange}(E^1)| \le 2^{|Y_{21}|}$, and $|R| \le 2^{|X_{21}| + |Y_{21}|} = 2^{k-1}$. Hence, Step 2
requires at most 2^{k-1} applications of the SAT algorithm for $[A^1/D^1]$, and
total effort for that step is $O(2^k \cdot \beta_1)$.

Using standard techniques, the set S of Step 4 and the vector f of

Step 5 are determined with $O(2^k \cdot k \cdot \beta_1)$ effort. Step 6 requires $O(\beta_2)$ effort, while the effort for Step 7 is clearly dominated by that for Step 4. Total effort is therefore $O(2^k \cdot k \cdot \beta_1 + \beta_2)$. $\qquad\square$

In general, Algorithm SOLVE AUGMENTED SUM SAT (11.4.10) solves several SAT instances involving B^1 and one SAT instance involving B^2. Hence, the augmented sum is of type II. We record this fact for future reference.

(11.4.11) Theorem. *The augmented sum is of type II.*

SAT Semicentrality

Recall from (5.2.1) that a class C of matrices A over $\mathbb{B}$ is SAT semicentral if

(11.4.12)
    (i)   If $A \in C$, then any submatrix of A is also in C.
    (ii)   There is a polynomial algorithm for solving the SAT instances given by the matrices of C.

Let C_1 and C_2 be two given classes of SAT semicentral matrices. We use the following process to construct a class C of matrices over $\mathbb{B}$ that later we prove to be SAT semicentral.

Initialize $C = C_1 \cup C_2$. In all possible ways, select matrices $B^1 \in C_1$ and $B^2 \in C_2$ of the form (11.4.2) such that the two matrices may be viewed as the components of an augmented sum B for which $|X_{21} \cup Y_{21}|$ is bounded by a constant. Add each B so constructed to C.

When C is constructed as described above, we say that C is *created* from C_1 and C_2 by augmented sums where each $|X_{21} \cup Y_{21}|$ is bounded by a constant.

We should mention that the construction rules guarantee that the length of B is larger than the length of B^1. However, the length of B may be less than that of B^2. In the latter case, the difference of the two lengths is well bounded. Let $k = |X_{21} \cup Y_{21}| + 1$. Then it is easily checked that the length of B plus $k + 2^{k-1}$ exceeds the length of B^2.

The reader may wonder why we do not define C via augmented k-sums from C_1 and C_2 instead of just augmented sums. The reason is that we need C to be closed under submatrix taking and that a construction via augmented k-sums does not satisfy that requirement.

We establish SAT semicentrality for the class C created from C_1 and C_2.

(11.4.13) Lemma. *Let C be the class of matrices created from given SAT semicentral classes C_1 and C_2 by augmented sums where each $|X_{21} \cup Y_{21}|$ is bounded by a constant. Then C is SAT semicentral.*

Proof. We show (11.4.12)(i) and (ii). For (i), let $B \in C$, and define $\overline{B}$ to be an arbitrary submatrix of B. We must show that $\overline{B}$ is in C.

If $B \in C_1 \cup C_2$, then by the SAT semicentrality of C_1 and C_2, $\overline{B} \in C_1 \cup C_2$, and hence $\overline{B} \in C$.

Assume that $B \notin C_1 \cup C_2$. By the construction of C, the matrix B is an augmented sum with components $B^1 \in C_1$ and $B^2 \in C_2$. Theorem (11.2.17) supplies the following conclusion. The submatrix $\overline{B}$ is contained in the submatrix A^1 of B^1, or is contained in the submatrix A^2 of B^2, or is an augmented sum whose components $\overline{B}^1$ and $\overline{B}^2$ are submatrices of B^1 and B^2, respectively. In the first two cases, $\overline{B}$ is in $C_1 \cup C_2$ and hence is in C. In the third case, we have $\overline{B}^1 \in C_1$ and $\overline{B}^2 \in C_2$, and Theorem (11.2.17) establishes that $|X_{21} \cup Y_{21}| \geq |\overline{X}_{21} \cup \overline{Y}_{21}|$. Hence, $\overline{B}$ is in C as well.

To prove (11.4.12)(ii), we construct a polynomial solution algorithm for C, using the assumed polynomial solution algorithms for C_1 and C_2. Let the latter algorithms have upper time bounds β_1 and β_2, respectively.

In the nontrivial case, the given $B \in C$ is an augmented sum with components in C_1 and C_2, and with $|X_{21} \cup Y_{21}|$ bounded by a constant. We invoke Algorithm SOLVE AUGMENTED SUM SAT (11.4.10) to solve the SAT problem for B, using the solution algorithms for C_1 and C_2 as subroutines. We argued earlier that the length of B is larger than the length of B^1 and that the length of B plus $k + 2^{k-1}$ exceeds the length of B^2. These observations and the fact that k is bounded by a constant prove Algorithm SOLVE AUGMENTED SUM SAT (11.4.10) to be polynomial.   □

We use Lemma (11.4.13) to prove SAT semicentrality for a class of C of matrices that is constructed recursively from a given SAT semicentral class C_0 by the following process.

Initialize $C = C_0$. In the recursive step, define two classes C_1 and C_2 by declaring C_1 to be C_0 and C_2 to be the current C. Then create the next class C from C_1 and C_2 by augmented sums where each $|X_{21} \cup Y_{21}|$ is bounded by a constant. The process is stopped after a bounded number of recursive steps.

We mean the above construction of C when we say that C is *created* from C_0 by augmented sums where each $|X_{21} \cup Y_{21}|$ as well as the number of recursive construction steps is bounded by a constant.

We establish SAT semicentrality for C just defined.

(11.4.14) Theorem. *Let C_0 be a SAT semicentral class of matrices. Define C to be a class created from C_0 by augmented sums where each $|X_{21} \cup Y_{21}|$ and the number of recursive construction steps are bounded by constants. Then C is SAT semicentral.*

Proof. The result follows by induction from Lemma (11.4.13).   □

We turn to extensions and references.

11.5   Extensions and References

One may specialize the notion of augmented sum to obtain stronger results. We present two cases.

In the first case, we consider augmented k-sums where $k \leq 3$. In that situation, the submatrix F of B^2 has at most $k - 1 \leq 2$ columns and thus corresponds to 2SAT clauses. Suppose B^1 belongs to a SAT central class. Further suppose that B^2 also is in that SAT central class or has an augmented k-decomposition with $k \leq 3$. Under suitable assumptions that support recursion, the SAT problem for B is then easily solved. Using different concepts and terminology, one such class is constructed by Knuth (1990) and then extended to a larger class by Hansen, Jaumard, and Plateau (1993). In the latter class, the submatrix $[A^1|E^1]$ of B represents 2SAT clauses plus at most one general CNF clause, the submatrix D^1 is trivial, and the submatrix E^1 consists of at most two columns. Then B can be decomposed into B^1 and B^2 where B^1 is a 2SAT matrix except for one row and where B^2 is obtained from a submatrix of B by the addition of 2SAT rows. The conclusions remain valid if one demands that just A^1, and not $[A^1|E^1]$, is a 2SAT matrix except for at most one general row.

In the second case, we assume that the submatrix $[A^1/D^1]$ of an augmented k-sum B can be column scaled such that A^1 becomes nearly negative and D^1 becomes nonpositive. To simplify the notation, we suppose that A^1 and D^1 are already of that form. If E^1 is zero, then B is a monotone sum and should be treated as discussed in Chapter 9. Assume that E^1 is nonzero.

We make repeated use of the inequalities of (11.4.3), which are

$$A^1 \odot s^1 \geq b^1 \ominus e^1$$
(11.5.1)
$$D^1 \odot s^1 \geq d^1$$

Let e^1 be any vector of subrange(E^1). Suppose the inequality $A^1 \odot s^1 \geq b^1 \ominus e^1$ has a solution. By Theorem (5.5.2), Algorithm SOLVE NEARLY NEGATIVE SAT OR MINSAT (5.5.1) finds one such solution, say, s^{1*}, that, in the terminology of Section 5.5, is minimum with respect to *True*. The latter feature and the nonpositivity of D^1 imply the following conclusions, where $d^{1*} = D^1 \odot s^{1*}$. First, $s^1 = s^{1*}$ satisfies both inequalities of (11.5.1) when d^1 is chosen as $d^1 = d^{1*}$. Second, for any d^1 obeying, for some row index i, $d_i^1 > d_i^{1*}$, the inequalities of (11.5.1) have no solution.

The conclusions support the following simplification of Steps 1, 2, and 7 of Algorithm SOLVE AUGMENTED SUM SAT (11.4.10) to Steps 1', 2', and 7', respectively, below. Validity follows from a suitably adapted Theorem (11.4.6).

1'. For the submatrix E^1 of B, use Algorithm RANGE (4.3.11) to determine subrange(E^1). Initialize $R = S = \emptyset$.

2'. Do for all vectors $e^1 \in$ subrange(E^1):
Solve with Algorithm SOLVE NEARLY NEGATIVE SAT OR MIN-SAT (5.5.1) the SAT instance $(A^1, b^1 \ominus e^1)$. If the instance has a solution, then store it as $s^1(e^1)$, compute $d^1 = D^1 \odot s^1(e^1)$, and add $[d^1/e^1]$ to R.

7'. Compute $e^1 = E^1 \odot s^{21*}$. Define $s^{1*} = s^1(e^1)$. Output $s^* = [s^{1*}/s^{2*}]$ as a satisfying solution for (B, b), and stop.

The next chapter concerns a sum called linear sum.

Chapter 12

Linear Sum

12.1 Overview

This chapter introduces a sum called linear sum. Such a sum may have any number of components and may be used to solve both the SAT and MINSAT problems. These facts, plus the ease with which one may detect computationally attractive linear sums, make the linear sum the most general and most versatile of the sums discussed in this book. In fact, linear sums can be defined for any ID-system and thus are useful for the solution of combinatorial problems whose instances can be formulated as inequality systems over some ID-system.

Section 4.7 classifies sums with two components B^1 and B^2 according to worst-case upper bounds on the number of b^1- and b^2-satisfiability problems for certain column submatrices $\overline{B}^1$ and $\overline{B}^2$ of B^1 and B^2 that may have to be solved by the SAT algorithm we have developed for that sum. If that upper bound is 1 for both $\overline{B}^1$ and $\overline{B}^2$, the sum is said to be of type I. If the upper bound is at least 2 for $\overline{B}^1$ and is 1 for $\overline{B}^2$, then the sum is of type II. In the remaining case, where both upper bounds are at least 2, the sum is of type III. It turns out that the linear sum with two components is of type III. We proceed as follows.

In Section 12.2, we define the linear sum and related notions.

In Section 12.3, we develop algorithms for detecting linear sums.

In Section 12.4, we present an algorithm for solving SAT and MINSAT instances involving linear sums. At that time, the linear sum with two components is shown to be of type III.

The final section, 12.5, includes extensions.

12.2 Definitions

This section defines linear sums with any number of components by a direct definition and also by a recursive construction. At the same time, a number of related concepts are introduced.

We need a convention about sets to simplify the presentation. Let $p \geq 1$ be given. Suppose $Z_1, Z_2, \ldots, Z_p$ are sets that have been introduced by some definition. Then we take any set Z_i for which $i < 1$ or $i > p$ to be empty.

Linear Separation

Let B be a matrix over $\mathbb{B}$, with row index set X and column index set Y. Suppose that, for some $p \geq 2$, X and Y have been partitioned into $X_1, X_2, \ldots, X_p$ and $Y_1, Y_2, \ldots, Y_p$, respectively, where for all i, $X_i \cup Y_i$ is nonempty. Let these partitions induce the following partition of B.

(12.2.1)

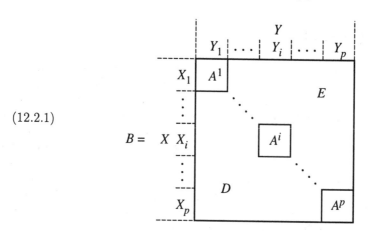

Matrix B with linear separation

Then $(X_1 \cup Y_1, X_2 \cup Y_2, \ldots, X_p \cup Y_p)$ is a *linear separation* of B.

Note the two areas of B labeled D and E. They contain the entries of B that are not part of any one of the submatrices $A^1, A^2, \ldots, A^p$.

Linear Decomposition

For $i = 1, 2, \ldots, p$, we introduce submatrices V^i and W^i of B. The matrix V^i (resp. W^i) is indexed by $\cup_{j<i} X_j$ and $\cup_{j<i} Y_j$ (resp. $\cup_{j>i} X_j$ and $\cup_{j>i} Y_j$). Thus,

(12.2.2)

$$V^i = \begin{array}{c|cccc} & Y_1 & \cdots & Y_{i-1} \\ \hline X_1 & A^1 & & & \\ \vdots & & \ddots & 0,\pm1 & \\ X_{i-1} & 0,\pm1 & & & A^{i-1} \end{array} \qquad W^i = \begin{array}{c|cccc} & Y_{i+1} & \cdots & Y_p \\ \hline X_{i+1} & A^{i+1} & & & \\ \vdots & & \ddots & 0,\pm1 & \\ X_p & 0,\pm1 & & & A^p \end{array}$$

Submatrices V^i and W^i

According to the earlier established convention about sets, both V^1 and W^p have empty row and column index sets and thus are empty matrices.

For $i = 1, 2, \ldots, p$, we derive a matrix B^i from B of (12.2.1) by replacing the submatrices V^i and W^i by zero matrices and by partitioning the unaffected portions of D and E into submatrices D^{i1}, D^{i2}, D^{i3} and E^{i1}, E^{i2}, E^{i3}, respectively. The precise form of B^i is

(12.2.3)

$$B^i = \begin{array}{c|c|c|c} & \multicolumn{3}{c}{Y} \\ & Y_1 \ldots & Y_i & Y_p \\ \hline X_1 \\ \vdots & 0 & E^{i1} & E^{i2} \\ \hline X\ X_i \\ \vdots & D^{i3} & A^i & E^{i3} \\ \hline \vdots \\ X_p & D^{i2} & D^{i1} & 0 \end{array}$$

Matrix B^i

We emphasize that any entry of B^i outside the explicitly shown zero matrices is equal to the corresponding entry of B. The matrices $B^1, B^2, \ldots, B^p$ are the *components* of a *linear decomposition* of B.

For the case $p = 2$, the matrix B is

(12.2.4)

$$B = \begin{array}{c|c|c} & Y_1 & Y_2 \\ \hline X_1 & A^1 & E \\ \hline X_2 & D & A^2 \end{array}$$

Matrix B producing two components

Comparing B^i of (12.2.3) with B of (12.2.4), we see that for this special case $D^{11} = D^{23} = D$ and $E^{13} = E^{21} = E$. Hence, the components B^1 and B^2 are

(12.2.5)

$$B^1 = \begin{array}{c|c|c} & Y_1 & Y_2 \\ \hline X_1 & A^1 & E \\ \hline X_2 & D & 0 \end{array} \qquad B^2 = \begin{array}{c|c|c} & Y_1 & Y_2 \\ \hline X_1 & 0 & E \\ \hline X_2 & D & A^2 \end{array}$$

Components B^1 and B^2

We list elementary facts about the components $B^1, B^2, \ldots, B^p$.

(12.2.6) Lemma. *For any $p \geq 2$, the components $B^1, B^2, \ldots, B^p$ satisfy (a) and (b) below.*

(a) *The following matrices are trivial or empty: the submatrices D^{12}, D^{13}, E^{11}, and E^{12} of B^1 and the submatrices D^{p1}, D^{p2}, E^{p2}, and E^{p3} of B^p.*

(b) *For $i = 1, 2, \ldots, p - 1$,*

(12.2.7)
$$[D^{i+1,3}/D^{i+1,2}] = [D^{i2}|D^{i1}]$$
$$[E^{i+1,1}|E^{i+1,2}] = [E^{i2}/E^{i3}]$$

Proof. Since V^1 and W^p are empty matrices, the submatrices of B^1 and B^p listed in part (a) must be trivial or empty.

A comparison of B^i of (12.2.3) with a correspondingly partitioned B^{i+1} proves part (b). □

Linear Composition

We may derive B of (12.2.1) from the matrices B^i of (12.2.3) in the obvious way. The matrix B is then obtained by a *linear composition* of $B^1, B^2, \ldots, B^p$.

Linear Sum

A matrix B is a *linear sum* of $B^1, B^2, \ldots, B^p$ if the latter matrices are the components of a linear decomposition of B or, equivalently, if B is created from $B^1, B^2, \ldots, B^p$ by a linear composition. We denote that situation by $B = B^1 \boxplus_l B^2 \boxplus_l \ldots \boxplus_l B^p$.

A linear sum is *proper* if, for $i = 1, 2, \ldots, p$, the submatrix A^i of B is nontrivial and nonempty, that is, if A^i has at least one entry.

Classification Using BG

We classify a subset of the proper linear sums using the system BG. The definitions are an extension of the concepts of matroid separation and matroid sum discussed in Sections 3.4 and 3.6. Suppose B is a proper linear sum given by (12.2.1). For $i = 1, 2, \ldots, p$, we use the submatrices D^{i1}, D^{i2}, D^{i3} and E^{i1}, E^{i2}, E^{i3} of B^i of (12.2.3) to define

$$(12.2.8) \quad \begin{aligned} \delta_i &= \text{BG-rank}([D^{i3}/D^{i2}]) + \text{BG-rank}([E^{i1}|E^{i2}]) + 1 \\ \epsilon_i &= \text{BG-rank}([D^{i2}|D^{i1}]) + \text{BG-rank}([E^{i2}/E^{i3}]) + 1 \end{aligned}$$

and

$$(12.2.9) \quad k = \max_i \{\delta_i, \epsilon_i\}$$

Let B be a proper linear sum; that is, B is a linear sum where each A^i contains at least one entry. If, for $i = 1, 2, \ldots, p$, the length of A^i is larger than both δ_i and ϵ_i, that is, if

$$(12.2.10) \quad |X_i \cup Y_i| \geq \max\{\delta_i, \epsilon_i\} + 1$$

then B has a *linear k-separation*. The linear decomposition, composition, and sum that correspond to a linear k-separation are called *linear k-decomposition*, *linear k-composition*, and *linear k-sum*, respectively. We stress that these definitions are based on a linear sum that is proper.

Suppose $k = 1$. By (12.2.8) and (12.2.9), for $i = 1, 2, \ldots, p$, $\delta_i = \epsilon_i = 1$, so D^{i1}, D^{i2}, D^{i3}, and E^{i1}, E^{i2}, E^{i3} must be zero matrices. Equivalently, the areas D and E in B of (12.2.1) contain zeros only. Hence, B is a block diagonal matrix with A^1, A^2, \ldots, A^p as blocks and may be constructed by repeated 1-sums from the blocks.

We establish some facts about δ_i and ϵ_i of (12.2.8) and about k of (12.2.9).

(12.2.11) Lemma. *For $i = 1, 2, \ldots, p-1$,*

$$(12.2.12) \quad \delta_{i+1} = \epsilon_i$$

Furthermore,

$$(12.2.13) \quad \delta_1 = \epsilon_p = 1$$

and

$$(12.2.14) \quad k = \max_i \{\delta_i\}$$

Proof. Lemma (12.2.6) plus the definition (12.2.8) of δ_i and ϵ_i establishes (12.2.12) and (12.2.13). The definition (12.2.9) of k plus (12.2.12) and (12.2.13) yields (12.2.14). ∎

Recursive Construction

One may derive any linear sum with $p \geq 3$ components by a recursive construction that uses linear decompositions with two components. Conversely, one may reduce any linear sum with $p \geq 3$ components to one with fewer components by recursive composition involving two components at a time.

We provide details for the latter process and thus implicitly for the former one as well. In a typical situation, B is the matrix of (12.2.1) and $p \geq 3$. We arbitrarily select i such that $1 \leq i \leq p-1$. In B, we replace the submatrices V^i and W^{i+1} by zero matrices to get the following matrix C.

(12.2.15)

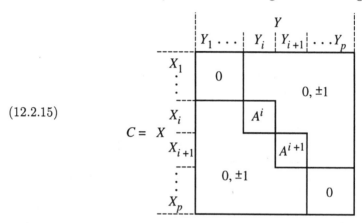

Matrix C

Comparing C with B^i and B^{i+1} defined via (12.2.3), we see that C is the linear sum of B^i and B^{i+1}. Furthermore, B is a linear sum with components $B^1, \ldots, B^{i-1}, C, B^{i+2}, \ldots, B^p$; if $i = 1$ (resp. $i = p-1$), then $B^1, \ldots, B^{i-1}$ (resp. $B^{i+2}, \ldots, B^p$) should be omitted from the list. We conclude that we may reduce a linear sum B with p components to one with $p-1$ components, by replacing two consecutive components B^i and B^{i+1} by the linear sum of those two components.

The next theorem extends the above observations and links them to linear k sums.

(12.2.16) Theorem. *For some $p \geq 3$, let B be a linear sum with components $B^1, B^2, \ldots, B^p$. Suppose that i and j satisfy $1 \leq i < j \leq p$ and that $i > 1$ or $j < p$. Then (a)–(c) below hold.*

(a) *The $j - i + 1$ component matrices $B^i, B^{i+1}, \ldots, B^j$ constitute by themselves the components of some linear sum, say, C.*

(b) *The matrix B is a linear sum with $p+i-j$ components $B^1, \ldots, B^{i-1}$, $C, B^{j+1}, \ldots, B^p$; if $i = 1$ (resp. $j = p$), $B^1, \ldots, B^{i-1}$ (resp. $B^{j+1}, \ldots, B^p$) should be omitted from the list.*

(c) *Suppose B is a linear k-sum of B^1, B^2, ..., B^p. Then, for some $k' \leq k$, B is a linear k'-sum of B^1, ..., B^{i-1}, C, B^{j+1}, ..., B^p; if $i = 1$ (resp. $j = p$), B^1, ..., B^{i-1} (resp. B^{j+1}, ..., B^p) should be omitted from the list.*

Proof. Parts (a) and (b) follow from the earlier observations and induction.

We show part (c). Since B is a proper linear sum of B^1, B^2, ..., B^p, it is also a proper linear sum of B^1, ..., B^{i-1}, C, B^{j+1}, ..., B^p.

Let δ_i and ϵ_i for B^i be given by (12.2.8), and, analogously, δ_j and ϵ_j for B^j. It is easy to see that the corresponding values for C, say, δ_C and ϵ_C, are equal to δ_i and ϵ_j, respectively. This implies that k of (12.2.14) for B^1, B^2, ..., B^p is at least as large as the corresponding k' for B^1, ..., B^{i-1}, C, B^{j+1}, ..., B^p. Furthermore, since the inequalities of (12.2.10) hold for B^1, B^2, ..., B^p, the corresponding inequalities are readily seen to hold for B^1, ..., B^{i-1}, C, B^{j+1}, ..., B^p.

The above arguments establish that B is a linear k'-sum of B^1, ..., B^{i-1}, C, B^{j+1}, ..., B^p with $k' \leq k$.   □

We call the linear sum B with components B^1, B^2, ..., B^p a *refinement* of the linear sum with components B^1, ..., B^{i-1}, C, B^{j+1}, ..., B^p. In addition, the linear separation of B corresponding to B^1, B^2, ..., B^p is a *refinement* of the linear separation corresponding to B^1, ..., B^{i-1}, C, B^{j+1}, ..., B^p.

Submatrix Taking

For the purpose of solving the SAT or MINSAT problem, one would desire that linear sums are in a certain sense maintained under submatrix taking. This is indeed the case.

(12.2.17) Theorem. *Let B be a linear sum with components B^1, B^2, ..., B^p. Then any nonempty submatrix $\overline{B}$ of B either is contained in the submatrix A^i of some component B^i or is a linear sum whose components are submatrices of the components B^i for which the intersection of A^i and $\overline{B}$ is a nonempty matrix.*

Proof. For $i = 1, 2, ..., p$, define $\overline{B}^i$ to be the submatrix of B^i having the same row and column index sets as $\overline{B}$. From $\overline{B}^1$, $\overline{B}^2$, ..., $\overline{B}^p$ so obtained, delete all matrices $\overline{B}^i$ for which the intersection of the submatrix A^i of B with $\overline{B}$ is an empty matrix.

If at most one matrix remains, the first case of the theorem applies. Otherwise, $\overline{B}$ is a linear sum having the remaining $\overline{B}^i$ as components.   □

In the next section, we develop decomposition algorithms for finding linear k-sums.

12.3 Decomposition Algorithms

According to Section 12.2, a linear 1-sum B is a block diagonal matrix where the blocks are the matrices A^1, A^2, ..., A^p. We ignore this simple case and concentrate on the situation where B is connected and where, for some $k \geq 2$, a linear k-sum with an arbitrary number of components is to be found.

We develop algorithms for finding such linear sums. The methods rely on the idea that linear sums may be obtained by a recursive construction. In the base case of that construction, a matrix B is given, and one must find a linear decomposition into two components. In the recursive step of the construction, a linear decomposition of B is given, and one must find a linear decomposition of a specified component C into two components to obtain a refined linear decomposition of B.

Each algorithm for the base case or the recursive step views the given matrix to be over BG and uses either Algorithm k-SEPARATION (3.5.20) or Heuristic BG-k-SEPARATION (3.5.34) as a subroutine. We review those two methods.

Let B be a matrix over BG, with row index set X and column index set Y. Define P_1, P_2 (resp. Q_1, Q_2) to be two disjoint subsets of X (resp. Y). Let m_1, m_2, and n be given integers. We want to find a BG-k-separation $(X_1 \cup Y_1, X_2 \cup Y_2)$ of B satisfying the following statements.

(12.3.1)
  (i) $(X_1 \cup Y_1, X_2 \cup Y_2)$ is an exact BG-k-separation of B with $k \leq n$.
  (ii) For $i = 1, 2$, $P_i \subset X_i$ and $Q_i \subset Y_i$.
  (iii) For $i = 1, 2$, $|X_i \cup Y_i| \geq |P_i \cup Q_i| + \max\{k, m_i\} + 1$.

Both Algorithm k-SEPARATION (3.5.20) and Heuristic BG-k-SEPARA-TION (3.5.34) search for such a separation while assuming that B does not have a BG-k-separation (12.3.1) for $k = 1$. Both methods are polynomial provided that m_1, m_2, and n are bounded by a constant. However, there is one important difference between the two methods. Algorithm k-SEPARATION (3.5.20) is theoretically satisfactory but practically not usable, while Heuristic BG-k-SEPARATION (3.5.34) is practically useful, but may fail to find a separation of the desired form even though one exists.

We are ready to present the algorithms that handle the base case of the construction.

Base Case of Construction

We want a linear k-separation $(X_1 \cup Y_1, X_2 \cup Y_2)$ for a given matrix B where, for given n, $k \leq n$. The associated partition of B and the corresponding components B^1 and B^2 are given by (12.2.4) and (12.2.5), respectively.

The conditions to be satisfied by the separation $(X_1 \cup Y_1, X_2 \cup Y_2)$ may be phrased as follows.

(12.3.2)

    (i) $(X_1 \cup Y_1, X_2 \cup Y_2)$ is an exact BG-k-separation of B with $k \leq n$.

    (ii)   For $i = 1, \overline{2}$, X_i and Y_i are nonempty.

    (iii)   For $i = 1, 2$, $|X_i \cup Y_i| \geq k + 1$.

Comparing (12.3.2) with (12.3.1), we see that the conditions (12.3.2)(i)–(iii) are the special case of (12.3.1)(i)–(iii) where for $i = 1, 2$, $P_i = Q_i = \emptyset$ and $m_i = 0$. Thus, we may solve the base case of the construction with Algorithm k-SEPARATION (3.5.20) or Heuristic BG-k-SEPARATION (3.5.34) using those values for P_i, Q_i, and m_i. Both methods impose the condition that B does not have a k-separation satisfying (12.3.1) for $k = 1$. That condition is clearly satisfied if B is a connected matrix.

We first give the method that solves the base case with Algorithm k-SEPARATION (3.5.20). The method is theoretically satisfactory, but practically unusable. Validity follows from the above discussion.

(12.3.3) Algorithm LINEAR k-SEPARATION. *Finds a linear k-separation of a matrix B over $\mathbb{B}$ or declares that such a separation does not exist.*

Input: Matrix B over $\mathbb{B}$, with row index set X and column index set Y. An integer n. The matrix B is connected.

Output: Either: A linear k-separation $(X_1 \cup Y_1, X_2 \cup Y_2)$ of B with minimal $k \leq n$. Or: "B does not have a linear k-separation $(X_1 \cup Y_1, X_2 \cup Y_2)$ with $k \leq n$."

Complexity: Polynomial if n is bounded by a constant.

Procedure:
1. Use Algorithm k-SEPARATION (3.5.20) with the following input. The input matrix is B, viewed to be over BG. For $i = 1, 2$, define $P_i = Q = \emptyset$ and $m_i = 0$. The input integer n is the given n.
2. If Algorithm k-SEPARATION (3.5.20) does not output a separation for B, then declare that B does not have a linear k-separation with $k \leq n$.
3. Output the separation $(X_1 \cup Y_1, X_2 \cup Y_2)$ of Algorithm k-SEPARATION (3.5.20) as the desired one, and stop.

When Heuristic BG-k-SEPARATION (3.5.34) is used instead of Algorithm k-SEPARATION (3.5.20), we get the following heuristic method.

(12.3.4) Heuristic LINEAR k-SEPARATION. *Finds a linear k-separation of a matrix B over $\mathbb{B}$ or declares that the method cannot find such a separation.*

Input: Matrix B over $\mathbb{B}$, with row index set X and column index set Y. An integer n. The matrix B is connected.

Output: Either: A linear k-separation $(X_1 \cup Y_1, X_2 \cup Y_2)$ of B with $k \leq n$. Or: "The heuristic algorithm cannot locate a linear k-separation of B with $k \leq n$."

Complexity: Polynomial if n is bounded by a constant.

Procedure:

1. Use Heuristic BG-k-SEPARATION (3.5.34) with the following input. The input matrix is B, viewed to be over BG. For $i = 1$, 2, define $P_i = Q = \emptyset$ and $m_i = 0$. The input integer n is the given n.
2. If Heuristic BG-k-SEPARATION (3.5.34) does not output a separation for B, then declare that the algorithm cannot find the desired separation of B, and stop.
3. Output the separation $(X_1 \cup Y_1, X_2 \cup Y_2)$ of Heuristic BG-k-SEPARATION (3.5.34) as the desired one, and stop.

We turn to the recursive step of the construction, where a given linear sum is to be refined.

Recursive Step of Construction

Let B be a given linear k-sum with component matrices $B^1, \ldots, B^{i-1}, C, B^{i+2}, \ldots, B^p$. Also given is an integer $n \geq k$. We want either to find a linear decomposition of C into two components, say, B^i and B^{i+1}, such that replacement of C by B^i and B^{i+1} yields a linear k'-decomposition with $k' \leq n$ and with components $B^1, B^2, \ldots, B^p$, or to conclude that such B^i and B^{i+1} do not exist.

The matrix C of (12.2.15) correctly shows the desired partition of C producing B^i and B^{i+1}. We display that matrix again.

(12.3.5)

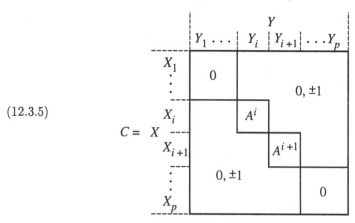

Matrix C

Define $P_1 = \cup_{j<i} X_j$, $P_2 = \cup_{j>i+1} X_j$, $Q_1 = \cup_{j<i} Y_j$, and $Q_2 = \cup_{j>i+1} Y_j$. Declare $X_C = X - (P_1 \cup P_2)$ and $Y_C = Y - (Q_1 \cup Q_2)$. Let δ_C and ϵ_C be defined for the component C of B using appropriately adapted formulas of (12.2.8).

We rephrase the task at hand. We want to partition X_C into X_i and X_{i+1}, and Y_C into Y_i and Y_{i+1}, such that the separation $(P_1 \cup Q_1 \cup X_i \cup Y_i, P_2 \cup Q_2 \cup X_{i+1} \cup Y_{i+1})$ of C satisfies the following three conditions.

First, to satisfy $k' \le n$, ϵ_i defined by (12.2.8) must obey $\epsilon_i \le n$.

Second, to assure that both submatrices A^i and A^{i+1} of C are nontrivial and nonempty, each one of the sets X_i, X_{i+1}, Y_i, and Y_{i+1} must be nonempty.

The third condition involves δ_i and ϵ_i of (12.2.8) as well as analogously defined δ_{i+1} and ϵ_{i+1}. We observe that $\delta_i = \delta_C$, $\epsilon_{i+1} = \epsilon_C$, and, by (12.2.12), $\delta_{i+1} = \epsilon_i$. The third condition consists of the cardinality requirements of (12.2.10) for the indices i and $i+1$; that is, $|X_i \cup Y_i| \ge \max\{\delta_i, \epsilon_i\}+1 = \max\{\delta_C, \epsilon_i\}+1$ and $|X_{i+1} \cup Y_{i+1}| \ge \max\{\delta_{i+1}, \epsilon_{i+1}\}+1 = \max\{\epsilon_i, \epsilon_C\} + 1$.

In summary, we want a separation $(P_1 \cup Q_1 \cup X_i \cup Y_i, P_2 \cup Q_2 \cup X_{i+1} \cup Y_{i+1})$ of C satisfying the following statements.

(12.3.6)

    (i) $(P_1 \cup Q_1 \cup X_i \cup Y_i, P_2 \cup Q_2 \cup X_{i+1} \cup Y_{i+1})$ is an exact BG-ϵ_i-separation of C with $\epsilon_i \le n$.

    (ii) X_i, X_{i+1}, Y_i, and Y_{i+1} are nonempty.

    (iii) $|X_i \cup Y_i| \ge \max\{\delta_C, \epsilon_i\} + 1$ and $|X_{i+1} \cup Y_{i+1}| \ge \max\{\epsilon_i, \epsilon_C\} + 1$.

We show that (12.3.6) is essentially (12.3.1) again. The sets P_1, P_2, Q_1, Q_2 of (12.3.6) correspond to the sets with same name in (12.3.1). The sets X_i, X_{i+1}, Y_i, and Y_{i+1} of (12.3.6) correspond to $X_1 - P_1$, $X_2 - P_2$, $Y_1 - Q_1$, and $Y_2 - Q_2$, respectively, in (12.3.1). The integers δ_C, ϵ_C, and n of (12.3.6) correspond to m_1, m_2, and n of (12.3.1).

Both Algorithm k-SEPARATION (3.5.20) and Heuristic BG-k-SEPARATION (3.5.34) require that the input matrix does not have a separation satisfying (12.3.1) with $k = 1$. For the situation at hand, a k-separation of C satisfying (12.3.6) with $k = 1$ is readily seen to be a 1-separation of B. Hence, if B is connected, such a separation of B or C does not exist.

Algorithm k-SEPARATION (3.5.20) and Heuristic BG-k-SEPARATION (3.5.34) are polynomial if m_1, m_2, and n are bounded by a constant. For the situation at hand, we have $m_1 = \delta_C \le k \le n$ and $m_2 = \epsilon_C \le k \le n$. Hence, if n is bounded by a constant, then both methods are polynomial.

We present the algorithms for refining linear k-separations. The first method uses Algorithm k-SEPARATION (3.5.20). Validity follows from the above discussion.

(12.3.7) Algorithm REFINE LINEAR k-SEPARATION. *Refines* a linear k-separation of a matrix B over $\mathbb{B}$ by determining a linear separation for a specified component, or concludes that such a refinement is not possible.

Input: Matrix B over $\mathbb{B}$, with a linear k-separation defining $p - 1 \geq 2$ components $B^1, \ldots, B^{i-1}, C, B^{i+2}, \ldots, B^p$. The matrix B (resp. any component B^j, component C) has row index set X (resp. X_j, X_C) and column index set Y (resp. Y_j, Y_C). An integer $n \geq k$. The matrix B is known to be connected.

Output: Either: For some $k' \leq n$, a linear k'-separation of B defining p components. (Necessarily, $k' \geq k$.) The components are obtained from the given components by replacing C by two components B^i and B^{i+1}. Subject to these conditions, k' is minimal. Or: "The linear k-separation of B cannot be refined to a linear k'-separation with $k' \leq n$, using a linear separation of C."

Complexity: Polynomial if n is bounded by a constant.

Procedure:

1. Use Algorithm k-SEPARATION (3.5.20) with the following input. The input matrix is C. The input sets P_1, P_2, Q_1, and Q_2 are $P_1 = \cup_{j<i} X_j$, $P_2 = \cup_{j>i+1} X_j$, $Q_1 = \cup_{j<i} Y_j$, and $Q_2 = \cup_{j>i+1} Y_j$. The input integers m_1 and m_2 are $m_1 = \delta_C$ and $m_2 = \epsilon_C$, where δ_C and ϵ_C are calculated for component C using appropriately adapted equations of (12.2.8). The input n is the integer n at hand.

2. If Algorithm k-SEPARATION (3.5.20) does not output a separation for C, then declare that the linear k-separation of B cannot be refined under the given restrictions, and stop.

3. Let the output separation for C be a BG-l-separation of the form $(P_1 \cup Q_1 \cup X_i \cup Y_i, P_2 \cup Q_2 \cup X_{i+1} \cup Y_{i+1})$, where X_i, X_{i+1} (resp. Y_i, Y_{i+1}) partition X_C (resp. Y_C). The index sets $X_1, X_2, \ldots, X_p$ and $Y_1, Y_2, \ldots, Y_p$ on hand provide the desired refined linear separation of B. Compute the value k' as $k' = \max\{k, l\}$. Output the refined k'-separation, and stop.

The second algorithm utilizes Heuristic BG-k-SEPARATION (3.5.34). Details are as follows.

(12.3.8) Heuristic REFINE LINEAR k-SEPARATION. *Refines a linear k-separation of a matrix B over $\mathbb{B}$ by determining a linear separation for a specified component, or concludes that the method cannot find such a refinement.*

Input: Matrix B over $\mathbb{B}$, with a linear k-separation defining $p - 1 \geq 2$ components $B^1, \ldots, B^{i-1}, C, B^{i+2}, \ldots, B^p$. The matrix B (resp. any component B^j, component C) has row index set X (resp. X_j, X_C) and column index set Y (resp. Y_j, Y_C). An integer $n \geq k$. The matrix B is known to be connected.

Output: Either: For some $k' \leq n$, a linear k'-separation of B defining p components. (Necessarily, $k' \geq k$.) The components are obtained from the given components by replacing C by two components B^i and B^{i+1}. Or: "The heuristic algorithm cannot locate a refinement of the linear k-separation of B to a linear k'-separation with $k' \leq n$, using a linear separation of C."

Complexity: Polynomial if n is bounded by a constant.

Procedure:

1. Use Heuristic BG-k-SEPARATION (3.5.34) with the following input. The input matrix is C. The input sets P_1, P_2, Q_1, and Q_2 are $P_1 = \cup_{j<i} X_j$, $P_2 = \cup_{j>i+1} X_j$, $Q_1 = \cup_{j<i} Y_j$, and $Q_2 = \cup_{j>i+1} Y_j$. The input integers m_1 and m_2 are $m_1 = \delta_C$ and $m_2 = \epsilon_C$, where δ_C and ϵ_C are calculated for component C using appropriately adapted equations of (12.2.8). The input n is the integer n at hand.

2. If Heuristic BG-k-SEPARATION (3.5.34) does not output a separation for C, then declare that the algorithm cannot refine the linear k-separation of B under the given restrictions, and stop.

3. Let the output separation for C be a BG-l-separation of the form $(P_1 \cup Q_1 \cup X_i \cup Y_i, P_2 \cup Q_2 \cup X_{i+1} \cup Y_{i+1})$, where X_i, X_{i+1} (resp. Y_i, Y_{i+1}) partition X_C (resp. Y_C). The index sets X_1, X_2, ..., X_p and Y_1, Y_2, ..., Y_p on hand provide the desired refined linear separation of B. Compute the value k' as $k' = \max\{k, l\}$. Output the refined k'-separation, and stop.

The next section provides an algorithm for solving SAT and MINSAT instances involving a given linear sum.

12.4 Solution Algorithm

In this section, we provide an algorithm that solves any SAT or MINSAT instance involving a linear sum B, say, with component matrices B^1, B^2, ..., B^p.

The algorithm proceeds as follows. Given is a SAT or MINSAT instance where a satisfying or least cost solution for an inequality system $B \odot s \geq b$ is to be found. The algorithm solves that SAT or MINSAT instance via p sets of SAT or MINSAT instances, where each SAT or MINSAT instance of the ith set involves a certain column submatrix $\overline{B}^i$ of component B^i and an inequality system of the form $\overline{B}^i \odot s^i \geq b^i$. The vector b^i is composed from the vector b and vectors of subrange sets of certain submatrices of B^i.

We present details following some preparations. For ease of reference, we display the linear sum B of (12.2.1), the matrices V^i and W^i of (12.2.2), and the component B^i of (12.2.3).

(12.4.1)

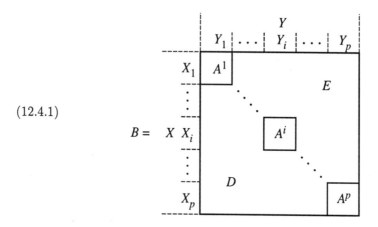

Matrix B with linear separation

(12.4.2)

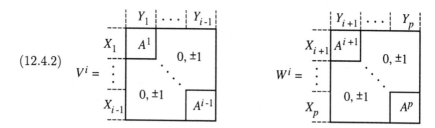

Submatrices V^i and W^i

(12.4.3)

Matrix B^i

For a fixed index i, $1 \leq i \leq p$, we obtain a comprehensive display of the relationships among the matrices when we partition B so that the submatrices A^i, D^{i1}, D^{i2}, D^{i3}, E^{i1}, E^{i2}, E^{i3} of B^i as well as V^i and W^i are exhibited. That is,

(12.4.4)

$$
B = \begin{array}{c|c|c|c|}
 & \multicolumn{3}{c}{Y} \\
 & Y_1 \ldots & Y_i & \ldots Y_p \\
\hline
X_1 \\
\vdots & V^i & E^{i1} & E^{i2} \\
\hline
X\ X_i & D^{i3} & A^i & E^{i3} \\
\hline
\vdots \\
X_p & D^{i2} & D^{i1} & W^i \\
\end{array}
$$

Matrix B

We extend the above notation to simplify the subsequent discussion. Specifically, we change the range of the index i from $1 \le i \le p$ to $1 \le i \le p+1$ and declare X_{p+1} and Y_{p+1} to be empty sets. Hence, A^{p+1} is an empty matrix, and the matrices V^{p+1}, W^{p+1}, B^{p+1}, as well as the partitioning of B in (12.4.4) for $i = p+1$, are well defined. The change permits the following extension of Lemma (12.2.6). The proof essentially is that of Lemma (12.2.6).

(12.4.5) Lemma. *For any $p \ge 2$, the components B^1, $B^2, \ldots,$ B^{p+1} satisfy (a) and (b) below.*

(a) *The following matrices are trivial or empty: the submatrices D^{12}, D^{13}, E^{11}, and E^{12} of B^1, the submatrices D^{p1}, D^{p2}, E^{p2}, and E^{p3} of B^p, and, for $j = 1, 2, 3$, the submatrices $D^{p+1,j}$ and $E^{p+1,j}$ of B^{p+1}.*
(b) *For $i = 1, 2, \ldots, p$,*

(12.4.6)
$$[D^{i+1,3}/D^{i+1,2}] = [D^{i2}|D^{i1}]$$
$$[E^{i+1,1}|E^{i+1,2}] = [E^{i2}/E^{i3}]$$

The solution algorithm to come is based on a reduction theorem that links the satisfying solutions of a given SAT or MINSAT instance involving B with the satisfying solutions of certain inequality systems involving the components B^i. We develop that result next.

Reduction Theorem

We need some definitions. Let b be a $\{0, 1\}$ vector indexed by X. For $i = 1$, $2, \ldots, p+1$, we partition b as $b = [b^{i1}/b^{i2}/b^{i3}]$, where b^{i1}, b^{i2}, and b^{i3} are indexed by $\cup_{j<i} X_j$, X_i, and $\cup_{j>i} X_j$, respectively.

We rely on the subrange of several submatrices of B of (12.4.4). Let i be a given index, where $1 \leq i \leq p+1$. For $j = 1, 2, 3$, we define d^{ij} (resp. e^{ij}) to be any vector in subrange(D^{ij}) (resp. subrange(E^{ij})). We denote by $[d^{i3}/d^{i2}]$ (resp. $[e^{i2}/e^{i3}]$) any vector in subrange($[D^{i3}/D^{i2}]$) (resp. subrange($[E^{i2}/E^{i3}]$)), where the partition of the vector is in agreement with that of the matrix. We use d^{i21} (resp. e^{i12}) for any vector in subrange($[D^{i2}|D^{i1}]$) (resp. subrange($[E^{i1}|E^{i2}]$)).

For $i = 2, 3, \ldots, p+1$, let S_i be the set of all triples $([d^{i3}/d^{i2}], e^{i12}, s^i)$ where $[d^{i3}/d^{i2}] \in$ subrange($[D^{i3}/D^{i2}]$) and $e^{i12} \in$ subrange($[E^{i1}|E^{i2}]$) and where s^i is a $\{\pm 1\}$ vector satisfying

$$
\begin{aligned}
V^i \odot s^i &\geq b^{i1} \ominus e^{i12} \\
D^{i3} \odot s^i &\geq d^{i3} \\
D^{i2} \odot s^i &\geq d^{i2}
\end{aligned}
$$

(12.4.7)

We give an alternate characterization of S_2 and relate S_{p+1} to the solutions of $B \odot s \geq b$.

(12.4.8) Lemma.

(a) For $i = 2$, the inequalities of (12.4.7) defining S_2 may be restated as $A^1 \odot s^2 \geq b^{12} \ominus e^{13}$ and $D^{11} \odot s^2 \geq d^{11}$, where $d^{11} \in$ subrange(D^{11}) and $e^{13} \in$ subrange(E^{13}).

(b) The set S_{p+1} consists precisely of the triples $(0, 0, s)$ for which $B \odot s \geq b$.

Proof. We show part (a). Lemma (12.4.5), the definition of V^i by (12.4.2), and the partitioning of b as $b = [b^{i1}/b^{i2}/b^{i3}]$ imply that $V^2 = A^1$, $b^{21} = b^{12}$, $[D^{23}/D^{22}] = D^{11}$, and $[E^{21}|E^{22}] = E^{13}$. When these equations are used in the above definition of S_i for $i = 2$, then the characterization of S_2 in part (a) results.

We prove part (b). Lemma (12.4.5)(a) says that, for $j = 1, 2, 3$, $D^{p+1,j}$ and $E^{p+1,j}$ are trivial or empty. Hence, both subrange($[D^{p+1,3}/D^{p+1,2}]$) and subrange($[E^{p+1,1}|E^{p+1,2}]$) consist of one zero vector each. In addition, $V^{p+1} = B$, and $b^{p+1,1} = b$. Let $s = s^{p+1}$. Then, for $i = p + 1$, (12.4.7) effectively is the inequality $B \odot s \geq b$, and the triples of S_{p+1} must have the claimed form.   □

For $i = 2, 3, \ldots, p$, we define R_i to be the set of all 5-tuples of the form $([d^{i3}/d^{i2}], d^{i21}, e^{i12}, [e^{i2}/e^{i3}], r^i)$ where $[d^{i3}/d^{i2}] \in$ subrange($[D^{i3}/D^{i2}]$), $d^{i21} \in$ subrange($[D^{i2}|D^{i1}]$), $e^{i12} \in$ subrange($[E^{i1}|E^{i2}]$), and $[e^{i2}/e^{i3}] \in$ subrange($[E^{i2}/E^{i3}]$) and where r^i is a $\{\pm 1\}$ vector satisfying

$$
\begin{aligned}
E^{i1} \odot r^i &\geq e^{i12} \ominus e^{i2} \\
A^i \odot r^i &\geq b^{i2} \ominus (d^{i3} \oplus e^{i3}) \\
D^{i1} \odot r^i &\geq d^{i21} \ominus d^{i2}
\end{aligned}
$$

(12.4.9)

The next lemma restates S_{i+1} in terms of vectors and matrices used in the definition of S_i and R_i.

(12.4.10) Lemma. For $i = 1, 2, \ldots, p$, the set S_{i+1} is the set of triples $(d^{i21}, [e^{i2}/e^{i3}], [s^i/r^i])$ where $d^{i21} \in$ subrange$([D^{i2}|D^{i1}])$ and $[e^{i2}/e^{i3}] \in$ subrange$([E^{i2}/E^{i3}])$ and where $[s^i/r^i]$ is a $\{\pm 1\}$ vector satisfying

$$(V^i \odot s^i) \oplus (E^{i1} \odot r^i) \geq b^{i1} \ominus e^{i2}$$
(12.4.11)
$$(D^{i3} \odot s^i) \oplus (A^i \odot r^i) \geq b^{i2} \ominus e^{i3}$$
$$(D^{i2} \odot s^i) \oplus (D^{i1} \odot r^i) \geq d^{i21}$$

Proof. Use (12.4.7) to state the definition of the triples of S_{i+1} in terms of V^{i+1}, s^{i+1}, and so on. Rewrite that definition using $[D^{i+1,3}/D^{i+1,2}] = [D^{i2}|D^{i1}]$ and $[E^{i+1,1}|E^{i+1,2}] = [E^{i2}/E^{i3}]$ of Lemma (12.4.5)(b), as well as the fact that V^{i+1} has by (12.4.2) and (12.4.4) a partition into two row submatrices $[V^i|E^{i1}]$ and $[D^{i3}|A^i]$. The resulting inequalities confirm the lemma. □

At long last, we are ready for the reduction theorem.

(12.4.12) Theorem. For $i = 1, 2, \ldots, p$, statements (a) and (b) below hold.

(a) If $([d^{i3}/d^{i2}], e^{i12}, s^i) \in S_i$ and $([d^{i3}/d^{i2}], d^{i21}, e^{i12}, [e^{i2}/e^{i3}], r^i) \in R_i$, then $(d^{i21}, [e^{i2}/e^{i3}], [s^i/r^i]) \in S_{i+1}$.

(b) If $(d^{i21}, [e^{i2}/e^{i3}], [s^i/r^i]) \in S_{i+1}$, where s^i is indexed by $\cup_{j<i}Y_j$ and r_i by Y_i, then there exist vectors $[d^{i3}/d^{i2}] \in$ subrange$([D^{i3}/D^{i2}])$ and $e^{i12} \in$ subrange$([E^{i1}|E^{i2}])$ such that $([d^{i3}/d^{i2}], e^{i12}, s^i) \in S_i$ and $([d^{i3}/d^{i2}], d^{i21}, e^{i12}, [e^{i2}/e^{i3}], r^i) \in R_i$.

Proof. For part (a), we add the inequalities of (12.4.7) pairwise to those of (12.4.9) and simplify the resulting system to the form (12.4.11). Lemma (12.4.10) then supports the desired conclusion. As an example for the pairwise addition process, we carry out details for the second inequality of (12.4.11). We add the second inequality of (12.4.7) to the second inequality of (12.4.9), getting, by (4.2.20), $(D^{i3} \odot s^i) \oplus (A^i \odot r^i) \geq d^{i3} \oplus [b^{i2} \ominus (d^{i3} \oplus e^{i3})]$. Using (4.2.11) and (4.2.12), we rewrite the right-hand side of the latter inequality as $[(b^{i2} \ominus e^{i3}) \ominus d^{i3}] \oplus d^{i3} \geq b^{i2} \ominus e^{i3}$ and thus have proved the second inequality of (12.4.11).

We turn to the proof of part (b). Lemma (12.4.10) says that the triple $(d^{i21}, [e^{i2}/e^{i3}], [s^i/r^i])$ of S_{i+1} satisfies the inequalities of (12.4.11). Using the vectors r^i and s^i of the triple, define $[d^{i3}/d^{i2}] = [D^{i3}/D^{i2}] \odot s^i$, $e^{i1} = E^{i1} \odot r^i$, and $e^{i12} = e^{i1} \oplus e^{i2}$. By these definitions, $[d^{i3}/d^{i2}] \in$ subrange$([D^{i3}/D^{i2}])$ and $e^{i12} \in$ subrange$([E^{i1}|E^{i2}])$. Appropriate substitutions using those equations in the inequalities of (12.4.11) produce the inequalities of (12.4.7) and (12.4.9). We demonstrate the process using the first inequality of (12.4.11), which is $(V^i \odot s^i) \oplus (E^{i1} \odot r^i) \geq b^{i1} \ominus e^{i2}$. Since $e^{i1} = E^{i1} \odot r^i$, we have $(V^i \odot s^i) \oplus e^{i1} \geq b^{i1} \ominus e^{i2}$. Using (4.2.11) and (4.2.13), the inequality becomes $V^i \odot s^i \geq b^{i1} \ominus (e^{i1} \oplus e^{i2})$. Since $e^{i12} = e^{i1} \oplus e^{i2}$, the right-hand side of the inequality is equal to $b^{i1} \ominus e^{i12}$. Thus, $V^i \odot s^i \geq b^{i1} \ominus e^{i12}$, which is the first inequality of (12.4.7).

Finally, since $e^{i1} = E^{i1} \odot r^i$ and $e^{i12} = e^{i1} \oplus e^{i2}$, we have, using (4.2.13), $E^{i1} \odot r^i \geq e^{i12} \ominus e^{i2}$, which is the first inequality of (12.4.9).   □

Theorem (12.4.12) has the following corollary.

(12.4.13) Corollary. *The following statements are equivalent.*

(i) *The inequality $B \odot s \geq b$ has a solution.*
(ii) *S_{p+1} is nonempty.*
(iii) *For $i = 2, 3, \ldots, p$, both R_i and S_i are nonempty.*

Proof. Lemma (12.4.8)(b) says that S_{p+1} consists of the triples $(0, 0, s)$ for which s satisfies $B \odot s \geq b$. Hence, (i)⇔(ii).

Theorem (12.4.12)(a) with $i = p$ establishes (iii)⇒(ii). Finally, Theorem (12.4.12)(b) and backward induction, with $i = p$ as base case, prove (ii)⇒(iii).

□

The solution algorithm for SAT and MINSAT instances with linear sums determines certain subsets of the sets S_i called S_i^*. The latter sets satisfy the following conditions.

SAT case: Consider all triples of S_i that are identical except for the solution vector s^i; the set S_i^* contains precisely one representative of such triples.

MINSAT case: Let c be the cost vector. Consider all triples of S_i that are identical except for the solution vector s^i; the set S_i^* contains precisely one representative of such triples that minimizes $\sum c_j$, where the summation is over the indices j for which $s_j^i = 1$.

We declare that any S_i^* observing the above conditions *represents* S_i.

Clearly, any S_i^* representing S_i is nonempty if and only if S_i is nonempty. That observation plus arguments almost identical to those proving Corollary (12.4.13) establishes the following result.

(12.4.14) Corollary. *The following statements are equivalent.*

(i) *The inequality $B \odot s \geq b$ has a solution.*
(ii) *Any S_{p+1}^* representing S_{p+1} is nonempty.*
(iii) *For $i = 2, 3, \ldots, p + 1$, any S_i^* representing S_i is nonempty.*

Solution Algorithm

The algorithm first computes a set S_2^*. Then, for $i = 2, 3, \ldots, p$, the method determines a set S_{i+1}^* using the set S_i^* on hand. The final set S_{p+1}^* supplies a solution vector for the SAT or MINSAT instance involving B.

(12.4.15) Algorithm SOLVE LINEAR SUM SAT OR MINSAT.
Solves SAT *instance* (B, b) *or* MINSAT *instance* (B, b, c) *where B is a linear sum, b is a $\{0, 1\}$ vector, and c is a rational nonnegative vector.*

Input: Matrix B over $\mathbb{B}$ of size $m \times n$, with row index set X and column index set Y. A $\{0, 1\}$ vector b with m entries. In the MINSAT case, a rational nonnegative vector c with n entries.
A linear decomposition of B with $p \geq 2$ components B^1, $B^2, \ldots, B^p$ as displayed by (12.4.1) and (12.4.3). For $i = 1, 2, \ldots, p$, consider b to be partitioned into b^{i1}, b^{i2}, and b^{i3}, where b^{i1}, b^{i2}, and b^{i3} are indexed by $\cup_{j<i}X_j$, X_i, and $\cup_{j>i}X_j$, respectively. Consider c to be partitioned into $c^1, c^2, \ldots, c^p$ according to the index sets $Y^1, Y^2, \ldots, Y^p$, respectively.
A total of p SAT or MINSAT algorithms. For $i = 1, 2, \ldots, p$, the ith SAT (resp. MINSAT) algorithm solves, for any $\{0, 1\}$ vector a^i of appropriate size, the SAT instance $([E^{i1}/A^i/D^{i1}], a^i)$ (resp. the MINSAT instance $([E^{i1}/A^i/D^{i1}], a^i, c^i)$ in at most β_i (resp. γ_i) time.

Output: Either: A solution s^* for (B, b) or (B, b, c), whichever applies. Or: "The given instance has no solution."

Complexity: $O(2^k \cdot p \cdot m \cdot n + 4^k \sum_{i=1}^{p} \beta_i)$ in the SAT case and $O(2^k \cdot p \cdot m \cdot n + 4^k \sum_{i=1}^{p} \gamma_i)$ in the MINSAT case, where $k = \max\{\delta_i\}$ using, for $i = 1, 2, \ldots, p$, $\delta_i = \text{BG-rank}([D^{i3}/D^{i2}]) + \text{BG-rank}([E^{i1}|E^{i2}]) + 1$.

Procedure:
1. For $i = 2, 3, \ldots, p+1$, initialize $S_i^* = \emptyset$.
2. For $i = 2, 3, \ldots, p$, use Algorithm RANGE (4.3.11) to compute subrange$([D^{i3}/D^{i2}])$ and subrange$([E^{i1}|E^{i2}])$. (According to Lemma (12.4.5), these calculations implicitly supply, for $i = 1, 2, \ldots, p-1$, subrange$([D^{i2}|D^{i1}])$ and subrange$([E^{i2}/E^{i3}])$.)
3. Using the appropriate given SAT or MINSAT algorithm, solve for each $d^{11} \in$ subrange(D^{11}) and for each $e^{13} \in$ subrange(E^{13}) the following problem.
   SAT case: The problem is $([A^1/D^{11}], [(b^{12} \ominus e^{13})/d^{11}])$.
   MINSAT case: The problem is $([A^1/D^{11}], [(b^{12} \ominus e^{13})/d^{11}], c^1)$.
   If the problem has a solution, say, s^{2*}, then add the triple (d^{11}, e^{13}, s^{2*}) to S_2^*.
4. If S_2^* is empty, declare that the original problem has no solution, and stop.
5. Do Steps 6–8 below for $i = 2, 3, \ldots, p$.
6. Using the appropriate given SAT or MINSAT algorithm, do Step 7 for each $d^{i21} \in$ subrange$([D^{i2}|D^{i1}])$ and $[e^{i2}/e^{i3}] \in$ subrange$([E^{i2}/E^{i3}])$.
7. SAT case: Find a triple $([d^{i3}/d^{i2}], e^{i12}, s^{i*}) \in S_i^*$ such that the SAT instance $([E^{i1}/A^i/D^{i1}], [(e^{i12} \ominus e^{i2})/(b^{i2} \ominus (d^{i3} \oplus e^{i3}))/(d^{i21} \ominus d^{i2})])$ has a solution, say, r^{i*}, or determine that no such triple exists in S_i^*. If r^{i*} exists, add the triple $(d^{i21}, [e^{i2}/e^{i3}], [s^{i*}/r^{i*}])$ to S_{i+1}^*.
   MINSAT case: Let $z^* = \infty$. Do for each triple $([d^{i3}/d^{i2}], e^{i12}, s^{i*}) \in$

S_i^*: Solve the MINSAT instance $([E^{i1}/A^i/D^{i1}], [(e^{i12} \ominus e^{i2})/(b^{i2} \ominus (d^{i3} \oplus e^{i3}))/(d^{i21} \ominus d^{i2})], c^i)$; if a solution, say, r^{i*}, is found, then compute $z = \sum c_j$ where the summation is over the indices j for which $s_j^{i*} = 1$ or $r_j^{i*} = 1$; if in addition $z < z^*$, then declare z^* to have value z and $s^{i+1,*}$ to be the vector $[s^{i*}/r^{i*}]$. Once all triples of S_i^* have been processed, and if $z^* < \infty$, then add the triple $(d^{i21}, [e^{i2}/e^{i3}], s^{i+1,*})$ to S_{i+1}^*.

8. If S_{i+1}^* is empty, then declare that the original problem has no solution, and stop.

9. Output the vector s^* of the single triple $(0, 0, s^*) \in S_{p+1}^*$ as a solution of the input SAT or MINSAT instance, and stop.

Proof of Validity. We first deal with the SAT case.

In Step 3, the SAT instance $([A^1/D^{11}], [(b^{12} \ominus e^{13})/d^{11}])$ involves the inequalities $A^1 \odot s^2 \geq b^{12} \ominus e^{13}$ and $D^{11} \odot s^2 \geq d^{11}$. By Lemma (12.4.8)(a), those inequalities define S_2. Hence, Step 3 determines a set S_2^* that represents S_2.

If S_2^* is empty, then, in agreement with Corollary (12.4.14), Step 4 declares the input problem to be unsatisfiable.

We validate Steps 5–8 by induction. Suppose that, for some $i \geq 2$, a nonempty set S_i^* is on hand that represents S_i. For the case $i = 2$, such S_i^* is supplied by Step 3.

Let d^{i21} and $[e^{i2}/e^{i3}]$ be given by Step 6. Step 7 either finds a triple $([d^{i3}/d^{i2}], e^{i12}, s^{i*}) \in S_i^*$ such that the SAT instance $([E^{i1}/A^i/D^{i1}], [(e^{i12} \ominus e^{i2})/(b^{i2} \ominus (d^{i3} \oplus e^{i3}))/(d^{i21} \ominus d^{i2})])$ has a solution, say, r^{i*}, or determines that no such triple exists in S_i^*. If r^{i*} exists, the triple $(d^{i21}, [e^{i2}/e^{i3}], [s^{i*}/r^{i*}])$ is added to S_{i+1}^*.

Evidently, the inequalities of the SAT instance $([E^{i1}/A^i/D^{i1}], [(e^{i12} \ominus e^{i2})/(b^{i2} \ominus (d^{i3} \oplus e^{i3}))/(d^{i21} \ominus d^{i2})])$ are precisely those of (12.4.9), which define R_i. Hence, if a solution r^{i*} is found, then the 5-tuple $([d^{i3}/d^{i2}], d^{i21}, e^{i12}, [e^{i2}/e^{i3}], r^{i*})$ is in R_i, and by Theorem (12.4.12)(a) the triple $(d^{i21}, [e^{i2}/e^{i3}], [s^{i*}/r^{i*}])$ is in S_{i+1}. Since Step 7 adds that triple to S_{i+1}^*, we conclude that the set S_{i+1}^* on hand in Step 8 is a subset of a set that represents S_{i+1}.

We prove that S_{i+1}^* of Step 8 indeed represents S_{i+1}. Let S_{i+1} contain a triple $(d^{i21}, [e^{i2}/e^{i3}], [s^i/r^i])$. We must exhibit one such triple, say, $(d^{i21}, [e^{i2}/e^{i3}], [s^{i*}/r^{i*}])$, in S_{i+1}^*.

By Theorem (12.4.12)(b), $(d^{i21}, [e^{i2}/e^{i3}], [s^i/r^i]) \in S_{i+1}$ implies that there exist $[d^{i3}/d^{i2}]$ and e^{i12} such that $([d^{i3}/d^{i2}], e^{i12}, s^i) \in S_i$ and $([d^{i3}/d^{i2}], d^{i21}, e^{i12}, [e^{i2}/e^{i3}], r^i) \in R_i$. By the inductive assumption, S_i^* represents S_i, so for some s^{i**} there is a triple $([d^{i3}/d^{i2}], e^{i12}, s^{i**})$ in S_i^*. Consider the moment when the case of d^{i21} and $[e^{i2}/e^{i3}]$ comes up in Step 6 and is processed in Step 7. Since $([d^{i3}/d^{i2}], d^{i21}, e^{i12}, [e^{i2}/e^{i3}], r^i) \in R_i$, the SAT instance $([E^{i1}/A^i/D^{i1}], [(e^{i12} \ominus e^{i2})/(b^{i2} \ominus (d^{i3} \oplus e^{i3}))/(d^{i21} \ominus d^{i2})])$ has a solution, say, r^{i**}. If that SAT instance, with the solution r^{i**}, was produced in Step 7, then the triple $(d^{i21}, [e^{i2}/e^{i3}], [s^{i**}/r^{i**}])$ would be placed into S_{i+1}^*. Of course, that SAT instance may not be processed by Step 7. In that case, a satisfiable SAT instance arising from some other triple in S_i^* is found in Step 7. So no matter which case applies, Step 7 does place, for some s^{i*} and r^{i*}, a triple $(d^{i21}, [e^{i2}/e^{i3}], [s^{i*}/r^{i*}])$ into S_{i+1}^*.

The above arguments validate the claim that the set S_{i+1}^* on hand in Step 8 represents S_{i+1}. They also prove that if S_{i+1}^* is empty, then Step 8 correctly concludes unsatisfiability of the input problem.

Finally, the single triple $(0, 0, s^*) \in S_{p+1}^*$ on hand in Step 9 must yield s^* as a solution for the input SAT instance.

The MINSAT case is argued almost identically, except that minimizing solutions are involved in the construction of S_2^* in Step 3 and of S_{i+1}^* in Steps 7. We leave the easy adaptation to the reader.

The complexity of the algorithm crucially depends on the effort for the subrange calculations in Step 2 and for the solution of the SAT or MINSAT instances in Steps 3 and 7. We prove that the given formulas bound the effort of those steps.

Let A be any matrix over $\mathbb{B}$ that, when viewed to be over BG, has BG-rank$(A) = l$. Theorem (4.4.12) says that the cardinality of the subrange of such A is at most 2^l. Theorem (4.4.19) establishes that Algorithm RANGE (4.3.11) determines the subrange of A with $O(2^l \cdot m \cdot n)$ effort.

We apply these formulas to the case at hand, using, for $i = 1, 2, \ldots, p$, $\delta_i = \text{BG-rank}([D^{i3}/D^{i2}]) + \text{BG-rank}([E^{i1}|E^{i2}]) + 1$ and using $k = \max\{\delta_i\}$. Thus, $2^{\delta_i - 1} \geq |\text{subrange}([D^{i3}/D^{i2}])| \cdot |\text{subrange}([E^{i1}|E^{i2}])|$, and the effort for the subrange calculations of Step 2 is $O(2^k \cdot p \cdot m \cdot n)$.

We estimate the complexity of the effort for Steps 3 and 7. Define $\delta_{p+1} = 1$. For $i = 1, 2, \ldots, p$, let $\alpha_i = 2^{\delta_i + \delta_{i+1} - 2}$. Since $k = \max\{\delta_i\}$, we have, for $i = 1, 2, \ldots, p$, $\alpha_i \leq 4^k$.

By (12.2.8), (12.2.12), and the definition $\delta_{p+1} = 1$, we conclude, for $i = 1, 2, \ldots, p$, $\delta_{i+1} = \text{BG-rank}([D^{i2}|D^{i1}]) + \text{BG-rank}([E^{i2}/E^{i3}]) + 1$, and $2^{\delta_{i+1} - 1} \geq |\text{subrange}([D^{i2}|D^{i1}])| \cdot |\text{subrange}([E^{i2}/E^{i3}])|$. For $i = 1, 2, \ldots, p$, $\alpha_i = 2^{\delta_i + \delta_{i+1} - 2}$, so we have $\alpha_i = 2^{\delta_i - 1} \cdot 2^{\delta_{i+1} - 1} \geq |\text{subrange}([D^{i3}/D^{i2}])| \cdot |\text{subrange}([E^{i1}|E^{i2}])| \cdot |\text{subrange}([D^{i2}|D^{i1}])| \cdot |\text{subrange}([E^{i2}/E^{i3}])|$. Hence, Step 3 solves at most α_1 inequality systems of type (12.4.7), and, for fixed i, Steps 7 solves at most α_i inequality systems of type (12.4.9). We conclude that the effort for Steps 3 and 7 is in the SAT case $O(\sum_{i=1}^p \alpha_i \cdot \beta_i)$ and in the MINSAT case $O(\sum_{i=1}^p \alpha_i \cdot \gamma_i)$. Using $\alpha_i \leq 4^k$, we simplify these formulas to $O(4^k \sum_{i=1}^p \beta_i)$ and $O(4^k \sum_{i=1}^p \gamma_i)$, respectively.

When the above bounds on the effort of Steps 2, 3, and 7 are combined, one obtains the overall complexity formulas claimed for the algorithm. $\square$

When Algorithm SOLVE LINEAR SUM SAT OR MINSAT (12.4.15)

solves SAT instances, then according to Steps 3, 6, and 7 several SAT instances involving each component of the linear sum may have to be solved. This implies that the linear sum with two components is of type III. The next theorem states this fact.

(12.4.16) Theorem. *The linear sum is of type* III.

SAT and MINSAT Semicentrality

Recall from (5.2.1) and (5.2.2) the following definitions. A class C of matrices A over $\mathbb{B}$ is SAT semicentral if the following holds.

(12.4.17) (i)   If $A \in C$, then any submatrix of A is also in C.
           (ii) There is a polynomial algorithm for solving the SAT instances given by the matrices of C.

A class C of matrix/vector pairs (A, c), where A is over $\mathbb{B}$ and c is a rational nonnegative vector, is SAT semicentral if the following holds.

(12.4.18) (i)   If $(A, c) \in C$, then any submatrix pair of (A, c) is also in C.
           (ii) There is a polynomial algorithm for solving the MINSAT instances given by the matrix/vector pairs of C.

We have the following SAT and MINSAT semicentrality result for linear sums.

(12.4.19) Theorem.
(a) Let C_0 be a SAT semicentral class of matrices. Enlarge C_0 to a class C by adding all possible linear sums where, in the notation of (12.4.3), the column submatrix $[E^{i1}/A^i/D^{i1}]$ of each component B^i is in C_0 and where k defined by (12.2.14) is bounded by a constant. Then C is SAT semicentral.

(b) Let C_0 be a MINSAT semicentral class of matrix/vector pairs. Enlarge C_0 to a class C by adding all pairs (A, c) satisfying the following conditions. The matrix A is a linear sum where, in the notation of (12.4.3), the column submatrix $[E^{i1}/A^i/D^{i1}]$ of each component B^i and the corresponding subvector c^i of c constitute a pair $([E^{i1}/A^i/D^{i1}], c^i)$ in C_0 and where k defined by (12.2.14) is bounded by a constant. Then C is MINSAT semicentral.

Proof. The arguments for parts (a) and (b) are essentially the same, so we just establish part (a).

To prove (12.4.17)(i), we must show that, for given $A \in C$, any submatrix $\overline{A}$ of A is also in C. If $A \in C_0$, then by the SAT semicentrality of C_0, we have $\overline{A} \in C_0$ and hence $\overline{A} \in C$. So assume $A \notin C_0$. By the construction of C, A is a linear sum where, for each component B^i, the column submatrix $[E^{i1}/A^i/D^{i1}]$ is in C_0 and where k defined by (12.2.14) is bounded by a constant.

We use Theorem (12.2.17), which roughly says that linear sums are maintained under submatrix taking. Specifically, that theorem implies that any submatrix $\overline{A}$ of A is a linear sum whose components are submatrices of some of the B^i, or that $\overline{A}$ is a submatrix of one of the $[E^{i1}/A^i/D^{i1}]$.

In the former case, the column submatrices of the components of the linear sum corresponding to the $[E^{i1}/A^i/D^{i1}]$ are in C_0, and $\overline{k}$ defined analogously to k of (12.2.14) for the linear sum satisfies $\overline{k} \le k$. Hence $\overline{A} \in C$.

In the latter case, by the SAT semicentrality of C, $\overline{A}$ is in C_0 and hence is in C as well.

To prove (12.4.17)(ii), we construct a polynomial solution algorithm for C, using the assumed polynomial solution algorithm for C_0. Suppose that the latter algorithm requires at most β effort. Let $A \in C$ be given. In the nontrivial case, A is a linear sum with components in C_0 where k of (12.2.14) is bounded by a constant. We use Algorithm SOLVE LINEAR SUM SAT OR MINSAT (12.4.15) to solve the SAT problem for A, with the solution algorithm for C_0 as subroutine. Using the complexity formula for Algorithm SOLVE LINEAR SUM SAT OR MINSAT (12.4.15), the complexity of the solution algorithm for A is $O(2^k \cdot p \cdot m \cdot n + 4^k \cdot p \cdot \beta)$. Since k is bounded by a constant, and since β is polynomially bounded, the bound proves the solution algorithm to be polynomial.   □

12.5 Extensions

We discuss extensions and improvements of the algorithms of this chapter, including their adaptation to inequality systems over ID-systems.

Section 12.2 defines the linear k-sum using BG-rank. Instead, one could define such sums using IB-rank. Correspondingly, one could change Algorithm LINEAR k-SEPARATION (12.3.3) and Algorithm REFINE LINEAR k-SEPARATION (12.3.7) by replacing Algorithm k-SEPARATION (3.5.20), which is used as a subroutine, by Algorithm IB-k-SEPARATION (4.6.16). The change produces decomposition algorithms that avoid the approximation via the system BG and seem appealing from a theoretical standpoint. However, just like Algorithm LINEAR k-SEPARATION (12.3.3) and Algorithm REFINE LINEAR k-SEPARATION (12.3.7), the new algorithms are too inefficient to be practically usable.

The performance of Algorithm SOLVE LINEAR SUM SAT OR MINSAT (12.4.15) can be significantly enhanced. We cover four improvements.

The first improvement applies only to the SAT problem. It rests on the simple observation that, for any $\{0, \pm 1\}$ matrix A and for any $\{0, 1\}$ vectors $a^1 \geq a^2$, any solution for $A \odot s \geq a^1$ is also a solution for $A \odot s \geq a^2$. Hence, if one has to decide satisfiability for both SAT instances, one should first solve the case $A \odot s \geq a^1$, and only if $A \odot s \geq a^1$ has no solution, one must solve the case $A \odot s \geq a^2$. The same consideration applies to the SAT cases of Steps 3 and 7, so a judicious sequencing of those instances may significantly reduce total computing effort.

The second improvement applies to both SAT and MINSAT instances. As specified, Step 7 solves (12.4.9) for 4-tuples $([d^{i3}/d^{i2}], d^{i21}, e^{i12}, [e^{i2}/e^{i3}])$ where $[d^{i3}/d^{i2}] \in \text{subrange}([D^{i3}/D^{i2}])$, $d^{i21} \in \text{subrange}([D^{i2}|D^{i1}])$, $e^{i12} \in$ subrange$([E^{i1}|E^{i2}])$, and $[e^{i2}/e^{i3}] \in$ subrange$([E^{i2}/E^{i3}])$, and where S_i^* contains a triple $([d^{i3}/d^{i2}], e^{i12}, s^{i*})$.

It turns out that these conditions may, and indeed typically do, admit a large number of 4-tuples that actually need not be considered. Let S be the set of $\{\pm 1\}$ vectors s satisfying $B \odot s \geq b$. For fixed i, $1 \leq i \leq p$, partition any $s \in S$ into s^{i1}, s^{i2}, and s^{i3} according to the index sets $\cup_{j<i} Y_j$, Y_i, and $\cup_{j>i} Y_j$, respectively. Then it is not difficult to prove that one only needs to consider 4-tuples in Step 7 that satisfy the above conditions plus the requirement that, for some $s \in S$,

$$
\begin{aligned}
[d^{i3}/d^{i2}] &= [D^{i3}/D^{i2}] \odot s^{i1} \\
d^{i21} &= [D^{i2}|D^{i1}] \odot [s^{i1}/s^{i2}] \\
e^{i12} &= [E^{i1}|E^{i2}] \odot [s^{i2}/s^{i3}] \\
[e^{i2}/e^{i3}] &= [E^{i2}/E^{i3}] \odot s^{i3}
\end{aligned}
$$

(12.5.1)

Of course, when Step 7 is executed, we do not know S. But the above observation remains valid when we enlarge S by adding any number of $\{\pm 1\}$ vectors of appropriate size. For example, we may take S to be the set of all $\{\pm 1\}$ vectors with $|Y|$ entries. Use of that S in Step 7 may result in a significant reduction of the number of 4-tuples to be considered. For SAT instances, the change can be combined with the one outlined at the beginning of this section.

A third improvement is possible when several SAT or MINSAT instances must be solved, each of which involves some submatrix of one given linear sum B. In that case, one may decide to replace the subrange computations of Step 2 plus the computations implementing the above reductions by simpler and faster calculations that rely on some precomputed information. Details are included in Chapter 13.

The fourth improvement assumes that one or more of the column submatrices $[E^{i1}/A^i/D^{i1}]$ of a linear sum B can be column scaled to become nearly negative, while one or both of the submatrices D^{i1} and E^{i1} become nonpositive. In the MINSAT case, such scaling must be restricted to the columns for which the corresponding entries of the cost vector are zero. The simplifications of Algorithm SOLVE LINEAR SUM SAT OR MIN-SAT (12.4.15) depend on which of the scaled submatrices D^{i1} and E^{i1} are nonpositive. The changes are analogous to those discussed in Section 11.5 for augmented sums, where it is assumed that the submatrix $[A^1/D^1]$ of an augmented sum B can be column scaled so that A^1 becomes nearly negative and D^1 becomes nonpositive. Given that similarity, we leave it to the reader to use the material of Section 11.5 as a guide and fill in the details.

The proof of validity of Algorithm SOLVE LINEAR SUM SAT OR MINSAT (12.4.15) rests on Theorem (12.4.12), which in turn holds due to several results for the operators $\odot$, $\oplus$, and $\ominus$ established by Lemmas (4.2.4), (4.2.8), and (4.2.14). In Section 4.2, some conclusions of these lemmas are extracted as axioms (4.2.21)–(4.2.27) and are used in Section 4.9 under (4.9.4)–(4.9.10) in the definition of ID-systems.

Given this link between Algorithm SOLVE LINEAR SUM SAT OR MINSAT (12.4.15) and ID-systems, one might expect that some extension of that algorithm should handle inequality systems over ID-systems whenever the underlying matrix has an appropriately defined linear decomposition.

This is indeed so. The needed changes mostly concern definitions and notation and are simple enough that we leave it to the reader to work out the details. Similarly, one may extend the decomposition algorithms of Section 12.3 so that the new methods locate appropriately defined linear decompositions for matrices over ID-systems.

Recall from Section 4.9 that a number of combinatorial problems can be formulated as inequality systems over some ID-system. Examples are covering and packing problems. The extensions of the decomposition algorithms of Section 12.3 and of Algorithm SOLVE LINEAR SUM SAT OR MINSAT (12.4.15) to ID-systems may thus be employed to solve such combinatorial problems.

The next chapter describes the analysis algorithm, which assembles solution algorithms for the SAT and MINSAT problems.

Chapter 13

Analysis Algorithm

13.1 Overview

In this chapter, we assemble an *analysis algorithm* for the SAT and MINSAT problems. In the SAT (resp. MINSAT) case, the analysis algorithm accepts as input a matrix A over $\mathbb{B}$ (resp. a matrix A over $\mathbb{B}$ and a rational nonnegative cost vector c). Given such input, the analysis algorithm produces a solution algorithm $\mathcal{M}$ that, for any $\{0,1\}$ vector a of appropriate size, for any column submatrix $\overline{A}$ of A, and, in the MINSAT case, for the corresponding subvector $\overline{c}$ of c, solves the SAT instance $(\overline{A}, a)$ or the MINSAT instance $(\overline{A}, a, \overline{c})$. Besides $\mathcal{M}$, the analysis algorithm also produces a rational number τ that is an upper bound on the run time of the method $\mathcal{M}$, no matter how the column submatrix $\overline{A}$ of A and the vector a are selected.

The analysis algorithm creates $\mathcal{M}$ and τ as follows. Given A, and c if applicable, the algorithm first finds a solution algorithm using the methods of Chapters 5 and 8. If that solution algorithm is fast, the desired $\mathcal{M}$ and τ are at hand. Otherwise, the analysis algorithm attempts to decompose A into component matrices using the decomposition methods of Chapters 9–12. If a decomposition is found, then the analysis algorithm essentially treats each component like another input matrix.

The analysis algorithm stops when, for each component, either a fast solution algorithm is at hand or further decomposition of the component is not possible. Regardless of the situation, a solution algorithm for the entire SAT or MINSAT problem is then at hand, together with a time bound. In addition, if that solution algorithm specifies a fast solution algorithm for each component, then the time bound is small.

For large subclasses of the SAT and MINSAT problems—in particular, for many classes of problems arising from real-world applications—the analysis algorithm creates a solution algorithm $\mathcal{M}$ with polynomial, indeed small, time bound τ. Whenever such a bound is obtained, $\mathcal{M}$ is certified to handle all SAT or MINSAT instances reliably fast, a crucial feature for real-world applications where a guaranteed fast response is needed.

In the language of computer science, the analysis algorithm is a *compiler* for the SAT and MINSAT problems. The matrix A and, if applicable, the vector c constitute the input for the compiler, while the solution algorithm $\mathcal{M}$ and the time bound τ are the output.

The presentation proceeds as follows.

Section 13.2 provides a summarizing description of the solution algorithms $\mathcal{M}$ that are created by the analysis algorithm.

Section 13.3 covers the selection of a solution algorithm for a given component matrix, using the algorithms of Chapters 5 and 8.

Section 13.4 describes the analysis algorithm, using the results of Section 13.3 and the decomposition methods and results of Chapters 9–12.

Section 13.5 describes an algorithm for the approximate solution of the MINSAT problem. The method relies on the integer programming heuristic method of Chapter 8 and on an assumed solution algorithm for the underlying SAT instances.

Section 13.6 presents pre- and postprocessing steps that may be added to the analysis algorithm to significantly increase the computational effectiveness of the solution algorithms.

The final section, 13.7, discusses extensions and lists references.

13.2 Structure of Solution Algorithms

This section describes the general structure of the solution algorithms that are produced by the analysis algorithm. We first review and extend some definitions of earlier chapters.

SAT and MINSAT Instances

Throughout this section, A is a matrix over $\mathbb{B}$, with row index set X and column index set Y. Define a to be any $\{0, 1\}$ vector indexed by X. Let c be any rational nonnegative vector indexed by Y. For fixed A, the possible pairs (A, a) are the SAT *instances* of A. Any such instance demands that one either finds a $\{\pm 1\}$ vector s satisfying $A \odot s \geq a$ or determines that no such s exists. For fixed A and c, the possible triples (A, a, c) are the

MINSAT *instances* of A and c. For a given (A, a, c), the total cost of a $\{\pm 1\}$ solution vector s for $A \odot s \geq a$ is $\sum c_j$, where the summation is over the indices j for which $s_j = 1$. The MINSAT instance (A, a, c) demands that one either finds a $\{\pm 1\}$ solution s for $A \odot s \geq a$ with minimum total cost or declares that $A \odot s \geq a$ has no solution.

The SAT instances *arising from*, or *involving*, the column submatrices $\overline{A}$ of A are all possible SAT instances of the form $(\overline{A}, a)$. Similarly, the MINSAT instances *arising from*, or *involving*, the column submatrices $\overline{A}$ of A and the corresponding subvectors $\overline{c}$ of c are all possible MINSAT instances of the form $(\overline{A}, a, \overline{c})$. In the subsequent discussion of MINSAT cases, just one cost vector is encountered for any given matrix. Accordingly, we typically simplify the above terminology by eliminating the explicit reference to the cost subvectors, and we refer, for example, to the MINSAT instances $(\overline{A}, a, \overline{c})$ as the MINSAT instances *arising from*, or *involving*, the column submatrices $\overline{A}$ of A.

General Form of Solution Algorithms

Given A and, in the MINSAT case, c, we want a solution algorithm that handles the SAT or MINSAT instances arising from the column submatrices of A. We create such an algorithm in three steps.

First, we decompose A into any number of component matrices, say, $B^1, B^2, \ldots, B^n$. In the MINSAT case, we also derive from c certain cost vectors for the components, say, $c^1, c^2, \ldots, c^n$. We permit the trivial case of decomposition where $n = 1$, $B^1 = A$, and, if applicable, $c^1 = c$.

Second, we construct SAT or MINSAT solution algorithms for the SAT or MINSAT instances involving certain column submatrices of the component matrices.

Third, we combine the SAT or MINSAT solution algorithms for the component matrices into an overall solution algorithm. The latter algorithm approximately is as follows.

(13.2.1) Algorithm SOLVE SAT OR MINSAT. (Summarizing Description) *Solves the SAT or MINSAT instance arising from any column submatrix $\overline{A}$ of a given matrix A over $\mathbb{B}$ and, in the MINSAT case, arising from the corresponding subvector $\overline{c}$ of a rational nonnegative vector c. Given are a decomposition of A into component matrices $B^1, B^2, \ldots, B^n$ and, in the MINSAT case, cost vectors $c^1, c^2, \ldots, c^n$. Also given are solution algorithms for the SAT or MINSAT instances arising from certain column submatrices of the component matrices.*

Input: Matrix A over $\mathbb{B}$, with row index set X and column index set Y. A column submatrix $\overline{A}$ of A. A $\{\pm 1\}$ vector a indexed by X, and, in the MINSAT case, a rational nonnegative vector c indexed by Y.

For some $n \geq 1$, component matrices B^1, B^2, ..., B^n and, in the MINSAT case, cost vectors c^1, c^2, ..., c^n derived from c.

A total of n SAT or MINSAT algorithms. For $i = 1, 2, \ldots, n$, the ith algorithm handles the SAT or MINSAT instances arising from certain column submatrices of B^i, using at most β_i time in the SAT case and at most γ_i time in the MINSAT case.

Integers $\alpha_1, \alpha_2, \ldots, \alpha_n$ such that, for $i = 1, 2, \ldots, n$, Step 2 below never processes more than α_i SAT or MINSAT instances involving column submatrices of B^i.

Output: Either: A $\{\pm 1\}$ solution vector s for the SAT instance $(\overline{A}, a)$ or the MINSAT instance $(\overline{A}, a, \overline{c})$, whichever applies. Or: "$(\overline{A}, a)$ is unsatisfiable."

Complexity: SAT case: $O(\sum_{i=1}^{n} \alpha_i \cdot \beta_i)$. MINSAT case: $O(\sum_{i=1}^{n} \alpha_i \cdot \gamma_i)$. Both formulas assume that the effort for Step 3 is dominated by that for Step 2. (The assumption is always satisfied for the cases considered later.)

Procedure:

1. Using $\overline{A}$, select a certain column submatrix $\overline{B}^1$ of B^1. In the MINSAT case, let $\overline{c}^1$ be the subvector of c^1 corresponding to $\overline{B}^1$. Derive certain $\{\pm 1\}$ vectors b^{1j} from the vector a.

2. Do for $i = 1, 2, \ldots, n$:
   Using the matrix $\overline{B}^i$, the vectors b^{ij}, and, in the MINSAT case, the vector $\overline{c}^i$, solve the SAT instances $(\overline{B}^i, b^{ij})$ or the MINSAT instances $(\overline{B}^i, b^{ij}, \overline{c}^i)$, whichever applies. If none of these SAT or MINSAT instances has a solution, declare that $(\overline{A}, a)$ is unsatisfiable, and stop. Otherwise, if $i < n$, derive from these solutions for the next iteration a certain column submatrix $\overline{B}^{i+1}$ of B^{i+1}, the corresponding subvector $\overline{c}^{i+1}$ of c^{i+1}, and certain vectors $b^{i+1,j}$.

3. Backtrack through the solutions for the SAT or MINSAT instances for $i = 1, 2, \ldots, n$ to assemble a solution for the input SAT or MINSAT instance. Output that solution, and stop.

The subsequent two sections describe how the decomposition of the given matrix A into the component matrices B^1, B^2, ..., B^n is determined, and how the SAT or MINSAT solution algorithms for these component matrices are found. The next section deals with the latter task.

13.3 Algorithm for Component Matrix

Algorithm SOLVE SAT OR MINSAT (13.2.1) requires solution algorithms for SAT or MINSAT instances involving certain column submatrices of the component matrices B^1, B^2, ..., B^n. In the MINSAT case, cost vectors c^1, c^2, ..., c^n are supplied with the components. In this section, we describe how these solution algorithms are found. Let B^i be an arbitrary component matrix. In the MINSAT case, let c^i be the cost vector for B^i.

In some decomposition cases, the column submatrices of B^i processed in Algorithm SOLVE SAT OR MINSAT (13.2.1) are proper submatrices of B^i. In such a situation, the column submatrices never involve certain columns of B^i. Hence, it makes sense that we define B^{i*} to be the column submatrix of B^i that contains all column submatrices of B^i that may possibly be processed by Algorithm SOLVE SAT OR MINSAT (13.2.1). In the MINSAT case, let c^{i*} be the corresponding subvector of c^i.

For the moment, we are not concerned how B^{i*} is derived from B^i. That task is described in the next section, where the decomposition of A into B^1, B^2, ..., B^n is covered.

We want a solution algorithm $\mathcal{M}_i$ that solves all SAT or MINSAT instances arising from the column submatrices of B^{i*}. Algorithm SELECT COMPONENT METHOD (13.3.1) below produces that solution algorithm, using the methods and techniques of Chapters 5 and 8. We summarize the steps of Algorithm SELECT COMPONENT METHOD (13.3.1).

Step 1 carries out certain reductions of the input matrix B^{i*}, obtaining another matrix, say, B.

Steps 2–4 check if one of the SAT or MINSAT solution algorithms of Chapter 5 can handle B.

If that search is successful, a fast solution algorithm for the SAT or MINSAT instances involving the column submatrices of B is at hand.

If that search is not successful, Step 5 uses algorithms of Chapter 8 to find a solution algorithm based on closed subregion decomposition. Chapter 8 contains three ways of selecting such a decomposition. Step 5 tries all of them and settles for a decomposition that yields a solution algorithm with the smallest bound on run time.

Regardless of the way a solution algorithm for the SAT or MINSAT instances arising from the column submatrices of B is found, Step 6 combines that solution algorithm with an algorithm that accounts for the reductions in Step 1 and thus obtains the desired solution algorithm $\mathcal{M}_i$.

Here are the details.

(13.3.1) Algorithm SELECT COMPONENT METHOD. *Determines a SAT or MINSAT solution algorithm $\mathcal{M}_i$ for a specified column submatrix B^{i*} of a component B^i and, in the MINSAT case, for the related cost subvector c^{i*} of B^{i*}. Algorithm $\mathcal{M}_i$ handles all SAT or MINSAT instances arising from the column submatrices of B^{i*}.*

Input: Matrix B^{i*} over $\mathbb{B}$. In the MINSAT case, rational nonnegative cost vector c^{i*} for B^{i*}.

Output: A SAT or MINSAT solution algorithm $\mathcal{M}_i$ that solves all SAT or MINSAT instances arising from the column submatrices of B^{i*}. An upper bound τ_i on the run time of that solution algorithm.

Complexity: Polynomial if a polynomial version of Heuristic SOLVE IP (8.3.3) is used. The scheme is invoked by the decomposition algorithms of Step 5.

Procedure:

1. (Reduce B^{i*} to a SAT or MINSAT simple submatrix B.) Apply Algorithm SIMPLE SUBMATRIX (5.3.3) to reduce B^{i*} to the maximum SAT or MINSAT simple submatrix. Call the latter submatrix B. In the MINSAT case, let c^B be the subvector of c^{i*} corresponding to B. If this is a MINSAT case, go to Step 3.

2. (SAT case only: Test for 2SAT property.) Check if each row of B has at most two nonzero entries. If this is so, declare Algorithm SOLVE 2SAT (5.4.1) to be the solution algorithm for B, use the complexity formula for that algorithm to compute an upper bound δ on the run time of the algorithm, and go to Step 6.

3. (Test for hidden near negativity.)
   SAT case: Define a matrix D by $D = B$.
   MINSAT case: Define D to be the column submatrix of B that contains all columns j of B for which $c_j = 0$.
   Let E be the matrix containing the columns of B that do not occur in D. Check with Algorithm TEST HIDDEN NEAR NEGATIVITY (5.6.1) if $B = [D|E]$ is hidden nearly negative relative to E. If this is so, declare Algorithm SOLVE HIDDEN NEARLY NEGATIVE SAT OR MINSAT (5.6.1) to be the solution algorithm for B, use the complexity formula for that algorithm to compute an upper bound δ on the run time of the algorithm, and go to Step 6.
   (Algorithm SOLVE HIDDEN NEARLY NEGATIVE SAT OR MINSAT (5.6.3) invokes Algorithm TEST HIDDEN NEAR NEGATIVITY (5.6.1) to determine the scaling factors that scale B to a nearly negative matrix. Since Algorithm TEST HIDDEN NEAR NEGATIVITY (5.6.1) has already been applied, one should pass these scaling factors to Algorithm SOLVE HIDDEN NEARLY NEGATIVE SAT OR MINSAT (5.6.3) instead of recomputing them in the latter algorithm.)

4. (Test for balancedness, total unimodularity, or network property.) Use Algorithm TEST BALANCEDNESS (5.7.3), Algorithm TEST TOTAL UNIMODULARITY (5.7.8), or Algorithm TEST NETWORK PROPERTY (5.7.9) to check if B is balanced, is totally unimodular, or has the network property. (The selection of the testing algorithm depends on the speed of the implementation of these algorithms. If speed is not important, then the balancedness test is preferred.)
   If B has any one of the listed properties, then it is balanced. In that case, declare Algorithm SOLVE BALANCED SAT OR MINSAT (5.7.25) to be the solution algorithm for B, use the complexity formula

for that algorithm to compute an upper bound δ on the run time of the algorithm, and go to Step 6.

5. (Find a closed subregion decomposition for B.)

   SAT case: Find three closed subregion decompositions of B using Heuristic DECOMPOSITION FOR 2SAT (8.4.5), Heuristic DECOMPOSITION FOR HIDDEN NEAR NEGATIVITY (8.5.5), and Heuristic DECOMPOSITION FOR NETWORK PROPERTY (8.6.12).

   MINSAT case: Find two closed subregion decompositions of B using Heuristic DECOMPOSITION FOR HIDDEN NEAR NEGATIVITY (8.5.5) and Heuristic DECOMPOSITION FOR NETWORK PROPERTY (8.6.12).

   Taking each of the just obtained decompositions in turn, let Algorithm SOLVE CLOSED SUBREGION DECOMPOSITION SAT OR MINSAT (8.2.6) be the solution algorithm, with the appropriate case of Algorithm SOLVE 2SAT (5.4.1), Algorithm SOLVE HIDDEN NEARLY NEGATIVE SAT OR MINSAT (5.6.1), or Algorithm SOLVE BALANCED SAT OR MINSAT (5.7.25) as subroutine.

   For each decomposition, use the complexity formula of Algorithm SOLVE CLOSED SUBREGION DECOMPOSITION SAT OR MINSAT (8.2.6) and of the applicable subroutine to get an upper bound on the run time of the solution algorithm. Select the decomposition producing the smallest time bound. Define δ to be that time bound.

   Declare Algorithm SOLVE CLOSED SUBREGION DECOMPOSITION SAT OR MINSAT (8.2.6) with the selected decomposition to be the solution algorithm for B. The algorithm uses as subroutine the appropriate case of Algorithm SOLVE 2SAT (5.4.1), Algorithm SOLVE HIDDEN NEARLY NEGATIVE SAT OR MINSAT (5.6.3), or Algorithm SOLVE BALANCED SAT OR MINSAT (5.7.25).

6. (Assemble solution algorithm $\mathcal{M}_i$.) In the SAT (resp. MINSAT) case, derive from Algorithm REDUCE SAT INSTANCE (5.3.4) (resp. Algorithm REDUCE MINSAT INSTANCE (5.3.5)) a reduction algorithm that carries out the following task. Input is any column submatrix $\overline{B}^i$ of B^{i*}, any $\{\pm1\}$ vector b of appropriate size, and, in the MINSAT case, the subvector $\overline{c}^i$ of c^{i*} corresponding to $\overline{B}^i$. The input defines a SAT instance $(\overline{B}^i, b)$ or a MINSAT instance $(\overline{B}^i, b, \overline{c}^i)$. The output is a SAT or MINSAT instance where the matrix is a submatrix of B.

   Combine the reduction algorithm with the solution algorithm selected in Step 2, 3, 4, or 5 for B to obtain the solution algorithm $\mathcal{M}_i$ for all SAT or MINSAT instances arising from the column submatrices of B^{i*}. Compute a time bound τ_i for $\mathcal{M}_i$ by combining the complexity bound for the reduction algorithm with the bound δ computed earlier. Output the solution algorithm $\mathcal{M}_i$ and the time bound τ_i, and stop.

We have the following result for the solution algorithms $\mathcal{M}_i$ constructed by Algorithm SELECT COMPONENT METHOD (13.3.1).

(13.3.2) Theorem. *Any solution algorithm $\mathcal{M}_i$ produced by Algorithm* SELECT COMPONENT METHOD (13.3.1) *is polynomial if the following conditions are satisfied.*

(a) *Suppose $\mathcal{M}_i$ is based on a closed subregion decomposition determined in Step 5 of Algorithm* SELECT COMPONENT METHOD (13.3.1). *Then the number of closed subregions of the decomposition must be bounded by a constant.*

(b) *Suppose $\mathcal{M}_i$ involves solution of linear programs. Then a polynomial algorithm for solving these linear programming problems must be used.*

Proof. Algorithm REDUCE SAT INSTANCE (5.3.4) and Algorithm REDUCE MINSAT INSTANCE (5.3.5) are polynomial. If conditions (a) and (b) hold, then any solution algorithm determined in Steps 2–5 of Algorithm SELECT COMPONENT METHOD (13.3.1) is polynomial as well.   □

Note that condition (b) of Theorem (13.3.2) applies only if $\mathcal{M}_i$ solves MINSAT instances involving balanced matrices via linear programming.

The next section provides the analysis algorithm.

13.4 Analysis Algorithm

This section describes the analysis algorithm. That scheme accepts as input a matrix A over $\mathbb{B}$ and, in the MINSAT case, a rational nonnegative vector c. The analysis algorithm produces a solution algorithm that handles the SAT or MINSAT instances arising from the column submatrices of A.

The main tools of the analysis algorithm are Algorithm SELECT COMPONENT METHOD (13.3.1) and the decomposition algorithms of Chapters 9–12.

SAT Case

We sketch the analysis algorithm, using the SAT case of a matrix A as an example.

First, we use Algorithm SELECT COMPONENT METHOD (13.3.1) to determine a solution algorithm for the SAT instances arising from the column submatrices of A, plus an upper time bound for the run time of the solution algorithm. If that time bound is acceptable, we are done. Otherwise, we decompose A in a recursive manner and apply Algorithm SELECT COMPONENT METHOD (13.3.1) to each component, until either we have an overall solution algorithm with an acceptable time bound or we cannot improve the overall solution algorithm by further decompositions.

The recursive decomposition process is guided by heuristic rules to ensure the decompositions lead to an attractive overall solution algorithm. The rules rely on some definitions and results for sums and decompositions of earlier chapters. We review that material.

Types of Sums and Decompositions

Section 4.7 classifies sums with two components B^1 and B^2 according to worst-case upper bounds on the number of b^1- and b^2-satisfiability problems for certain column submatrices $\overline{B}^1$ and $\overline{B}^2$ of B^1 and B^2 that may have to be solved by the SAT algorithm we have developed for that sum. If that upper bound is 1 for both $\overline{B}^1$ and $\overline{B}^2$, the sum is said to be of type I. If the upper bound is at least 2 for $\overline{B}^1$ and is 1 for $\overline{B}^2$, then the sum is of type II. In the remaining case, where both upper bounds are at least 2, the sum is of type III.

We apply the classification of sums as type I, II, or III to the related decompositions and separations in the obvious way.

The sums of Chapters 9–12 are called monotone, closed, augmented, and linear. There also is an elementary sum called 1-sum, which corresponds to a block decomposition.

The types of these sums are as follows.

(13.4.1) Theorem.
(a) *The monotone sum and the 1-sum are of type I.*
(b) *The closed sum and the augmented sum are of type II.*
(c) *The linear sum with two components is of type III.*

Proof. The conclusion for the 1-sum is trivial. The remaining statements represent Theorems (9.4.9), (10.4.19), (11.4.11), and (12.4.16).   □

Ranking of Decompositions

The analysis algorithm prefers type I sums to type II sums and, in turn, prefers type II sums to type III sums. This heuristic rule is based on the consideration that, all other things being equal, a SAT or MINSAT instance of a type I (resp. type II) sum can be more efficiently solved than a type II (resp. type III) sum.

Using the definitions of closed k-sum, augmented k-sum, and linear k-sum in Chapters 10–12, it is easy to verify that any closed 1-sum, augmented 1-sum, or linear 1-sum essentially is a 1-sum and that any augmented 2-sum or linear 2-sum can be replaced by a closed 2-sum. Accordingly, we only need to consider closed k-sums for $k \geq 2$ and augmented k-sums as well as linear k-sums for $k \geq 3$.

Chapters 10–12 provide effective decomposition algorithms for closed k-sums (resp. augmented k-sums, linear k-sums) for values of k satisfying

$2 \leq k \leq 3$ (resp. $3 \leq k \leq 4$, any $k \geq 3$). For a given k-sum of the above variety, we prefer cases with small k values.

We combine the earlier ranking according to type I, II, and III with the above observations concerning k-sums to the following overall ordering of sums, where (i) is ranked highest and preferred and (v) is ranked lowest. For ready reference, we add the sum type in parentheses.

(13.4.2)

	(i)	Monotone sum (type I)
	(ii)	1-sum (type I)
	(iii)	Closed k-sum, $2 \leq k \leq 3$ (type II)
	(iv)	Augmented k-sum, $3 \leq k \leq 4$ (type II)
	(v)	Linear k-sum, $k \geq 3$ (type III)

It may seem odd that the monotone sum is ranked higher than the 1-sum. That decision is based on the fact that, generally, it is more efficient to first decompose a matrix according to a monotone sum, and then to decompose the components of the monotone sum by repeated 1-sum decompositions. Indeed, that approach is equivalent to the following, more elaborate process. First, decompose by repeated 1-sum decompositions. Next, decompose each component according to monotone decompositions. Finally, decompose each component using 1-sum decompositions. Equivalence of the two processes can be proved via Theorem (9.2.3), which asserts uniqueness of maximal monotone decompositions. We leave it to the reader to provide details of the straightforward proof.

Restrictions Imposed on Decompositions

The analysis algorithm selects decompositions in the order specified by (13.4.2), but in addition imposes restrictions depending on the situation. For the discussion of these restrictions, let B either be the original matrix A or be a component obtained from A by any number of recursively applied decompositions. We assume that the upper time bound for the solution algorithm on hand for B is not attractive and that we want to decompose B.

If $B = A$, then we search for a sum decomposition of B into two components, in the order given by (13.4.2).

If $B \neq A$, then the decomposition cases considered for B depend on the type of decomposition that produced B. The rules are as follows.

If a type I decomposition produced B, then the rules are the same as for the case $B = A$.

Assume that a type II decomposition, say, of a matrix C, produced B. Let $(X_1 \cup Y_1, X_2 \cup Y_2)$ be the corresponding separation of C, and let B^1 and B^2 be the components of the decomposition. Thus, B is equal to B^1 or B^2.

If B is the second component B^2, then the rules are the same as for $B = A$.

If B is the first component B^1, we also treat it like the case $B = A$, except that we do not search for any type II decomposition of B. However, we do search for a type II separation $(X_1' \cup Y_1', X_2' \cup Y_2')$ of C that satisfies $X_1' \cup Y_1' \subset X_1 \cup Y_1$. If that search is successful, we replace B (resp. B^2) by the first (resp. second) component of the new decomposition of C. The process may be viewed as a two-step method where B^1 and B^2 are composed to C, which then is decomposed again according to $(X_1' \cup Y_1', X_2' \cup Y_2')$. Since $X_1' \cup Y_1' \subset X_1 \cup Y_1$, the first component of the new decomposition, which is the new B, is smaller than the original B.

The above modification for the case $B = B^1$ is introduced so that recursively nested type II decompositions cannot occur. Such decompositions are undesirable, since they may result in an exponential number of SAT subproblems in the solution algorithm for A. One could permit a nonrecursive nesting. We have not done so to simplify the exposition. But we do allow nesting of a type III decomposition into a type II decomposition, since otherwise the analysis algorithm would not be effective for large classes of practical problems.

We turn to the final case, where B is a component of a type III decomposition. There is just one such decomposition, the linear one. We demand that one may decompose B only by a refinement of that linear decomposition. We impose that rule to avoid nested type III decompositions, since such nesting may produce an exponential number of SAT subproblems in the solution algorithm for A.

MINSAT Case

We have covered the decomposition rules for the SAT problem. The MINSAT problem is handled in essentially the same manner, except that sums that apply just to SAT are ruled out. Thus, closed sums and augmented sums are excluded, and the list of candidate sums (13.4.2) becomes (13.4.3) below.

(13.4.3)
	(i)	Monotone sum (type I)
	(ii)	1-sum (type I)
	(v)	Linear k-sum, $k \geq 2$ (type III)

Structure of Analysis Algorithm

The analysis algorithm consists of a main routine, called Algorithm ANALYSIS (13.4.4), and three subroutines, called Algorithm SELECT TYPE I METHOD (13.4.5), Algorithm SELECT TYPE II METHOD (13.4.6), and Algorithm SELECT TYPE III METHOD (13.4.7).

The three subroutines involve two global Boolean variables called Allow_type_II and Allow_type_III. These variables control the use of type II and type III decompositions.

The three subroutines employ the term "small" in connection with the time bounds τ of solution algorithms $\mathcal{M}$ for the SAT or MINSAT instances arising from the column submatrices of a given matrix B. We define the use of that term.

Let B, $\mathcal{M}$, and τ be given. Define B' to be any matrix over $\mathbb{B}$ that has the same size as B, has the same number of nonzeros as B, and is nearly negative. Algorithm SOLVE NEARLY NEGATIVE SAT OR MINSAT (5.5.1) can solve the SAT instance B'. Indeed, the complexity formula of that algorithm supports a linear time bound τ' for checking satisfiability of B'. That bound is valid regardless of the distribution of the nonzeros in B'.

We use τ' to classify the time bound τ of the solution algorithm $\mathcal{M}$. Specifically, we say that the time bound τ for $\mathcal{M}$ is *small* if, for a fixed positive integer α of moderate size, say, $\alpha = 8$, we have $\tau \leq \alpha \cdot \tau'$.

In the summarizing description of solution algorithms given by Algorithm SOLVE SAT OR MINSAT (13.2.1), a forward and a backward pass are sketched. The analysis algorithm avoids explicit construction of these two passes by nesting solution algorithms taken from Chapters 5 and 8–12.

Main Routine of Analysis Algorithm

We list Algorithm ANALYSIS (13.4.4) and the three subroutines next, followed by the proof of validity.

(13.4.4) Algorithm ANALYSIS. *Analyzes the structure of a matrix A or a matrix/vector pair (A, c) where A is over $\mathbb{B}$ and c is a rational nonnegative vector. Constructs a solution algorithm $\mathcal{M}$ that handles all SAT or MINSAT instances arising from the column submatrices of A. Supplies an upper time bound τ on the run time of $\mathcal{M}$.*

Input: Matrix A over $\mathbb{B}$, with row index set X and column index set Y. In the MINSAT case, a rational nonnegative vector c indexed by Y.

Output: A SAT or MINSAT solution algorithm $\mathcal{M}$, and an upper time bound τ on its run time. Algorithm $\mathcal{M}$ handles all SAT or MINSAT instances arising from the column submatrices of A. Algorithm $\mathcal{M}$ is polynomial if the number of closed subregions of each closed subregion decomposition used in the construction of $\mathcal{M}$ is bounded by a constant and if solution of any linear programs by $\mathcal{M}$ is done by a polynomial method.

Complexity: Polynomial.

Procedure:

1. (Initialize two Boolean variables Allow_type_II and Allow_type_III that control the use of type II and type III decompositions.) In the SAT (resp. MINSAT) case, declare a Boolean variable Allow_type_II to be *True* (resp. *False*). Also declare a Boolean variable Allow_type_III to be *True*.

2. (Determine solution algorithm $\mathcal{M}$.) Do Algorithm SELECT TYPE I METHOD (13.4.5). The input consists of the matrix A and, if applicable, the vector c. The output of Algorithm SELECT TYPE I METHOD (13.4.5) is the desired solution algorithm $\mathcal{M}$ and the time bound τ. Output $\mathcal{M}$ and τ, and stop.

First Subroutine of Analysis Algorithm

Algorithm SELECT TYPE I METHOD (13.4.5) below is the first subroutine. It proceeds as follows.

Step 1 finds a monotone decomposition of the input matrix into B^1 and B^2.

Step 2 decomposes B^2 into blocks.

Step 3 calls the subsequently listed Algorithm SELECT TYPE II METHOD (13.4.6) and Algorithm SELECT TYPE III METHOD (13.4.7) to get solution algorithms for the blocks.

Finally, Steps 4 and 5 combine the solution algorithms for the blocks with Algorithm SOLVE MONOTONE SUM SAT OR MINSAT (9.4.8) to obtain the desired solution algorithm for the input matrix.

(13.4.5) Algorithm SELECT TYPE I METHOD. *Determines for an input matrix B and, if applicable, for an input vector c a SAT or MINSAT solution algorithm $\mathcal{M}$ that handles all SAT or MINSAT instances arising from the column submatrices of B. Computes an upper time bound τ on the run time of $\mathcal{M}$. The construction of $\mathcal{M}$ is based on type I decompositions.*

Input: Matrix B over $\mathbb{B}$, with row index set X and column index set Y. In the MINSAT case, a rational nonnegative vector c indexed by Y.

Output: A SAT or MINSAT solution algorithm $\mathcal{M}$ and an upper time bound τ on its run time. Algorithm $\mathcal{M}$ handles all SAT or MINSAT instances arising from the column submatrices of B. Algorithm $\mathcal{M}$ is polynomial if the number of closed subregions of each closed subregion decomposition used in the construction of $\mathcal{M}$ is bounded by a constant and if solution of any linear programs by $\mathcal{M}$ is done by a polynomial method.

Complexity: Polynomial.

Procedure:

1. (Monotone decomposition of B)
   SAT case: Let $C = B$.

MINSAT case: Let C be the matrix containing the columns j of B for which $c_j = 0$.

Do Algorithm MONOTONE DECOMPOSITION (9.3.3) with B and C as input. In the notation of (9.4.3), the output consists of the components $B^1 = A^1$ and $B^2 = [D|A^2]$, plus the $\{\pm 1\}$ scaling factors that were used in the decomposition to achieve near negativity for B^1 and $D \leq 0$.

(We take Algorithm SOLVE MONOTONE SUM SAT OR MINSAT (9.4.8) as the solution algorithm $\mathcal{M}$. Algorithm SOLVE MONOTONE SUM SAT OR MINSAT (9.4.8) requires a solution algorithm, say, $\mathcal{M}_2$, for the SAT or MINSAT instances involving the column submatrices of a matrix B^{2*} derived from B^2 as follows.)

Declare B^{2*} to be the column submatrix A^2 of $B^2 = [D|A^2]$. In the MINSAT case, let c^{2*} be the subvector of c corresponding to A^2.

2. (1-Sum decomposition of B^{2*}) Use Algorithm 1-SEPARATION (3.5.1) to determine 1-sum decompositions that derive from B^{2*} the blocks of B^{2*}, say, for some $p \geq 1$, $B^{21}, B^{22}, \ldots, B^{2p}$.

3. (Construct solution algorithms for $B^{21}, B^{22}, \ldots, B^{2p}$.) Do for $i = 1, 2, \ldots, p$:

   Define $B^{2i*} = B^{2i}$. In the MINSAT case, let c^{2i*} be the subvector of c^{2*} corresponding to B^{2i*}. Find a solution algorithm $\mathcal{M}_{2i}$ and a time bound τ_{2i} using the algorithm specified below, with B^{2i*} and, if applicable, with c^{2i*} as input.

   If Allow_type_III is *False*, use Algorithm SELECT COMPONENT METHOD (13.3.1).

   If Allow_type_II is *True* and Allow_type_III is *True*, use Algorithm SELECT TYPE II METHOD (13.4.6).

   If Allow_type_II is *False* and Allow_type_III is *True*, use Algorithm SELECT TYPE III METHOD (13.4.7).

4. (Determine $\mathcal{M}_2$.) Construct $\mathcal{M}_2$ using $\mathcal{M}_{21}, \mathcal{M}_{22}, \ldots, \mathcal{M}_{2p}$ of Step 3, plus the straightforward reduction of any SAT or MINSAT instance arising from a column submatrix of B^{2*} to one SAT or MINSAT instance each for $\mathcal{M}_{21}, \mathcal{M}_{22}, \ldots, \mathcal{M}_{2p}$. The upper time bound τ_2 for $\mathcal{M}_2$ is equal to $\sum_{i=1}^{p} \tau_{2i}$ plus the time required for the reduction effort.

5. (Determine $\mathcal{M}$.) Define $\mathcal{M}$ to be Algorithm SOLVE MONOTONE SUM SAT OR MINSAT (9.4.8), with $\mathcal{M}_2$ of Step 4 as subroutine. Compute the time bound τ for $\mathcal{M}$ using the time bound formula for Algorithm SOLVE MONOTONE SUM SAT OR MINSAT (9.4.8) and the time bound τ_2 for $\mathcal{M}_2$. Output $\mathcal{M}$ and τ, and stop.

Second Subroutine of Analysis Algorithm

The second subroutine is Algorithm SELECT TYPE II METHOD (13.4.6) presented next. It applies to SAT only. Since it is invoked by Step 3 of

Algorithm SELECT TYPE I METHOD (13.4.5), it is known that the input matrix is connected and does not have a proper monotone decomposition. Algorithm SELECT TYPE II METHOD (13.4.6) proceeds as follows.

Step 1 uses the later given Algorithm SELECT TYPE III METHOD (13.4.7) to construct a solution algorithm $\mathcal{M}_0$ that does not involve type II decompositions.

If the time bound τ_0 for $\mathcal{M}_0$ is small, a satisfactory solution algorithm has been found, and the algorithm stops. Otherwise, Steps 2 and 3 search for a type II decomposition.

If a type II decomposition is not found, the solution algorithm $\mathcal{M}_0$ of Step 1 is accepted, and the algorithm stops.

If a type II decomposition is found, Step 4 constructs a solution algorithm $\mathcal{M}_1^{II}$ for the first component of the decomposition using Algorithm SELECT TYPE I METHOD (13.4.5). The latter solution algorithm is not allowed to involve type II decompositions, but may involve type III decompositions. If the time bound τ_1^{II} for $\mathcal{M}_1^{II}$ algorithm is not small, we search for another type II decomposition where the first component is smaller than that of the decomposition on hand. If that search is successful, we repeat Steps 3 and 4 to determine a new $\mathcal{M}_1^{II}$, then continue as described above. The iterative process stops when the $\mathcal{M}_1^{II}$ on hand has a small time bound τ_1^{II}, or when another type II decomposition with smaller first component cannot be found. At that time, we accept the current $\mathcal{M}_1^{II}$ as the solution algorithm for the first component of the type II decomposition on hand.

In Step 5, we find a solution algorithm $\mathcal{M}_2^{II}$ for the second component of the type II decomposition, using Algorithm SELECT TYPE I METHOD (13.4.5).

Step 6 uses $\mathcal{M}_1^{II}$, $\mathcal{M}_2^{II}$, and the solution algorithm for the selected type II decomposition to construct a solution algorithm $\mathcal{M}^{II}$ for the input matrix. Let τ^{II} be the time bound for $\mathcal{M}^{II}$. Based on a comparison of τ^{II} and τ_0 of Step 1, Step 6 outputs $\mathcal{M}_0$ or $\mathcal{M}^{II}$, plus the corresponding time bound, as the solution algorithm and time bound for the input matrix.

(13.4.6) Algorithm SELECT TYPE II METHOD. *Determines for an input matrix B a SAT solution algorithm $\mathcal{M}$ that handles all SAT instances arising from the column submatrices of B. Computes an upper time bound τ on the run time of $\mathcal{M}$. The construction of $\mathcal{M}$ is based on type II decompositions.*

Input: Matrix B over $\mathbb{B}$, with row index set X and column index set Y. The matrix B is connected and does not have a proper monotone decomposition.

Output: A SAT solution algorithm $\mathcal{M}$ and an upper time bound τ on its run time. Algorithm $\mathcal{M}$ handles all SAT instances arising from the column submatrices of B. Algorithm $\mathcal{M}$ is polynomial if the number of closed subregions of each closed subregion decomposition used in the construction of $\mathcal{M}$ is bounded by a constant and if solution of any linear programs by $\mathcal{M}$ is done by a polynomial method.

Complexity: Polynomial.

Procedure:

1. (Find solution algorithm $\mathcal{M}_0$ without type II decompositions.) Do Algorithm SELECT TYPE III METHOD (13.4.7) with B as input. Declare the output solution algorithm to be $\mathcal{M}_0$, and declare the time bound to be τ_0. If the time bound τ_0 is small, output $\mathcal{M} = \mathcal{M}_0$ and $\tau = \tau_0$, and stop.

2. Using Algorithm CLOSED 2-SEPARATION (10.3.8), Algorithm CLOSED 3-SEPARATION (10.3.9), or Algorithm AUGMENTED k-SEPARATION (11.3.1), search for a closed k-separation with $2 \le k \le 3$ or for an augmented k-separation with $3 \le k \le 4$, satisfying the following two conditions. First, any closed separation must be preferred to any augmented separation. Second, the value of k must be as small as possible.

   If a k-separation, say, $(X_1 \cup Y_1, X_2 \cup Y_2)$, with the desired features is found, go to Step 3. Otherwise, output $\mathcal{M} = \mathcal{M}_0$ and $\tau = \tau_0$, and stop.

3. (Select B^{1*} and B^{2*} from the type II decomposition.)

   If $(X_1 \cup Y_1, X_2 \cup Y_2)$ is a closed separation: Two decomposition cases are possible, given by (10.4.5) and (10.4.11). If $|X_1 \cup X_2 \cup Y_1| \le |X_2 \cup Y_2|$, select case (10.4.5), and define $B^{1*} = [A^1/D]$ and $B^{2*} = [D|A^2]$; otherwise, select case (10.4.11), and define $B^{1*} = A^2$ and $B^{2*} = [A^1/D]$. (The selection and definition of B^{1*} and B^{2*} are based on Algorithm SOLVE CLOSED SUM SAT (10.4.18) and produce B^{1*} of least length.) Define $\mathcal{M}^{II}$ to be Algorithm SOLVE CLOSED SUM SAT (10.4.18).

   If $(X_1 \cup Y_1, X_2 \cup Y_2)$ is an augmented separation: In the notation of (11.4.2), define $B^{1*} = [A^1/D]$ and $B^{2*} = B^2$. Define $\mathcal{M}^{II}$ to be Algorithm SOLVE AUGMENTED SUM SAT (11.4.10).

   (Regardless of how $\mathcal{M}^{II}$ is chosen, that algorithm requires two subroutines, say, $\mathcal{M}_1^{II}$ and $\mathcal{M}_2^{II}$, for solving SAT instances involving column submatrices of B^{1*} and B^{2*}, respectively.)

4. (Construct $\mathcal{M}_1^{II}$.) Set Allow_type_II to *False*. Do Algorithm SELECT TYPE I METHOD (13.4.5) with B^{1*} as input. Declare the output solution algorithm to be $\mathcal{M}_1^{II}$, and declare the time bound to be τ_1^{II}. Reset Allow_type_II to *True*. If τ_1^{II} is not small, go to Step 8 (to search for a better type II decomposition).

5. (Construct $\mathcal{M}_2^{II}$.) Do Algorithm SELECT TYPE I METHOD (13.4.5) with B^{2*} as input. Declare the output solution algorithm to be $\mathcal{M}_2^{II}$, and declare the time bound to be τ_2^{II}.

6. (Construct time bound for $\mathcal{M}^{II}$.) Using τ_1^{II} and τ_2^{II} in the complexity

formula for $\mathcal{M}^{II}$, compute a time bound τ^{II} for the latter algorithm.

7. (Compare the time bound τ_0 for $\mathcal{M}_0$ with τ^{II} for $\mathcal{M}^{II}$ to select the solution algorithm $\mathcal{M}$.) If $\tau_0 \leq \tau^{II}$, then let $\mathcal{M} = \mathcal{M}_0$ and $\tau = \tau_0$. Otherwise, let $\mathcal{M} = \mathcal{M}^{II}$ and $\tau = \tau^{II}$. Output $\mathcal{M}$ and τ, and stop.

8. (Time bound τ_1^{II} is not small. Search for an improved type II decomposition.) Select the appropriate decomposition algorithm from those cited in Step 2, and search for a k'-separation $(X_1' \cup Y_1', X_2' \cup Y_2')$ of B of the same kind as $(X_1 \cup Y_1, X_2 \cup Y_2)$, that is, closed or augmented, whichever applies, while enforcing $X_1' \cup Y_1' \subset X_1 \cup Y_1$. If such a separation is found, update $X_1 = X_1'$, $X_2 = X_2'$, $Y_1 = Y_1'$, $Y_2 = Y_2'$, and go to Step 3 with the new $(X_1 \cup Y_1, X_2 \cup Y_2)$. Otherwise go to Step 5. (If an improved separation is not found, we accept the current $(X_1 \cup Y_1, X_2 \cup Y_2)$ and the solution algorithm $\mathcal{M}_1^{II}$ on hand.)

Third Subroutine of Analysis Algorithm

The third subroutine is Algorithm SELECT TYPE III METHOD (13.4.7) below. It applies to both SAT and MINSAT. Since it is invoked by Step 3 of Algorithm SELECT TYPE I METHOD (13.4.5) or by Step 1 of Algorithm SELECT TYPE II METHOD (13.4.6), it is known that the input matrix is connected and does not have a proper monotone decomposition. We sketch the steps of the algorithm.

Step 1 uses Algorithm SELECT COMPONENT METHOD (13.3.1) to determine a solution algorithm $\mathcal{M}_0$ that does not involve a type III decomposition. If the time bound for $\mathcal{M}_0$ is small, $\mathcal{M}_0$ is taken as the desired solution algorithm. Otherwise, the remaining steps search for a linear decomposition and solution algorithms for its components. Details are as follows.

Steps 2 and 3 search for a linear decomposition with two components and determine a solution algorithm for each component.

Steps 4 and 5 proceed recursively. If the time bound for the solution algorithm of any one of the components is not small, then these steps search for a refined linear decomposition where the component in question is replaced by two components. The recursive process stops when each component on hand has a small time bound or does not admit a linear decomposition into two components.

The results of the recursive process of Steps 4 and 5 are recorded in a binary tree T where each node represents a component and where the linear decomposition of a component into two components is depicted by a parent node with two immediate descendants. The name or label for each node of T specifies the component matrix it represents, the solution algorithm obtained for that component, and the related time bound.

Step 6 is a record keeping step that we need not discuss here.

Step 7 uses the time bounds and the complexity formula for Algorithm SOLVE LINEAR SUM SAT OR MINSAT (12.4.15) to extract a linear decomposition from the tree T that in a certain sense produces a best solution algorithm.

Step 8 assembles that best solution algorithm and computes a time bound for it. That solution algorithm and time bound constitute the output.

(13.4.7) Algorithm SELECT TYPE III METHOD. *Determines for an input matrix* B *and, if applicable, for an input vector* c *a SAT or MIN-SAT solution algorithm* $\mathcal{M}$ *that handles all SAT or MINSAT instances arising from the column submatrices of* B. *Computes an upper time bound* τ *on the run time of* $\mathcal{M}$. *The construction of* $\mathcal{M}$ *is based on type III decompositions.*

Input: Matrix B over $\mathbb{B}$, with row index set X and column index set Y. In the MINSAT case, a rational nonnegative vector c indexed by Y. The matrix B is connected and does not have a proper monotone decomposition.

Output: A SAT or MINSAT solution algorithm $\mathcal{M}$ and an upper time bound τ on its run time. Algorithm $\mathcal{M}$ handles all SAT or MINSAT instances arising from the column submatrices of B. Algorithm $\mathcal{M}$ is polynomial if the number of closed subregions of each closed subregion decomposition used in the construction of $\mathcal{M}$ is bounded by a constant and if solution of any linear programs by $\mathcal{M}$ is done by a polynomial method.

Complexity: Polynomial.

Procedure:
1. (Find solution algorithm without type III decomposition.) Do Algorithm SELECT COMPONENT METHOD (13.3.1) with B and, in the MINSAT case, with vector c as input. Declare the output solution algorithm to be $\mathcal{M}_0$ and the time bound to be τ_0. If τ_0 is small, output $\mathcal{M} = \mathcal{M}_0$ and $\tau = \tau_0$, and stop.
2. (Find initial type III decomposition.) Use Algorithm LINEAR k-SEPARATION (12.3.3) or Heuristic LINEAR k-SEPARATION (12.3.4) to search for a linear k-separation of B where k is bounded by a given constant n; for example, $k \le n \le 8$. If no decomposition is found, output $\mathcal{M} = \mathcal{M}_0$ and $\tau = \tau_0$, and stop. Otherwise, let B^1 and B^2 be the components of the decomposition, as given by (12.2.5).
3. (Find solution algorithms for the components of the initial type III decomposition, and start the tree T.) Using the notation of (12.2.5) for B^1 and B^2, define $B^{1*} = [A^1/D]$ and $B^{2*} = [E/A^2]$. In the MINSAT case, define c^{1*} and c^{2*} to be the subvectors of c corresponding to B^{1*} and B^{2*}, respectively. Set Allow_type_III to *False*. For $i = 1$, 2, do Algorithm SELECT TYPE I METHOD (13.4.5) with B^{i*} and, in the MINSAT case, with c^{i*} as input. Declare the output solution

algorithm to be $\mathcal{M}_i^{\mathrm{III}}$, and declare the time bound to be τ_i^{III}. Reset Allow_type_III to *True*.

Construct a rooted T with three nodes. The root node is labeled $(B, \mathcal{M}_0, \tau_0)$. The two descendant nodes are labeled $(B^1, \mathcal{M}_1^{\mathrm{III}}, \tau_1^{\mathrm{III}})$ and $(B^2, \mathcal{M}_2^{\mathrm{III}}, \tau_2^{\mathrm{III}})$. Declare the root node to be scanned, and declare the two descendant nodes to be unscanned.

4. (The tip nodes of T correspond to a linear decomposition of B. Refine that decomposition if needed and possible.) If all nodes of the tree are scanned, go to Step 6. Otherwise, select one unscanned node, say, $(C, \mathcal{M}_C^{\mathrm{III}}, \tau_C^{\mathrm{III}})$, and declare it to be scanned.

   If the time bound τ_C^{III} is small, go to the beginning of this step to process another unscanned node. Otherwise, use Algorithm REFINE LINEAR k-SEPARATION (12.3.7) or Heuristic REFINE LINEAR k-SEPARATION (12.3.8) to search for a refinement of the linear decomposition given by the tip nodes of T where the component matrix C is decomposed into, say, B^i and B^{i+1}. The value k for any such decomposition must be bounded by a given constant n—for example, by $k \leq n \leq 8$.

   If a refinement is not found, go to the beginning of this step to process another unscanned node. Otherwise, let the decomposition correspond to the partition of C given by (12.3.5), and define B^i and B^{i+1} via (12.4.3).

5. (Update tree T.) Declare B^{i*} to be the matrix $[E^{i1}/A^i/D^{i1}]$, and declare B^{i+1*} to be the matrix $[E^{i+1,1}/A^{i+1}/D^{i+1,1}]$. In the MINSAT case, let c^{i*} and c^{i+1*} be the subvectors of c corresponding to B^{i*} and B^{i+1*}.

   Set Allow_type_III to *False*. Do SELECT TYPE I METHOD (13.4.5) twice: once with input matrix B^{i*} and once with B^{i+1*}, using cost vectors c^{i*} and c^{i+1*} if applicable. Declare the respective output solution algorithms to be $\mathcal{M}_i^{\mathrm{III}}$ and $\mathcal{M}_{i+1}^{\mathrm{III}}$, with time bounds τ_i^{III} and $\tau_{i+1}^{\mathrm{III}}$. Reset Allow_type_III to *True*.

   Enlarge T by creating two descendant nodes $(B^i, \mathcal{M}_i^{\mathrm{III}}, \tau_i^{\mathrm{III}})$ and $(B^{i+1}, \mathcal{M}_{i+1}^{\mathrm{III}}, \tau_{i+1}^{\mathrm{III}})$ of $(C, \mathcal{M}_C^{\mathrm{III}}, \tau_C^{\mathrm{III}})$. Declare the two new nodes to be unscanned. Go to Step 4.

6. (All nodes of the tree T are scanned. Prepare for the selection of the linear decomposition.) Define all nontip nodes of T to be unscanned.

7. (Select linear decomposition.) If all nodes of T have been scanned, go to Step 8. Otherwise, select any unscanned node, say, $(C, \mathcal{M}_C^{\mathrm{III}}, \tau_C^{\mathrm{III}})$, that is the parent of two scanned nodes.

   Declare $(C, \mathcal{M}_C^{\mathrm{III}}, \tau_C^{\mathrm{III}})$ to be scanned. Let T_C be the subtree of T that has $(C, \mathcal{M}_C^{\mathrm{III}}, \tau_C^{\mathrm{III}})$ as root node and that contains all descendants of that node. Consider two linear decompositions of B. The first decomposition has as components the matrices of the tip nodes of T. The second decomposition is derived from the first one by replacing the components corresponding to the tip nodes of the subtree T_C by the matrix C.

Compute a time bound τ' (resp. τ'') for the first (resp. second) decomposition, using the complexity formula for Algorithm SOLVE LINEAR SUM SAT OR MINSAT (12.4.15) and the time bounds of the nodes defining the components of the decomposition. (If $\tau' \geq \tau''$, the second decomposition is deemed superior to the first one.) If $\tau' \geq \tau''$, redefine T by deleting all descendants of node $(C, \mathcal{M}_C^{\text{III}}, \tau_C^{\text{III}})$. Go to the beginning of this step to process another unscanned node.

8. (The tip nodes of T give the desired decomposition. Construct $\mathcal{M}$ using that decomposition.) If T consists of just the root node, output $\mathcal{M} = \mathcal{M}_0$ and $\tau = \tau_0$, and stop. Otherwise, for some $p \geq 2$, let $(B^1, \mathcal{M}_1^{\text{III}}, \tau_1^{\text{III}})$, $(B^2, \mathcal{M}_2^{\text{III}}, \tau_2^{\text{III}})$, ..., $(B^p, \mathcal{M}_p^{\text{III}}, \tau_p^{\text{III}})$ be the tip nodes of T, indexed in such a way that B^1, $B^2, \ldots,$ B^p are the components of a linear decomposition of B. Declare $\mathcal{M}^{\text{III}}$ to be Algorithm SOLVE LINEAR SUM SAT OR MINSAT (12.4.15), using the algorithms $\mathcal{M}_1^{\text{III}}$, $\mathcal{M}_2^{\text{III}}, \ldots,$ $\mathcal{M}_p^{\text{III}}$ as subroutines. Compute a time bound τ^{III} for $\mathcal{M}^{\text{III}}$ using the complexity formula for Algorithm SOLVE LINEAR SUM SAT OR MINSAT (12.4.15) and the time bounds τ_1^{III}, $\tau_2^{\text{III}}, \ldots,$ τ_p^{III}. Output $\mathcal{M} = \mathcal{M}_{\text{III}}$ and $\tau = \tau^{\text{III}}$, and stop.

We note that the calculations of τ' and τ'' in Step 7 are very similar. Accordingly, the test whether $\tau' \geq \tau''$ can be simplified. We leave it to the reader to fill in details.

Proof of Validity of Algorithm ANALYSIS (13.4.4). Given the prior discussion, we only need to confirm that the complexity of the analysis algorithm is polynomial and that the algorithm produces polynomial solution algorithms under the stated assumptions.

Each subroutine of the analysis algorithm by itself is polynomial. The variables Allow_type_II and Allow_type_III are manipulated so that the following effects are achieved. The first component of a type II decomposition cannot undergo another type II decomposition. The components of a type III decomposition cannot undergo a type II decomposition, and they can undergo a type III decomposition only if the latter decomposition is a refinement of the given type III decomposition. These restrictions imply that the analysis algorithm is polynomial.

We turn to the polynomiality claim for solution algorithms. The analysis algorithm is so designed that any solution algorithm constructed by it is polynomial if Algorithm SELECT COMPONENT METHOD (13.3.1) supplies polynomial solution algorithms. Theorem (13.3.2) guarantees that this is so under the assumptions stated in Algorithm ANALYSIS (13.4.4) and its subroutines.   □

One may employ Algorithm ANALYSIS (13.4.4) to create approximation algorithms for MINSAT. The next section presents details.

13.5 Approximate Minimization

Section 8.3 contains Heuristic SOLVE IP (8.3.3) for the approximate solution of integer programming problems. In this section, we specialize that heuristic method to obtain an approximate solution algorithm for MINSAT.

Outline of Approach

Let the matrix A over $\mathbb{B}$ and the rational nonnegative cost vector c^A for A be given. Suppose that the time bound of the solution algorithm constructed by Algorithm ANALYSIS (13.4.4) for the MINSAT instances arising from the column submatrices of A is too large to be acceptable. Assume that we are willing to consider solution algorithms that solve these MINSAT instances approximately. We construct a solution algorithm of the latter kind in a two-step process.

First, we apply Algorithm ANALYSIS (13.4.4) to the matrix A and obtain a solution algorithm, say, Q, for the SAT instances arising from the column submatrices of A. If the time bound for Q is too large to be acceptable, then the subsequent results of this section do not apply. So assume that the time bound for Q is acceptable; that is, the bound is small or at least reasonable.

Second, we use the SAT solution algorithm Q in Heuristic SOLVE IP (8.3.3) to obtain an approximate solution algorithm that solves the MINSAT instances arising from the column submatrices of A. The latter solution algorithm proceeds as follows.

MINSAT Instance as Integer Program

Let the given MINSAT instance involve a column submatrix B of A, the corresponding subvector c of c^A, and a $\{0, 1\}$ vector b. Suppose that B has row index set X and column index set Y. We are to find a $\{\pm 1\}$ vector s satisfying $B \odot s \geq b$ and minimizing $\sum_j c_j$ where the summation is over the indices $j \in Y$ for which $s_j = 1$, or conclude that $B \odot s \geq b$ has no solution.

We reformulate the MINSAT instance as an integer program (IP). Let $q(B)$ be the integer vector with elements indexed by X where, for each $x \in X$, the element $q(B)_x$ is equal to the number of -1s in row x of B. Let r be a $\{0, 1\}$ vector indexed by Y. The vector r is related to s as follows. For each $j \in Y$, $r_j = 0$ (resp. $r_j = 1$) if $s_j = -1$ (resp. $s_j = 1$).

According to the discussion of Section 5.7, we may formulate the MINSAT instance involving B as the following IP, where B is viewed to be over the integers.

$$
\begin{aligned}
\min \quad & c^t \cdot r \\
\text{s. t.} \quad & B \cdot r \geq b - q(B) \\
& r \text{ is a } \{0, 1\} \text{ vector}
\end{aligned}
$$

(13.5.1)

Heuristic Solution Algorithm for MINSAT

We use Heuristic SOLVE IP (8.3.3) to solve (13.5.1) approximately. That method requires as input the arrays B, $b - q(B)$, and c, a positive integer k, a subset J of Y, and two subroutines Q and R. We select k, J, Q, and R as follows.

The positive integer k controls the extent of the enumerative effort by Heuristic SOLVE IP (8.3.3) and may be arbitrarily chosen.

We do not restate the conditions concerning the choice of J, but take $J = Y$ and note that this selection trivially satisfies those conditions.

Given $J = Y$, we may rephrase the required features of the subroutine Q as follows. Suppose arbitrary $\{0, 1\}$ values have been assigned to the variables r_j with index j in some subset of Y. Then subroutine Q is to decide whether one can assign $\{0, 1\}$ values to the remaining variables r_j such that the resulting $\{0, 1\}$ vector r is a feasible solution for the IP (13.5.1).

Given the link between the IP (13.5.1) and the MINSAT instance, subroutine Q must effectively be able to solve all SAT instances arising from the column submatrices of B. We have an algorithm for the latter task, the SAT algorithm Q constructed by Algorithm ANALYSIS (13.4.4) for B. Hence, we use that Q here. Let the time bound for that solution algorithm Q be σ.

For the discussion of the subroutine R, we derive the following linear program (LP) from the IP (13.5.1).

$$
\begin{array}{ll}
\min & c^t \cdot r \\
\text{s. t.} & B \cdot r \geq b - q(B) \\
& 0 \leq r \leq 1
\end{array}
$$

(13.5.2)

The subroutine R is to find an optimal extreme point solution for any one of the following modified versions of the LP (13.5.2). Each version is obtained from the LP (13.5.2) by fixing the variables r_j with index j in some subset of Y to some $\{0, 1\}$ values such that the modified LP still has a feasible solution. Subroutine R is assumed to require at most λ effort.

Heuristic SOLVE IP (8.3.3) either provides a good but not necessarily optimal solution for the IP (13.5.1), plus a rational number β that is a lower bound on the optimal objective function value of the IP (13.5.1), or concludes that the IP (13.5.1) has no feasible solution.

In MINSAT terminology, Heuristic SOLVE IP (8.3.3) either provides a good but not necessarily optimal solution for the MINSAT instance defined by B, b, and c or concludes that the MINSAT instance is unsatisfiable. In addition, the value of β is a lower bound on the optimal objective function value of the MINSAT instance. Thus, if the difference between β and the objective function value of the solution is small (resp. 0), then that solution is close to optimal (resp. is indeed optimal).

The effort of Heuristic SOLVE IP (8.3.3) is $O(2^k \cdot (|J| + 1) \cdot (\sigma + \lambda))$ and thus is polynomial if k is bounded by a constant and if both σ and λ are polynomially bounded.

We summarize the above heuristic method for MINSAT.

(13.5.3) Heuristic SOLVE MINSAT. *Finds a good but not necessarily optimal solution for the MINSAT instance arising from any column submatrix of a matrix A and from the corresponding subvector of a rational nonnegative cost vector c^A.*

Input: Matrix A over $\mathbb{B}$. A column submatrix B of A. The matrix B has row index set X and column index set Y. A $\{0,1\}$ vector b with entries indexed by X. A rational nonnegative cost vector c^A for A. Let c be the subvector of c^A corresponding to B.
A positive integer k.
A solution algorithm Q obtained by Algorithm ANALYSIS (13.4.4) that solves the SAT instances arising from the column submatrices of A. Let the time bound for Q be σ.
A subroutine R that for any column submatrix $\overline{A}$ of A, for any $\{0,1\}$ vector a of appropriate size, and for the subvector $\overline{c}$ of c^A corresponding to $\overline{A}$, finds an optimal extreme point solution for the LP

$$
(13.5.4) \qquad \begin{aligned} \min \quad & \overline{c}^t \cdot r \\ \text{s. t.} \quad & \overline{A} \cdot r \geq a - q(\overline{A}) \\ & 0 \leq r \leq \underline{1} \end{aligned}
$$

as well as for modified versions of that LP where some variables have been fixed to some $\{\pm 1\}$ values. In each case, it is known that the LP has a feasible solution. Subroutine R is assumed to require at most λ effort.

Output: Either: A good but not necessarily optimal solution for the MIN-SAT instance (B, b, c), plus a rational number β that is a lower bound on the optimal objective function value of that instance. (If the difference between β and the objective function value of the solution is small (resp. 0), then that solution is close to optimal (resp. is indeed optimal)). Or: "The MINSAT instance (B, b, c) is unsatisfiable."

Complexity: $O(2^k \cdot (|Y| + 1) \cdot (\sigma + \lambda))$. The effort is polynomial if σ and λ are polynomially bounded and if k is bounded by a constant.

Procedure:

1. Apply Heuristic SOLVE IP (8.3.3) to the IP

$$
(13.5.5) \qquad \begin{aligned} \min \quad & c^t \cdot r \\ \text{s. t.} \quad & B \cdot r \geq b - q(B) \\ & r \text{ is a } \{0,1\} \text{ vector} \end{aligned}
$$

The input consists of B, $b - q(B)$, c, k, a set J defined to be $J = Y$, and the input subroutines Q and R. The use of Q requires the translation of vectors r for the IP to solution vectors s for SAT instances, and vice versa. The relationship is $r_j = 0$ (resp. $r_j = 1$) if and only if $s_j = -1$ (resp. $s_j = 1$).

2. If Heuristic SOLVE IP (8.3.3) does not produce a solution, declare that the MINSAT instance (B, b, c) is unsatisfiable, and stop. Otherwise, let r^* be the solution vector. Define for all $j \in Y$, $s_j^* = -1$ (resp. $s_j^* = 1$) if $r_j^* = 0$ (resp. $r_j^* = 1$). Output s^* as a good but not necessarily optimal solution for the MINSAT instance (B, b, c), together with the lower bound β, and stop.

The next section provides pre- and postprocessing steps that improve the solution algorithms produced by Algorithm ANALYSIS (13.4.4).

13.6 Pre- and Postprocessing

When Algorithm ANALYSIS (13.4.4) processes the matrix A of a SAT problem or the matrix/vector pair (A, c) of a MINSAT problem arising from real-world applications, one typically has additional information about the SAT or MINSAT instances to be solved. In this section, we show how such information can sometimes be utilized to improve the solution algorithms produced by Algorithm ANALYSIS (13.4.4).

The improvements we have in mind are of two types. Improvements of the first type reduce or simplify the matrix A before Algorithm ANALYSIS (13.4.4) is applied and thus are considered *preprocessing*.

Improvements of the second type streamline the solution algorithms generated by Algorithm ANALYSIS (13.4.4) and thus constitute *postprocessing*.

Preprocessing

Let A be a matrix over $\mathbb{B}$. Consider the SAT instances arising from the column submatrices B of A. Any such instance is defined by B and a $\{0, 1\}$ vector b and requires solution of the inequality system $B \odot s \geq b$. We reformulate that instance. We drop from B the rows i for which $b_i = 0$, getting a submatrix $\overline{A}$ of B, and declare that $\overline{A}$ is a SAT instance demanding solution of the inequality $\overline{A} \odot s \geq \underline{1}$. Under this viewpoint, the SAT instances are completely specified by the submatrices $\overline{A}$ of A. Analogously, one may restate MINSAT instances. For the discussion below, we consider SAT or MINSAT instances to be so reformulated.

Suppose we know in advance that certain rows and columns of A will be part of every SAT or MINSAT instance that is to be solved. Specifically, define $V \subseteq X$ (resp. $W \subseteq Y$) to be the index set of these rows (resp. columns). We utilize V and W for two reductions of A.

Reduction Using Resolution

The first reduction applies to SAT only. Suppose a column y of A has its 1s (resp. -1s) in rows indexed by a subset X_+ (resp. X_-) of X. Further suppose that for any $x \in X_+$ and any $z \in X_-$, there exists a column $w \in W$, $w \neq y$, that has, for some $\{\pm 1\}$ value α, an entry equal to α in row x and an entry equal to $-\alpha$ in row z.

Assume that we must solve a SAT instance given by a submatrix $\overline{A}$ of A. Let $\overline{A}$ have row index set $\overline{X}$ and column index set $\overline{Y}$. By assumption, $W \subseteq \overline{Y}$. We invoke Algorithm RESOLUTION FOR MATRIX (5.4.4) to eliminate column y from $\overline{A}$. We skip details of that algorithm and simply claim that, due to the assumption about nonzero entries in the submatrix of $\overline{A}$ indexed by $(X_+ \cup X_-) \cap \overline{X}$ and $\overline{Y} - \{y\}$, the algorithm eliminates column y and all rows indexed by $(X_+ \cup X_-) \cap \overline{X}$ and does not add any new rows. Since this conclusion is independent of the particular form of $\overline{A}$, we might as well proceed as follows.

We delete column y from A and find for the reduced matrix a solution algorithm with Algorithm ANALYSIS (13.4.4). If a SAT instance $\overline{A}$ arising from A and indexed by $\overline{X}$ and $\overline{Y}$ does not involve column y, then that solution algorithm is appropriate. Otherwise, we delete column y and all rows of $(X_+ \cup X_-) \cap \overline{X}$ from $\overline{A}$ and solve the reduced SAT instance, say, $\overline{A}'$, with the solution algorithm. If the latter problem is unsatisfiable, so is the SAT instance $\overline{A}$. Otherwise, we extend the satisfying solution for $\overline{A}'$ to one for $\overline{A}$ by assigning an appropriate *True/False* value to column y.

Reduction Using Special Row Submatrices

The second reduction applies to both SAT and MINSAT. Let A, X, Y, V, and W be as before. We determine constraints imposed by row submatrices of A that have 2SAT form, are nearly negative, or have the network property.

Collect in a matrix $\overline{A}$ all rows $v \in V$ of A satisfying the following conditions. Each such row must have at most two nonzero entries, and these entries must occur in columns indexed by W.

By repeated use of Algorithm SOLVE 2SAT (5.4.1), we determine which values s_y of the solution vector s for $\overline{A} \cdot s \geq 1$ are unique.

The analogous process can be carried out for the largest submatrix $\overline{A}$ of A where each row $v \in V$ has all nonzero entries in columns with index in W and where $\overline{A}$ is nearly negative or has the network property. This time, we repeatedly use Algorithm SOLVE NEARLY NEGATIVE SAT OR MINSAT (5.5.1) or Algorithm SOLVE BALANCED SAT OR MINSAT (5.7.25) to determine the s_y with unique solution value for $\overline{A} \cdot s \geq 1$.

As soon as an s_y with unique value has been found for $\overline{A}$, we record that value, reduce A accordingly, and repeat the above process.

Postprocessing

We revert to our customary way of stating SAT and MINSAT instances. That is, each SAT instance involves a column submatrix B of the given matrix A and a $\{0,1\}$ vector b and demands solution of the inequality system $B \odot s \geq b$. MINSAT instances are formulated correspondingly.

Let A have row index set X and column index set Y. As in the preprocessing case, we assume to have a set $W \subseteq Y$ that indexes the columns of A that will be part of every SAT or MINSAT instance to be solved. Note that we do not need the set V of the preprocessing case.

Postprocessing is done once Algorithm ANALYSIS (13.4.4) has produced a solution algorithm. The improvements apply to the use of Algorithm SOLVE CLOSED SUM SAT (10.4.18), Algorithm SOLVE AUGMENTED SUM SAT (11.4.10), and Algorithm SOLVE LINEAR SUM SAT OR MINSAT (12.4.15) as subroutines in the solution algorithm. We discuss each case.

Closed Sum Case

Algorithm SOLVE CLOSED SUM SAT (10.4.18) determines in Step 1 one subrange set and may compute in Step 3 sets J_0, J_+, J_-, and $J_\pm$. According to Algorithm SELECT TYPE II METHOD (13.4.6) invoked by Algorithm ANALYSIS (13.4.4), the underlying closed k-sum has $k \leq 3$. Hence, we may compute the subrange set and the sets J_0, J_+, J_-, and $J_\pm$ in advance for all possible cases, and later just look up the appropriate sets.

Augmented Sum Case

Steps 1, 4, and 5 of Algorithm SOLVE AUGMENTED SUM SAT (11.4.10) compute two subrange sets, a set S, and a vector f. According to Algorithm SELECT TYPE II METHOD (13.4.6) invoked by Algorithm ANALYSIS (13.4.4), the underlying augmented k-sum has $k \leq 4$. Hence, one may compute the subrange sets, the set S, and the vector f in advance for all possible cases and may replace the computations of Steps 1, 4, and 5 by straightforward lookups.

Linear Sum Case

Step 2 of Algorithm SOLVE LINEAR SUM SAT OR MINSAT (12.4.15) computes, for $i = 2, 3, \ldots, p$, subrange($[D^{i3}/D^{i2}]$) and subrange($[E^{i1}|E^{i2}]$). If k is not small, computation of these subrange sets may require considerable effort. Hence, one would want to compute the subrange sets one time, prior to solving SAT or MINSAT instances. This is possible if we know that the column submatrix B of A of any SAT or MINSAT instance is A itself, that is, if $W = Y$.

Suppose we do not have such assurance, but that the set $Y - W$, which contains the indices of the columns of A that might be deleted, is small. We then precompute certain sets that take the place of the sets subrange($[D^{i3}/D^{i2}]$) and subrange($[E^{i1}|E^{i2}]$).

We present details in a moment, but first note that validity of the change crucially depends on the fact that Algorithm SOLVE LINEAR SUM SAT OR MINSAT (12.4.15) remains valid if any subrange set computed in Step 2 is replaced by some set containing it. For a proof, one only needs to check that all results supporting Algorithm SOLVE LINEAR SUM SAT OR MINSAT (12.4.15) remain valid upon such a substitution. We leave it to the reader to confirm the correctness of this claim.

We describe the sets to be precomputed. Let $2 \leq i \leq p$. We partition A like B of (12.4.4), except that we add the hat symbol to each submatrix to differentiate it from the corresponding submatrix of B. For example, $\hat{D}^{i3}$ and $\hat{D}^{i2}$ are the submatrices of A that correspond to the submatrices D^{i3} and D^{i2}, respectively, of B.

We first create the set that contains subrange($[D^{i3}/D^{i2}]$). By the above definitions, the submatrix $[\hat{D}^{i3}/\hat{D}^{i2}]$ of A has $[D^{i3}/D^{i2}]$ as a column submatrix. Let Z_i be the column index set of $[\hat{D}^{i3}/\hat{D}^{i2}]$, and $J_i = Z_i - W$. Thus, J_i contains the indices of the columns of $[\hat{D}^{i3}/\hat{D}^{i2}]$ that are not present in $[D^{i3}/D^{i2}]$.

Define S_i to be the set

$$(13.6.1) \qquad S_i = \{s \mid s_j \in \{0, \pm 1\}, j \in J_i;\ s_j \in \{\pm 1\}, j \in Z_i - J_i\}$$

and let

$$(13.6.2) \qquad \text{range}([\hat{D}^{i3}/\hat{D}^{i2}], J_i) = \{b \mid b = [\hat{D}^{i3}/\hat{D}^{i2}] \odot s;\ s \in S_i\}$$

Evidently, range($[\hat{D}^{i3}/\hat{D}^{i2}], J_i$) $\supseteq$ subrange($[D^{i3}/D^{i2}]$), regardless of the choice of B. Thus, we take range($[\hat{D}^{i3}/\hat{D}^{i2}], J_i$) as the desired set.

For $i = 2, 3, \ldots, p$, we compute range($[\hat{D}^{i3}/\hat{D}^{i2}], J_i$) with Algorithm RANGE (4.3.11) as follows. First, we handle the case $i = 2$ by one application of that algorithm. Inductively, assume that, for some $2 \leq i < p$, we have range($[\hat{D}^{i3}/\hat{D}^{i2}], J_i$). Since (12.2.7) says that $[D^{i+1,3}/D^{i+1,2}] = [D^{i2}|D^{i1}]$, we analogously have

$$(13.6.3) \qquad [\hat{D}^{i+1,3}/\hat{D}^{i+1,2}] = [\hat{D}^{i2}|\hat{D}^{i1}]$$

Thus, range($[\hat{D}^{i+1,3}/\hat{D}^{i+1,2}], J_{i+1}$) is equal to range($[\hat{D}^{i2}|\hat{D}^{i1}], J_{i+1}$) and may be obtained from range($[\hat{D}^{i3}/\hat{D}^{i2}], J_i$) by projecting out the entries corresponding to $\hat{D}^{i3}$ and then extending the resulting set of vectors to range($[\hat{D}^{i2}|\hat{D}^{i1}]$, J_{i+1}). Algorithm RANGE (4.3.11) is designed to perform these operations efficiently.

Computation of the sets that contain the sets subrange($[E^{i1}|E^{i2}]$) is handled in an analogous manner, except that we start with $i = p$ and reduce the index i iteratively until $i = 2$.

The final section presents extensions and references.

13.7 Extensions and References

A number of improvements of Algorithm ANALYSIS (13.4.4) and its subroutines are possible.

One can enhance the subregion decomposition approach of Algorithm SELECT COMPONENT METHOD (13.3.1), using the ideas of Section 8.7. In addition, for MINSAT cases, one may consider Heuristic SOLVE MINSAT (13.5.3) as a possible solution algorithm in Algorithm SELECT COMPONENT METHOD (13.3.1).

One may add other, possibly specialized, decompositions of type I, II, or III and related subroutines to Algorithm ANALYSIS (13.4.4), without changing the basic structure of that algorithm.

The resolution-based preprocessing of Section 13.6, which deals with one variable at a time, can be extended so that links between any two variables are discovered and used to simplify the given matrix. Related material is discussed in Hansen (1976), Johnson and Padberg (1982), and Hansen, Jaumard, and Minoux (1986).

The postprocessing may, indeed should, include the improvements for linear decompositions presented in Section 12.5.

One can speed up the solution algorithms produced by Algorithm ANALYSIS (13.4.4) by inserting various heuristic methods—for example, the reduction scheme of Section 9.5.

Chapter 14

Central and Semicentral Classes

14.1 Overview

In this chapter, we construct large matrix classes that are SAT central or semicentral, as well as large matrix/vector classes that are MINSAT central or semicentral. Using an abbreviated terminology that simultaneously covers both the SAT and MINSAT cases, we refer to these classes simply as central or semicentral classes.

The chapter proceeds as follows. Section 14.2 provides an overview over the centrality and semicentrality results of the preceding chapters.

Section 14.3 assembles from the results of Section 14.2 two central and two semicentral classes.

Section 14.4 offers empirical evidence that the classes of Section 14.3 are of practical importance.

Finally, Section 14.5 provides extensions and references.

14.2 Review of Centrality and Semicentrality Results

A number of centrality and semicentrality results are contained in the preceding chapters. We collect them here for ready reference.

Definitions

Recall from Section 5.2 that a class of matrices A over $\mathbb{B}$ is SAT central if the following holds.

(14.2.1)

    (i)   If $A \in C$, then any submatrix of A is also in C.
    (ii)   There is a polynomial algorithm for solving the SAT instances given by the matrices of C.
    (iii) There is a polynomial algorithm for recognizing the matrices of C.

The class C is SAT semicentral if it observes (14.2.1)(i) and (ii).

A class C of matrix/vector pairs (A, c), where A is over $\mathbb{B}$ and c is a rational nonnegative vector, is MINSAT central if the following holds.

(14.2.2)

    (i)   If $(A, c) \in C$, then any submatrix pair of (A, c) is also in C.
    (ii)   There is a polynomial algorithm for solving the MINSAT instances given by the matrix/vector pairs of C.
    (iii) There is a polynomial algorithm for recognizing the matrix/vector pairs of C.

The class C is MINSAT semicentral if it observes (14.2.2)(i) and (ii).

Subclasses, Union, and Intersection

Centrality and semicentrality are preserved under certain reductions and under finite union and intersection of classes.

(14.2.3) Lemma. (See Lemma (5.3.1).) *Let C be a class of matrices or matrix/vector pairs. Define $\overline{C}$ to be a subclass of C that is maintained under submatrix taking. Then (a) and (b) below hold.*

(a) *If C is SAT or MINSAT semicentral, then $\overline{C}$ also has that property.*
(b) *Suppose membership in $\overline{C}$ can be tested in polynomial time provided that membership in C is known. If C is SAT or MINSAT central, then $\overline{C}$ also has that property.*

(14.2.4) Lemma. (See Lemma (5.3.2).) *For given $n \geq 2$, let $C_1, C_2, \ldots, C_n$ be classes of matrices or matrix/vector pairs. Assume that the classes have a given centrality property, that is, SAT or MINSAT centrality or semicentrality. Then the union and the intersection of these classes also have that property.*

Elementary Extensions

The next result concerns extensions of central or semicentral classes. According to Section 5.3, a matrix A over $\mathbb{B}$ is SAT simple if A has no rows with less than two nonzeros, no duplicate rows, and no parallel or monotone columns. The matrix A is MINSAT simple if A has no rows with less than two nonzeros, no duplicate rows, and no nonpositive columns.

A maximum SAT or MINSAT simple submatrix of a given matrix A is the submatrix derived by Algorithm SIMPLE SUBMATRIX (5.3.3) from A.

(14.2.5) Theorem. (See Theorem (5.3.6).) *Let C be a class of matrices that is maintained under submatrix taking, and let C' be a subclass of C. If C' is SAT central (resp. semicentral) and if C consists precisely of the matrices whose maximum SAT simple matrix is in C', then C is SAT central (resp. semicentral) as well. The MINSAT version of the above statements also holds provided that C and C' are classes of matrix/vector pairs.*

Special Matrix Classes

Sections 5.4–5.7 define the following special matrices and related classes.

A 2SAT matrix has at most two nonzeros in each row.

A nearly negative matrix has at most one $+1$ in each row. A hidden nearly negative matrix can be column scaled to become nearly negative.

A balanced matrix does not have any cycle submatrix whose entries sum to 2(mod 4). In a totally unimodular matrix, the determinant of any square submatrix is 0, $+1$, or -1. A network matrix is, up to transposition, a totally unimodular 2SAT matrix.

Chapter 5 establishes the following centrality results for these matrix classes.

(14.2.6) Theorem. (See Theorem (5.4.2).) *The class of 2SAT matrices is SAT central.*

(14.2.7) Theorem. (See Theorem (5.5.4).)
(a) *The class of nearly negative matrices is SAT central.*
(b) *The class of matrix/vector pairs (A, c) where A is nearly negative and c is a rational nonnegative vector is MINSAT central.*

(14.2.8) Theorem. (See Theorem (5.6.4).)
(a) *The class of hidden nearly negative matrices A is SAT central.*
(b) *The class of the following matrix/vector pairs (A, c) is MINSAT central. The vector c is rational nonnegative, and the matrix A is hidden nearly negative relative to the column submatrix E of A whose columns correspond to the zero entries of c.*

(14.2.9) Theorem. (See Theorem (5.7.28).)

(a) *The three classes consisting of the balanced matrices, the totally uni-modular matrices, and the matrices with the network property are SAT central.*

(b) *The three classes of matrix/vector pairs where the matrices are balanced, are totally unimodular, or have the network property are MIN-SAT central.*

Chapters 8–12 describe a number of decompositions and sums. The centrality and semicentrality results of those chapters are listed next.

Classes Based on Closed Subregion Decomposition

We begin with a theorem for the closed subregion decompositions of Chapter 8.

(14.2.10) Theorem. (See Theorem (8.2.11).) *Let C be a class of matrices A (resp. matrix/vector pairs (A, c)) each of which belongs to a given SAT (resp. MINSAT) semicentral class C' or has, for some q bounded by a constant, a closed subregion decomposition into A^0, $A^1, \ldots, A^q$ where A^0 (resp. (A^0, c)) is in C'. Then C is SAT or MINSAT semicentral, whichever applies.*

Classes Based on Monotone Sums

For the monotone sums of Chapter 9, we have the following result.

(14.2.11) Theorem. (See Theorem (9.4.12).)

(a) *Let C_0 be a class of SAT central (resp. semicentral) matrices. Then the class C of monotone sums $B = B^1 \boxplus_m B^2$ where the submatrix A^2 of B^2 is in C_0 is SAT central (resp. semicentral).*

(b) *Let C_0 be a class of MINSAT central (resp. semicentral) matrix/vector pairs. Then the class C of matrix/vector pairs (B, c) for which B is a monotone sum $B = B^1 \boxplus_m B^2$ and for which the submatrix A^2 of B^2 and the related subvector c^2 of c form a matrix/vector pair (A^2, c^2) of C_0, is MINSAT central (resp. semicentral).*

Classes Based on Closed Sums

The closed sums of Chapter 10 are treated next.

(14.2.12) Theorem. (See Theorem (10.5.4).) *Let C_0 be a SAT central class of matrices. Define C to be the class of matrices created from C_0 by repeated closed k-sum steps where k is bounded by a constant. Then C is SAT central.*

Classes Based on Augmented Sums

We turn to the augmented sums of Chapter 11.

(14.2.13) Theorem. (See Theorem (11.4.14).) *Let C_0 be a SAT semicentral class of matrices. Define C to be a class created from C_0 by augmented sums where each $|X_{21} \cup Y_{21}|$ and the number of recursive construction steps are bounded by constants. Then C is SAT semicentral.*

Classes Based on Linear Sums

Finally, we cover the linear sums of Chapter 12.

(14.2.14) Theorem. (See Theorem (12.4.19).)
(a) *Let C_0 be a SAT semicentral class of matrices. Enlarge C_0 to a class C by adding all possible linear sums where, in the notation of (12.4.3), the column submatrix $[E^{i1}/A^i/D^{i1}]$ of each component B^i is in C_0 and where k defined by (12.2.14) is bounded by a constant. Then C is SAT semicentral.*
(b) *Let C_0 be a MINSAT semicentral class of matrix/vector pairs. Enlarge C_0 to a class C by adding all pairs (A, c) satisfying the following conditions. The matrix A is a linear sum where, in the notation of (12.4.3), the column submatrix $[E^{i1}/A^i/D^{i1}]$ of each component B^i and the corresponding subvector c^i of c constitute a pair $([E^{i1}/A^i/D^{i1}], c^i)$ in C_0 and where k defined by (12.2.14) is bounded by a constant. Then C is MINSAT semicentral.*

14.3 Construction of Central and Semicentral Classes

We construct several central and semicentral classes.

SAT Central Classes

Let S_0 be the union of the classes of 2SAT matrices, hidden nearly negative matrices, and balanced matrices.

Extend S_0 to a class S_1 by enlarging each $A \in S_0$ by the following operations, applied in all possible ways. Add to A duplicate rows, rows with at most one nonzero entry, and monotone or parallel columns.

Extend S_1 to S_2 by composing matrices of S_1 in repeated 1-sum composition steps, in all possible ways.

Extend S_2 to S_3 by monotone sum compositions, in all possible ways where, in the notation of (9.2.1), the matrix A^2 is taken from S_2.

In subsequent constructions, we repeatedly use the 1-sum and monotone compositions that extend S_1 to S_3. To describe this process, we then simply say that a given class is extended to another class by 1-sum and monotone sum compositions.

Create S_4 from S_3 by repeated closed k-sum steps where k is bounded by a constant.

Extend S_4 to S_5 by 1-sum and monotone sum compositions.

(14.3.1) Theorem. *The classes S_0–S_5 are SAT central.*

Proof. The applicable results of Section 14.2 prove the claim. □

MINSAT Central Classes

We use a more restricted construction for MINSAT central classes. Let M_0 be the class of pairs (A, c) where A is balanced or hidden nearly negative relative to the column submatrix of A whose columns correspond to the zero entries of c.

Extend M_0 to M_1 by enlarging each $(A, c) \in M_0$ by the following operations, applied in all possible ways. Add to A duplicate rows, rows with at most one nonzero entry, and nonpositive columns. Extend c to another nonnegative rational vector.

Extend M_1 to M_2 by 1-sum and monotone sum compositions analogously to the extension of S_1 to S_3.

(14.3.2) Theorem. *The classes M_0–M_2 are MINSAT central.*

Proof. The applicable results of Section 14.2 prove the theorem. □

We proceed to larger matrix or matrix/vector classes, this time aiming at semicentrality.

SAT Semicentral Classes

Define $T_0 = S_0$.

Extend T_0 to T_1 by adding all possible matrices with closed subregion decompositions $A^0, A^1, \ldots, A^q$ where A^0 is taken from T_0 and where q is bounded by a constant.

Extend T_1 to T_2 by enlarging each $A \in T_1$ by the following operations, applied in all possible ways. Add to A duplicate rows, rows with at most one nonzero entry, and monotone or parallel columns.

Extend T_2 to T_3 by 1-sum and monotone sum compositions.

Extend T_3 to T_4 by adding linear sums B where, in the notation of (12.2.3), the column submatrix $[E^{i1}/A^i/D^{i1}]$ of each component B^i is in T_3 and where k defined by (12.2.14) is bounded by a constant.

Extend T_4 to T_5 by 1-sum and monotone sum compositions.

Create T_6 from T_5 by repeated closed k-sum steps and augmented sum steps. The construction process is nothing but a mixing of the two recursive construction processes described in Chapters 10 and 11. Thus, in each step one either carries out a closed k-sum composition or an augmented sum composition. The parameter k of the closed sum case and $|X_{21} \cup Y_{21}|$ of the augmented sum case must be bounded by a constant. In addition, the number of augmented sum compositions used to create any matrix must be bounded by a constant.

Extend T_6 to T_7 by 1-sum and monotone sum compositions.

(14.3.3) Theorem. *The classes T_0–T_7 are SAT semicentral.*

Proof. The applicable results of Section 14.2 prove the claim for T_0–T_5 and T_7. The SAT semicentrality of T_6 follows easily from the proofs of Theorems (10.5.4) and (11.4.14). □

MINSAT Semicentral Classes

We restrict the above construction of T_0–T_7 to obtain MINSAT semicentral classes. Let $N_0 = M_0$.

Extend N_0 to N_1 using closed subregion decompositions analogously to the extension of T_0 to T_1.

Extend N_1 to N_2 using duplicate rows, rows with at most one nonzero entry, and nonpositive columns, as in the extension of M_0 to M_1.

Extend N_2 to N_3 by 1-sum and monotone compositions as in the extension of M_1 to M_2.

Extend N_3 to N_4 using linear sums analogously to the extension of T_3 to T_4.

Extend N_4 to N_5 using 1-sum and monotone sum compositions.

(14.3.4) Theorem. *The classes N_0–N_5 are MINSAT semicentral.*

Proof. The applicable results of Section 14.2 supply the conclusion. □

14.4 Link to Analysis Algorithm

The construction of the central and semicentral classes of Section 14.3 is inverse to the way in which decomposition and recognition algorithms are employed in Algorithm ANALYSIS (13.4.4). It would be pleasing if we could show that the matrices or matrix/vector pairs for which Algorithm ANALYSIS (13.4.4) produces attractive solution algorithms correspond to classes of Section 14.3. But that is not possible, since, for example, Algorithm ANALYSIS (13.4.4) relies on some heuristics and restricts the search for certain k-sum decompositions to small values of k. However, one can prove a somewhat weaker result, which says that the SAT or MINSAT instances for which Algorithm ANALYSIS (13.4.4) produces fast solution algorithms belong to the semicentral classes of Section 14.3. In this section, we derive that result and use it along with some empirical evidence to argue that the semicentral classes of Section 14.3 are of practical importance.

We begin with a definition. A solution algorithm $\mathcal{M}$ produced by Algorithm ANALYSIS (13.4.4) is *attractive* if the time bound computed for $\mathcal{M}$ is small enough to guarantee fast or at least reasonable execution times. The definition implies that a solution algorithm $\mathcal{M}$ that today is judged to be unattractive may become attractive in the future due to an improved performance of computers. This is perfectly reasonable.

The constructions of the semicentral classes T_0–T_7 and N_0–N_5 assume bounds on the parameter q of closed subregions decompositions, on the parameter k of closed k-sums and linear sums, and on the value of $|X_{21} \cup Y_{21}|$ of augmented sums. For present purposes, we assume that the bounds on k and $|X_{21} \cup Y_{21}|$ are sufficiently large so that they do not exclude cases for which the corresponding decomposition case is accepted by Algorithm ANALYSIS (13.4.4). We also assume that the bound on q is of reasonable size.

Since T_7 contains S_0–S_5 and T_0–T_6 and since N_5 contains M_0–M_2 and N_0–N_4, we just link T_7 and N_5 to the SAT and MINSAT instances for which Algorithm ANALYSIS (13.4.4) generates attractive solution algorithms $\mathcal{M}$.

(14.4.1) Theorem. *Suppose Algorithm ANALYSIS (13.4.4) produces an attractive solution algorithm $\mathcal{M}$ for a given SAT or MINSAT instance. If it is a SAT case, assume that the number of augmented sum decompositions employed by $\mathcal{M}$ is bounded by the same constant used for such decompositions in the definition of T_7. Then that instance is in T_7 or N_5, whichever is applicable.*

Proof. We sketch the arguments, using the material of Sections 13.3 and 13.4. In the general case, $\mathcal{M}$ utilizes monotone, 1-sum, closed, augmented, linear, and closed subregion decompositions and the related solution algorithms of Chapters 8–12, as well as the solution algorithms for the special matrix classes of Chapter 5.

Regardless of the case of $\mathcal{M}$, it is straightforward to check that the decomposition sequences utilized by $\mathcal{M}$ correspond to composition sequences used in the construction of T_7 or N_5, whichever is applicable.

Since $\mathcal{M}$ is attractive, the values of q of the closed subregion decompositions must be small or at least reasonable. The proof of the theorem utilizes that fact; the earlier made assumption on the bounds on q, k, and $|X_{21} \cup Y_{21}|$ used in the construction of T_7 or N_5; the assumption of the theorem on the number of augmented sums used in SAT cases; and the above observation linking the sequences of decompositions in the construction of

$\mathcal{M}$ to the composition sequences creating T_7 or N_5.   ☐

A number of years ago, we began to implement Algorithm ANALY-SIS (13.4.4). We call the resulting software the *Leibniz System*, to honor G. W. Leibniz (1646–1716), who first proposed that logic computations should be employed to solve real-world problems. The implementation effort is not yet finished and is continuing as we write this book. Nevertheless, the present version of the Leibniz System is sufficiently powerful to be practically useful. That version has been used to create solution algorithms for hundreds of SAT and MINSAT instances. In almost all cases, the solution algorithm produced by the Leibniz System has turned out to have a small upper time bound and thus to be attractive. According to Theorem (14.4.1), the underlying matrices or matrix/vector pairs of these cases are in T_7 or N_5, respectively, assuming in the SAT case that an appropriate bound on the number of augmented sums is used in the construction of T_7.

These results constitute empirical evidence that T_7 and N_5 contain a large number of SAT and MINSAT instances of real-world applications and that Algorithm ANALYSIS (13.4.4) produces effective solution algorithms for these applications.

14.5 Extensions and References

Section 13.7 mentions that Algorithm ANALYSIS (13.4.4) may be enhanced by additional, possibly specialized, decompositions and related subroutines, without changing the basic structure of that algorithm. Similarly, one may enlarge the central and semicentral classes of this chapter by adding the compositions corresponding to such additional decompositions.

A number of previously published SAT matrix classes are contained in some of the classes described in this chapter—in particular, the matrices with *bounded bandwidth* of Monien and Sudborough (1985), where the rows are indexed by $1, 2, 3 \ldots$, and where any two nonzero entries of any column, say, in rows indexed by i and j, must have $|i - j|$ bounded by some constant; the instances of Gallo and Scutellà (1988), which are based on a class of Yamasaki and Doshita (1983); the instances given indirectly by Gallo and Urbani (1989) according to a rewrite rule; and the q-Horn instances of Boros, Crama, and Hammer (1990). Except for the first reference, details about the cited classes are included in Sections 8.7 and 9.5.

Related to, but not contained in, the classes of this chapter are the classes of Dalal and Etherington (1992) and of Pretolani (1993a, 1996). The cited classes are discussed in Section 8.7.

References

Agarwal, S., Sharma, P., and Mittal, A. K. (1982), An extension of the edge covering problem, *Mathematical Programming* 23 (1982) 353–356.

Ahuja, R. K., Magnanti, T. L., and Orlin, J. B. (1993), *Network Flows*, Prentice-Hall, Englewood Cliffs, New Jersey, 1993.

Aigner, M. (1979), *Combinatorial Theory*, Springer-Verlag, Berlin, 1979.

Ajtai, M. (1994), The complexity of the pigeonhole principle, *Combinatorica* 14 (1994) 417–433.

Applegate, D., and Cook, W. (1993), Solving large-scale matching problems, in: *Network Flows and Matching: First DIMACS Implementation Challenge* (D. S. Johnson and C. C. McGeoch, eds.), DIMACS Series in Discrete Mathematics and Theoretical Computer Science, Vol. 12, American Mathematical Society, Providence, Rhode Island, 1993, pp. 557–576.

Arora, S., Lund, C., Motwani, R., Sudan, M., and Szegedy, M. (1992), Proof verification and hardness of approximation problems, in: *Proceedings of 33rd Symposium on Foundations of Computer Science*, Pittsburgh, Pennsylvania, 1992, IEEE Computer Society Press, Los Alamitos, California, 1992, pp. 14–23.

Arvind, V., and Biswas, S. (1987), An $O(n^2)$ algorithm for the satisfiability problem of a subset of propositional sentences in CNF that includes all Horn sentences, *Information Processing Letters* 24 (1987) 67–69.

Aspvall, B. (1980), Recognizing disguised NR(1) instances of the satisfiability problem, *Journal of Algorithms* 1 (1980) 97–103.

Aspvall, B., Plass, M. F., and Tarjan, R. E. (1979), A linear-time algorithm for testing the truth of certain quantified Boolean formulas, *Information Processing Letters* 8 (1979) 121–123.

Avron, A. (1993), Gentzen-type systems, resolution and tableaux, *Journal of Automated Reasoning* 10 (1993) 265–281.

Bagchi, A., Servatius, B., and Shi, W. (1995), 2-satisfiability and diagnosing faulty processors in massively parallel computing systems, *Discrete Applied Mathematics* 60 (1995) 25–37.

Bartholdi, J. J., III (1982), A good submatrix is hard to find, *Operations Research Letters* 1 (1982) 190–193.

Bartholdi, J. J., III, Orlin, J. B., and Ratliff, H. D. (1980), Cyclic scheduling via integer programs with circular ones, *Operations Research* 28 (1980) 1074–1085.

Bellman, R. (1957), *Dynamic Programming*, Princeton University Press, Princeton, New Jersey, 1957.

Bellman, R. E., and Dreyfus, S. E. (1962), *Applied Dynamic Programming*, Princeton University Press, Princeton, New Jersey, 1962.

Ben-Ari, M. (1980), A simplified proof that regular resolution is exponential, *Information Processing Letters* 10 (1980) 96–98.

Berge, C. (1972), Balanced matrices, *Mathematical Programming* 2 (1972) 19–31.

Berge, C. (1973), *Graphs and Hypergraphs*, North-Holland, Amsterdam, 1973.

Berman, K. A., Franco, J. V., and Schlipf, J. S. (1995), Unique satisfiability of Horn sets can be solved in nearly linear time, *Discrete Applied Mathematics* 60 (1995) 77–91.

Bertsekas, D. P. (1987), *Dynamic Programming*, Prentice-Hall, Englewood Cliffs, New Jersey, 1987.

Bibel, W. (1990), Short proofs of the pigeonhole formulas based on the connection method, *Journal of Automated Reasoning* 6 (1990) 287–297.

Bibel, W. (1993), *Deduction*, Academic Press, London, 1993.

Billionnet, A., and Sutter, A. (1992), An efficient algorithm for the 3-satisfiability problem, *Operations Research Letters* 12 (1992) 29–36.

Blair, C. E., Jeroslow, R. G., and Lowe, J. K. (1986), Some results and experiments in programming techniques for propositional logic, *Computers and Operations Research* 13 (1986) 633–645.

Blass, A., and Gurevich, Y. (1982), On the unique satisfiability problem, *Information and Control* 55 (1982) 80–88.

Böhm, M. (1996), *Verteilte Lösung harter Probleme: Schneller Lastausgleich*, thesis, University of Cologne, published by Shaker-Verlag, Aachen, Germany, 1996.

Böhm, M., and Speckenmeyer, E. (1996), A fast parallel SAT-solver – efficient workload balancing, *Annals of Mathematics and Artificial Intelligence* 9 (1996) 1–20.

Bondy, J. A., and Murty, U. S. R. (1976), *Graph Theory with Applications*, Macmillan, London, 1976.

Boros, E., and Čepek, O. (1994), On the complexity of Horn minimization, RUTCOR Research Report RRR 1-94, Rutgers University, 1994.

Boros, E., and Čepek, O. (1995), Perfect $0, \pm 1$ matrices, *Discrete Mathematics*, to appear.

Boros, E., Crama, Y., and Hammer, P. L. (1990), Polynomial-time inference of all valid implications for Horn and related formulae, *Annals of Mathematics and Artificial Intelligence* 1 (1990) 21–32.

Boros, E., Crama, Y., Hammer, P. L., and Saks, M. (1994), A complexity index for satisfiability problems, *SIAM Journal on Computing* 23 (1994) 45-49.

Boros, E., and Hammer, P. L. (1992), A generalization of the pure literal rule for the satisfiability problem, RUTCOR Research Report RRR 20-92, Rutgers University, 1992.

Boros, E., Hammer, P. L., and Sun, X. (1994), Recognition of q-Horn formulae in linear time, *Discrete Applied Mathematics* 55 (1994) 1–13.

Brown, C. A., and Purdom, P. W., Jr. (1981), An average time analysis of backtracking, *SIAM Journal on Computing* 10 (1981) 583–593.

Bryant, R. E. (1986), Graph-based algorithms for Boolean function manipulation, *IEEE Transactions on Computers* C-35 (1986) 677–691.

Bugrara, K. M., Pan, Y., and Purdom, P. W., Jr. (1989), Exponential average time for the pure literal rule, *SIAM Journal on Computing* 18 (1989) 409–418.

Bugrara, K. M., and Purdom, P. W. (1988), An exponential lower bound for the pure literal rule, *Information Processing Letters* 27 (1988) 215–219.

Buro, M., and Kleine Büning, H. (1993), Report on a SAT competition, *Bulletin of the European Association for Theoretical Computer Science* 49 (1993) 143–151.

Buss, S. R. (1987), Polynomial size proofs of the propositional pigeonhole principle, *Journal of Symbolic Logic* 52 (1987) 916–927.

Buss, S. R., and Turán, G. (1988), Resolution proofs of generalized pigeonhole principles, *Theoretical Computer Science* 62 (1988) 311–317.

Carraresi, P., Gallo, G., and Rago, G. (1993), A hypergraph model for constraint logic programming and applications to bus drivers' scheduling, *Annals of Mathematics and Artificial Intelligence* 8 (1993) 247–270.

Čepek, O. (1995), *Structural properties and minimization of Horn Boolean functions*, thesis, Rutgers University, 1995.

Chandrasekaran, R. (1970), A special case of the complementary pivot problem, *Opsearch* 7 (1970) 263–268.

Chandrasekaran, R. (1984), Integer programming problems for which a simple rounding type algorithm works, in: *Progress in Combinatorial Optimization* (W. R. Pulleyblank, ed.), Proceedings of the first week of the Silver Jubilee Conference on Combinatorics, University of Waterloo, 1982, Academic Press Canada, Toronto, 1984, pp. 101–106.

Chandrasekaran, R., Kabadi, S. N., and Lakshminarayanan, S. (1996), An extension of a theorem of Fulkerson and Gross, *Linear Algebra and Its Applications* 246 (1996) 23–29.

Chandru, V., Coullard, C. R., and Montañez, M. (1988), On Horn and related structures in propositional logic, working paper CC-88-32, Purdue University, 1988.

Chandru, V., Coullard, C. R., Hammer, P. L., Montañez, M., and Sun, X. (1990), On renamable Horn and generalized Horn functions, *Annals of Mathematics and Artificial Intelligence* 1 (1990) 33–47.

Chandru, V., and Hooker, J. N. (1991), Extended Horn sets in propositional logic, *Journal of the Association for Computing Machinery* 38 (1991) 205–221.

Chandru, V., and Hooker, J. N. (1992), Detecting embedded Horn structure in propositional logic, *Information Processing Letters* 42 (1992) 109–111.

Chandru, V., and Hooker, J. N. (1997), *Optimization Methods for Logical Inference*, in preparation.

Chang, C. L. (1970), The unit proof and the input proof in theorem proving, *Journal of the Association for Computing Machinery* 17 (1970) 698–707.

Chang, C.-L., and Lee, R. C.-T. (1973), *Symbolic Logic and Mechanical Theorem Proving*, Academic Press, New York, 1973.

Cheriyan, J., Cunningham, W. H., Tunçel, L., and Wang, Y. (1996), A linear programming and rounding approach to Max 2-SAT, in: *Cliques, Coloring, and Satisfiability: Second DIMACS Implementation Challenge* (D. S. Johnson and M. A. Trick, eds.), DIMACS Series in Discrete Mathematics and Theoretical Computer Science, Vol. 26, American Mathematical Society, Providence, Rhode Island, 1996, pp. 395–414.

Chvátal, V. (1983), *Linear Programming*, Freeman, New York, 1983.

Chvátal, V., and Szemerédi, E. (1988), Many hard examples for resolution, *Journal of the Association for Computing Machinery* 35 (1988) 759–768.

Cohn, P. M. (1982), *Algebra*, Vol. 1 (2nd edition), Wiley, Chichester, 1982.

Conforti, M., and Cornuéjols, G. (1992), A class of logic problems solvable by linear programming, in: *Proceedings of 33rd Symposium on Foundations of Computer Science*, Pittsburgh, Pennsylvania, 1992, IEEE Computer Society Press, Los Alamitos, California, 1992, pp. 670–675. Also, *Journal of the Association for Computing Machinery* 42 (1995) 1107–1113.

Conforti, M., and Cornuéjols, G. (1995), Balanced 0, ±1 matrices, bicoloring and total dual integrality, *Mathematical Programming (A)* 71 (1995) 249–258.

Conforti, M., Cornuéjols, G., and de Francesco, C. (1997), Perfect 0, ±1 matrices, *Linear Algebra and Its Applications* 253 (1997) 299–309.

Conforti, M., Cornuéjols, G., Kapoor, A., and Vušković, K. (1994a), Balanced {0, ±1} matrices. I. Decomposition, working paper, Carnegie-Mellon University, 1994.

Conforti, M., Cornuéjols, G., Kapoor, A., and Vušković, K. (1994b), Balanced {0, ±1} matrices. II. Recognition Algorithm, working paper, Carnegie-Mellon University, 1994.

Conforti, M., Cornuéjols, G., Kapoor, A., and Vušković, K. (1996), Perfect, ideal and balanced matrices, *Ricerca Operativa* 26 (1996) 66–80. Also appeared in *Annotated Bibliographies in Combinatorial Optimization* (M. Dell'Amico, F. Maffioli, and S. Martello, eds.), 1997, pp. 81–94.

Conforti, M., Cornuéjols, G., Kapoor, A., Vušković, K., Rao, M. R. (1994), Balanced matrices, in: *Mathematical Programming: State of the Art 1994* (J. R. Birge and K. G. Murty, eds.), Proceedings of 15th International Symposium on Mathematical Programming, University of Michigan, 1994, pp. 1–33.

Conforti, M., Cornuéjols, G., and Rao, M. R. (1997), Decomposition of balanced matrices, working paper, Carnegie-Mellon University, 1997.

Conforti, M., Cornuéjols, G., and Truemper, K. (1994), From totally unimodular to balanced {0, ±1} matrices: A family of integer polytopes, *Mathematics of Operations Research* 19 (1994) 21–23.

Cook, S. A. (1971), The complexity of theorem-proving procedures, in: *Proceedings of Third Annual ACM Symposium on Theory of Computing*, Shaker Heights, Ohio, 1971. ACM, New York, 1971, pp. 151–158.

Cook, S. A. (1976), A short proof of the pigeon principle using extended resolution, *ACM SIGACT News* 8 (1976) 28–32.

Cook, S., and Pitassi, T. (1990), A feasibly constructive lower bound for resolution proofs, *Information Processing Letters* 34 (1990) 81–85.

Cook, S. A., and Reckhow, R. A. (1979), The relative efficiency of propositional proof systems, *Journal of Symbolic Logic* 44 (1979) 36–50.

Cook, W., Coullard, C. R., and Turán, G. (1987), On the complexity of cutting-plane proofs, *Discrete Applied Mathematics* 18 (1987) 25–38.

Cottle, R. W., Pang, J.-S., and Stone, R. E. (1992), *The Linear Complementarity Problem*, Academic Press, Boston, 1992.

Crama, Y., Ekin, O., and Hammer, P. L. (1997), Variable and term removal from Boolean formulae, *Discrete Applied Mathematics* 75 (1997) 217–230.

Crawford, J. M., and Auton, L. D. (1996), Experimental results on the crossover point in random 3-SAT, *Artificial Intelligence* 81 (1996) 31–57.

Cunningham, W. H., and Edmonds, J. (1980), A combinatorial decomposition theory, *Canadian Journal of Mathematics* 32 (1980) 734–765.

Dalal, M., and Etherington, D. W. (1992), A hierarchy of tractable satisfiability problems, *Information Processing Letters* 44 (1992) 173–180.

Davis, M., and Putnam, H. (1960), A computing procedure for quantification theory, *Journal of the Association for Computing Machinery* 7 (1960) 201–215.

Davis, M., Logemann, G., and Loveland, D. W. (1962), A machine program for theorem-proving, *Communications of the Association for Computing Machinery* 5 (1962) 394–397.

Derigs, U., and Metz, A. (1991), Solving (large scale) matching problems combinatorially, *Mathematical Programming (A)* 50 (1991) 113–121.

Dilworth, R. P. (1950), A decomposition theorem for partially ordered sets, *Annals of Mathematics* 51 (1950) 161–166.

Dorfman, R., Samuelson, P. A., and Solow, R. M. (1958), *Linear Programming and Economic Analysis*, McGraw-Hill, New York, 1958.

Dowling, W. F., and Gallier, J. H. (1984), Linear-time algorithms for testing the satisfiability of propositional Horn formulae, *Journal of Logic Programming* 1 (1984) 267–284.

Dreyfus, S. E., and Law, A. M. (1977), *The Art and Theory of Dynamic Programming*, Academic Press, New York, 1977.

Dubois, O. (1991), Counting the number of solutions for instances of satisfiability, *Theoretical Computer Science* 81 (1991) 49–64.

Dubois, O., and Carlier, J. (1991), Probabilistic approach to the satisfiability problem, *Theoretical Computer Science* 81 (1991) 65–75.

Edmonds, J. (1965a), Maximum matching and a polyhedron with $\{0,1\}$ vertices, *Journal of Research of the National Bureau of Standards (B)* 69B (1965) 125–130.

Edmonds, J. (1965b), Paths, trees, and flowers, *Canadian Journal of Mathematics* 17 (1965) 449–467.

Edmonds, J. (1967), Systems of distinct representatives and linear algebra, *Journal of Research of the National Bureau of Standards (B)* 71B (1967) 241–245.

Eiter, T., Kilpeläinen, P., and Mannila, H. (1995), Recognizing renamable generalized propositional Horn formulas is NP-complete, *Discrete Applied Mathematics* 59 (1995) 23–31.

Ekin, O., Hammer, P. L., and Peled, U. N. (1997), Horn functions and submodular Boolean functions, *Theoretical Computer Science* 175 (1997) 257–270.

Elias, P., Feinstein, A., and Shannon, C. E. (1956), A note on the maximum flow through a network, *IRE Transactions on Information Theory* IT-2 (1956) 117-119.

Evan, S. A., Itai, A., and Shamir, A. (1976), On the complexity of timetable and multicommodity flow problems, *SIAM Journal on Computing* 5 (1976) 691–703.

Faddeev, D. K., and Faddeeva, V. N. (1963), *Computational Methods of Linear Algebra*, Freeman, San Francisco, 1963.

Feige, U., and Goemans, M. X. (1995), Approximating the value of two prover proof systems, with applications to MAX2SAT and MAXDICUT, in: *Proceedings of the 3rd Israel Symposium on the Theory of Computing and Systems*, 1995, pp. 182–189.

Fitting, M. (1990), *First-Order Logic and Automated Theorem Proving*, Springer-Verlag, New York, 1990.

Ford, L. R., Jr., and Fulkerson, D. R. (1956), Maximal flow through a network, *Canadian Journal of Mathematics* 8 (1956) 399–404.

Ford, L. R., Jr., and Fulkerson, D. R. (1962), *Flows in Networks*, Princeton University Press, Princeton, New Jersey, 1962.

Fouks, J.-D. (1992), Tseitin's formulas revisited, *Theoretical Computer Science* 99 (1992) 315–326.

Franco, J. (1983), Probabilistic analysis of the Davis Putnam procedure for solving the satisfiability problem, *Discrete Applied Mathematics* 5 (1983) 77–87.

Franco, J. (1986), On the probabilistic performance of algorithms for the satisfiability problem, *Information Processing Letters* 23 (1986) 103–106.

Franco, J. (1991), Elimination of infrequent variables improves average case performance of satisfiability algorithms, *SIAM Journal on Computing* 20 (1991) 1119–1127.

Franco, J. (1993), On the occurrence of null clauses in random instances of satisfiability, *Discrete Applied Mathematics* 41 (1993) 203–209.

Franco, J., and Ho, Y. C. (1988), Probabilistic performance of a heuristic for the satisfiability problem, *Discrete Applied Mathematics* 22 (1988/89) 35–51.

Franco, J., and Paull, M. (1983), Probabilistic analysis of the Davis Putnam procedure for solving the satisfiability problem, *Discrete Applied Mathematics* 5 (1983) 77–87.

Franco, J., and Swaminathan, R. P. (1997a), Average case results for satisfiability algorithms under the random clause width model, *Annals of Mathematics and Artificial Intelligence* 20 (1997) 357–391.

Franco, J., and Swaminathan, R. P. (1997b), Toward a good algorithm for determining unsatisfiability of propositional formulas, *Journal of Combinatorial Optimization*, to appear.

Freeman, J. W. (1996), Hard random 3-SAT problems and the Davis–Putnam procedure, *Artificial Intelligence* 81 (1996) 183–198.

Fulkerson, D. R., Hoffman, A. J., and Oppenheim, R. (1974), On balanced matrices, *Mathematical Programming Study* 1 (1974) 120–132.

Galil, Z. (1977a), On the complexity of regular resolution and the Davis–Putnam procedure, *Theoretical Computer Science* 4 (1977) 23–46.

Galil, Z. (1977b), On resolution with clauses of bounded size, *SIAM Journal on Computing* 6 (1977) 444–459.

Gallo, G., Gentile, C., Pretolani, D., and Rago, G. (1997), Max Horn SAT and the minimum cut problem in directed hypergraphs, *Mathematical Programming (A)*, to appear.

Gallo, G., and Pretolani, D. (1995), A new algorithm for the propositional satisfiability problem, *Discrete Applied Mathematics* 60 (1995) 159–179.

Gallo, G., and Rago, G. (1994), The satisfiability problem for the Schöenfinkel-Bernays fragment: Partial instantiation and hypergraph algorithms, working paper TR-4/94, Department of Computer Science, University of Pisa, Italy, 1994.

Gallo, G., and Scutellà, M. G. (1988), Polynomially Solvable Satisfiability Problems, *Information Processing Letters* 29 (1988) 221–227.

Gallo, G., and Urbani, G. (1989), Algorithms for testing the satisfiability of propositional formulae, *Journal of Logic Programming* 7 (1989) 45–61.

Garey, M. R., and Johnson, D. S. (1979), *Computers and Intractability: A Guide to the Theory of NP-Completeness*, Freeman, San Francisco, 1979.

Garey, M. R., Johnson, D. S., and Stockmeyer, L. (1976), Some simplified NP-complete graph problems, *Theoretical Computer Science* 1 (1976) 237–267.

Garfinkel, R. S., and Nemhauser, G. L. (1972), *Integer Programming*, Wiley, New York, 1972.

van Gelder, A. (1988), A satisfiability tester for non-clausal propositional calculus, *Information and Computation* 79 (1988) 1–21.

Genesereth, M. R., and Nilsson, N. J. (1987), *Logical Foundations of Artificial Intelligence*, Morgan Kaufmann, Los Altos, California, 1987.

Gent, I. P., and Walsh, T. (1996), The satisfiability constraint gap, *Artificial Intelligence* 81 (1996) 59–80.

Ghallab, M., and Escalada-Imaz, E. (1991), A linear control algorithm for a class of rule-based systems, *Journal of Logic Programming* 11 (1991) 117–132.

Goemans, M. X., and Williamson, D. P. (1994), New $\frac{3}{4}$-approximation algorithms for the maximum satisfiability problem, *SIAM Journal on Discrete Mathematics* 7 (1994) 656–666.

Goemans, M. X., and Williamson, D. P. (1995), Improved approximation algorithms for maximum cut and satisfiability problems using semidefinite programming, *Journal of the Association for Computing Machinery* 42 (1995) 1115–1145.

Goerdt, A. (1992a), Davis–Putnam resolution versus unrestricted resolution, *Annals of Mathematics and Artificial Intelligence* 6 (1992) 169–184.

Goerdt, A. (1992b), Unrestricted resolution versus N-resolution, *Theoretical Computer Science* 93 (1992) 159–167.

Goerdt, A. (1993), Regular resolution versus unrestricted resolution, *SIAM Journal on Computing* 22 (1993) 661–683.

Goldberg, A., Purdom, P. W., Jr., and Brown, C. A. (1982), Average time analysis of simplified Davis–Putnam procedures, *Information Processing Letters* 15 (1982) 72–75. Corrigendum, *Information Processing Letters* 16 (1983) 213.

Gomory, R. E. (1965), On the relation between integer and noninteger solutions to linear programs, *Proceedings of the National Academy of Sciences* 53 (1965) 260–265.

Gomory, R. E. (1967), Faces of an integer polyhedron, *Proceedings of the National Academy of Sciences* 57 (1967) 16–18.

Gomory, R. E. (1969), Some polyhedra related to combinatorial problems, *Linear Algebra and Its Applications* 2 (1969) 451–558.

Grötschel, M., Lovász, L., and Schrijver, A. (1993), *Geometric Algorithms and Combinatorial Optimization* (2nd edition), Springer-Verlag, Heidelberg, 1993.

Guan, J. W., and Bell, D. A. (1991), *Evidence Theory and Its Applications*, Vol. 1, (Studies in Computer Science and Artificial Intelligence No. 7), North-Holland, Amsterdam, 1991.

Guenin, B. (1997), Perfect and ideal $0, \pm 1$ matrices, *Mathematics of Operations Research*, to appear.

Gusfield, D., and Pitt, L. (1992), A bounded approximation for the minimum cost 2-Sat problem, *Algorithmica* 8 (1992) 103–117.

Hailperin, T. (1986), *Boole's Logic and Probability* (2nd edition), (Studies in Computer Science and Artificial Intelligence No. 85), North-Holland, Amsterdam, 1986.

Haken, A. (1985), The intractability of resolution, *Theoretical Computer Science* 39 (1985) 297–308.

Hall, P. (1935), On representatives of subsets, *Journal of the London Mathematical Society* 10 (1935) 26–30.

Hamilton, A. G. (1988), *Logic for Mathematicians*, Cambridge University Press, Cambridge, 1988.

Hammer, P. L., and Kogan, A. (1992), Horn functions and their DNFs, *Information Processing Letters* 44 (1992) 23–29.

Hammer, P. L., and Kogan, A. (1993), Optimal compression of propositional Horn knowledge bases: Complexity and approximation, *Artificial Intelligence* 64 (1993) 131–145.

Hammer, P. L., and Kogan, A. (1995), Quasi-acyclic propositional Horn knowledge bases: Optimal compression, *IEEE Transactions on Knowledge and Data Engineering* 7 (1995) 751–762.

Hammer, P. L., and Kogan, A. (1996), Essential and redundant rules in Horn knowledge bases, *Decision Support Systems* 16 (1996) 119–130.

Hansen, P. (1976), A cascade algorithm for the logical closure of a set of binary relations, *Information Processing Letters* 5 (1976) 50–54.

Hansen, P., and Jaumard, B. (1985), Uniquely solvable quadratic Boolean equations, *Discrete Applied Mathematics* 12 (1985) 147–154.

Hansen, P., and Jaumard, B. (1990), Algorithms for the maximum satisfiability problem, *Computing* 44 (1990) 279–303.

Hansen, P., Jaumard, B., and Minoux, M. (1986), A linear expected-time algorithm for deriving all logical conclusions implied by a set of Boolean inequalities, *Mathematical Programming* 34 (1986) 223–231.

Hansen, P., Jaumard, B., and Plateau, G. (1993), An extension of nested satisfiability, GERAD Research Report G-93-27, McGill University, 1993, and RUTCOR Research Report RRR 29-93, Rutgers University, 1993.

Harary, F. (1969), *Graph Theory*, Addison-Wesley, Reading, Massachusetts, 1969.

Harche, F., Hooker, J. N., and Thompson, G. L. (1994), A computational study of satisfiability algorithms for propositional logic, *ORSA Journal on Computing* 6 (1994) 423–435.

Harche, F., and Thompson, G. L. (1994), The column subtraction algorithm: An exact method for solving weighted set covering, packing, and partitioning problems, *Computers and Operations Research* 21 (1994) 689–705.

Hébrard, J.-J. (1994), A linear algorithm for renaming a set of clauses as a Horn set, *Theoretical Computer Science* 124 (1994) 343–350.

Hébrard, J.-J. (1995), Unique Horn renaming and unique 2-satisfiability, *Information Processing Letters* 54 (1995) 235–239.

Heller, I., and Tompkins, C. B. (1956), An extension of a theorem of Dantzig's, in: *Linear Inequalities and Related Systems* (H. W. Kuhn and A. W. Tucker, eds.), Princeton University Press, Princeton, New Jersey, 1956, pp. 247–254.

Henschen, L., and Wos, L. (1974), Unit refutation and Horn sets, *Journal of the Association for Computing Machinery* 21 (1974) 590–605.

Heusch, P. (1994), *Implikation der Implikation*, thesis, University of Düsseldorf, Germany, 1994.

Hochbaum, D. S., Megiddo, N., Naor, J., and Tamir, A. (1993), Tight bounds and 2-approximation algorithms for integer programs with two variables per inequality, *Mathematical Programming* 62 (1993) 69–83.

Hoffman, A. J., and Kruskal, J. B. (1956), Integral boundary points of convex polyhedra, in: *Linear Inequalities and Related Systems* (H. W. Kuhn and A. W. Tucker, eds.), Princeton University Press, Princeton, New Jersey, 1956, pp. 223–246.

Hooker, J. N. (1988a), Resolution vs. cutting plane solution of inference problems: Some computational experience, *Operations Research Letters* 7 (1988) 1–7.

Hooker, J. N. (1988b), A quantitative approach to logical inference, *Decision Support Systems* 4 (1988) 45–69.

Hooker, J. N. (1988c), Generalized resolution and cutting planes, *Annals of Operations Research* 12 (1988) 217–239.

Hooker, J. N. (1989), Input proofs and rank one cutting planes, *ORSA Journal on Computing* 1 (1989) 137–145.

Hooker, J. N. (1992), Generalized resolution for 0-1 linear inequalities, *Annals of Mathematics and Artificial Intelligence* 6 (1992) 271–286.

Hooker, J. N. (1993), Solving the incremental satisfiability problem, *Journal of Logic Programming* 15 (1993) 177–186.

Hooker, J. N. (1996), Resolution and the integrality of satisfiability polytopes, *Mathematical Programming* 74 (1996) 1–10.

Hooker, J. N., and Fedjki, C. (1990), Branch-and-cut solution of inference problems in propositional logic, *Annals of Mathematics and Artificial Intelligence* 1 (1990) 123–139.

Hooker, J. N., and Vinay, V. (1995), Branching rules for satisfiability, *Journal of Automated Reasoning* 15 (1995) 359–383.

Hopcroft, J. E., and Tarjan, R. E. (1973), Dividing a graph into triconnected components, *SIAM Journal on Computing* 2 (1973) 135–158.

Horn, A. (1951), On sentences which are true of direct unions of algebras, *Journal of Symbolic Logic* 16 (1951) 14–21.

Hu, T. C. (1969), *Integer Programming and Network Flows*, Addison-Wesley, Reading, Massachusetts, 1969.

Hunt, H. B., III, and Stearns, R. E. (1990), The complexity of very simple Boolean formulas with applications, *SIAM Journal on Computing* 19 (1990) 44–70.

Itai, A., and Makowsky, J. A. (1982), On the complexity of Herbrand's theorem, working paper 243, Department of Computer Science, Israel Institute of Technology, 1982.

Itai, A., and Makowsky, J. A. (1987), Unification as a complexity measure for logic programming, *Journal of Logic Programming* 4 (1987) 105–117.

Iwama, K. (1989), CNF satisfiability test by counting and polynomial average time, *SIAM Journal on Computing* 18 (1989) 385–391.

Jacobson, N. (1985), *Basic Algebra*, Vol. 1 (2nd edition), Freeman, New York, 1985.

Jaumard, B., Marchioro, P., Morgana, A., Petreschi, R., and Simeone, B. (1990), On-line 2-satisfiability, *Annals of Mathematics and Artificial Intelligence* 1 (1990) 155–165.

Jaumard, B., and Simeone, B. (1987), On the complexity of the maximum satisfiability problem for Horn formulas, *Information Processing Letters* 26 (1987/88) 1–4.

Jeroslow, R. G. (1989), *Logic-Based Decision Support - Mixed Integer Model Formulation*, monograph, published as vol. 40 of *Annals of Discrete Mathematics*, 1989.

Jeroslow, R. G., Martin, K., Rardin, R. L., Wang, J. (1992), Gainfree Leontief substitution flow problems, *Mathematical Programming* 57 (1992) 375–414.

Jeroslow, R. G., and Wang, J. (1989), Dynamic programming, integral polyhedra, and Horn clause knowledge bases, *ORSA Journal on Computing* 1 (1989) 7–19.

Jeroslow, R. G., and Wang, J. (1990), Solving propositional satisfiability problems, *Annals of Mathematics and Artificial Intelligence* 1 (1990) 167–187.

Johnson, D. S. (1974), Approximation algorithms for combinatorial problems, *Journal of Computer and Systems Sciences* 9 (1974) 256–278.

Johnson, E. L., and Padberg, M. W. (1982), Degree-two inequalities, clique facets, and biperfect graphs, *Annals of Discrete Mathematics* 16 (1982) 169–187.

Kamath, A. P., Karmarkar, N. K., Ramakrishnan, K. G., and Resende, M. G. C. (1990), Computational experience with an interior point algorithm on the satisfiability problem, *Annals of Operations Research* 25 (1990) 43–58.

Kamath, A. P., Karmarkar, N. K., Ramakrishnan, K. G., and Resende, M. G. C. (1992), A continuous approach to inductive inference, *Mathematical Programming* 57 (1992) 215–238.

Karloff, H. (1991), *Linear Programming*, Birkhäuser, Boston, 1991.

Kleine Büning, H. (1990), Existence of simple propositional formulas, *Information Processing Letters* 36 (1990) 177–182.

Kleine Büning, H. (1993), On generalized Horn formulas and k-resolution, *Theoretical Computer Science* 116 (1993) 405–413.

Kleine Büning, H., and Lettmann, T. (1994), *Aussagenlogik: Deduktion und Algorithmen*, Teubner, Stuttgart, Germany, 1994.

Kleine Büning, H., and Löwen, U. (1989), Optimizing propositional calculus formulas with regard to questions of deducibility, *Information and Computation* 80 (1989) 18–43.

Kneale, W., and Kneale, M. (1984), *The Development of Logic*, Clarendon Press, Oxford, 1984.

Knuth, D. E. (1990), Nested satisfiability, *Acta Informatica* 28 (1990) 1–6.

König, D. (1936), *Theorie der endlichen und unendlichen Graphen*, Akademische Verlagsgesellschaft, Leipzig, 1936 (reprinted: Chelsea, New York, 1950, and Teubner, Leipzig, 1986).

Koutsoupias, E., and Papadimitriou, C. H. (1992), On the greedy algorithm for satisfiability, *Information Processing Letters* 43 (1992) 53–55.

Kratochvíl, J. (1994), A special planar satisfiability problem and a consequence of its NP-completeness, *Discrete Applied Mathematics* 52 (1994) 233–252.

Kratochvíl, J., and Křivánek, M. (1993), Satisfiability of co-nested formulas, *Acta Informatica* 30 (1993) 397–403.

Kratochvíl, J., Savický, P., and Tuza, Z. (1993), One more occurrence of variables makes satisfiability jump from trivial to NP-complete, *SIAM Journal on Computing* 22 (1993) 203–210.

Kullmann, O. (1997a), A generalization of extended resolution, *Discrete Applied Mathematics*, to appear.

Kullmann, O. (1997b), Worst-case analysis, 3-SAT decision and lower bounds: Approaches for improved SAT algorithms, in: *The Satisfiability (SAT) Problem* (D. Du, J. Gu, and P. Pardalos, eds.), DIMACS Series in Discrete Mathematics and Theoretical Computer Science, Vol. 35, American Mathematical Society, Providence, Rhode Island, 1997, to appear.

Kullmann, O. (1997c), A systematical approach to 3-SAT-decision, yielding 3-SAT-decision in less than 1.5045^n steps, *Theoretical Computer Science*, to appear.

Kullmann, O., and Luckhardt, H. (1997), Deciding propositional tautologies: Algorithms and their complexity, working paper, University of Frankfurt, Germany, 1997.

Kung, J. P. S. (1986), *A Source Book in Matroid Theory*, Birkhäuser, Boston, 1986.

Lagarias, J. C. (1985), The computational complexity of simultaneous diophantine approximation problems, *Siam Journal on Computing* 14 (1985) 196–209.

Lakshminarayanan, S. and Chandrasekaran, R. (1994), A rounding algorithm for integer programs, *Discrete Applied Mathematics* 50 (1994) 267–282.

Lancaster, P., and Tismenetsky, M. (1985), *The Theory of Matrices with Applications*, Academic Press, Orlando, Florida, 1985.

Lang, S. (1984), *Algebra*, Addison-Wesley, Reading, Massachusetts, 1984.

Larrabee, T. (1992), Test pattern generation using Boolean satisfiability, *IEEE Transactions on Computer-Aided Design* 11 (1992) 4–15.

Lawler, E. L. (1976), *Combinatorial Optimization: Networks and Matroids*, Holt, Rinehart and Winston, New York, 1976.

Leontief, W. (1986), *Input-Output Economics* (2nd edition), Oxford University Press, New York, 1986.

Letz, R., Schumann, J., Bayerl, S., and Bibel, W. (1992), SETHEO: A high-performance theorem prover, *Journal of Automated Reasoning* 8 (1992) 183–212.

Lewis, H. R. (1978), Renaming a set of clauses as a Horn set, *Journal of the Association for Computing Machinery* 25 (1978) 134–135.

Lewis, J. M., and Yannakakis, M. (1980), The node-deletion problem for hereditary properties is NP-complete, *Journal of Computer and System Sciences* 20 (1980) 219–230.

Lichtenstein, D. (1982), Planar formulae and their uses, *SIAM Journal on Computing* 11 (1982) 329–343.

Lieberherr, K. J. (1982), Algorithmic extremal problems in combinatorial optimization, *Journal of Algorithms* 3 (1982) 225–244.

Lieberherr, K. J., and Specker, E. (1981), Complexity of partial satisfaction, *Journal of the Association for Computing Machinery* 28 (1981) 411–421.

Lindhorst, G., and Shahrokhi, F. (1989), On renaming a set of clauses as a Horn set, *Information Processing Letters* 30 (1989) 289–293.

Lloyd, J. W. (1987), *Foundations of Logic Programming* (2nd edition), Springer-Verlag, Berlin, 1987.

Lovász, L., and Plummer, M. D. (1986), *Matching Theory*, Akadémiai Kiadó, Budapest, 1986.

Loveland, D. W. (1978), *Automated Theorem Proving: A Logical Basis*, North-Holland, Amsterdam, 1978.

Loveland, D. W. (1984), Automated theorem-proving: A quarter-century review, *Contemporary Mathematics* 29 (1984) 1–45.

MacLane, S., and Birkhoff, G. (1988), *Algebra* (3rd edition), Macmillan, New York, 1988.

Mannila, H., and Mehlhorn, K. (1985), A fast algorithm for renaming a set of clauses as a Horn set, *Information Processing Letters* 21 (1985) 269–272.

Mayer, J., Mitterreiter, I., and Radermacher, F. J. (1995), Running time experiments on some algorithms for solving propositional satisfiability problems, *Annals of Operations Research* 55 (1995) 139–178.

Meltzer, B. (1965), Theorem-proving for computers: Some results on resolution and renaming, *Computer Journal* 8 (1965/66) 341–343.

Minoux, M. (1988), LTUR: A simplified linear-time unit resolution algorithm for Horn formulae and computer implementation, *Information Processing Letters* 29 (1988) 1–12.

Minoux, M. (1992), The unique Horn-satisfiability problem and quadratic Boolean equations, *Annals of Mathematics and Artificial Intelligence* 6 (1992) 253–266.

Mitchell, D. G., and Levesque, H. J. (1996), Some pitfalls for experimenters with random SAT, *Artificial Intelligence* 81 (1996) 111–125.

Monien, B., and Speckenmeyer, E. (1985), Solving satisfiability in less than 2^n steps, *Discrete Applied Mathematics* 10 (1985) 287–295.

Monien, B., and Sudborough, I. H. (1985), Bandwidth constrained NP-complete problems, *Theoretical Computer Science* 41 (1985) 141–167.

Nemhauser, G. L., and Trotter, L. E., Jr. (1975), Vertex packings: Structural properties and algorithms, *Mathematical Programming* 8 (1975) 232–248.

Nemhauser, G. L., and Wolsey, L. A. (1988), *Integer and Combinatorial Optimization*, Wiley, New York, 1988.

Nerode, A., and Shore, R. (1993), *Logic for Applications*, Springer-Verlag, New York, 1993.

Newman, J. R. (1956), *The World of Mathematics*, Vols. 3 and 4, Simon and Schuster, New York, 1956.

Nobili, P., and Sassano, A. (1997), $(0, \pm 1)$ ideal matrices, *Mathematical Programming (A)* 80 (1997) 253–270.

Ore, O. (1962), *Theory of Graphs* (American Mathematical Society Colloquium Publications, Vol. 38), American Mathematical Society, Providence, Rhode Island, 1962.

Oxley, J. G. (1992), *Matroid Theory*, Oxford University Press, Oxford, 1992.

Paris, J. B., Wilkie, A. J., and Woods, A. R. (1988), Provability of the pigeonhole principle and the existence of infinitely many primes, *Journal of Symbolic Logic* 53 (1988) 1235–1244.

Pearl, J. (1988), *Probabilistic Reasoning in Intelligent Systems: Networks of Plausible Inference*, Morgan Kaufmann, San Mateo, California, 1988.

Petreschi, R., and Simeone, B. (1980), A switching algorithm for the solution of quadratic Boolean equations, *Information Processing Letters* 11 (1980) 193–198.

Petreschi, R., and Simeone, B. (1991), Experimental comparison of 2-satisfiability algorithms, *Operations Research* 25 (1991) 241–264.

Poljak, S., and Turzík, D. (1982), A polynomial algorithm for constructing a large bipartite graph, with an application to a satisfiability problem, *Canadian Journal of Mathematics* 34 (1982) 519–524.

Pretolani, D. (1993a), *Satisfiability and Hypergraphs*, thesis TD-12/93, Department of Computer Science, University of Pisa, Italy, 1993.

Pretolani, D. (1993b), A linear time algorithm for unique Horn satisfiability, *Information Processing Letters* 48 (1993) 61–66.

Pretolani, D. (1994), Hierarchies of polynomially solvable satisfiability problems, *Annals of Mathematics and Artificial Intelligence* 17 (1996) 339–357.

Pretolani, D. (1996), Efficiency and stability of hypergraph SAT algorithms, in: *Cliques, Coloring, and Satisfiability: Second DIMACS Implementation Challenge* (D. S. Johnson and M. A. Trick, eds.), DIMACS Series in Discrete Mathematics and Theoretical Computer Science, Vol. 26, American Mathematical Society, Providence, Rhode Island, 1996, pp. 479–498.

Purdom, P. W., Jr. (1984), Solving satisfiability with less searching, *IEEE Transactions on Pattern Analysis and Machine Intelligence* PAMI-6 (1984) 510–513.

Purdom, P. W., Jr. (1990), A survey of average time analysis of satisfiability algorithms, *Journal of Information Processing* 13 (1990) 449–455.

Purdom, P. W., Jr., and Brown, C. A. (1985a), The pure literal rule and polynomial average time, *SIAM Journal on Computing* 14 (1985) 943–953.

Purdom, P. W., Jr., and Brown, C. A. (1985b), *The Analysis of Algorithms*, Holt, Rinehart and Winston, New York, 1985.

Purdom, P. W., Jr., and Brown, C. A. (1987), Polynomial-average-time satisfiability problems, *Information Sciences* 41 (1987) 23–42.

Rago, G. (1994), *Optimization, Hypergraphs and Logical Inference*, thesis TD-4/94, Department of Computer Science, University of Pisa, Italy, 1994.

Recski, A. (1989), *Matroid Theory and Its Applications in Electrical Networks and Statics*, Springer-Verlag, Heidelberg, 1989.

Robinson, J. A. (1965a), A machine-oriented logic based on the resolution principle, *Journal of the Association for Computing Machinery* 12 (1965) 23–41.

Robinson, J. A. (1965b), Automatic deduction with hyper-resolution, *International Journal of Computer Mathematics* 1 (1965) 227–234.

Rodošek, R. (1996), A new approach on solving 3-satisfiability, in: *Artificial Intelligence and Symbolic Mathematical Computation* (J. Calmet, J. A. Campbell, and J. Pfalzgraf, eds.), Proceedings of AISMC-3 International Conference, Steyr, Austria, 1996, published as vol. 1138 of *Lecture Notes in Computer Science*, Springer-Verlag, Berlin, 1996, pp. 197–212.

Rodošek, R., and Schiermeyer, I. (1997), Binary decisions for solving 3-satisfiability problems, *Discrete Applied Mathematics*, to appear.

Salkin, H. M. (1975), *Integer Programming*, Addison-Wesley, Reading Massachusetts, 1975.

Schiermeyer, I. (1993), Solving 3-satisfiability in less than $1,579^n$ steps, in: *Computer Science Logic* (E. Börger, G. Jäger, H. Kleine Büning, S. Martini, M. M. Richter, eds.), Selected Papers of CSL '92 6th Workshop, San Miniato, Italy, 1992, published as vol. 702 of *Lecture Notes in Computer Science*, Springer-Verlag, Berlin, 1993, pp. 379–394.

Schiermeyer, I. (1996), Pure literal look ahead: An $O(1,497^n)$ 3-satisfiability algorithm, in: *Workshop on the Satisfiability Problem* (J. Franco, G. Gallo, H. Kleine Büning, E. Speckenmeyer, and C. Spera, eds.), Technical Report 96-230, University of Cologne, Germany, 1996, pp. 127–136.

Schlipf, J. S., Annexstein, F. S., Franco, J. V., and Swaminathan, R. P. (1995), On finding solutions for extended Horn formulas, *Information Processing Letters* 54 (1995) 133–137.

Schrag, R., and Crawford, J. M. (1996), Implicates and prime implicates in random 3-SAT, *Artificial Intelligence* 81 (1996) 199–222.

Schrijver, A. (1986), *Theory of Linear and Integer Programming*, Wiley, Chichester, 1986.

Scutellà, M. G. (1990), A note on Dowling and Gallier's top-down algorithm for propositional Horn satisfiability, *Journal of Logic Programming* 8 (1990) 265–273.

Selman, B., and Kirkpatrick, S. (1996), Critical behavior in the computational cost of satisfiability testing, *Artificial Intelligence* 81 (1996) 273–295.

Selman, B., Mitchell, D. G., and Levesque, H. J. (1996), Generating hard satisfiability problems, *Artificial Intelligence* 81 (1996) 17–29.

Seymour, P. D. (1980), Decomposition of regular matroids, *Journal of Combinatorial Theory (B)* 28 (1980) 305–359.

Simeone, B. (1985), Consistency of quadratic Boolean equations and the König-Egerváry property for graphs, *Annals of Discrete Mathematics* 25 (1985) 281–290.

Speckenmeyer, E., Böhm, M., and Heusch, P. (1997), On the imbalance of distributions of solutions of CNF-formulas and its impact on satisfiability solvers, in: *The Satisfiability (SAT) Problem* (D. Du, J. Gu, and P. Pardalos, eds.), DIMACS Series in Discrete Mathematics and Theoretical Computer Science, Vol. 35, American Mathematical Society, Providence, Rhode Island, 1997, to appear.

Speckenmeyer, E., Monien, B., and Vornberger, O. (1988), Superlinear speedup for parallel backtracking, in: *Proceedings of 1st International Conference on Supercomputing* (E. N. Houstis, T. S. Papatheodorou, C. D. Polychronopoulos, eds.), Athens, Greece, 1987, published as vol. 297 of *Lecture Notes in Computer Science*, Springer-Verlag, Berlin, 1988, pp. 985–993.

Strang, G. (1980), *Linear Algebra and Its Applications* (2nd edition), Academic Press, New York, 1980.

Swaminathan, R. P., and Wagner, D. K. (1995), The arborescence-realization problem, *Discrete Applied Mathematics* 59 (1995) 267–283.

Tanaka, Y. (1991), A dual algorithm for the satisfiability problem, *Information Processing Letters* 37 (1991) 85–89.

Tarjan, R. E. (1972), Depth-first search and linear graph algorithms, *SIAM Journal on Computing* 1 (1972) 146–160.

Tovey, C. A. (1984), A simplified NP-complete satisfiability problem, *Discrete Applied Mathematics* 8 (1984) 85–89.

Truemper, K. (1976), An Efficient Scaling Procedure for Gain Networks, *Networks* 6 (1976) 151–159.

Truemper, K. (1982), Alpha-balanced graphs and matrices and GF(3)-representability of matroids, *Journal of Combinatorial Theory (B)* 32 (1982) 112–139.

Truemper, K. (1990), A decomposition theory for matroids. V. Testing of matrix total unimodularity, *Journal of Combinatorial Theory (B)* 49 (1990) 241–281.

Truemper, K. (1992), *Matroid Decomposition*, Academic Press, Boston, 1992, and Leibniz, Plano, Texas, 1998.

Truemper, K., and Chandrasekaran, R. (1978), Local unimodularity of matrix-vector pairs, *Linear Algebra and Its Applications* 22 (1978) 65–78.

Tseitin, G. S. (1968), On the complexity of derivations in the propositional calculus, in: *Structures in Constructive Mathematics and Mathematical Logic* (A. O. Slisenko, ed.), Part II (translated from Russian), Consultants Bureau, New York, 1968, pp. 115–125.

Tutte, W. T. (1966), Connectivity in matroids, *Canadian Journal of Mathematics* 18 (1966) 1301–1324.

Tutte, W. T. (1971), *Introduction to the Theory of Matroids*, American Elsevier, New York, 1971.

Urquhart, A. (1987), Hard examples for resolution, *Journal of the Association for Computing Machinery* 34 (1987) 209–219.

Vlach, F. (1993), Simplification in a satisfiability checker for VLSI applications, *Journal of Automated Reasoning* 10 (1993) 115–136.

Wang, J. (1993), Inference flexibility in Horn clause knowledge bases and the simplex method, *Journal of Automated Reasoning* 11 (1993) 269–288.

Whitney, H. (1935), On the abstract properties of linear dependence, *American Journal of Mathematics* 57 (1935) 509–533.

Wilson, R. J. (1972), *Introduction to Graph Theory*, Longman Group Limited, London, 1972.

Wos, L., Overbeek, R., Lusk, E., and Boyle, J. (1992), *Automated Reasoning* (2nd edition), McGraw-Hill, New York, 1992.

Yamasaki, S., and Doshita, S. (1983), The satisfiability problem for a class consisting of Horn sentences and some non-Horn sentences in propositional logic, *Information and Control* 59 (1983) 1–12.

Yannakakis, M. (1981), Node-deletion problems on bipartite graphs, *SIAM Journal on Computing* 10 (1981) 310–327.

Yannakakis, M. (1992), On the Approximation of Maximum Satisfiability, in: *Proceedings of the Third Annual ACM-SIAM Symposium on Discrete Algorithms*, Association for Computing Machinery and Society for Industrial and Applied Mathematics, Orlando Florida, 1992, pp. 1–9. Also, *Journal of Algorithms* 17 (1994) 475–502.

Zhang, H. (1993), Sato: A decision procedure for propositional logic, *Association of Automated Reasoning Newsletter* No. 22 (1993) 1–3.

Zhang, W. (1996), Number of models and satisfiability of sets of clauses, *Theoretical Computer Science* 155 (1996) 277–288.

Author Index

Subject Index

457

W

Z